Aircraft Aerodynamic Design with Computational Software

This modern text presents the aerodynamic design of aircraft with realistic applications using computational fluid dynamics software, as well as guidance on its use. The tutorials, exercises, and mini-projects provided involve the design of real aircraft, ranging from straight to swept to slender wings and from low speed to supersonic. Supported by online resources and supplements, this tool kit covers topics including shape optimization to minimize drag and collaborative design. It prepares senior-year undergraduate students and first-year graduate students for design and analysis tasks in aerospace companies. In addition, it is a valuable resource for practicing engineers, aircraft designers, and entrepreneurial consultants.

Arthur Rizzi is Professor of Aeronautics at the KTH Royal Institute of Technology. He is the recipient of the Royal Aeronautical Society's Busk Prize and the Swedish Aeronautics and Astronautics Society's Thulin Medal.

Jesper Oppelstrup is Professor of Numerical Analysis at the KTH Royal Institute of Technology. He has extensive experience in industry of applying computational mathematics to practical engineering problems.

Aircraft Aerodynamic Design with Computational Software

ARTHUR RIZZI
KTH Royal Institute of Technology

JESPER OPPELSTRUP
KTH Royal Institute of Technology

CAMBRIDGE
UNIVERSITY PRESS

University Printing House, Cambridge CB2 8BS, United Kingdom

One Liberty Plaza, 20th Floor, New York, NY 10006, USA

477 Williamstown Road, Port Melbourne, VIC 3207, Australia

314–321, 3rd Floor, Plot 3, Splendor Forum, Jasola District Centre, New Delhi – 110025, India

79 Anson Road, #06–04/06, Singapore 079906

Cambridge University Press is part of the University of Cambridge.

It furthers the University's mission by disseminating knowledge in the pursuit of education, learning, and research at the highest international levels of excellence.

www.cambridge.org
Information on this title: www.cambridge.org/9781107019485
DOI: 10.1017/9781139094672

First published 2021

A catalogue record for this publication is available from the British Library.

ISBN 978-1-107-01948-5 Hardback

Contents

Figures

Tables

Preface

The KTH Royal Institute of Technology is the prime seat of aeronautical education in Scandinavia, having granted degrees in aeronautics continuously for over a century. Generations of engineers have been trained for careers in the aerospace industry, designing and producing outstanding aircraft at Saab AB as well as at other manufacturers.

The origins of this book date back to the late 1980s when the authors participated in the Center for Computational Mathematics and Mechanics at KTH under the tutelage of Professors Heinz Kreiss and Mårten Landahl. Our contribution was a series of courses on aircraft aerodynamics and computational fluid dynamics (CFD). The starting points for the aerodynamic design lectures were the classical tracts on the subject in the 1970s by, for example, Ryle and Küchemann. In our research work, we connected CFD, a then emerging tool, with design tasks.

Our long-term experience of teaching aerodynamic design tasks with computational software has proven this to be an effective teaching approach for a semester-long course. It has met with student approval, improving our approach with each year's constructive feedback. The maturing of computational aerodynamic tools and computer hardware over the years has driven this teaching approach. Today students can run meaningful CFD on their laptops and apply it to aerodynamic design.

The examples we chose span a large part of the design space for conventional aircraft with straight or slender wings at low or high speeds. The intention is to whet students' curiosity and incite them to explore on their own and learn through active computation. With the software at hand, students can explore the design space around these points and understand quantitatively the mapping from flight shape to performance. Students can try out their own ideas and see the results with reasonable response times, thereby learning through their own actions instead of just reading about what someone else has done.

CFD is a process that includes a sequence of techniques, and any tool is only as good as the user's ability to handle it with skill. The twofold aim of this book, therefore, is to *inform* students about CFD applications to aerodynamic design (what we term *user awareness* of applying the tools to the design tasks) and to *explain* CFD *due diligence* in wielding the tools.

It is our firm conviction that CFD should not be taught as a *spectator sport* with dazzling, eye-catching examples computed by professionals who know their codes intimately and have extensive experience in generating grids. Instead, we encourage beginners, students, and amateurs to be inspired by what the pros can do, and, continuing the sports analogy, get themselves a ball and a pair of shoes and go out on the practice field to *learn by doing*. We want them to mimic the professional aerodynamicists in how they use CFD and start the learning process with basic tools! Practically all of the examples shown in this book are results that the students, with the software available on the book's website and starting from a three-view drawing of the configuration, can produce on their own.

In this sense, our approach is less listening to the "sage on the stage" and more of the "coach on the sidelines" guiding the team to higher performance. This learning-by-doing approach to teaching aerodynamic design is accomplished by working with exercises, tutorials, and extended projects and using the computational tools under guidance. Experience gained in carrying out these exercises will help the student when completing a term project or a capstone design course or writing a senior-year, masters, or PhD thesis. The hands-on assignments are presented not in the book itself, but on the book's website: www.cambridge.org/rizzi. Useful public-domain software, which has been used for computing many of the examples in this book, can be found at http://airinnova.se/education/aerodynamic-design-of-aircraft/. This website provides downloads and also links to the home and developer pages for the different packages.

The material in the book is suitable for a final-year undergraduate course or a first-year graduate course. Students should have entry-level knowledge from a basic course in the fundamentals of flight and of elementary numerical methods. We hope the book will remain a guide on the side, even in future work.

Acknowledgments

Recognition of all of the people and events that expanded our intellectual horizons as we traveled the long journey that led to the writing of this book would be a Herculean effort. At the risk of being unfair, we highlight a few of those we have worked closely together with in numerous projects over the years.

Heinz Kreiss, Björn Engquist, Bertil Gustafsson, Per Lötstedt, and Bernhard Müller at the Uppsala University School of Scientific Computing lit up our theoretical path with friendship and mathematization.

Roger Larsson and Yngve Sedin, among others at Saab AB, provided firsthand views into the industrial process of designing aircraft, as well as good companionship in European collaborative projects, bringing academia and industry together.

Always ready to help, our long-term friend and associate Jan Vos at CFS Engineering in Lausanne has worked with us for over 30 years in developing and applying computational fluid dynamics methods and software, both industrial and academic. Many years ago, Daniel Raymer at Conceptual Research Corp. introduced us to aircraft conceptual design and has set the standard for teaching it. Denis Darracq, Thierry Poinsot, and Jean-Francois Boussuge at CERFACS, in close conjunction with Eric Chaput and Loic Tourrette at Airbus France in Toulouse, partnered with us in many European projects developing Navier–Stokes solvers and bringing them into civil aircraft design.

Likewise, Ernst Hirschel at EADS Military Aircraft in Ottobrunn has been our long-term colleague on aerodynamic topics of military aircraft design, especially concerning separated vortex flow. With his AGILE project, Björn Nagel at DLR led us to the world of multidisciplinary optimization and, together with colleagues Klaus Becker and Markus Fischer at Airbus Bremen, to applications to civil aircraft design.

For over 30 years, we have had numerous exchanges with Mark Drela at MIT, in particular on teaching topics of aerodynamic design, and we are most grateful to him for sharing his MSES software with us to the benefit of KTH students. Over the years, James Luckring at NASA Langley has lent his hand in guiding our doctoral students in a number of STO-NATO task groups and the NASA-sponsored CAWAPI projects on the F-16XL flight-test aircraft, adding an extra dimension to their doctoral studies.

Working with students is a challenge, but at the same time a source of energy. Their passion, curiosity, engagement, and dreams are contagious, and they have rubbed off on us, propelling us to higher levels than we could have reached without them.

Tangible evidence of this synergy is that PhD and MSc students have very generously made large parts of their theses available to us. A strong initial motivation for compiling this material into a textbook is our felt responsibility to pass on the results of their efforts to benefit younger generations. Some appear by name in the text and remembering every one, we owe them *all* sincere gratitude for making our lives as teachers so fulfilling.

More specifically for their direct input to the book, we first of all are very grateful to Peter Eliasson and Mengmeng Zhang for the many new computations they ran at our request. This book would have been impossible without them.

Our thanks also go to all of those who so graciously gave us permission to use their results from other publications or shared them in private communication.

Fritz Bark, Mark Voskuijl, Aaron Dettman, Kenneth Nilsson, Tord Jonsson, and Bernhard Müller reviewed selected parts of our manuscript and provided valuable comments. The feedback from this supportive group brought forth scores of improvements and corrections, and our thanks go to all of you. Any remaining errors are ours alone.

Ellen Rizzi added some professional touches to a number of our illustrations for which we are highly appreciative. Applying her sharp eye, Kerstin Assarsson-Rizzi proofread the manuscript and improved its readability to the benefit of all.

Our gratitude goes out to the editorial team at Cambridge University Press: to Peter Gordon, for seeing in our course compendium the makings of this book; to Steven Elliot, who even at those darkest of moments never lost belief in us; and to Julia Ford, for holding us true to the task.

Finally, our spouses deserve special thanks for putting up with the many long nights and weekends that writing this book entailed, testing their patience over a much too long time.

Abbreviations

(C,E,I,T)AS	(Calibrated, equivalent, indicated, true) airspeed
AC	Aerodynamic center
ADODG	AIAA Aerodynamic Design Optimization Discussion Group
AFWAL	Air Force Wright Aeronautical Laboratory (Wright-Patterson Air Force Base, OH)
AIAA	American Institute of Aeronautics and Astronautics
AMR	Adaptive mesh refinement
AMR	Automatic mesh refinement
AoA	Angle of attack
API	Application program interface
AR	Aspect ratio
ARSM	Algebraic RSM
BAE	British Aerospace
BSL	Baseline (turbulence model)
CAD	Computer-aided design
CFD	Computational fluid dynamics
CFL	Courant–Friedrichs–Lewy
CFSE	Computational Fluid and Structures Engineering
CG	Center of gravity
CGNS	CFD General Notation System
CP	Center of pressure
CPACS	Common Parametric Aircraft Configuration Schema
CRM	NASA Common Research Model
CSM	Computational structural mechanics
CST	Class–shape function transformation
DATCOM	Data compendium
DES	Detached-eddy simulation
DLM	Doublet lattice model
DLR	German Aerospace Center
DNS	Direct numerical simulation
DOC	Direct operating cost
DOF	Degree of freedom
DoE	Design of experiment
DRSM	Differential Reynolds stress model

DVL	Deutsche Versuchsanstalt für Luftfahrt
EARSM	Explicit algebraic Reynolds stress model
FAR	Federal aviation regulations
FAS	Full approximation scheme
FCS	Flight control system
FDS	Flux-difference splitting
FE	Finite element
FFA	Swedish Aeronautical Research Establishment
FFD	Free-form deformation
FOI	Swedish Defence Research Agency
FV	Finite volume
GD	General Dynamics
GMRES	Generalized minimal-residual algorithm
GPL	General public license
GUI	Graphical user interface
HCST	High-speed civilian transport
Hi-Fi	High fidelity
HSR	High-speed research program
ICAO	International civil aviation organization
IGES	Initial Graphics Exchange Specification
JAR	Joint Aviation Requirements
JAS	Jakt Attack Spaning (Swedish: Intercept Attack Recon)
KTH	Royal Institute of Technology
LCO	Limit cycle oscillation
LES	Large-eddy simulation
LEX	Leading-edge extension
LU–SGS	Lower-upper symmetric Gauss–Seidel
Lo-Fi	Low fidelity
MAC	Mean aerodynamic chord
MDO	Multidisciplinary optimization
Me	Messerschmitt
MIT	Massachusetts Institute of Technology
MTOW	Maximum takeoff weight
MUSCL	Monotone upstream scheme for conservation laws
NAA	North American Aviation
NACA	National Advisory Committee on Aeronautics
NASA	National Air and Space Administration
NLF	National Laminar Flow program
NLR	Netherlands Aerospace Center
NURBS	Nonuniform rational B-spline
ODE	Ordinary differential equation
OEI	One engine inoperative
ONERA	French national aerospace research center
PG	Prandtl–Glauert

PDE	Partial differential equation
RAE	Royal Aircraft Establishment
RAF	Royal Air Force
RANS	Reynolds-averaged Navier–Stokes
RSM	Reynolds stress model
RMS(E)	Root mean square (error)
SA	Spalart–Allmaras
SAS	Stability augmentation system
SBLI	Shock–boundary layer interaction
SCID	Streamline curvature iterative displacement
SDSA	Static and dynamic stability analyzer
SGS	Sub-grid scale
SST	Shear stress transport (turbulence model)
STEP	Standard for the exchange of product data
Sumo	Surface modeler
TCR	Transonic Cruiser
TKE	Turbulence kinetic energy
TVD	Total variation diminishing
U(C)AV	Unmanned (combat) aerial vehicle
USAF	US Air Force
VLM	Vortex lattice method
WWI	World War I
WW II	World War II
XML	Extensible markup language

Nomenclature

Table 0.1 Nomenclature.

Symbol	Verbal definition	Formula definition	Reference
$D(.)/Dt$	Material derivative	$\frac{\partial(.)}{\partial t} + (\mathbf{u} \cdot \nabla)(.)$	p. 68
S	Wing area	–	Figure 0.1, p. xxix
AR	Aspect ratio	$AR = b^2/S$	Ditto
λ	Taper ratio	c_t/c_r	Ditto
p	Pressure or angular rate around x-axis	–	Figure 0.2, p. xxix
C_p	Pressure coefficient	$C_p = \frac{p-p_\infty}{q}$	Equation (2.3), p. 52
C_f	Friction coefficient	$C_f = \tau/q$	Equation (2.2), p. 52
Re_l	Reynolds number	$Re_l = \frac{\rho V l}{\mu}$	p. 51
W	Weight	$W = mg$	–
V, V_∞, TAS	True airspeed, velocity	–	–
EAS	Equivalent airspeed	$EAS = TAS\sqrt{\frac{\rho_{air}(Alt)}{\rho_{air}(0)}}$	p. 10
L	Lift force	–	Figure 1.2, p. 6
D	Drag force	–	Ditto
q	Dynamic pressure or angular rate around y	$q = \frac{1}{2}\rho_{air}V^2$	pp. xxx, xxix
C_L	Lift coefficient	$C_L = \frac{L}{qS}$	Ditto
C_D	Drag coefficient	$C_D = \frac{D}{qS}$	Ditto
M, M_∞	Mach number	$M = V/a, a$ speed of sound	p. 9
M_{crit}	Critical Mach number	—	p. 26
M_{dd}	Drag divergence Mach number	—	Ditto
MAC	Mean aerodynamic chord	See ref.	p. xxxi

Wing Planform Geometry

The wing planform geometry and its characteristic quantities are given in Figure 0.1.

Coordinate Systems

There are several in use, all right handed, as shown in Figure 0.2:

- The *body axes (aerodynamics)*: used for geometry definitions: x runs from nose to tail, z up, and y out the starboard wing.
- The *body axes (western flight mechanics)*: (x, y, z) x from tail to nose, z down and y (still) out the starboard wing.

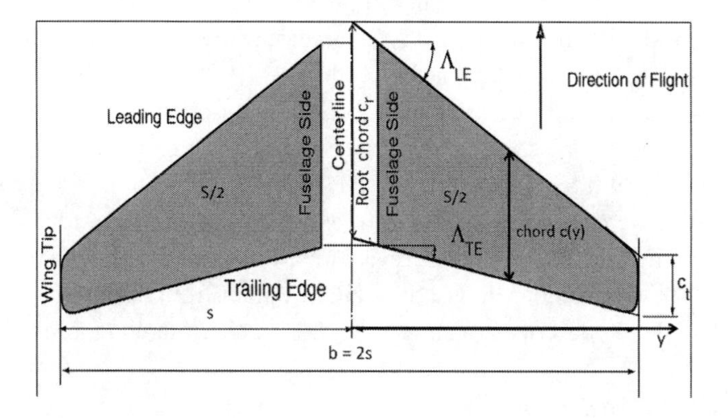

Figure 0.1 Wing planform and its parameters.

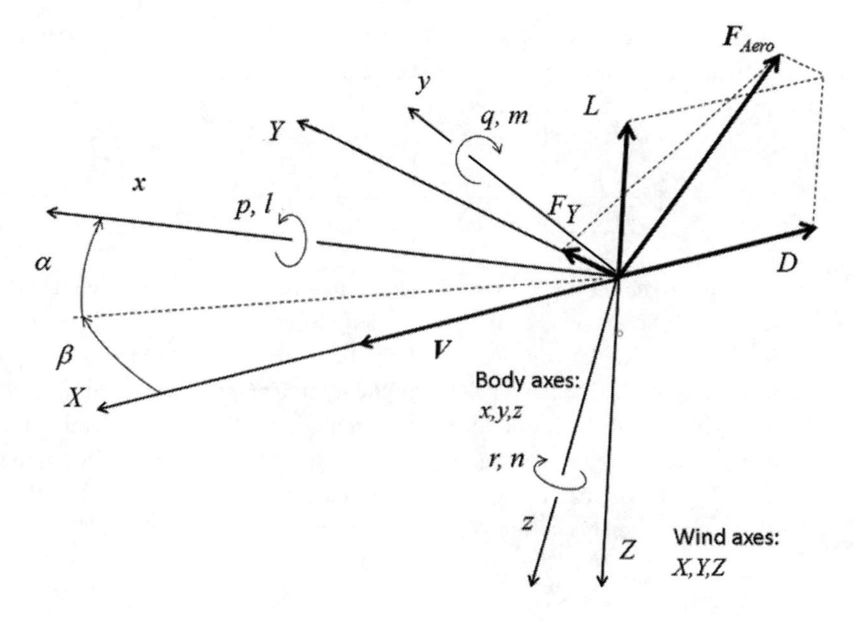

Figure 0.2 Wind (X, Y, Z) and body (x, y, z) axes; forces and moments; angles and angular rates.

- The *wind axes*: $(x, y, z$; here $X, Y, Z)$. X points in the direction of the aircraft velocity, so drag D is in the opposite direction. Lift L is aligned with the cross product $D \times y$, so Z points in the opposite direction. Side force F_Y and Y are aligned with $Y = L \times D$.

The aircraft velocity is described in the body axes coordinates, with u, v, and w representing velocities in the x, y, and z directions, respectively. p, q, and r indicate rotational rates about the x, y, and z axes, respectively, counted positive by the right-hand screw rule: $p > 0$ indicates a roll right wing down, $q > 0$ indicates pitch up, and $r > 0$ indicates yaw to the right.

Angle of attack (AoA) is taken as $\alpha = \arctan(w/u)$ and angle of sideslip is taken as $\beta = \arcsin(v/\sqrt{u^2 + v^2 + w^2})$. Aircraft position in a global reference system G is needed in flight simulation, usually with X_G increasing toward the north, Y_G increasing toward the east, and Z_G indicating negative altitude.

Euler angles are used to describe aircraft rigid body orientation in global coordinates. The attitude can be represented as a sequence of three rotations from a reference around defined directions. Note that the order of application of the rotations is significant. Positive directions are chosen to be consistent with right-handed coordinate systems. The yaw angle, Ψ, describes aircraft heading from due north, positive toward east. The pitch angle, θ, describes aircraft pitch from nose level, positive up, and the roll angle, Φ, describes rotation about the x-axis from wing level.

Force and Moment Coefficients

The six force and moment components are usually defined by six nondimensional *force and moment coefficients* $C_{...}$, with definitions of angles and angular rates as in Figure 0.2. This isolates the dependence on shape from speed and overall size. The 3D coefficients referring to the wind axes are as follows.

$$C_L = L/(q_\infty S), C_D = D/(q_\infty S), C_Y = F_Y/(q_\infty S) \tag{0.1}$$

$$C_m = m/(q_\infty S c_{ref}), \text{ etc.} \tag{0.2}$$

$$q_\infty = 1/2 \, \rho_\infty \, {V_\infty}^2$$

q_∞ is the dynamic pressure of the free stream, S is the chord length or the wing reference area. and c_{ref} is a reference length, usually the mean aerodynamic chord (MAC; see below) or wingspan b. Often, lower-case c is used for the two-dimensional flow coefficients. The actual values of the reference quantities are immaterial as long as the same numbers are used in all computations. There are several definitions for reference wing area in common use, with differences in the treatment of the fuselage cover. It must be remembered that what matters in the end is lift and drag, not the values of coefficients, although they are very helpful.

CP and AC

The center of pressure (CP) X_{cp}, Y_{cp} is the centroid of the pressure distribution. Consider now the total moment on the starboard half wing around a point $AC = (X_{ac}, Y_{ac})$.

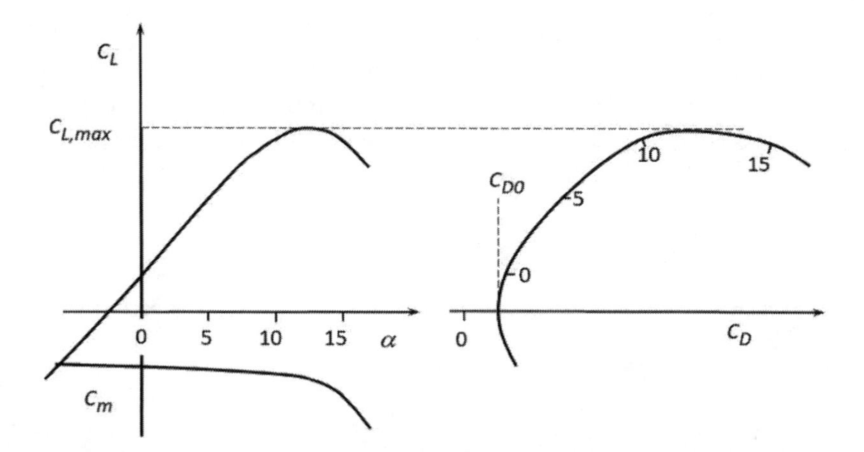

Figure 0.3 Lift and moment coefficients vs α and lift coefficient vs drag coefficient for a low-speed cambered wing.

It is the (half) wing aerodynamic center, also called the neutral point, if the moment variation with α vanishes. Let the aerodynamic center of the section lift be $x_{ac}(y)$, similarly defined. To first order, the moment change with α vanishes when its α-derivative does, which gives the following.

$$X_{ac} \cdot S/2 = \int_0^{b/2} c(y)x_{ac}(y)dy, Y_{ac} \cdot S/2 = \int_0^{b/2} yc(y)dy \tag{0.3}$$

The section-lift coefficient slope $c_{L,\alpha}$ has been assumed constant across the span. For thin sections, the aerodynamic center is at quarter-chord, $x_{ac} = x_{LE}(y) + 1/4c(y)$, so the reference wing aerodynamic center is determined by planform geometry alone. We define also the MAC by the following.

$$MAC \cdot S/2 = \int_0^{b/2} c(y)^2 dy \tag{0.4}$$

The aerodynamic center of a whole *symmetric* wing has the same X_{ac} and, of course, $Y_{ac} = 0$.

Drag Polars

The variation of lift and drag coefficients with AoA is customarily plotted as lift vs α and lift vs drag in a *drag polar*. The moment is adjoined as a third curve with α as abscissa. Figure 0.3 shows an example. The lift curve is linear for small angles, then bends over to exhibit a maximum. For higher angles, the lift decreases as the wing stalls while drag continues to rise. $C_{L,max}$ is about 1–2, and is higher for wings with high-lift devices deployed. The drag coefficient, measured in drag counts (i.e. 10^{-4}), may range from a few tens to a thousand. The drag polar C_L vs C_D is close to parabolic, as an effect of lift-induced drag (see Chapter 2), and the minimal drag occurs for C_L close to 0. But if this were a plot for a wing section, there would be no lift-induced drag. $C_{D0} > 0$ is due to skin-friction and separation losses (Chapter 2). The wing section of this plot has positive camber (i.e. is convex upwards), so there is nonzero lift at zero AoA.

Thermodynamic Properties of Air

For atmospheric flight at moderate Mach numbers, air does not dissociate and behaves like an ideal gas, which means that its state is determined by two quantities, such as pressure p (N/m^2) and absolute temperature T (K). The specific internal energy e (J/kg) depends on temperature alone as follows.

$$e = c_v T \tag{0.5}$$

where c_v $\left(\frac{\text{J}}{\text{K·kg}}\right)$ is the specific heat at constant volume. The perfect gas law states the following.

$$p = \bar{R}\rho T \tag{0.6}$$

where ρ (kg/m^3) is density and $\bar{R}$ is the specific gas constant (for air $\bar{R} = 287.058$ $\left(\frac{\text{J}}{\text{K·kg}}\right)$. The ratio γ (−) of specific heats at constant pressure c_p and volume c_v is the following.

$$c_p/c_v = \gamma = 7/5, \text{ and the difference is } c_p - c_v = \bar{R} \tag{0.7}$$

Total specific energy is the sum of specific internal and kinetic energies, $E = e + 1/2V^2$. Specific total enthalpy is $H = E + p/\rho$. When heat exchange can be neglected, the change of state is adiabatic, the specific total enthalpy of a moving packet of air is constant, and the flow is *isenthalpic*: $D/Dt[H] = 0$. A flow with the same total enthalpy in the whole flow field is called *homenthalpic*.

If the change is smooth, it is also isentropic, so $D/Dt[p/\rho^\gamma] = 0$. For flows both isenthalpic and isentropic, the state is determined by the value of one quantity. "Stagnation" states $()_0$ have $M = 0$ and "critical" ones $()^*$ have $M = 1$, and free stream properties are $()_\infty$.

As an example, C_p^* is the critical pressure coefficient:

$$C_p^* = \frac{2}{\gamma M_\infty^2}\left\{\left[\frac{1 + \frac{\gamma-1}{2}M_\infty^2}{1 + \frac{\gamma-1}{2}}\right]^{\frac{\gamma}{\gamma-1}} - 1\right\} \tag{0.8}$$

Shock waves are characterized by discontinuities in state and across them entropy is discontinuous. Chapter 4 discusses shock waves in more detail.

Sound waves are low-amplitude isentropic pressure waves traveling at the speed

$$a = \sqrt{\gamma\frac{p}{\rho}} \text{ (m/s)} \tag{0.9}$$

The dissipative processes are viscous diffusion of momentum and heat conduction. Viscosity is characterized by the coefficient of dynamic viscosity μ $\left(\frac{\text{kg}}{\text{ms}}\right)$. The kinematic coefficient of viscosity is $\nu = \mu/\rho$ $\left(\frac{\text{m}^2}{\text{s}}\right)$. The coefficient of thermal conductivity appearing in Fourier's law is κ $\left(\frac{\text{W}}{\text{mK}}\right)$.

Boundary-Layer Parameters

Chapter 2 discusses mathematical models for boundary layers and Chapter 7 discusses numerical models. A boundary layer, laminar or turbulent, is characterized by its thickness δ, momentum thickness θ, and shape factor H, all of which are defined in Chapter 7. These provide a rough description of the velocity profile from wall out to free stream.

In turbulence models, the following are also of importance:

- The friction velocity $u_\tau = \sqrt{\tau_w/\rho}$, where τ_w is the wall stress.
- Nondimensional wall distance $y^+ = yu_\tau/\nu$.

y^+ is a Reynolds number based on wall distance y and friction velocíty. For $y^+ > 50$, the effect on shear stress τ of mean flow viscosity $\mu d\overline{u}/dy$ is negligible; the dominant contribution to the stress budget is the Reynolds stress (see Chapter 2). $\overline{u}$ is the time average of the wall parallel velocity.

1 Introduction to Aircraft Aerodynamic Design

A scientist studies what is, whereas an engineer creates what never was.

Theodore von Kármán[1]

The successful aeroplane, like many other pieces of mechanism, is a huge mass of compromise.

Howard T. Wright, early British aircraft builder and designer

Preamble

The subject of this book, the aerodynamic design of aircraft, is an integral task within the entire aircraft design, and a prime focus is the shaping and lay out of the aircraft's lifting surfaces. Introducing the subject matter of the book, this chapter also conveys some appreciation for, and fundamental insight into, how and why wings evolve toward the geometric configurations we see in reality. An intrinsic characteristic of the development of a new type of aircraft is that it evolves from a *succession of design cycles*. This chapter describes and explains three of these cycles occurring in the early design process. As Theodore von Kármán implies, creativity lies at the heart of any engineering activity such as aircraft design. Belonging to the cognitive aspects of the human brain, creativity is not the realm of technology, but we do indicate how and where it enters into the three design cycles, and we encourage students to "think outside the box."

The fundamental aerodynamic quantities, lift and drag, are key to performance. Sizing the wing surface to the mission of the design is a crucial step in determining the baseline configuration, which then develops further in the succession of Cycles 2 and 3. This chapter introduces the tools, tasks, and workflows of the three design cycles and explains how computational fluid dynamics (CFD) and optimization procedures are involved, and it maps out where in the coming chapters each of these is treated in depth.

[1] https://doi.org/10.17226/10566, with permission of The National Academies Press.

1.1 Introduction

The prime task of the aerodynamics team in aircraft design is to suggest the aircraft's *flight shape* (i.e. the shape of its outer skin exposed to the airflow). But the overall shape of the aircraft must also reflect considerations such as structural integrity, weight, engine characteristics, and performance. We call this design process *configuration development*.

This chapter describes the aircraft development process and the role of aerodynamics, focusing on computational tools. In particular, we consider the iterations where a proposed aerodynamic shape is modified to better suit its requirements. Such a step can be approached and formulated as an optimization problem for which the computational multidisciplinary analysis and optimization (MDAO) tools clearly cross disciplinary boundaries, and these are in very rapid development as we write. The acronym MDAO is often shortened to MDO. This book is *not* about MDO in its entirety, but rather only the part that CFD plays within it. Examples are given on how the aircraft's mission requirements influence basic features such as the shape of the wing's horizontal projection: its *planform* (Figure 0.1).

The science of aerodynamics involves two apparently separate, but in fact related, studies. *Fundamental aerodynamics* is concerned with the qualitative and quantitative examination of air in motion – with its displacement, velocity, and acceleration. *Applied aerodynamics* concerns the physical forces exerted by air on the bodies immersed therein through the motion of the air relative to the body. There are four major questions to be addressed:

(1) How is the aerodynamic force created to keep an aircraft in the air, and how does this force *vary* with *shape*, *attitude*, and *speed*? This is the problem of *lift*.
(2) What is the propulsive force necessary to keep the aircraft moving through the air? This problem is associated with the air resistance or *drag*, which is fundamental to the general study of aircraft performance.
(3) How does the force and its distribution on the aircraft *vary* in *flight*? This is the problem of the stability and control of aircraft.
(4) How do the airloads during flight deform the airplane into the *flight shape*? This is the engineering field of (static) aero-elasticity.

Aerodynamics is seen by some as a branch of applied mathematics; others consider it largely an experimental subject. Mathematical analysis alone, however, is ineffective, as its necessary simplifying assumptions prove useful only in some situations, but they are invalid in others. On the other hand, to proceed only by experiment limits one's knowledge to very specific situations and inhibits the making of reliable predictions.

The aerodynamicist, therefore, needs good enough theories to combine both of these approaches, using analysis to deepen and extend their knowledge. Continuous experimenting is required to check the validity of the assumptions and to improve understanding of the physics. Answers are always to some extent approximate, and the conclusions drawn are often limited to certain classes of situations.

To the novice, this all becomes a mixture of engineering experience, with models being constructed by guidance from theory but completed by curve fits, all producing a forest of formulas, each with limited applicability. Indeed, the approaches to solutions range from mostly statistical and empirically based models to fully physics-based methods. The limits to applicability of the physics-based methods have been pushed back significantly by high-performance computing machinery and software for CFD. But these computational tools are, and will be for the near future, unable to solve problems through first-principles models that are universally valid for typical-flight Reynolds numbers. Thus, many assumptions are still made in applied aerodynamics to facilitate a computational approach, and the caveat above still stands. Analogy with medical science is appropriate here, where engineering experience in the field plays the role of *clinical experience*.

Theory of Ideal Fluids

Three simplifying assumptions about airflow that are very useful at times are as follows.

(1) *Incompressible* flow: The assumption that fluid density is constant leads to a major simplification: the thermodynamics decouples from the kinematics. This gives very good results, provided that the fluid velocity is not too great, but is totally invalid at high speeds.
(2) *Inviscid* flow: Here, it is assumed that the viscosity of the fluid vanishes. A useful theory may be developed that gives good answers to the problem of lift. On the other hand, drag cannot be accounted for at all on this basis.
(3) *Irrotational* flow: Here, fluid particles do not rotate, being mathematically expressed as the vanishing of the vorticity of the velocity field $\omega = \nabla \times \mathbf{v}$.

Flow of an "ideal fluid" satisfies all three of these assumptions and leads to the d'Alembert paradox that the net force on a body vanishes. It would seem then to be completely useless, but the theory was modified and made into the Prandtl–Glauert wing flow engineering mathematical tool by Ludwig Prandtl and his followers.

1.1.1 Aerodynamic Design Is Part of Aircraft Design

The task of designing an aircraft is among the most complex in engineering. Not counting the smallest components such as nuts, bolts, and rivets, an aircraft may have hundreds of thousands of components, with over a million important design parameters and many more that are less important. Complex and advanced simulation and data management software systems are needed to support the design teams, both in the tasks each design team undertakes and for putting together the data and design parameters for the configuration as it evolves through (many) iterations and redesigns to arrive at a satisfactory solution. By the term *configuration* we mean the general layout and external shape, dimensions, and other relevant characteristics of the design.

Every aerospace company has its own structure and process for design, reflecting the diverse and complex nature of conceiving a new aircraft.

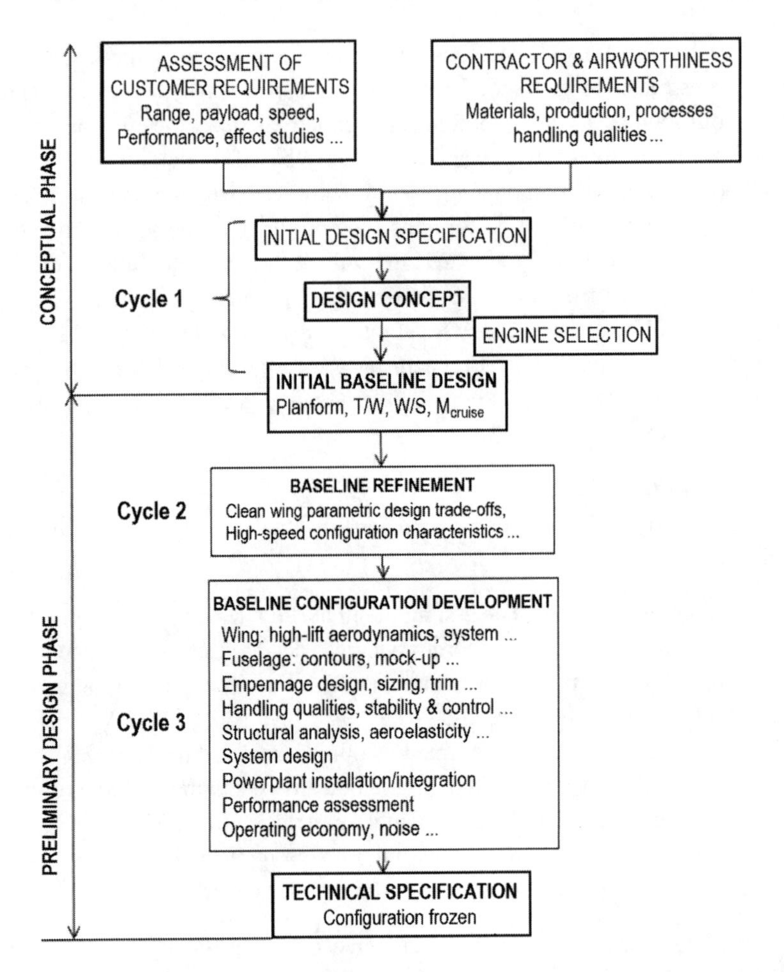

Figure 1.1 Configuration design and development of a typical transport aircraft evolving after a succession of Cycles 1, 2, and 3. (Adapted from Torenbeek [29], reprinted with permission)

For example, the preliminary design of the Boeing 777 was carried out by 3000 people. Coordination was facilitated through weekly design meetings of 25 lead engineers, each representing 100+ engineers in their specialty. Aerodynamic design is only one part of this vast enterprise.

There are a number of good textbooks on aircraft design (e.g. [19, 23, 29]) that spell out how the many disciplines work together to synthesize an aircraft in a process that is subdivided into conceptual, preliminary, and detailed design *stages*. Figure 1.1 presents a flowchart of the conceptual and preliminary design stages. It provides an overview of where and how aerodynamic design enters into the overall synthesis. The entire development of a new aircraft takes place in a *succession of design cycles*.

In the course of each of these cycles, the aircraft is designed in its entirety. Investigation is carried out into all of the main groups, airframe systems, and pieces of equipment to a similar level of detail. The extent of this detailing steadily increases as the design cycles succeed each other, until finally the entire aircraft is defined in every detail.

Using Torenbeek's [29] terminology, we further designate the subsequent basic design stages as follows.

(1) Cycle 1: conceptual design, also called speculative design, explores a large basic parameter space.
(2) Cycle 2: baseline refinement design, which demonstrates the feasibility of the speculative design.
(3) Cycle 3: baseline configuration development, which determines the best conceivable design among the feasible ones regarded as sufficiently mature. It finishes with the decision to freeze the configuration, ending the preliminary design.
(4) Detailed design comprises the final hardware design cycles of the configuration, which goes into production after flight-testing of prototypes. These cycles are beyond the scope of this book.

A number of aspects of the first three design cycles will be further elaborated and discussed in this book. The Cycle 1 conceptual design concludes with an initial baseline configuration, and this is covered in the present chapter. Chapters 8 and 9 present the Cycle 2 procedures followed in evolving an initial baseline configuration, and they constitute a major portion of this book.

Chapters 9, 10, and 11 present surveys of several topics related to the further Cycle 3 elaboration and development of the baseline configuration.

To give some idea of the magnitude of work in preliminary design, the initial baseline design of a transport aircraft will require several thousand person-hours. The subsequent design phase of variants and parametric studies will demand multiples of this effort.

Torenbeek [29] gives some idea of the scope of configuration development, citing the Lockheed L-1011 program. In the two years of its configuration development, over two million person-hours were expended on investigating various configurations and approaches in order to determine the optimum design.

Clearly, a textbook cannot hope to elucidate every aspect of what an aerospace company carries out. Instead, this text

(1) provides an overview of how modern software for analysis and optimization is used;
(2) outlines the mapping from shape to aerodynamic forces in some detail; and
(3) discusses the techniques and design tasks that can be carried out with academic software tools.

Hierarchical Breakdown of Aircraft: The Cayley Paradigm

Viewing the configuration as a hierarchy of its constituent components helps to manage the complexity of the design task, both in handling the design space (the set of design parameters) and in modeling its function.

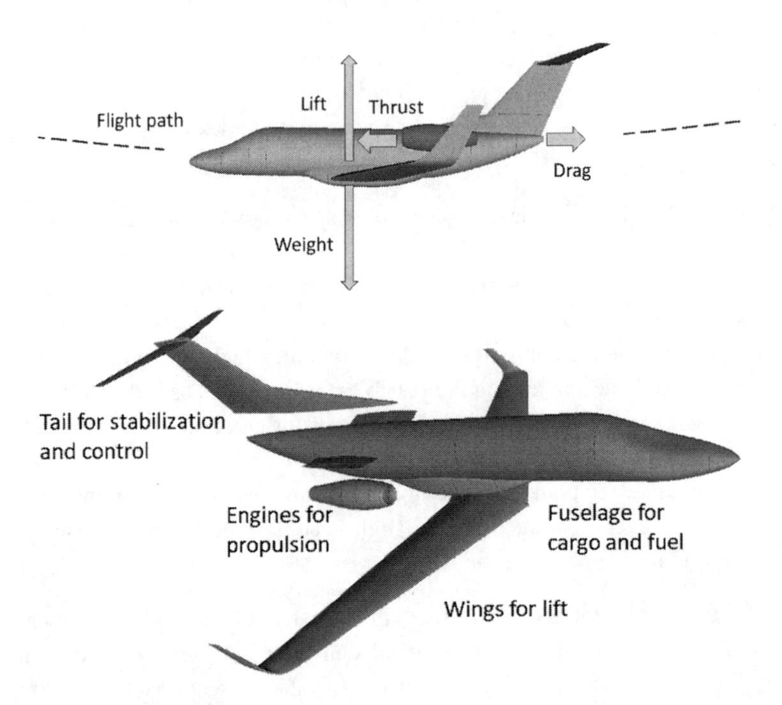

Figure 1.2 Forces in un-accelerated flight and aircraft components. (Top) Lift balances weight and thrust balances drag. (Bottom) Cayley's principle: each component of the aircraft has its distinct function.

The traditional hierarchical approach to design follows Sir George Cayley's design paradigm (see Figure 1.2, bottom). It assigns functions such as lift, propulsion, trim, pitch-and-yaw stabilization and control, etc., exclusively to corresponding subsystems such as the wing, the engine, the tail unit, etc. The top of Figure 1.2 shows weight (the gravitational force on the aircraft), which the lift must overcome, and drag, which must be balanced by thrust to stay aloft.

If these subsystems and their functions influence each other only *weakly* in well-understood ways, one is able to treat and optimize each subsystem with its functions more or less independently. Implicit in the paradigm is the decomposition not only of the subsystems and functions but also of the engineering disciplines into aerodynamics, structures, flight control, etc. This *decomposition* of the engineering disciplines has led to the established practice of *sequential and iterative design cycles*. However, the decomposition may limit the design space by neglecting potentially beneficial couplings between subsystems. If the design can be carried out with more concurrency between the design activities of the different teams, the number of design cycles can be reduced and the configurations can become more efficient. This is the goal of the MDO techniques enabled by high-performance computing.

1.1.2 Lift and Drag: Keys to Performance

When performance of the aircraft is discussed, the context is important: an airline executive looks at the bottom line in servicing the needs of the company's clients,

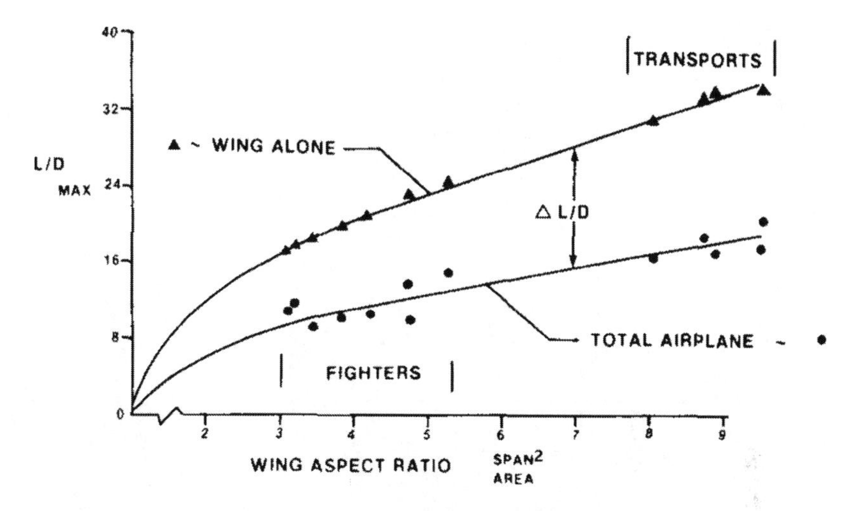

Figure 1.3 Maximum lift-to-drag ratio for total configuration and wing alone vs wing aspect ratio. (From Chuprun [3], AFWAL, public domain)

a stunt pilot is interested in quick response to control inputs, while top speed and turn rate are important qualities for fighter pilots. All performance metrics require data at least on airplane velocity and acceleration, and these must be derived from the forces exerted on the airplane by the surrounding air. While it is important to understand the airflow patterns, as discussed in Chapter 2, it is really only the resulting forces that matter for the aerodynamic designer.

1.1.3 Wings, Lift, and Drag

Wings provide lift that is much greater than their own weight. Fighter wings lift about 90 times their own weight, while for transport aircraft the ratio is about 22, where a prime focus is given to high cruise efficiency, hence maximum *lift-to-drag* ratio is the objective. In contrast to transport aircraft, fighters need high lift to maneuver at the expense of higher drag. Using data representative of modern fighter and transport aircraft [3], Figure 1.3 highlights the powerful leverage that wings possess in fulfilling their prime function of generating lift.

The plot in Figure 1.3 shows $(L/D)_{max}$ growing strongly with increasing aspect ratio. The wing contributes all of the lift and only half of the drag for the whole configuration, so its $(L/D)_{max}$ is about half of that of the total airplane. With some hyperbole, Chuprun coined the phrase "wing is king" in aircraft design, in the sense that the wing is the *backbone* of an airplane. Certainly, there is no heavier-than-air flight without aerodynamic lift. This fact also motivates why a flying wing, unburdened by the drag of other airplane components, potentially has high aerodynamic efficiency.

At its most elementary level, Figure 1.4 symbolizes and summarizes the description of the first steps in aircraft design. It also indicates the process that repeats itself in the subsequent cycles to further develop the design in more detail using an increasing number of parameters.

Table 1.1 Primary parameters for baseline planform design.

Planform	Area S
	Wing span b
	Aspect ration AR
	Taper ratio λ
	Sweep angle Λ
	Average thickness $(t/c)_{ave}$
Performance	Cruise Mach M_{cruise}
Propulsion	Thrust-to-weight ratio T/W
Structures	Weight-to-wing area ratio W/S

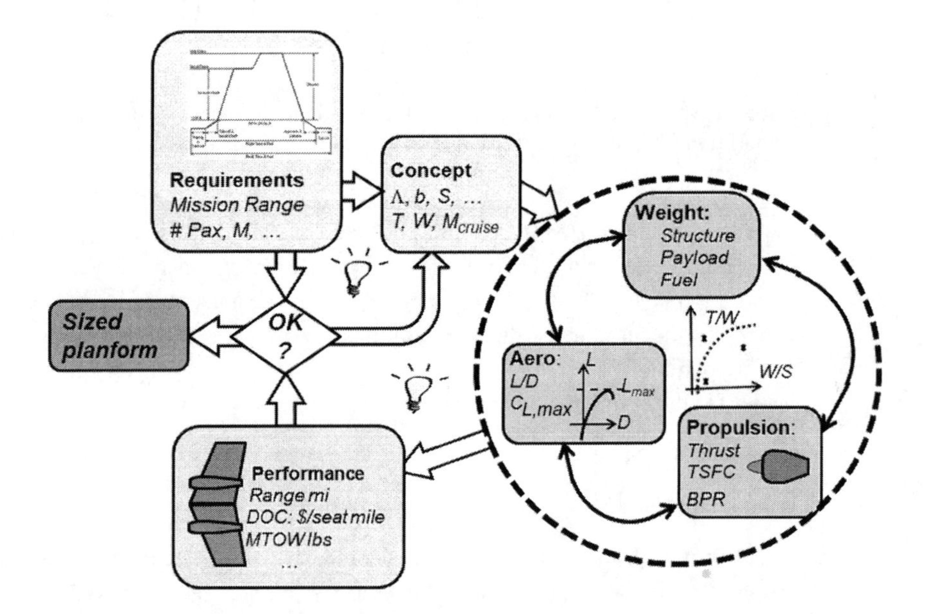

Figure 1.4 Initial sizing process in the Cycle 1 aircraft design process, with outcome of wing planform and size, showing the role of aerodynamics.

To start the process, the concept – usually hand-drawn – must be transformed into a geometry (i.e. its *flight shape* specified by a handful of primary parameters, such as those in Table 1.1).

That so many primary variables concern wing shape gives credence to Chuprun's claim that wing is king.

Specific Range

For most airplanes, range is one of the most important measures of performance. This is certainly the case for commercial aircraft, and for many military airplanes the maximum combat radius is of major importance. The differential form shown in Eq. (1.1) of Breguet's celebrated range equation relates the specific range SR of an aircraft in cruise (here, miles traveled per gallon of fuel at weight W) to properties of propulsion, weight, and aerodynamics:

$$SR = Const \underbrace{1/TSFC}_{engine} \quad \underbrace{ML/D}_{aerodynamics} \quad \underbrace{1/W}_{structure} \tag{1.1}$$

The equation is a simple consequence of the fact that, for steady flight at constant altitude, $L = W$ and $T = D$ (top of Figure 1.2), with M being the flight Mach number. The weight of the aircraft W diminishes as it burns its fuel. We see three important quantities:

(1) Choice of engine and its thrust-specific fuel consumption $TSFC$
(2) Total aircraft weight W
(3) Aerodynamic cruise efficiency $M\ L/D$

Measured in, for instance, $\frac{Gal}{N \cdot hr}$, $TSFC$ decreases when newer engines bring better fuel economy. For propeller propulsion, *power-specific fuel consumption* (= PSFC) is a more relevant measure across the speed range. Advanced materials allow for a lighter-weight aircraft structure. Equation (1.1) tells us that, to improve specific range, the *aerodynamic efficiency* $M\ L/D$ – the product of the flight Mach number (dependent on the planform) and the aerodynamic *quality* (the lift-to-drag ratio) – should be as high as possible.

Compound Benefits
Such improvements compound the benefits through interdependencies. If less fuel is needed for a given mission, the takeoff weight is reduced, less lift is required, so the wing area can be smaller (hence less drag), the cruise Mach can be increased, etc., in a virtuous circle. On the other hand, when weight *increases*, one unfortunately encounters the opposite: the vicious circle of compounding weight penalties rushing toward poorer performance.

For an innovative aircraft concept with better fuel efficiency, Eq. (1.1) points design efforts in the direction of a configuration with optimized aero-structural sizing to yield less weight and drag. Design represents the search for the optimization of these innovative aero-structural concepts for maximum aerodynamic and structural efficiency as well as safe and controlled flight during normal operation and in critical conditions.

1.1.4 Sizing the Wing Planform: Initial Parametric Design Cycle 1

The design process starts with the main mission requirements, desired performance, and cost goals, as Figure 1.4 suggests.

Requirements for commercial airline services follow from the analysis of the intended flight route, including data on expected traffic volume and desired frequency, typically between city pairs. This sets the desired payload-range characteristics.

The *performance* requirements would typically include factors such as maximum takeoff weight, start and landing distances, maneuverability, rate of climb, service ceiling, speed, fuel economy given a maximum size and weight, etc. In addition, the aircraft must be certified by regulatory authorities, which adds requirements that are discussed in more detail below and in Section 1.1.5.

Airworthiness Requirements

The suggested configuration must satisfy further conditions to ensure it is *airworthy* (i.e. can be operated with adequate safety in the air as well as on the ground). Airworthiness requirements govern performance, control, stability, trim, structural and mechanical design and a host of other aspects. Published by the International Civil Aviation Organization, the International Standards of the Chicago Convention (1944) spell out what is required for an aircraft in civil aviation to be deemed airworthy and certifiable by the regulatory authorities.

Structural design is of critical importance to aircraft safety and plays a key role in aircraft cost, weight, and performance. In addition, the aircraft structural weight affects performance through the compounding effect explained above. In order to predict the aircraft cost and empty weight, we must estimate the weight of each of the components. Thus, we need to calculate the loads that they will have to support in flight and on the ground, wing bending moments due to aerodynamic lift, the weight of the structures, landing and taxi-bump loads, etc. For certification of an aircraft structure, one might examine tens of thousands of loading conditions, several hundred of which may be critical for some structural element.

The definition of strength requirements for commercial aircraft is specified in Federal Aviation Regulations (FAR), Part 25. Many of the load requirements are defined in terms of the *load factor*, n. This is defined as the effective transversal acceleration of the airplane in units of g, $n \approx L/W$ when angles of attack and sideslip are small.

Flight Envelope

The flight envelope is usually depicted in altitude (H) – Mach (M) (or Placard) diagrams and $V-n$ diagrams with speed V and load factor n (Figure 1.5). Speeds are given as *equivalent airspeed* (EAS): the speed at sea level that would give the same dynamic pressure as the true airspeed (TAS) at altitude.

$$EAS = TAS\sqrt{\rho(H)/\rho(0)} \tag{1.2}$$

The diagrams below are taken from the design of the TCR discussed in Section 1.1.5.

Mach-Altitude Envelope

The top of Figure 1.5 shows a Placard diagram. The left boundary is the low-speed stall limit, set by weight, wing area, and $C_{L,max}$ for the configuration. The right boundary is set by the maximum allowable dynamic pressure and is of the form $M^2 < const./p(H)$, with pressure $p(H)$ taken from the standard atmosphere. The dashed circle surrounding the design cruise condition – the black dot – indicates the necessary region where a stable "healthy" flow pattern must obtain and becomes part of the design strategy.

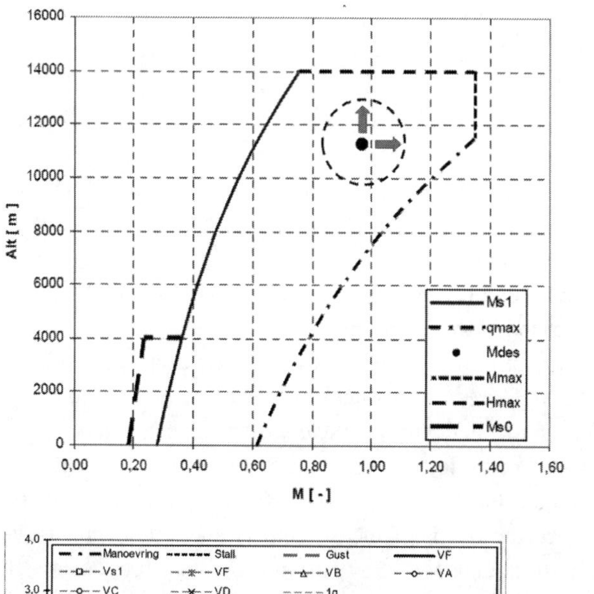

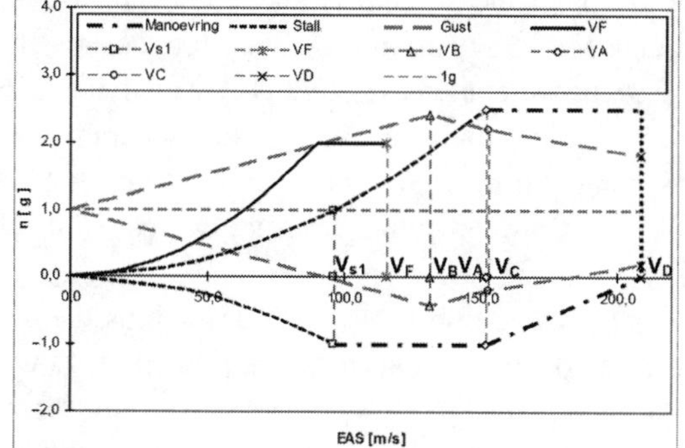

	Name	Speed [m/s]
V_{S1}	1-g stall speed	96
V_F	Max flap speed	115
V_A	Manoeuvring speed	152
V_B	Speed for max gust	131
V_C	Cruise speed	153
V_D	Dive speed	209

Figure 1.5 Diagrams describing the aircraft flight-state envelope, with a dashed circle indicating the region of stable "healthy" flow surrounding the design point. Top: Mach-altitude envelope; bottom-left: $V - n$ diagram; bottom-right: speed definitions table. (Courtesy of R. Larsson, private communication, from EU project SimSAC [20])

Speeds

Details of how speeds are determined are too technical for this textbook (see the table in Figure 1.5). The $1g$ stall speed V_{S1}, for instance, is the minimum airspeed for level flight (i.e. with load factor 1). It is determined by the maximal lift obtainable and weight.

$$V_{S1} = \sqrt{\frac{2W}{\rho(0)C_{L,max} \cdot S}}$$

Cruise speed V_c is a performance requirement, and dive speed V_D is determined from the q_{max} boundary in the Mach-altitude envelope: $V_D \approx 0.61 \cdot 340 = 207$ m/s.

The maneuvering speed V_a is the maximal speed for maximal control surface deployment: larger deflections could result in load factors exceeding the maximum allowed.

Maneuver Diagram

This $V - n$ diagram illustrates the variation in load factor with EAS for maneuvers. At low speeds, the maximum load factor is constrained by stall (i.e. the aircraft $C_{L,max}$), with high-lift devices deployed up to speed V_F and retracted up to V_D. At higher speeds, the maneuver load factor may be restricted as specified by FAR Part 25. The maximum maneuver load factor is usually +2.5.

The negative value of n is -1.0 at speeds up to V_c, decreasing linearly to 0 at V_D. Maximum elevator deflection at V_A and pitch rates from V_A to V_D must also be considered. Loads associated with vertical gusts are evaluated over the range of speeds. The dashed "Gust" line in Figure 1.5 shows estimated gust loads that are probably encountered no more than once per 100-h flight time at that EAS. The gust speeds employed are the result of measurements and statistical models, hence the "probably." The "buffet boundary" is akin to the boundary for "healthy flow." Its limit is not sharp, since the severity of buffet increases gradually with dynamic pressure and mild buffet is tolerable. Low-speed buffet is caused by flow separation as the aircraft approaches stall and the separated flow becomes unsteady. At higher Mach numbers, high-speed buffet is caused by flow separation from the wings by shock–boundary layer interaction.

The flight envelope is not given a priori, but must develop as the design studies reveal more and more of the aerodynamic and structural properties. Low-speed properties like the stall boundary are related to $C_{L,max}$, as discussed in Chapters 8 and 9. Airworthiness certification imposes other limits, such as the fact that a load factor between -1 and +2.5 must be tolerated at all speeds up to V_D. These numbers, which are typical for airliners, depend on the type of aircraft.

Parametric Cycle

At this point, the *parametric cycle* of conceptual design starts, evaluating the initial idea and concepts with *many variants* studied. Past experience along with simplified aerodynamic and structural estimates and statistical databases, which we will refer to as "fidelity level L0 Handbook Methods," then synthesize these characteristics into the first concept, an initial sized configuration, together with preliminary performance levels. Note that range is given, whereas aircraft weight (and thus aircraft physical size) is to be calculated. The initial concept may prove, or cast doubt on, the feasibility of the proposed mission. It may appear too ambitious, requiring an excessively large and expensive aircraft, or the indications are that it can be carried out by an aircraft of acceptable size. Thus, the design process takes another iteration, usually followed by yet more iterations, to arrive at a mission and corresponding design that can be expected to capture the desired market share and give sufficient return on investment.

For example, Figure 1.3 indicates how $(L/D)_{max}$ increases with aspect ratio because induced drag decreases. It would seem that you should design the wing with

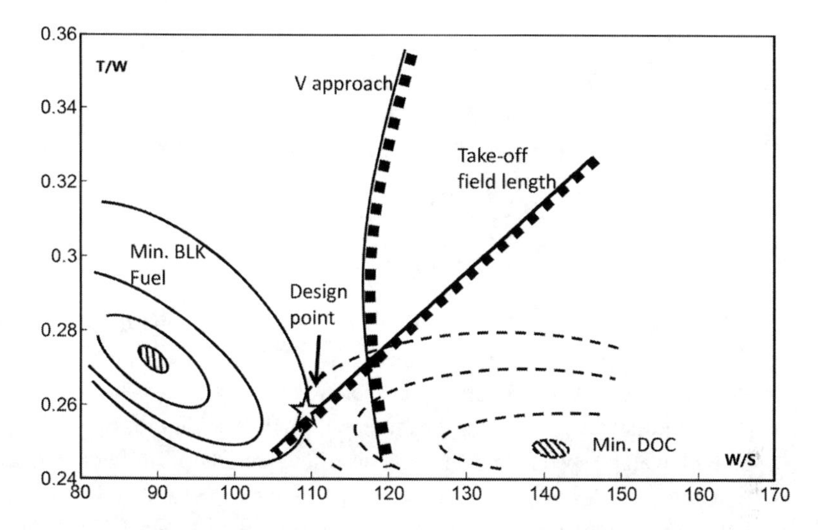

Figure 1.6 Constraints and isocurves of measures of merit in the W/S–T/W diagram.

a very high aspect ratio. However, due to the higher strength required and thus heavier structure, wing weight increases with increasing aspect ratio and area. Since the goal is a wing with the necessary lift and the least weight and drag, trade-offs between the structural and aerodynamic benefits are necessary to find the best compromise.

With the planform described mathematically, the analysis procedures can estimate the concept's functional properties, such as its:

1. Weight and balance, including payload, fuel, systems and structural weight predicted by structural design
2. Thrust and fuel needed to carry out the mission to fly at a specified cruise Mach number
3. Lift and drag forces during the mission

These procedures in turn generate the data to estimate performance and flight dynamics in order to determine whether the concept fulfills the requirements and goals for the design.

Outcome: Sized Baseline Configuration

Once converged, the initial iteration loop yields a *baseline configuration*, a three-view drawing, including the wing planform sized to fulfill the mission and represented by a small number of parameters, given in Table 1.1 and displayed in Figure 0.1. It is common to display a T/W versus W/S diagram such as Figure 1.6 with the various requirements and criteria as boundaries for feasible designs. Usually, the *set* of design parameters $\mathbf{X}$ is fixed, so the various quantities of interest, such as takeoff field length and direct operating cost (DOC), can be worked out for some given *values* of $\mathbf{X}$. Such parametric studies also produce *sensitivities* of performance measures of parameter changes, for small excursions in the design space from the design point (i.e. within the dashed circle in Figure 1.5). The constraints are shown in Figure 1.6 as curves with cross-hatches on the forbidden side. The landing speed limit is almost independent of T/W: the limit on landing speed $V_{approach} < limit$ is determined essentially by wing loading W/S,

$$W/S = L/S \leq 1/2\rho_{air}limit^2 \cdot C_{L,max}, \tag{1.3}$$

since the attainable maximal lift coefficient $C_{L,max}$, defined in Eq. (0.1), is determined by the wing shape and the aerodynamics "laws." Takeoff length is also influenced by the thrust, hence its sloping constraint line.

The block fuel "BLK Fuel" is the amount of fuel necessary to fly the mission, and it depends on weight, speed, drag, and engine efficiency. DOC is measured in, say, dollars per seat mile. The design point must be in the feasible subset allowed by the constraints, most probably on the boundary of it. Now consider the iso-curves for DOC and BLK Fuel, a design point, and moving it in some direction. It is clear that for the point to be good there must be no directions that improve both measures. This is the definition of the Pareto front for the two objectives (see any standard textbook on optimization techniques for an explanation of the Pareto front). So the trade-off is between points on the Pareto front for DOC and BLK Fuel.

We now have at the design point T/W and W/S. For cruise $T = D$ and $W = L$, so

$$\frac{T}{W} = \frac{1}{L/D} \text{ and } \frac{W}{S} = q\, C_L, \tag{1.4}$$

which gives requirements for the wing aerodynamics to fulfill where $q = 1/2\rho(alt)V_\infty^2$ is dynamic pressure, dependent on altitude and airspeed.

Designs Vary Widely

Figure 1.7 indicates how greatly the resulting design can vary depending on mission requirements, from human-powered aircraft at the bottom-left to jet fighters at the top-right. Note that the ordinate is *power*-to-weight, and that the scales are logarithmic to cover the vast expanse of this design space.

Sizing

"Sizing" is the design process that determines the aircraft takeoff gross weight, empty weight, and fuel weight required for the configuration to perform a specified mission with a specified payload weight. This calculated size is used to estimate a revised wing area, fuselage length, etc., appropriate to the determined weight. For a detailed description, see Raymer [19].

Initial sizing determines the specific geometric data needed for the baseline layout by analyzing the aerodynamics, weights, and propulsion characteristics in order to check all performance requirements for the design mission. Decisions and trade-offs are traditionally made on the basis of compilations of carpet plots of the parameter

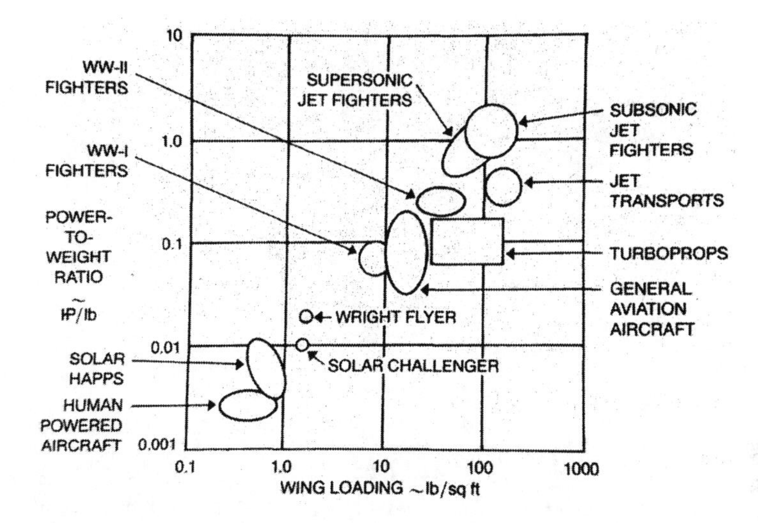

Figure 1.7 Mapping of aircraft classes onto the thrust–weight landscape. (Courtesy of David Hall [11])

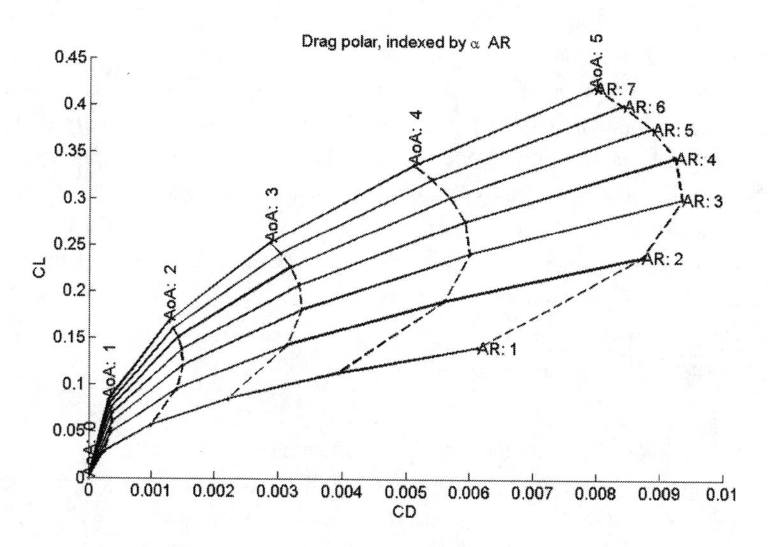

Figure 1.8 Carpet plot: $C_L, C_D, \alpha = AoA$, and AR.

studies: a carpet plot shows graphs of four variables constrained by two relations. Figure 1.8 is an example of this, showing lift and induced drag coefficients, angle of attack α, and aspect ratio AR for rectangular wings at low speed.

Creativity

The design process just summarized involves much *creativity*, as symbolized by the light bulb symbol below, and it is not easily taught, as is explained in the panel below.

Panel: Creativity – when the *light bulb* glows
The "light bulb" symbol is used in our figures to indicate creativity – an important part of the design process.

Creativity is not in the realm of technology; rather, it belongs to the cognitive aspects of the human brain. How we create is assessed in neuroscientific studies, which indicate that human beings innovate by absorbing the best *existing ideas* and making them *better*. "Whether inventing an iPhone, manufacturing the next-gen aircraft, or launching modern art, creators remodel what they inherit," write the authors Brandt and Eagleman in their book *The Runaway Species*. We do so by engaging in three basic strategies by which all ideas evolve: "bending, breaking, and blending." We take the raw materials of experience and then bend, break, and blend them to spawn outbursts of creative activity symbolized by a bulb lighting up, leading to new outcomes. Unlike wild creatures, who operate largely on autopilot, humans usually avoid repetition and seek novelty, whence the dictum *think outside of the box*. Brandt and Eagleman's narrative is filled with tips on how to produce successful ideas: practice, experiment, generate many ideas, and let most die. A cornerstone of the creative process is to generate *options*, which is equally applicable to aerodynamic design, as in Hemingway's reputed 39 endings to *A Farewell to Arms*.

Handbook Methods to Size the Baseline Configuration

The design and analysis methods discussed in the book assume that a sized baseline proposal exists. This begs the question of *how* to produce a sized baseline if we do not have the design tools for all disciplines shown in Figure 1.4.

Handbook methods capable of producing sizing information on weights, wing planform, and propulsion from basic mission requirements are presented in aircraft design textbooks and their accompanying software. They belong to fidelity level L0. Raymer's classic textbook on conceptual design [19] is an excellent starting point, complemented by its RDS software package. In addition, Roskam's book [23] and associated commercial AAA software from the DAR corporation is recommended.

The USAF Data Compendium (DATCOM) [13] is a collection of algorithms, formulas, and design rules based on physics, statistics from data for many existing aircraft, and curve fits, obtained through the Air Force's long-standing cooperation with aerospace companies. The Digital DATCOM software [30] is capable of producing weights, inertias, and force and moment data from subsonic to supersonic speeds from the DATCOM formulas and algorithms and scant geometric information on the configuration. Its fidelity is limited to (and for) conventional configurations.

ESDU (www.esdu.com) is a high-end commercial product providing validated engineering analysis tools for aerospace engineering. The tools include methodologies, design guides, equations, and software.

With some experience of and access to such L0 tools, the design team will look for useful templates from, for example:

1. Raymer's webpage (www.aircraftdesign.org)
2. the SUMO aerospace CAD package [14]
3. European collaborative projects such as
   SimSAC [20, 22]
   AGILE [4], and its Academy [6], etc.

Then the work begins, as the next section exemplifies.

1.1.5 Example: Sized Baseline Configuration – The TransCruiser

High-Speed Civil Transport and the Saab TransCruiser

The US High-Speed Research program was started in 1990 aiming at Mach 2 transoceanic flight. The High-Speed Civilian Transport concept presented in 1995 had a "cranked" wing. Its inner part was highly swept, with a leading edge extending far forward along the fuselage, and the outboard panel was less swept. The cranked wing concept is similar to the F16-XL research aircraft built for high-speed research as a stretch of the standard F16 (Figure 1.9). The wing modification aimed at improved supersonic and preserved transonic performance. The F16-XL was instrumented to provide a wealth of flight data, including pressure maps through the transonic speed range. The availability of the data for a wing similar to the HSCT concept provided a unique opportunity for CFD correlation and code validation with flight and wind tunnel data.

The HSCT project was abandoned due to economic and environmental issues, such as sonic boom concerns. In 2001, Boeing announced the Sonic Cruiser concept, also

Figure 1.9 The F16-XL in flight. (From Elegance in Flight [18], NASA, public domain)

with a cranked wing. With a range of 10,000 nautical miles and a cruise Mach number above 0.95, it was thought to be a competitive configuration. Its cranked leading-edge allowed leading edge high-lift devices, so take-off and landing would not require extreme angles of attack. The Concorde, with its curved leading edge, had no slats, needed a high angle of attack at take-off and landing, and resorted to drooping the nose for better pilot vision.

Hepperle [12] carried out a concept study on a possible aircraft configuration. One result of the study is that, due to the near-sonic speed and high drag, the concept encounters demanding engineering challenges within a very narrow design space. John F. Kennedy's "We choose to go to the moon" address comes to mind:

> We choose to …do the other things, not because they are easy, but because they are hard.

Similar to the Boeing Sonic Cruiser, Saab proposed the cranked wing TransCruiser (TCR), cruising at Mach 0.97, for one of the Design, Simulate, and Evaluate exercises in the SimSAC Project [20–22]. Cycle 1 work starts with the development of an initial baseline design, using L0 in-house sizing methods. It is inefficient to fly very close to the sonic wall, such as Mach 0.97. Saab purposely selected this case to *stress* the shortcomings of their in-house methods for predicting characteristics in transonic flight.

Specification and Baseline Configuration

Table 1.2 spells out the requirements for the TCR design.

Using in-house design methods, the Saab team produced the sized baseline configuration in Figure 1.10. The sizing procedure includes many aspects, among them estimates of weights, engine performance, and aerodynamics. The landing and cruise phases define constraints for thrust-to-weight and wing loading. The predicted aircraft weight is also an output of the procedure. The outcome from this sizing is as follows.

- Maximum take-off weight (W_0): 227 metric tons × g = 2,225 kN
- Required static thrust (T_0): 890 kN
- Required net wing area (S_{net}): 377 m^2

The design C_L varies from 0.48 at Mach 0.8 to 0.3 at Mach 0.97

Table 1.2 Nominal TCR specification.

Parameter	Requirement
MTOW	180 metric tons × g = 1766 kN
No. Pax	200
Range	10,000 km (5500 nm)
Design cruise speed	$M_c = 0.97$
Baggage and freight	LD3-46W containers

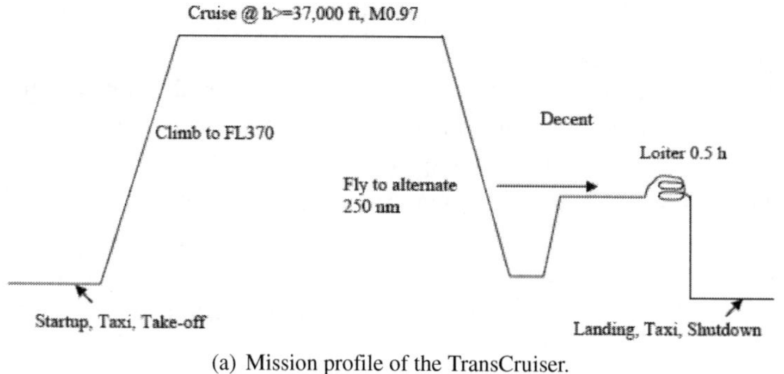

(a) Mission profile of the TransCruiser.

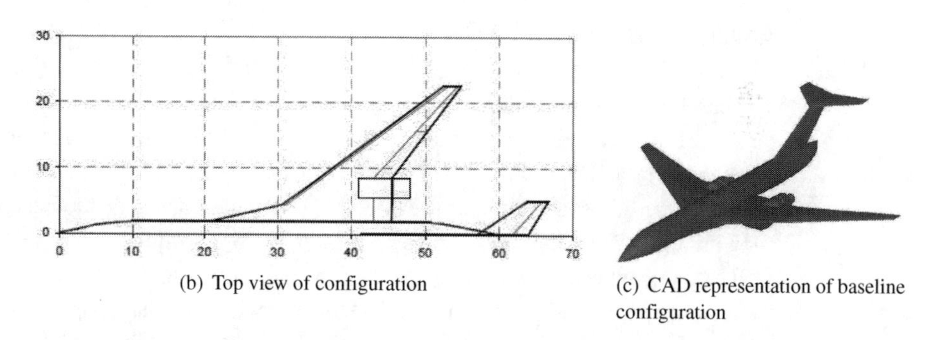

(b) Top view of configuration

(c) CAD representation of baseline configuration

Figure 1.10 The baseline T-tail configuration for the TCR exercise. (From EU project SimSAC [20, 21], courtesy of Roger Larsson)

The baseline is a 'mid-to-low'-winged T-tail configuration with two wing-mounted engines. Ailerons and rudder are used together with an all-moving horizontal tail for control. Flaps and slats are used as high-lift devices. The landing gear is a conventional tri-cycle type with main gears mounted in the wing (see Figure 1.10).

The Saab team concluded their Cycle 1 design with the following conclusions and suggestions:

> Classical flight-mechanics analysis indicates that the base-line is stable in pitch at all time, and increasingly so in the transonic region, but the elevator effectiveness is low. The TCR aircraft is both laterally and directionally stable throughout the whole Mach envelope. The flying characteristics are generally good, and the control system can easily improve the undesirable Short period and Phugoid-mode characteristics. The main problem with the baseline is the excessive elevator angles needed for trim. The elevator angle should be not more than 5 to 10 ° for most flight conditions, we estimate it to be nearer to 15 °. It follows that the elevator effectiveness together with the pitch stability, especially in the transonic region, should be improved in Cycle 2. An option is to re-position the main wing and modify the horizontal tail layout.

For the purposes of this book, the TCR transonic cruiser has all the ingredients to make an excellent case study for using the aerodynamic tools to take the initial aircraft specification through Cycle 2 to a feasible concept, and then through Cycle 3 for closed-loop control system design in the following steps. Chapter 1 presents the baseline TCR configuration.

A votex lattice method (VLM) model (L1) is created by the tools of Chapter 3.

Chapter 5 produces a CAD model and grid for L2 analysis.

Chapter 8 studies the airfoil used to shape the wing.

Chapter 9 describes the aerodynamic redesign from variants TCR-1 to TCR-4 to TCR-C15.

Chapter 10 covers the iterative process that solved the trim problem, presents the prediction of flying qualities, and draws overall conclusions.

1.2 Advanced Wing Design: Cycle 2

The baseline sizing procedure, Cycle 1, is a broad-ranging exploration over the parameter space embracing much cross-disciplinary-type analysis ranging from aerodynamics through manufacturing and costs. The work is usually carried out in the Future Projects Office of the aerospace company. A complete layout consists of the baseline design showing three view drawings and some principal cross-sections.

If the concept is to be developed further, the next step is to check the extent to which the characteristics and performance of the design will meet the design requirements.

The Future Projects Office hands this work over to the Advanced Design Office, where further development (Cycle 2) begins with experts in each respective discipline (e.g. aerodynamics) to establish the technically consistent feasibility of the design. Figure 1.1 indicates this "hand-off" of the "Initial Baseline Design" to the "Advanced Design" teams to further explore the parameter space. The object is twofold: first, to improve the design where it does not meet the requirements; and second, to investigate the most likely possibilities and to determine whether other variants may prove a better proposal. A protracted and labor-intensive effort involving more advanced trade studies develops the baseline in greater detail to become a realistic candidate. A family of designs thus emerges that are easily comparable with each other as well as with the baseline design.

Although the fidelity of the methods used should be as high as possible, there always remains uncertainty. It is necessary to carry out further detailed investigations in Cycle 3 of (at least) *two* alternative configurations to differentiate between the evolving designs before an irrevocable choice is made.

Design Examples

Chapter 9 presents a number of historical examples beginning in the 1940s with a study of one of the first swept-wing jet fighters, the Saab J 29. The Mach 2 Concorde was designed in the 1960s when CFD was incapable of producing flow solutions across the speed range. Its design uses purposely separated vortical flow over the highly swept delta wing to give reasonable lift at low speed. The 1970s saw Saab developing the SF340 commuter turboprop with modern NASA airfoils. The Common

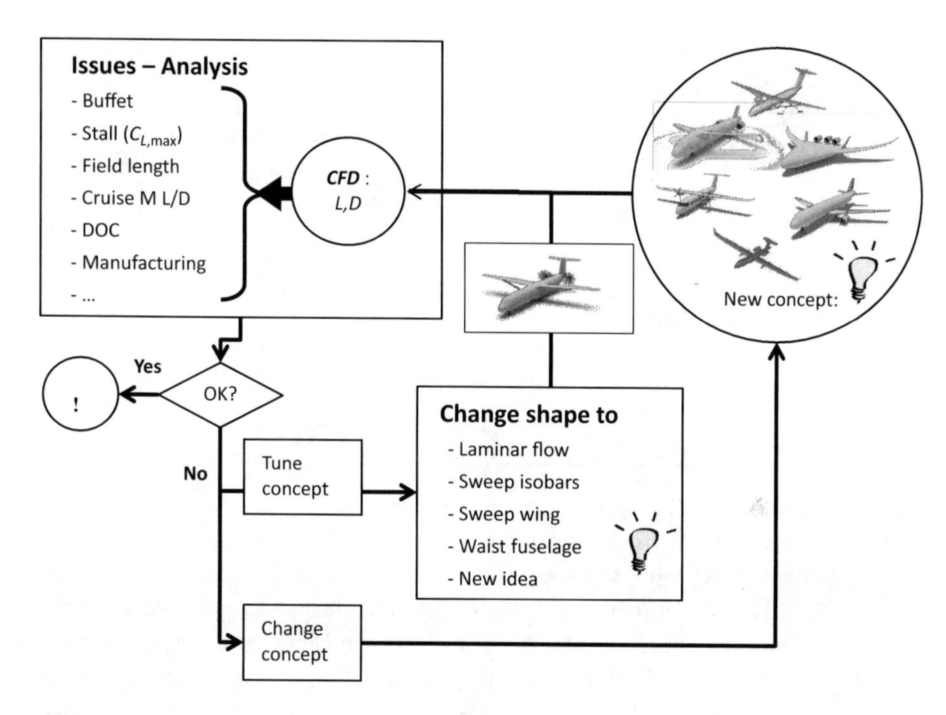

Figure 1.11 The design loop, Cycle 2, intersperses tuning of the selected concept with changing focus onto another – perhaps only just invented – concept.

Research Model is a configuration study that is typical of transonic airliners today, with details of shape published to encourage further optimization studies using CFD.

Tasks and Tools

This section goes further in identifying more precisely the aerodynamic tasks to improve the baseline performance, as exemplified by Figure 1.11. In a quantitative way, it spells out the protocol for the design parameters, shape variables, and performance measures.

Low-fidelity tools are traditionally used by the Future Projects Office in the concept design Cycle 1, where many alternatives need to be analyzed in a short period. Higher-fidelity tools are used by the Advanced Design Office in the following Cycles 2 and 3 as the multiple concepts narrow down to a few that are mature enough for such tools. The term "mature" refers here to the representation of the aircraft's outer-skin geometry and its internal structure, where applicable. This in turn influences the fidelity of the aerodynamical modeling that determines the aircraft's behavior and performance. Today, this is the existing practice for developing new aircraft.

We focus on the CFD-related tasks in Baseline Refinement. Clean-wing design is treated in Chapters 8 and 9 and must produce a light wing with high lift-to-drag in cruise and good low-speed properties. The secondary parameters for detailed aerodynamic wing-shape design are shown in Table 1.3. This sets the stage for refinement of

Table 1.3 Secondary parameters for clean-wing design.

Section shape	Parameter
	Thickness chordwise
	Camber line chordwise
	Leading-edge radius R_n
	Trailing-edge angle τ_{te}
Spanwise shape	**Parameter**
	Twist variation spanwise
	Camber variation spanwise
	$(t/c)_{max}$ variation spanwise

the whole configuration through analysis of the flying qualities. Chapter 10 covers the sizing of control surfaces, rudders, elevators, elevons, canard, spoilers, and ailerons, if they are present. Compilation of the tables of the aerodynamic forces over the flight envelope, which is necessary for flight simulation, is discussed. Finally, Chapter 11 presents a low-fidelity model for investigating the effects of the deformations of the airframe in flight and maneuvers.

1.2.1 Aerodynamic Wing-Shape Design

All airplanes must perform satisfactorily over a range of speeds. Look at Figure 1.5 and Figure 1.11. To achieve acceptable operation across the flight envelope, airplanes must adapt the geometry of their lifting surfaces to different operation points. A transport airplane could be said to have three main design points in the flight state envelope: take-off and climb-out, cruise, and landing. In the first condition, the flaps are set for take-off conditions (i.e. at not too large angles to ensure low drag). The wing is designed for cruise with the flaps retracted. In landing, drag is beneficial and the flaps are set at large angles to increase lift and reduce landing speed. For each of these flight conditions, effects of deviations in speed and attitude as well as of gusts must be considered. The aerodynamic design must give the airplane a well-ordered, stable flow in all of these conditions.

Clean-Wing design

Clean-Wing design is one of the cornerstones of aircraft manufacture, and details of the procedures used are well-guarded intellectual properties of the manufacturers. It is safe to assume that creation of an efficient wing from scratch for a new configuration requires significant industrial effort and one or more years of devoted time. Ryle [24] gives us a detailed account of wing design in the 1970s, but leaves out much of the relevant mathematics. Obert's book [17] also describes the process.

At the time of Ryle's writing, CFD was in its infancy, and accurate computation from first principles of the transonic flow around a suggested wing was all but impossible. Thus, experimental data provided the main guidance, together with potential flow as the workhorse mathematical model (Chapter 3), corrected in various ways such as the Prandtl–Glauert Mach number scaling for compressibility effects and shape modification by estimated boundary-layer displacement thickness.

It is also safe to assume that there is much to gain by copying an existing wing that is known to be efficient and modifying it to satisfy the new performance requirements. Wing design for unconventional configurations is an even greater challenge, since there is no obvious baseline to start from.

Conceptual design is a very repetitive process, with the design evolving with each iteration. New ideas come forth in response to problems as they are uncovered when the design is investigated in ever-increasing detail, as illustrated in Figure 1.11, which highlights aerodynamics. Similar analysis/synthesis loops appear for the design of the structures, the propulsion, the control system, etc. The analysis step is structured with quantitative tools to put numbers on the features and properties once a shape has been defined. The shape change to achieve the desired effect may be a small nudge along well-known directions. But when the well-known ideas have been exhausted, a creative step is called for. We maintain that such *creativity* is best learned on the job, by trying and sometimes failing, and by studying the masters. But software can help unfetter creativity in the engineering teams. It can take care of menial tasks such as file formatting and running computational tools in "workflows" involving several computational analyses over and again. Computers do this without the errors typical of humans in repetitive jobs with many details to get correct. They can also, without fail, log all analysis steps for backtracking and documentation.

Real Wing Design and Optimization

The shape of a real wing must also satisfy non-flow-related geometric and structural constraints. The wing must have sufficient volume to contain the fuel and control surface actuators, be thick enough to carry a wing box that can withstand the aerodynamic loads, and be simple enough to allow economic manufacture. Aero-elastic effects must also be taken into account. The designer must *create a shape* with low drag at cruise, safe stall characteristics, and satisfactory divergence, flutter, and buffet speeds.

This is much more complicated than choosing the shape from a given family that minimizes the drag at a given lift coefficient and Mach number by running an optimization algorithm. It is reasonable to ask what help solutions of such simple tasks can be to the design office. There is, of course, the long-term hope that, with *much* faster computation, the complexity of the optimization task can be increased to become more realistic, and that the algorithms will become less myopic in their search of the design space.

But in the near term, optimization exercises are already very useful in at least two different ways. First, they show *patterns* of shape modification and the resulting flow changes, so that the engineers learn more about the mapping from *shape to performance*. Second, they are useful in shape modifications to shave the last drag counts off an already well-designed wing by weakening shocks or reducing adverse pressure gradients locally. It follows from these deliberations that software tools for aerodynamic design should provide efficient solutions to tasks formulated as optimization problems where the objectives as well as constraints can be easily changed.

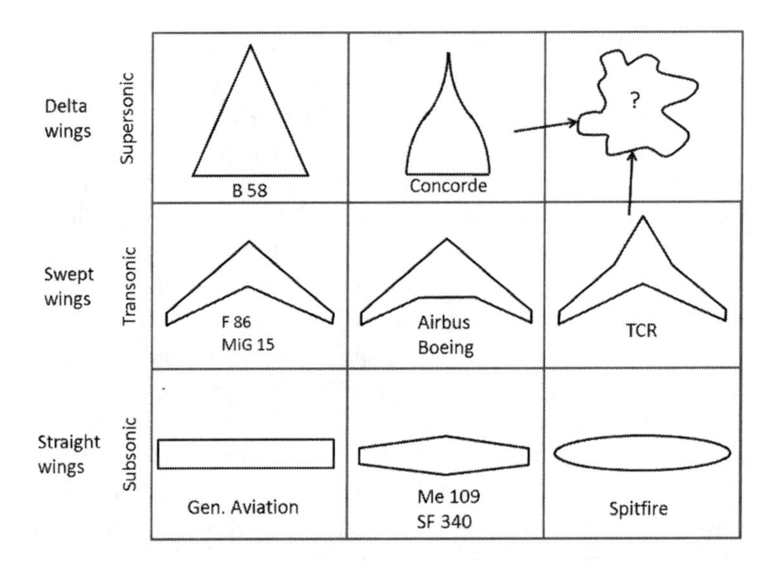

Figure 1.12 Variations of the three planform classes. Straight wings (bottom) for subsonic flight; swept wings (middle) for transonics; and delta or arrow wings (top) for supersonic speeds.

In the following, we consider only the aerodynamic shape optimization subtask in the *decomposed overall design task*, epitomized in Figure 1.11.

Aircraft conceptual design processes can be improved by proper application of multidisciplinary optimization (MDO). MDO techniques can reduce the weight and cost of an aircraft design concept by making fairly minor changes to the key design variables, and with no additional downstream costs. In effect, we can get a better airplane for next to nothing. An overview is presented in Section 1.3.1.

Speed Regimes and Planforms

Figure 1.12 shows several variants of the basic three planforms: straight, swept, and delta. The main wing planform variants for high speed are the swept wing (backward and forward), the swept wing with strake, the delta wing, the delta wing with canard, and the hybrid wing. The hybrid wing is a trapezoidal wing combined with a strake, also called a leading-edge extension (LEX). The trapezoidal wing typically has a well-rounded leading edge with rather small sweep angle, $\Lambda_{LE} = 25\text{–}45^\circ$. The strake is a slender, far forward-reaching, highly swept inner wing with a small leading-edge radius. The leading edge may be straight or slightly curved. The cruise speed increases, as does the sweep angle, from subsonic straight wings in the bottom row of Figure 1.12 to transonic swept ones in the middle row, and then to supersonic with highly swept wings in the top row. The determining factor is the Mach number, M_{cruise}, the ratio of airspeed to the speed of sound. Supersonic craft all are slender, with small wingspan-to-fuselage length ratios. Many more variations have evolved from the three basic ones. The US fighters have hybrid small aspect ratio swept wings with various LEXs. In the EU, the later generations of manned fighters instead have delta

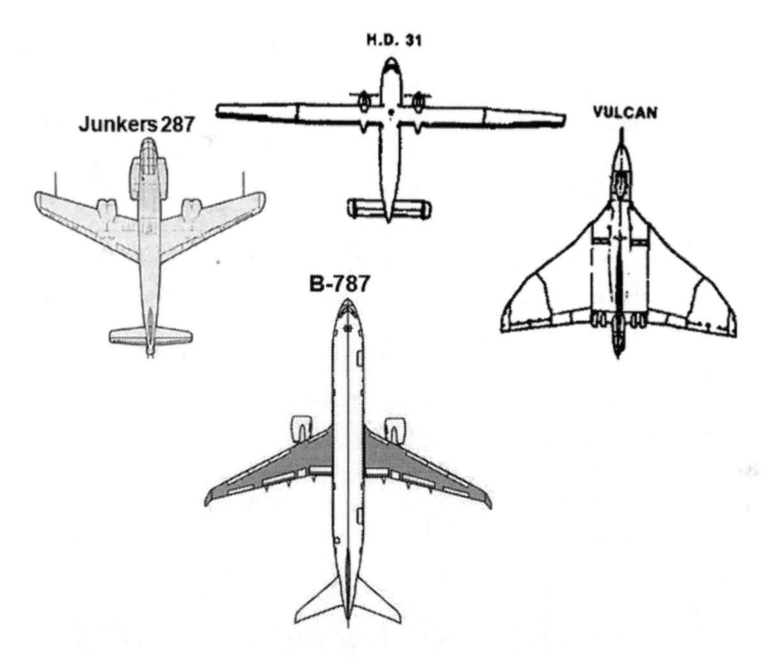

Figure 1.13 Four different planforms. (Top) The very high aspect ratio Hurel–Dubois HD.31. (Center left) The Ju 287 jet World War II bomber with forward sweep. (Center right) The delta wing Avro Vulcan strategic bomber. (Bottom) The Boeing 787 Dreamliner, a typical modern long-range airliner. (Adapted from Chuprun [3], AFWAL, public domain)

(or double-delta) wings with canards. Unconventional configurations such as flying wings and Prandtl box wings evolved as combinations of these elements. For instance, winglets of many different shapes now appear on modern planes designed for low cruise drag. More detailed discussions on transonic and supersonic planforms can be found in Chapters 3 and 9.

Examples: Advanced Design Work

Reflecting the wide variety of aspect ratios and sweep angles in Figure 1.12, Figure 1.13 shows four very different wing planforms that have found their way onto flying aircraft. One is a classical wing with extreme aspect ratio, and the other three are different solutions to the challenge of high-speed flight. It would seem that the high aspect ratio swept wing is the current best practice for economical close-to-Mach-1 flight.

The Junkers Ju 287 was a German World War II aerodynamic test bed built to develop the technology required for a multi-engine jet bomber. It was powered by four Junkers Jumo 004 engines and featured a revolutionary forward-swept wing. Later prototypes were captured by the Red Army in the closing stages of World War II and the design was further developed in the Soviet Union after the end of the war. Flight tests displayed extremely good handling characteristics, but also revealed some of the problems of the forward-swept wing. The most notable of these drawbacks was "wing warping," or excessive in-flight flexing of the main spar and wing assembly. Forward sweep makes the wing tip twist up, increasing the lift on the outer part, which further increases the twist. Swept-back wings are free of this vicious cycle, but, as we shall see later, they have other problems.

The Hurel–Dubois HD.31 was produced in France in the 1950s, based on Maurice Hurel's high aspect ratio wing designs. Tests with the Hurel–Dubois HD.10 research aircraft had validated Hurel's ideas. Two prototypes of a medium-range airliner utilizing this principle were built, conventional designs in all respects other than their unorthodox wings with an aspect ratio of 20.2!

The Avro Vulcan, later named the Hawker Siddeley Vulcan, a jet-powered delta-wing high-altitude strategic bomber, represents a highly successful "Flying Wing" type with compound leading-edge sweep. Several scale aircraft, designated Avro 707, were produced to test and refine the delta-wing design principles. Influential in the design was wind tunnel testing performed by the Royal Aircraft Establishment at Farnborough. This indicated the need for a wing redesign with breaks in the leading-edge sweep to avoid the onset of compressibility drag, which would have restricted the maximum speed.

Representing a current airliner, the Boeing 787 Dreamliner first flew in 2009, with high aspect ratio wings and winglets. It is built with carbon fiber composites. The chevrons on the nacelles reduce nozzle exhaust noise by influencing the mixing layers between the high-speed core and bypass flows and the exterior flow.

1.2.2 Aerodynamic Cruise Efficiency and Wing Planform

Thus, the task of aerodynamic design becomes that of *aerodynamic shape optimization*, and just as in Figure 1.13, the planform can vary greatly to achieve optimum aerodynamic cruise efficiency. Figure 1.14 illustrates the three "sweet spots" for speeds of the DC-3 at the low end, the Concorde at the high end, and a typical transonic jet airliner in-between. As we shall investigate in Chapter 9, the TCR cruises inefficiently just below Mach 1. The parameter most often used to assess the aerodynamic cruise efficiency E of a transport configuration is the product of the cruise lift-to-drag ratio, L/D, and the cruise Mach number, $\mathsf{E} = M(L/D)$. This parameter is directly related to the range of an airplane (Eq. (1.3)).

Critical Flow and Transonic Drag Rise

For a "classical" type of aircraft, L/D is approximately constant until the *critical* Mach number M_{crit} of the wing section, the flight Mach number at which supersonic flow first appears, is reached.

At higher speed, drag increases significantly due to compressibility effects: local pockets of supersonic flow terminated by shocks produce wave drag- and/or shock-induced separation. Therefore, E will vary approximately linearly with Mach number until M_{crit} is reached, and then decrease quite suddenly because the drag increases rapidly above M_{crit}. The slope of the E curve increases with wingspan as described by

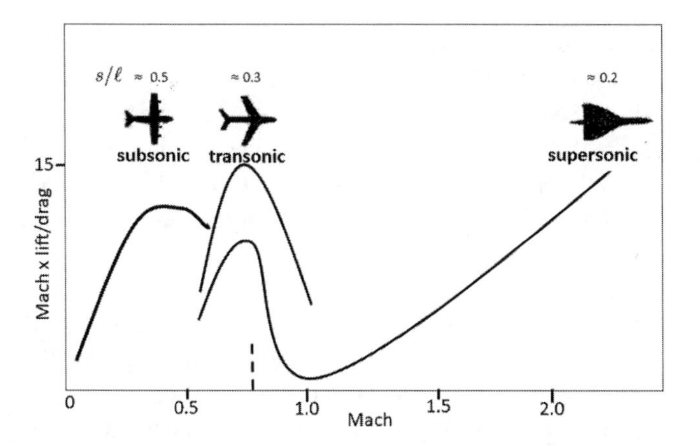

Figure 1.14 Wing cruise efficiency for three classes of aircraft in three different speed regimes.

wingspan load efficiency e (see Eq. (1.5)) and decreases with increased wetted area of the configuration.

The airfoil technology sets M_{crit}, and it decreases when the lift coefficient or the wing thickness increases. Improvement in airfoil technology typically produces a slight change in the slope of the E curve and a continuation of the straight portion to higher M_{crit}. The gain in E with advanced airfoil technology is then proportional to the increase in the critical Mach number.

A few relationships clearly and simply reveal the way in which the several measures of airplane characteristics affect airplane performance. The lift-induced drag of an airplane typically varies parabolically with lift C_L,

$$C_D = C_{D,0} + k' C_L^2/(\pi AR) = C_{D,0} + C_L^2/(e\pi AR), \tag{1.5}$$

a relation predicted by, for example, lifting line theory (Chapter 3). Another general relation is the specific range equation (Eq. (1.1)). The basic credo is that *the most economic aircraft travels farthest per gallon of fuel burned.* The TSFC appearing in the specific range is approximately a constant for jet propulsion,

$$dW/dt = TSFC \cdot T \text{ Jet propulsion,}$$

but for propeller propulsion with a piston engine, fuel consumption is proportional to the power, VT,

$$dW/dt = PSFC \cdot VT \text{ Propeller propulsion.}$$

From this, Smith [26] derives a relation for propeller aircraft showing that $(L/D)_{max}$ varies with $AR^{1/2}$, hence gains from increased aspect ratio are fairly substantial. $(L/D)_{max}$ improves as $C_D^{-1/2}$, so reductions in friction drag help greatly. For the jet airplane, the effects of these parameters are somewhat different. Now the gains from the aspect ratio vary only as $AR^{1/4}$, while the gains from drag reduction vary as $C_D^{-3/2}$. Thus, compared with piston-propeller airplanes, the jet will tend to have a lower aspect ratio, and its aerodynamic cleanliness, above all low-wave drag, must be at a premium.

What Supersonic Cruise Speed? Concorde Example

During the initial planning of the Concorde design concept (the speculative Cycle 1) in the 1960s, the prospects for favorable L/D ratios at supersonic speed were not promising, owing to the shock-wave drag and the tendency for the lift coefficient to fall with Mach number. There was thought to be a chance of achieving shock-free flows up to Mach 1.2 by increasing sweep, carefully shaping the body/engine-pod junctions and wingtips, and using judicious twist and camber.

Thus, fairly straightforward extensions of contemporary Mach 0.8 swept-wing transports might offer cruising at 50% higher speed. However, it proved difficult to evolve shapes that gave sufficiently high L/D to counterbalance the adverse effects on economy at low supersonic speed and low engine efficiency.

The speed range Mach 1.8–3.0 looked more attractive. The fact that supersonic L/D ratios, albeit low, did not worsen with Mach number meant that the speed should be as high as possible. Long, *slender shapes* could produce good aerodynamics. When combined with confident predictions of greater engine efficiency, the higher speed gave acceptable productivity and encouraging cost figures.

At the lower end of the speed range, it was thought possible to evolve shapes that also had reasonable landing performance. The optimum appeared to lie between Mach 2.0 and 2.5, and the choice within this band rested as much on structural and power-plant considerations as on aerodynamics, since kinetic heating difficulties mount very rapidly. While L/D is not much different from 7.5, being roughly half as good as a transonic airliner, the doubled M compensates in terms of cruise efficiency. In addition, the greater intake compression improves the engine efficiency, which could mean greater range or more payload. Chapter 9 examines further the aerodynamic design of Concorde carried out in Cycle 2.

1.2.3 Why the Slender Delta?

The designers faced a number of concerns in their initial configuration studies. Compared with skin friction and vortex drag due to lift, the major obstacle to economic supersonic flight is the large wave drag associated with lift and volume. Ever since supersonic flight was first seriously considered, it was understood that wave drag could be kept low by using slender shapes. Elongating the fuselage was the obvious line to take, but elongation of the wing required much more careful consideration.

As explained in Chapter 3, it makes sense to keep the wing's leading edge behind the Mach lines, and certainly to keep the whole aircraft well within the Mach cone from its nose, simultaneously distributing the volume and lift fairly evenly.

During the development of aircraft for high-speed flight after the World War II, sweepback steadily increased and thickness and aspect ratios decreased, tending toward the delta planform. Supersonic aircraft demanded even thinner wings and more elongated fuselages, so greater extremes of aspect ratio and planform began to be explored. It appeared that, due to their small wave drag, long, slender aerodynamic shapes, advocated by Dietrich Küchemann at the Royal Aircraft Establishment (RAE), promised to be serious contenders in generating an L/D enabling economic cruising at Mach 2. In addition to thickness ratio, the other geometric parameter that greatly influences supersonic drag is the *slenderness ratio* s/ℓ, the ratio of semi-span s to total length ℓ of the aerodynamic shape. For a lift coefficient of 0.1 at Mach 2, the drag is at a minimum when $s/\ell = 0.2$. This helps explain why slender delta planforms looked attractive, although, at the time, the possibility of landing them satisfactorily seemed rather remote.

1.3 Integrated Aircraft Design and MDO

For more than 50 years, conceptual aircraft design has involved parametric studies by the traditional carpet plot technique that poses trade-offs among sets of four variables constrained by two relations. The perspective on the design process is widened here to include additional discussion of optimization.

1.3.1 Protocol for Aircraft Design Optimization

Conceptual design is primarily a search process. Its goal is to find a set of design variable quantities **X**, which produces a vehicle that fulfills some desired list of minimum requirements (see Figure 1.15).

The mechanism behind this search is mathematics, and the core utilities required to conduct the design process can be itemized as follows, starting with the design

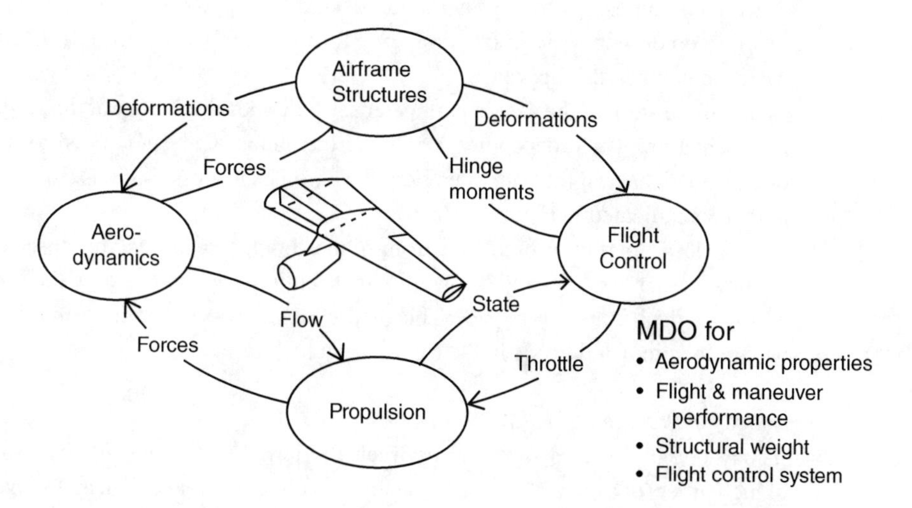

Figure 1.15 Integrated MDO design, Cycle 3, where subfields such as flight control, propulsion, aerodynamics and airframe structures are weakly coupled to each other through constraints imposed by the coupling.

specifications or requirements **R** that define the success of any aircraft design candidate. If a requirement is strict, it is incorporated into the set of constraints; if it can be relaxed, it may be incorporated into the figure of merit (FoM).

- The design parameters **X**, a set of abscissa values intentionally selected to define the airplane in a physically tangible sense, describing the vehicle's physical features or characteristics. The dimension of the *design space* is the number of parameters, which can run to tens of thousands.
- **P**, a set of vehicle properties that satisfy functional relationships, say $\mathbf{P} = prop(\mathbf{X})$, with the design parameters through physical principles or statistical correlations, with fidelity levels L0–L3.

  Examples are the flow prediction methods, such as VLM, to be discussed in Chapter 3.
- $\mathbf{FoM} = fom(\mathbf{X}, \mathbf{P})$ quantifies performance or compliance with regard to the design specifications. Examples could be specific fuel consumption, cruise speed, or range.
- Design constraints $\mathbf{C} = constr(\mathbf{X}, \mathbf{P})$ are typically the aircraft's required performance values such as takeoff distance, climb rates, or cruise speed. There are geometric constraints such as limits to wingspan and fuselage diameter and environmental restrictions such as limits to emissions of noise and greenhouse gases or ozone-reducing substances.

Conspicuously missing is the flow variable **F**, which appears in flow simulations. **F** describes the whole flow field and is huge. But **P** is computed from **F**, whose computation is abstracted into $cfd(\mathbf{X}, \mathbf{F}) = 0$, so **F** is just an intermediary without intrinsic value in the design. However, management of **F** is, of course, a technical challenge. The search involves analysis to deduce the properties of a product from its characteristics, as well as synthesis to create a particular design **X** that yields some desired properties **P**(**X**); in other words, optimization in one form or another. Figure 1.16 illustrates the basic elements in this approach.

The dimension of the design parameter set **X** defines the *fidelity* of the configuration representation. Then, depending on the aerodynamic analysis applied to it (L1, L2, or L3), the *fidelity* of the computed flow **F**, as well as its dimension, is correspondingly higher, as indicated in Figure 1.16.

Vehicle properties **P** predicted from higher-fidelity analyses are closer to reality, so they have more localized probability distributions, as indicated by the sizes of the ellipses in the lift-and-drag space. This in turn implies less uncertainty and shorter error bars in the predicted FoMs to the right in Figure 1.16.

Notions of Order and Fidelity

It may be helpful to distinguish between the terms *order* and *fidelity*. An analysis of high order (of accuracy) describes a physical effect in great detail, and a high-fidelity analysis predicts properties accurately. In fact, a low-order analysis might return results of high fidelity. For example, if a primitive geometry – just the planform and camber distribution (no thickness) – were computed in an Euler simulation, the predictions may

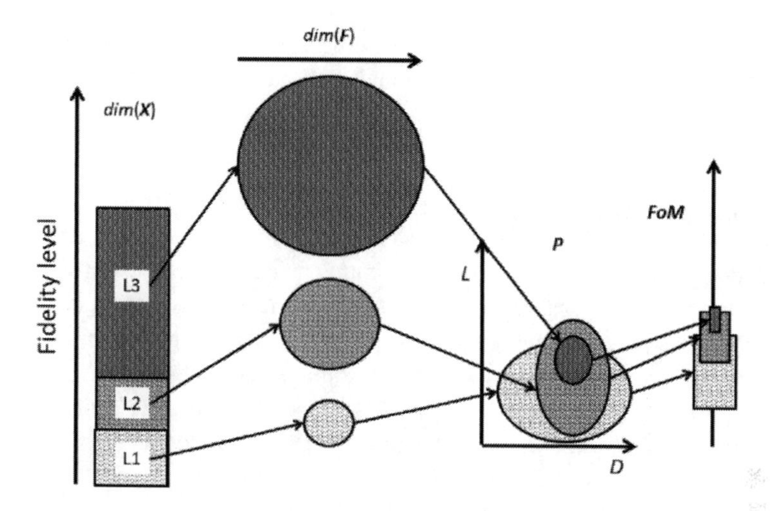

Figure 1.16 Higher-fidelity models give larger data and reduced uncertainty.

well be less accurate than those made by a vortex–lattice method because there will be unrealistic vortices and other separation phenomena from this primitive geometry. The lesson here, of course, is that the fidelity of the prediction method $prop$ should match the fidelity of $\mathbf{X}$.

As an example, the first pass through the design concept loop (Cycle 1) in Figure 1.1 synthesizes the first baseline configuration $\mathbf{X}_{bl}$. With the requirements taken as an initial set of properties $\mathbf{P}$, it derives the basic configuration parameters (e.g. thrust to weight, wing loading, the wing reference area, etc.).

Then, the refinement loop applies these characteristics $\mathbf{X}$ to derive again by analysis properties $\mathbf{P}$, albeit at a higher level of detail (e.g. lift-induced drag from the given wing planform), and then reenters those properties into the configuration development (synthesis) loop to determine a more mature set of characteristics $\mathbf{X}$. Several configurations with different characteristics may share similar properties $\mathbf{P}$, so the evolution of the design to a more mature state also requires down-selection. This means a choice between different candidates that, based on the requirements considered so far, look equally good. Thus, as the design evolves, the requirements are augmented or made stricter or, if no feasible candidates are found, relaxed.

An essential feature of MDO is the presence of design constraints and FoMs of system-level concern. In a typical aircraft conceptual design application, the FoM is either cost or its surrogate, weight, where the aircraft is sized to some specified mission that includes the range and payload requirements. The design constraints are typically the aircraft's required performance such as takeoff distance and climb rates, plus geometric or operational constraints such as a wingspan limit.

A typical application for MDO is the simultaneous aerodynamic and structural optimization of a wing. The wing is defined in terms of geometric variables, and the effects on aerodynamics and structural strength are determined as the geometry is varied. Results are assessed versus a defined FoM, with constraints based on performance, safety, operability, and practicality. The crucial issue in either classical optimization or MDO is the number of design parameters because the workload quickly becomes immense as this number grows.

1.3.2 Problem Complexity and Optimization

The design problem can be posed as an optimization problem, but it is so complex that the formulation is a challenge in itself, and numerical solution will remain beyond the reach of computers in the near future. Therefore, it must be approached by decomposition. Decomposition and software technology is under intense development, as exemplified by papers by M. Giles [8, 9] and the Stanford group around I. Kroo [16]. Different types of software architectures have been proposed, such as collaborative optimization (CO) [2, 15], bi-level integrated system synthesis (BLISS) [27], concurrent subspace optimization (CSO) [31], and many others.

Traditional decomposition leads to subproblems of aerodynamic shape optimization coupled with structural design by simplified constraints such as on wing thickness, limits on wing root bending moment, etc. An example of this will be shown in Chapter 3, indicating that quite detailed constraints must be included for the optimal wing to look like those of contemporary airliners. Even purely aerodynamic shape optimization for clean wings is complex and is in practice carried out using a combination of mathematical tools and engineering know-how. The tools developed and analysed in this work are described in Chapter 9. Thus, although technically a single-discipline optimization, it is still complex enough that decomposition and other methods employed in MDO may be useful.

The Curse of Dimensionality

Consider the basic set of six design parameters commonly used in sizing the wing (see Figures 1.1 and 1.4): thrust-to-weight T/W, wing loading W/S, and wing aspect ratio AR, taper ratio λ, sweep Λ, and thickness t/c. In the language of experimental design, a full factorial design requires a minimum of $3^6 = 729$ data points to span the design space, and $5^6 = 15,625$ would improve accuracy. Each data point represents a different airplane and requires analysis for aerodynamics, propulsion, weights, sizing, and performance. To better optimize an aircraft in the next loops in Figure 1.1, one includes additional design parameters to more precisely specify the actual shape, including wing planform breaks, nacelle locations, tail locations, fuselage fineness ratio, wing design lift coefficient (or camber), wing thickness distribution, engine bypass ratio or propeller diameter, etc. As additional design variables are added, the number of required data points (i.e. aircraft parametric evaluations) rapidly spirals out of control. This brings into focus the issues of information and time, sometimes labeled the "curse of dimensionality," addressed in D. Böhnke's PhD dissertation [1]. Some relief for small-dimensional design spaces is brought by sampling schemes developed for experimental design, such as Latin hypercube, but not enough to cast off the curse.

Decomposition of the design space offers potential reduction of the computational workload. In practice, engineers know that some analysis results may be rather insensitive to some selected design variables. Many design variables are not very interdependent ("weakly coupled") and can be optimized separately. The designer partitions a large engineering design optimization problem into a number of smaller subproblems, formulated so that their couplings can be relaxed only to be ever better satisfied as the candidates converge to the optimum.

This decomposition allows separate optimizations with private variables in the subproblems. The subproblems in turn may also be broken into weakly interconnected models, creating a hierarchical structure to the optimization process.

As an example of a particular decomposition, consider an aerodynamic wing-shape optimization for minimizing drag, where the weight of a candidate shape is minimized by a structural optimization tool. This realizes a "Stackelberg equilibrium" in a game played by the shape optimizer and the structural optimizer.

$$\min_x Drag(x) \tag{1.6}$$

$$s.t. \quad g(x, y(x)) \leq 0 \text{ where}$$

$$y(x) = arg \min_y Weight(x, y) \tag{1.7}$$

$$s.t. \quad \text{structure tolerates airloads}$$

Here, x represents the public parameters governing wing surface shape and y represents the private parameters of the structural mechanics (position, thickness, etc., of ribs, spars, skin). g represents shape constraints such as maximal wingspan. Each evaluation of y requires the resolution of a structural mechanics optimization problem.

1.3.3 Basic Elements of Weakly Coupled MDO

The usual development approach of MDAO tools uses established disciplinary codes, often referred to as "legacy codes," in a "weak" coupling mode. For example, in the present treatment of aero-elastic and structural dynamics problems, predominantly linear aerodynamic methods are employed, dynamic structure problems are treated in the frequency domain, and the flow/structure coupling is made by passing analysis results back and forth between aerodynamics and structural analyses. The couplings with flight dynamics and flight control are made unidirectionally. The new treatment basically must evolve out of the present capabilities in order to reduce risks and to ensure the acceptance of users in the aircraft definition and development processes. However, many of the required strategies for dealing with such complex systems have yet to be developed.

The property variables **P** from the corresponding multidisciplinary analysis problem for the design variables **X** are subject to the optimizer that minimizes an objective function, a single FoM, composed from the whole set of **FoM** considered.

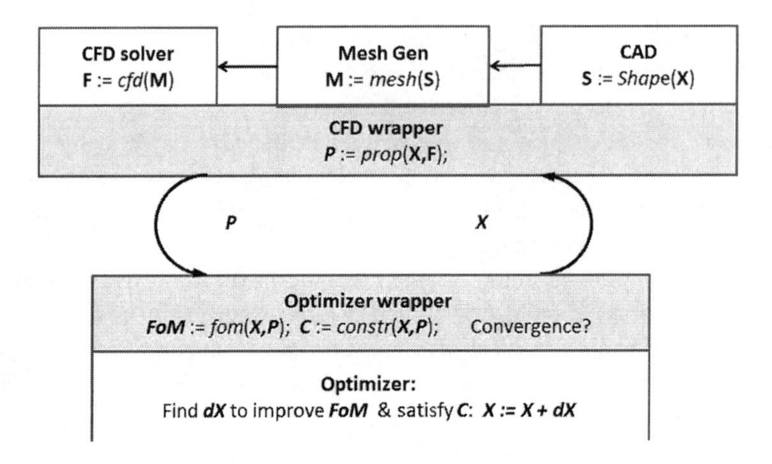

Figure 1.17 Shape optimization, with mesh regeneration.

$$\min_{\mathbf{X}} FoM = fom(\mathbf{X}, \mathbf{P}) \tag{1.8}$$

$$\text{subject to constraints } \mathbf{P} = prop(\mathbf{X})$$

$$\mathbf{R} = constr(\mathbf{X}, \mathbf{P}) \leq \mathbf{0}$$

Note that the inequality constraints also include equalities from a construction such as $f \leq 0, -f \leq 0$. Figure 1.17 shows a common aerodynamic shape optimization scheme with loosely coupled modules, the CFD module comprising a CAD program, a mesh generator, and the CFD solver proper. The CAD program creates the shape **S** of the wing from the geometric parameters in **X**. The mesh generator makes the computational mesh **M** around **S**, and finally the flow solver computes the flow **F** on **M**. The CFD module is driven by an application program interface (API) exposed by the "wrapper" to external modules. The CFD "wrapper" also computes FoMs and constraints from **F** and exports them to the optimizer, which finds an increment **dX** to improve the FoM and satisfy the constraints. The new parameters are exported to the CFD module for the next iteration. To operate efficiently, the optimizer needs gradients of the FoMs and constraints. The gradients can be computed by numerical differencing, with one run for each parameter, so for Np parameters, the flow solver must be run $Np+1$ times. Modern flow solvers often provide calculation of derivatives by solution of an adjoint partial differential equation (PDE) problem, which reduces the work significantly for large Np. See the book's website for an overview of the mathematics involved. Chapter 3 gives an example of induced drag minimization where the aero-analysis is done by the vortex lattice method.

Current State-of-the-Art Develpoment

The current state of the art is exemplified in Section 9.3.3, which reports on the results of mathematical shape optimization for drag of the Common Research Model wing.

The baseline geometry has been made available by Boeing and AIAA as a realistic test case for developers of MDO. The group of J. Martins at the University of Michigan produced the results we quote with a state-of-the-art software approach applicable to real multidisciplinary design exercises. Their gallery of applications includes such diverse systems as optimization of the shape and structure of wings, structural mechanics of satellites, and lithium-ion batteries [25]. The speed of optimization algorithms depends crucially on the availability of gradients of the objective with respect to the design variables. For aerodynamic shape optimization, the differentiation must involve the effect of surface shape on the volume grid, which in turn affects the flow solution, etc. Algorithms for the differentiation of computer codes [5] have been around for decades and are important ingredients in the overall software system. Progress in MDO requires flexible and general APIs to allow differentiation through the whole chain of analysis modules; the MSES airfoil analysis and design system described in Chapter 7 uses analytically computed derivatives for efficient solution of the flow problem and as ingredients in the shape design task. The Martins' group `OpenMDAO` software bundle [10] takes this approach to a level where it is applicable to large complex systems.

1.4 Aerodynamic Design and CFD

Let us summarize the discussion so far. At its earliest stage, aircraft design begins with the specifications of the flight mission and a request that starts with a blank-sheet design. With an idea for a concept expressed as a sketch, generalists set in motion the parametric process using the L0 empirical/handbook methods described in Section 1.1.4. These are the most suitable for this boot-strapping process that, using only a sketch or even a sized planform, must "guess" what the actual weight is that must be lifted into the air. The best tools of this type are industry-owned and proprietary because they have been honed through many years of industrial know-how and experience.

This book focuses on the aerodynamic tools and tasks used to evolve a baseline into a mature and viable shape – a feasible design – with enough detail that could, for example, be machined into a wind tunnel model or a remotely flyable demonstrator. CFD is the tool for this task.

1.4.1 Tasks and CFD Tools

Thus, the major task of aerodynamics is to define the overall shape of the vehicle in order to fulfill the performance demands of the flight envelope. At its extremes are found load cases that are critical for structural dimensioning. In addition, fly-ability and controllability of the vehicle must be analyzed (see Figure 1.18). More than a pure aerodynamic optimization, the shape definition process must take into account structural constraints in an aero-structural optimization, as well as stability

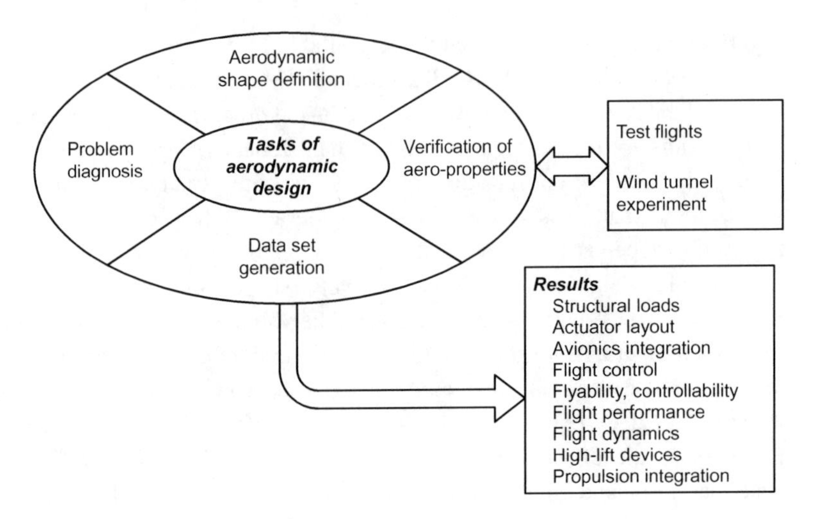

Figure 1.18 The tools, tasks, and results of the aerodynamic design process.

and control characteristics. Hence, aerodynamic data set generation is a task that enables integration with other disciplines and problem diagnosis when performance requirements are not met. Once the design detail have advanced sufficiently far, models can be built. Other equally important tasks and processes are, in concert with wind tunnel and flight testing, the verification of the aerodynamic design and its guaranteed performance.

Senior Technical Fellow at Boeing Commercial Airplanes Philippe Spalart [28] gives the industry perspective on the role and challenges of CFD in building aircraft. It resonates with the issues of aerodynamic design that we spell out here at a pedagogical level. With much in common with Spalart's views, Figure 1.18 outlines our perspective on the overall tasks of aerodynamic design. Let us enumerate them in some detail. Typical design applications where CFD can make major contributions are as follows.

1. *Aerodynamic shape definition* (i.e. its lofting and analysis).
   - Select optimum airfoils, wing lofting, and empennage configuration for performance. Chapters 8 and 9 cover these topics.
   - Define the outer surface shape of the airframe to fulfill the aerodynamic requirements regarding lift, drag, loads, stability criteria, etc., by sizing, shaping, and positioning its various geometric components, taking into account propulsion integration, structures, and their volumetric demands.
2. *Performance data set*: compute force and moment data to estimate lift, drag, and moments over the flight envelope of the vehicle.
   - *Control surfaces*: determine aerodynamic coefficients, force, and moment data to evaluate stability and control derivatives.

- Verify performance, control, and handling characteristics in a flight simulator. Chapter 10 covers these topics.

3. *Air-loads data set*: compute surface pressure to determine structural loading and aero-elastic impacts, including the likely loss of aerodynamic performance and control surface effectiveness due to structural deformation under load (static aero-elasticity). Chapter 11 covers this topic.
4. *Problem diagnosis*: understand and solve *problems* discovered in wind tunnel or flight testing.
   - Poor flying qualities that cannot be cured by control laws implemented in the flight control system must be remedied through aerodynamic redesign. Chapter 10 covers this topic.

Some of the applications concern *attached flow* (e.g. cruise shape definition). Many of the others involve *separated flow*, such as at the extremes of the flight envelope, where only Reynolds-averaged Navier–Stokes (RANS) is a viable flow model. For example, the loads that determine the sizing of the structures usually occur at separated flow conditions well away from the cruise design point. Chapter 2 explains in detail the physics of attached and separated airflow and describes the various appropriate mathematical models.

1.4.2 CFD Workflow and User Awareness

CFD is the science of producing numerical solutions, by computational methods, to a system of PDEs that describe fluid flow, or airflow in the case of aerodynamics. The purpose is to better understand, qualitatively and quantitatively, flow phenomena in order to improve upon engineering design.

Teaching User Awareness: The Informed User

As is shown in Figure 1.19, CFD draws upon elements from six related disciplines: fluid mechanics (of which aerodynamics is a part), numerical analysis, programming principles, computer science, grid generation, and scientific visualization.

The code developers described in Figure 1.19 are hard at work to improve their products, but it is still necessary for the CFD engineer to *understand* what goes on under the hood. For example, CFD codes do not always work. They may give up and produce no solution at all, or they may produce a flow field in significant disagreement with the real flow. The user must assess the *quality* of the solution, and modify the parameters that control the code to give a *reliable* solution. This is what an *informed* CFD user must do, and we call it *CFD due diligence*. Philippe Spalart [28] calls for more user awareness in industrial CFD:

Code Developer

- Define flow phenomenon to model
- Construct PDE model
- Analyze well-posedness, boundary conditions
- Discretize in space and time
- Analyze discretized model:
  - Boundary conditions, stability, ...
- Develop computer code

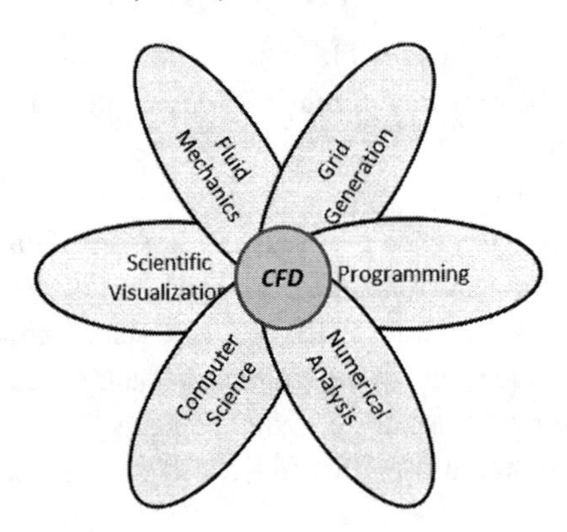

Student = Aerodynamicist User

- Define shape design problem
- Define flow regime
- Define configuration
  - CAD model, 3-view drawing, ...
- Run CFD:
  - Model geometry, grid
  - Flow model
  - Boundary conditions
  - Solver parameters
  - Assess fidellty
- Quantify configuration performance
- If unacceptable: Modify shape

Figure 1.19 Disciplines in CFD and tasks of the code developer and typical user – the student as aerodynamicist.

> *The quality of CFD answers depends at least on three factors: the code, the available computer resources (which limit the grid resolution and number of iterations), and the user.*

In addition, codes should provide more actionable information of direct importance to the designer:

> *Another support for awareness of physics ...will be available if and when a rigorous definition of induced drag, wave drag, and parasitic drag from viscous flow fields is established. These concepts are constantly used by designers ...*

Greater automation is called for in the overall workflow. Even more important is automation of the numerical solution process. Grid adaptation is a case in point, which is a challenge for the resolution of boundary layers, shocks, and vortices. But user know-how is key, no matter how advanced the code and how fast the computer. Over-confidence and undercompetence in CFD must be addressed through education. Sufficient emphasis must be placed on the flow models, the numerical methods, and hands-on experience in CFD codes. The student must become an informed CFD user with awareness of the power and limitations of their tools. Spalart proposes that, at the same time, curricula should

...temper the erosion of classical aerodynamic knowledge in the younger generations.

The European Research Community on Flow, Turbulence and Combustion (ERCOFTAC) has put out a set of very helpful *Best Practice Guidelines for CFD* [7], available via the web. Since a CFD code offers a variety of turbulence models at the touch of a button (see Chapter 6), it may be tempting to believe that bad results are usually caused by inadequate choice of turbulence model. However, as made very clear by the Guidelines, as often as not the problem lies with the basic ingredients of grid quality, convergence level of iterative schemes, and choice of boundary conditions.

An *informed user*, applying *due diligence*, will ascertain – with a significant degree of certainty – that the computed flow field agrees with the real flow. This calls for:

1. Choosing the CFD fidelity level and numerical modeling appropriate to the shock waves and vortical features of the problem at hand (see Chapters 3 and 4, and hands-on tutorials on VLM and DEMOFLOW).
2. Modeling the geometry and generating a grid of sufficient quality and resolution to represent the vehicle and the flow under investigation adequately (see Chapter 5 and hands-on modeling and automatic grid generation tutorials).
3. Verifying convergence to steady state, if it exists, and if not, computing the unsteady flow with sufficient time accuracy (see Chapter 6 and its associated hands-on tutorials).
4. Visualization and analysis of the computed solution to verify reliably that the chosen level of physical modeling is appropriate (see Chapters 4 and 6 with hands-on examples).

Workflow

Once competent in CFD due diligence, the user can proceed to apply CFD to problems in aircraft aerodynamics, as outlined in Figure 1.19, as follows.

1. Shape design and performance analysis of *airfoils*: Chapters 7 and 8 focus on transonic airfoils, with hands-on running of MSES and RANS for active learning of mapping of airfoil shapes to performance.
2. Shape design and performance analysis of *clean wings*: Chapters 9 and 10 focus on the wing shape secondary parameters in Table 1.3, with hands-on running of RANS for active learning of wing geometry mapping to performance. The studies include twist, camber, and thickness scheduling for the CRM wing to improve flow over root and tip.
3. Development of a *full configuration* with acceptable flying qualities and aero-elastic behavior: Chapter 10 develops the TCR design proposal further in Cycle 3, including analysis of stability and flying qualities in flight simulation using multi-fidelity aerodynamics modeling. Chapter 11 discusses how the wing shape in flight can be predicted from its jig shape.

CFD is the main tool in all of the above actions, which we call the *CFD workflow*. The text uses the EDGE code as generic examples of concepts that apply to most CFD

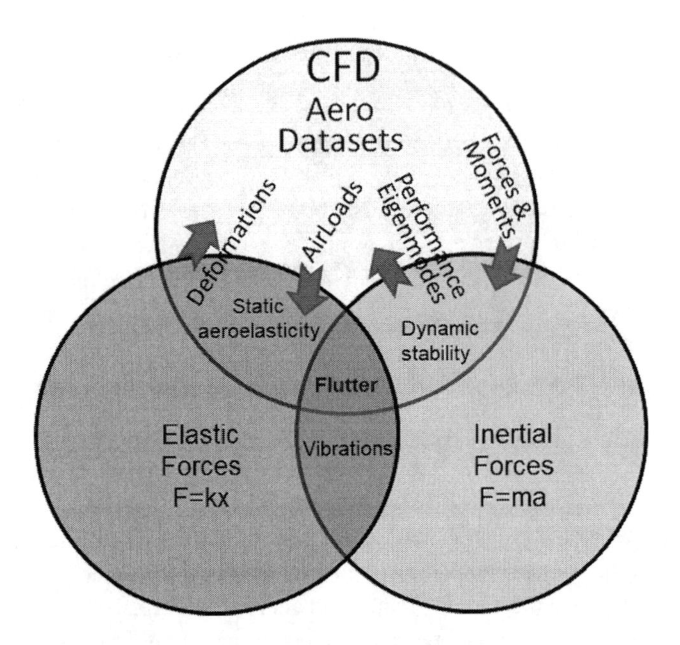

Figure 1.20 CFD results as used in several design tasks.

codes in the hope that the similarities outweigh the differences in the code(s) available to the student. We abstain from discussing the peculiarities of any single CFD computer program.

Data Exchange with Other Disciplines

A successful design must guarantee that the airplane behaves as expected for all possible flight states (i.e over the entire *flight envelope*). It must be controllable by the pilot, its structural integrity must be guaranteed, and its performance in terms of speed, climb rate, payload capacity, fuel consumption, field requirements for take-off and landing, etc., must agree with specifications. Only a multidisciplinary analysis can provide such a guarantee, as is suggested in Figure 1.20. The airplane behavior can be predicted from the *aerodynamic data set*, with its tables of the forces and moments exerted on it from the air, inertia, and gravity, in the different flight states. The data can be obtained through measurements or CFD computations. The first simplification is to assume that the aircraft is rigid. The aerodynamic forces and moments computed by CFD are applied to the six degrees of freedom model of a rigid body, thus allowing the study of dynamic stability. This enables investigation of flying qualities via the eigenmodes of motion, as explained in Chapter 10. The next approximation is to allow flexibility and to take into account static aero-elastic deformation, where structural mechanics interacts with CFD through the airloads (surface pressure, possibly unsteady) on the wings and control surfaces. This is the *loads data set*. As the design matures, increasingly rich data sets generated by the analyses are used for evaluating the performance of the concept.

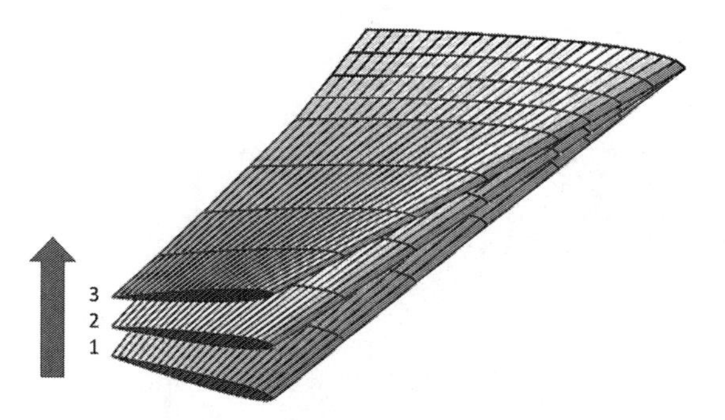

Figure 1.21 Deformations of a wing under aero-loads.

Static Aero-Elastic Effects

CFD and structural analysis need to be integrated, because the aerodynamic loads size the structure and influence the mass of the aircraft, and therefore, its cost and performance.

Through the design phase of an aircraft, the fidelity of the structural model could increase from a simple approximation, such as a beam model, to a sophisticated finite-element model later on. Typically, for the static deflection of the wing from its jig shape to its flight shape, the aero-structural coupling is done in a loose fashion and often converges in just a few iterations, as is indicated in Figure 1.21 and explained in Chapter 11. Dynamic load interactions, including flutter and vibrations, call for time-dependent or time-harmonic CFD analysis to be coupled with a structural module, and this is beyond the scope of this book.

1.4.3 Challenges to Improving Performance

Improvements in aerodynamics (see Figure 1.22), as well as materials and propulsion, can contribute modest gains in the overall productivity and quality of the aircraft.

Fuel reduction is a long-standing goal in improved performance. It can result from concurrent adaption of advanced technologies in aerodynamics by increasing the lift-to-drag ratio, in structures and materials by decreasing the overall weight, and in systems and engine design by reducing the specific fuel consumption. The specific range (Eq. (1.1)) illustrates quantitatively how the different aircraft subsystems influence fuel consumption.

In this book, we focus on the aerodynamics, and Figure 1.22 highlights some aerodynamic design tasks involved with five different classes of aircraft: propeller-driven commuter, conventional jet airliner, blended wing–body transport, business jet, and jet fighter. Many of the tasks are common to all classes, but some are specific to particular classes. A number of these tasks will be delved into in Chapters 9 and 10.

For example, improvements in airframe technology, such as high-speed airfoil and wing design, can extend the operational boundaries of the aircraft by delaying the onset of buffet and increasing the lift at maximum cruise speed. This can mean reducing wave and friction drag and enhancing the maximum operating Mach number, thus improving flight performance and allowing more flexible conditions for operating the aircraft.

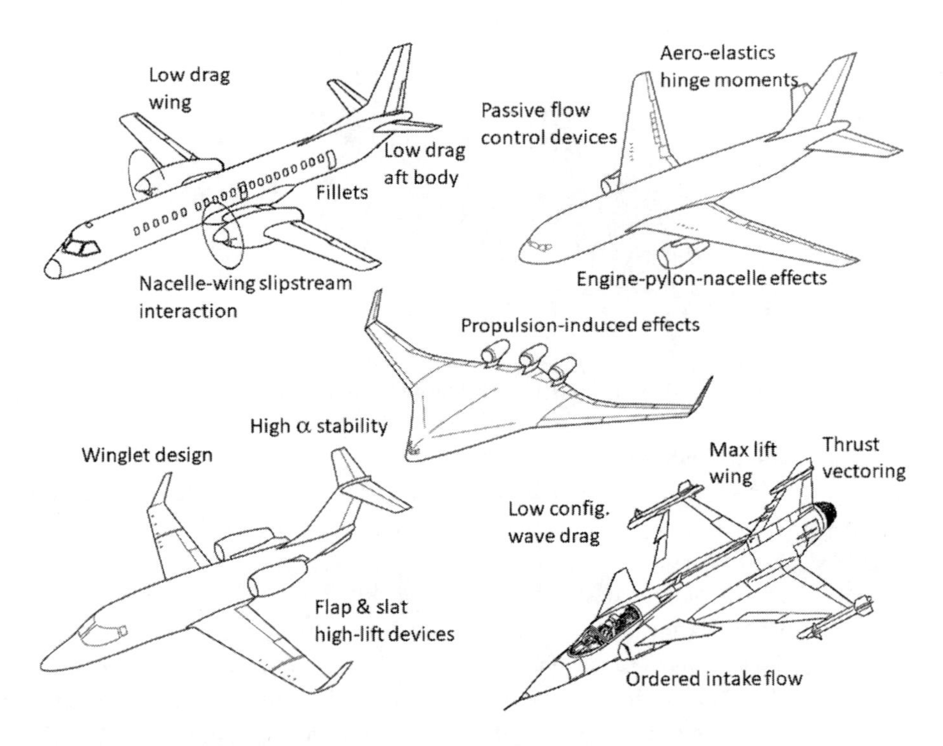

Figure 1.22 Five different classes of aircraft pose challenging aerodynamic design tasks.

Advanced wing-tip devices decrease drag, increase the maximum range, and relax the limitations on take-off weight. Better integration of propulsion into the airframe reduces interference drag. Means to enhance lift and decrease lift-induced drag include using wings with higher aspect ratio, variable camber, and smart wings that change shape with flight regime. Better shock-wave boundary-layer control delays the onset of buffet and drag divergence. Greater regions of laminar flow over the wing, the tail, and nacelles translate into reduced friction drag, as does overall better management of the turbulent flow.

With this introduction to the subjects in the book, we are ready to look in the next chapter at the types of flow fields that the aerodynamic designer strives to exploit.

1.5 Learn More by Computing

Gain hands-on experience of the computational tools for the topics in this chapter by working with the on-line resources. Exercises, tutorials, and project suggestions are found on the book website www.cambridge.org/rizzi. In particular, work the tutorial introducing the mathematical shape optimization. Software used to compute many of the examples shown is available from http://airinnova.se/education/aerodynamic-design-of-aircraft

References

[1] D. Böhnke. *A Multi-Fidelity Workflow to Derive Physics-Based Conceptual Design Methods*. Doctoral thesis, DLR, August 2014.

[2] R. Braun. *Collaborative Optimization: An Architecture for Large-Scale Distributed Design*. PhD thesis, Stanford University, May 1996.

[3] J. Chuprun. Wings. Technical report, 1980. AIAA-Paper-80-3031.

[4] P. D. Ciampa and B. Nagel. AGILE the next generation of collaborative MDO: Achievements and open challenges. 2018 Multidisciplinary Analysis and Optimization Conference, June 2018. AIAA-Paper-2018-3249.

[5] G. Corliss, C. Faure, A. Griewank, L. Hascoët, and U. Naumann, editors. *Automatic Differentiation of Algorithms: From Simulation to Optimization*. Springer, 2002.

[6] P. DellaVecchia, P. Ciampa, P. S. Prakasha, B. Aigner, and I. vanGent. MDO framework for university research collaboration: AGILE academy initiatives and outcomes. 2018 Multidisciplinary Analysis and Optimization Conference, June 2018. AIAA-Paper-2018-3254.

[7] Ercoftac. www.ercoftac.org/publications/.

[8] M. B. Giles. Aerospace design: A complex task. Technical Report, Computing Laboratory, Oxford University, July 1997.

[9] M. B. Giles and N. A. Pierce. An introduction to the adjoint approach to design. *Flow, Turbulence and Combustion*, 65: 393–415, 2000.

[10] J. S. Gray, J. T. Hwang, J. R. R. A. Martins, K. T. Moore, and B. A. Naylor. OpenMDAO: An open source framework for multidisciplinary design, analysis, and optimization. *Structural and Multidisciplinary Optimization*, 2019. Available from http://mdolab.engin.umich.edu.

[11] D. W. Hall. To fly on the wings of the sun: A study of solar-powered aircraft. Technical Report Vol. 18, Lockheed Horizons, Lockheed Corp., Burbank, CA, January 1985.

[12] M. Hepperle. The sonic cruiser – a concept analysis. Presented at Aviation Technologies of the XXI Century: New Aircraft Concepts and Flight Simulation. Aviation Salon, ILA Berlin Air Show, Berlin, May 2002.

[13] D. E. Hoak et al. The USAF Stability and Control DATCOM. Technical Report TR-83-3048, Air Force Wright Aeronautical Laboratories, October 1960 (revised 1978).

[14] SUMO homepage. Available from www.larosterna.com/sumo.html.

[15] I. Kroo and V. Manning. Collaborative optimization: Status and directions. AIAA Paper 2000-4721, September 2000.

[16] I. M. Kroo. MDO for Large-Scale Design. In *Multidisciplinary Design Optimization: State-of-the-Art* SIAM, 1997, pp. 22–44.

[17] E. Obert. *Aerodynamic Design of Transport Aircraft*. IOS Press, 2009.

[18] A. C. Piccirillo. *Elegance in Flight: A Comprehensive History of the F-16XL Experimental Prototype and Its Role in NASA Flight Research*. NASA Aeronautics Book Series, 2014. Available from www.nasa.gov/ebooks.

[19] D. P. Raymer. *Aircraft Design: A Conceptual Approach*. American Institute of Aeronautics and Astronautics, 6th edition, 2018.

[20] A. Rizzi. Modeling and simulating aircraft stability and control – the simsac project. *Progress in Aerospace Sciences*, 47(8): 573–588, 2011.

[21] A. Rizzi, P. Eliasson, T. Goetzendorf-Grabowski, J. B. Vos, M. Zhang, and T. S. Richardson. Design of a canard configured TransCruiser using CEASIOM. *Progress in Aerospace Sciences*, 47(8): 695–705, 2011.

[22] A. Rizzi and J. B. Vos. Preamble to special issue – modeling and simulating aircraft stability and control. *Progress in Aerospace Sciences*, 47(8): 571–572, 2011.

[23] J. Roskam. *Airplane Design Parts I–VIII.* DAR Corporation, 1985.

[24] D. M. Ryle. *High Reynolds number subsonic aerodynamics*, chapter Paper No. 6, pages 6–1 – 6–115. AGARD Lecture Series LS-37-70. AGARD, Neuilly-sur-Seine, France, June 1970.

[25] A. Sgueglia, P. Schmollgruber, N. Bartoli, E. Benard, J. Morlier, J. Jasa, J. R. R. A. Martins, J. T. Hwang, and J. S. Gray. Multidisciplinary design optimization framework with coupled derivative computation for hybrid aircraft. *Journal of Aircraft*, 2020 (in press).

[26] A. M. O. Smith. Wing design and analysis – your job. In T. Cebeci, editor, *Numerical and Physical Aspects of Aerodynmic Flows*. Springer Science + Business Media, 1984, pp. 41–59.

[27] J. Sobeiszczanski-Sobieski, J. Agte, and R. Sandusky. Bi-level integrated system synthesis (BLISS). Presented at *AIAA Symposium on Multidisciplinary Analysis and Optimization*, St. Louis, MO, September 1998. AIAA-1998-4916.

[28] P. R. Spalart and V. Venkatakrishnan. On the role and challenges of CFD in the aerospace industry. *Aeronautical Journal*, 120(1223): 209–232, 2016.

[29] E. Torenbeek. *Synthesis of Subsonic Airplane Design*. Springer Netherlands, 1982.

[30] J. E. Williams and S. R. Vukelich. *The USAF Stability and Control Digital Datcom.* Air Force Wright Aeronautical Laboratories, 1979.

[31] B. Wujek, J. Renaud, and S. Batill. A concurrent engineering approach for multidisciplinary design in distributed computing environment. In N. Alexandrov and M. Y. Hussaini, editors, *Multidisciplinary Design Optimization: State of the Art*. SIAM, 1997, pp. 189–208.

2 Airflow Physics and Mathematical Models

...know your flow and cultivate "healthy" flow patterns suited to design performance
...do not spend your time seeking patent medicines to cure flows with chronic disorders.

Dietrich Küchemann[1]

All models are wrong, but some are useful.

George E. P. Box, British statistician[2]

Preamble

The shape of the aircraft determines how its movement creates the aerodynamic forces on it, and hence its performance: fuel economy, cruise speed, and maneuverability and controllability. The designers use their knowledge on the mapping from shape to flow to forces to improve on existing designs or create new ones. The mapping happens to be explicitly known as a set of partial differential equations (PDEs), but the human mind cannot operate with PDEs. It must create a finite set of mental models of the flow phenomena with reasonable causality producing quantitative conclusions. This chapter discusses such models and how the originally intractable PDEs can be turned into computationally feasible tools that enable the engineer/designer to move from cause–effect understanding to numbers and shapes.

The usefulness of the models is *not* determined by mathematical estimates of their errors, although that might be desirable. Like in other sciences dealing with complex systems, the quality of a model is judged by extensive experience and comparison of its predictions with reality. Thus, the models now in common use have been selected by theoretical soundness and by best "clinical" experience. Mechanisms for the creation of lift and drag are fundamental to the building of useful models.

The crucial phenomenon is *flow separation*. When the flow is stable at the trailing edges of the configuration, we have *streamlined flow* and all is well. But when the angle of attack is too large or the airspeed approaches the speed of sound, the flow may suffer abrupt changes, and so too the forces. The flow may suddenly separate over much of the wing area in a catastrophic loss of lift called "stall"; shocks may appear and move suddenly or interact with the boundary-layer flow to create large-scale separated flow that is no longer streamlined.

[1] [8] with permission Cambridge University Press.
[2] [2] with permission John Wiley & Sons.

On the other hand, a flow separating from the *leading edge* of a thin wing of low aspect ratio develops into a stable *coherent* vortex that produces additional lift in a benign fashion. In addition, flow separation in the form of a thin bubble on the wing suction side is tolerable and a feature of many successful low-speed wing designs.

2.1 Introduction to Wing Flow Physics

Section 1.2.2 in the previous chapter indicated very generally the three types of planforms characterized by their slenderness ratio s/ℓ that achieve optimal cruise efficiency over the speed spectrum, illustrated in Figure 1.13. Now we look to understand the flow physics that makes these three classes of wing planforms, and their associated flow patterns, yield L/D values that are so beneficial.

To this end, we redraw Figure 1.13 in the form of Figure 2.1, showing the various flow features expected with the three typical wing planforms in terms of aspect ratio, sweep angle, and cruise speed. The primary aerodynamic wing parameters (Table 1.1) control the planform shape in our discussion.

The various flow features that we see involved in these three flight categories are: boundary-layer separation at all speeds, resulting in either chaotic wakes or coherent vortices; and shock waves at transonic and supersonic speeds. These flow phenomena

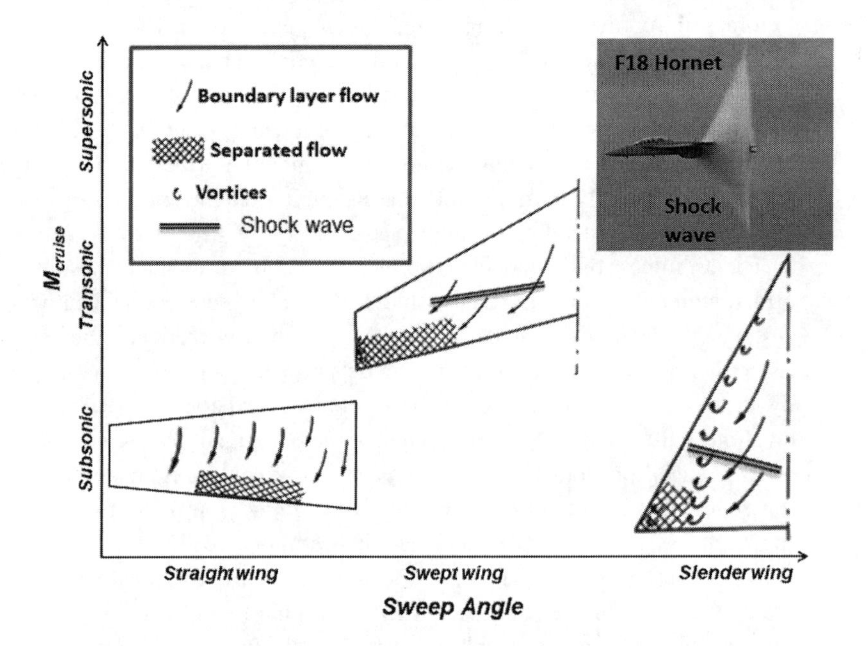

Figure 2.1 Factors affecting "healthy" flow patterns over wings (i.e. planform (AR, sweep angle) and cruise speed M_{cruise}). (F-18 photo reprinted from Wikipedia Creative Commons)

can strongly interact with each other. Each of them is discussed at some length in this chapter. Deeper and more detailed treatment of high aspect ratio straight and swept-wing attached flow is given in Hirschel et al. [12]. Another Hirschel et al. [13] monograph investigates vortex-separated flow over slender wings very thoroughly. A discussion on the evolution of our computational fluid dynamics (CFD) capabilities, in particular related to separated flows, is provided in Rizzi and Luckring [26].

An F-18 Hornet is shown in close to sonic flight as an example. It has a low aspect ratio swept wing with a leading-edge extension (LEX) to provide a lifting vortex, typical of current US fighter designs. The transonic expansion wave "vapor cone" is visible by condensation in the temperature drop where flow expands rapidly (i.e. the expansion fan). The heating through the terminating shock wave vaporizes the water drops again, hence the sharp localization of the vapor cone.

2.1.1 Classical Straight Wings

Before the jet propulsion era, high-speed propeller-driven aircraft were typically characterized by slenderness ratio $s/\ell \approx 0.5$ and straight wings having thickness ratios in the range from 14% to 18%, defined as the thickness of the wing divided by its chord. The aerodynamic design of such aircraft could, to a first approximation, be considered in terms of the *addition* of various elements of the aircraft, seen as a further step of linearizing the Cayley paradigm. For example, the drag of the wing, fuselage, and tail, measured separately, could be added together, with only minimal consideration of interference effects, to obtain the drag of the entire aircraft. Moreover, the wing shape was a simple ruled surface given by two airfoil sections at the root and the tip, implying that the wing flow behaves much as the flow does over these airfoils. Thus, a major part of wing shaping was the choice of airfoils with optimal lift-to-drag characteristics. The airfoil should have a round leading edge to give high suction on the upper side, followed by gentle pressure recovery to the trailing edge where the flow separates and pressure is again almost as high as the stagnation value.

Examples of such designs during the 1920s are Lindbergh's Ryan monoplane, Spirit of St. Louis, the Lockheed Vega, the Heinkel He-70, the DC-3 during the 1930s, and the fighter aircraft in the 1940s. Obtaining a streamlined flow pattern is the driving principle behind these successful designs. It had been known since the time of Aristotle that a moving body passing through air encounters resistance (i.e. aerodynamic drag). During the 1920s, Melvil Jones developed the ideas of Louis Breguet into a refined theory to demonstrate emphatically the importance of drag to the performance of aircraft, and hence the need for streamlined shaping.

Jones's 1929 paper, "The Streamline Aeroplane" [21], proposed an ideal aircraft that would have minimal drag. This led to the concepts of a "clean" (i.e. streamlined) monoplane and retractable undercarriage that became the genus for this class of aircraft.

Lift-Induced Drag

Lanchester evolved his vortex theory of lift in 1894, but owing to the opaque language and lack of rigorous mathematics, his work was not accepted. Eventually, Prandtl evolved a clear mathematical theory showing that $C_D \propto C_L^2/AR$. It also showed that, for a given aspect ratio, the induced drag is minimized if the spanwise distribution of lift is elliptical. This was a major breakthrough, but because at the time induced drag was only around 5% of the total drag, the emphasis that Lanchester and Prandtl put on a high aspect ratio was neglected. Inevitably, however, the Lanchester–Prandtl vortex theory gained acceptance, initially in Germany, where it was applied to the Heinkel He-70, and then more widely in the USA and the UK.

Attached Boundary Layer: Aerodynamic Streamlining

In terms of aerodynamic efficiency, as measured by L/D, the best that had been achieved by the late 1920s was a L/D of about 10, typical for an aircraft with a water-cooled engine. Those with air-cooled engines were slightly worse.

Jones computed the performance of an ideal streamlined aircraft in which the drag was only the sum of the induced drag corresponding to the span loading of the aircraft, as predicted by Lanchester and Prandtl, and the skin-friction drag. He was able to show that many contemporary aircraft had two to three times more drag than such an ideal aircraft. Furthermore, even the best contemporary aircraft would have needed only a third of the power with which they were provided, if they conformed to his ideal aeroplane. Put another way, if they were streamlined so that the boundary layer remained attached, they could fly 60 mph (95 km/h) faster on the same power.

Clean-up operations were mounted: undercarriages were made retractable, wings were made thinner and their surfaces made smoother by flush riveting, wing–body junctions were filleted, flaps became common, and wing loadings were increased.

2.1.2 Transonic Swept Wings

The potential for even higher speed offered by the revolutionary new jet propulsion systems that had their beginnings in the 1940s, however, could in no way be realized by the typical aircraft designs of the World War II era. As jet propulsion ushered in prospects of flight in the transonic and supersonic flight regimes, these wing-design procedures, along with other time-honored tenets of airplane design, were to undergo radical change that came about through much trial-and-error development.

At some high flight speed, though lower than the speed of sound, say $M_\infty \approx 0.8$, there occur regions over the wing where the flow velocities reach supersonic values. Such flight cases are said to be *transonic*. Before reaching the wing trailing edge where the velocity is subsonic, the local flow velocity retards from supersonic to a subsonic value across a *shock wave*. The wing experiences an increase in drag, called *wave drag*. The flight Mach number at which wave drag first appears is the critical Mach number M_{crit}. The shock wave also acts on the boundary layer through the discontinuous increase in pressure across it. Such *shock–boundary layer interaction* is a substantial cause of flow separation. Modern airliners fly fast enough that wave drag must be kept low. Chapters 6 and 8 discuss all of this in more detail.

Wing Sweep

Now, in addition to keeping the flow attached to the wing, the designer has to shape the wing to allow a high critical Mach number. A small amount of wave drag is tolerable so that the aircraft cruises at as high a speed as possible at the expense of that wave drag. The top speed of an aircraft in level flight was determined by the drag divergence Mach number M_{dd}, beyond which the rapid increase in drag occurred. Research concentrated on means to increase M_{dd} rather than to reduce the level of the drag in this condition. This led to the development of wing sweepback: the classic wing shape of the late 1940s was swept to permit efficient and safe operation, bringing the slenderness ratio to $s/\ell \approx 0.35$. The design of such swept wings for today's jet airlines to fly economically at transonic speed is an enormously important activity. Chapter 3 explains wing sweep in further detail.

Shock Wave Acting on the Boundary Layer

When localized supersonic flow, terminated by a weak shock wave, first appears on a wing, the dominating parameter is the adverse pressure gradient resulting from the abrupt rise in static pressure across the shock. This will cause at least a thickening of the boundary layer, if not flow separation. Just above the critical Mach number, the shock wave is not strong enough to cause separation, and the effects on drag are barely perceptible. Further increase in speed causes the supersonic region to enlarge and the strength of the shock wave to grow. The interaction between the shock wave and boundary layer will eventually cause boundary-layer separation. This prevents the static pressure at the wing's trailing edge from recovering to the free stream value, leaving the aftmost, rearward-facing surface of the wing at a lower-than-desired pressure. This increases the pressure drag and is one of the main contributors to wave drag. Thicker airfoils were found to aggravate this undesirable behavior, causing it to occur at lower speeds. It was conclusively established in the late 1940s that thinner sections, typically about 10% thickness, could greatly alleviate these effects and postpone the drag rise to higher Mach numbers. See [12] for a thorough exposition on this topic.

2.1.3 Slender Supersonic Wings

Transonic aircraft usually do not cruise much beyond M_{crit} because of the excessive wave drag; their optimum cruise efficiency is transonic. For supersonic flight, however, the aircraft must have sufficient power to overcome the high drag in the transonic speed range and be capable of controlled flight through this capricious Mach number range. (Chapter 6 shows just how capricious this range actually is.)

For many years prior to the jet propulsion era, reducing the airfoil thickness ratio was the only known method of increasing the wing critical Mach number by any significant amount. Further study and development led to thin, slender wings of low aspect ratio with slenderness ratios around 0.2 and thickness 3%–4% as the most suitable, and these have become part of the aerodynamic ingredients of a low-drag supersonic aircraft. High sweep produces lower high-speed drag, but such wings also generate less lift, a decided disadvantage. A thin, low-*AR* wing yields relatively low maximum lift-to-drag ratio $(L/D)_{max}$ at subsonic speeds, as well as certain undesirable handling characteristics in the low-speed, high-angle-of-attack regime.

Vortex Lift

Thin wings also mean small leading-edge nose radii that are prone to flow separation. However, it was discovered that separation from a sufficiently swept leading edge created a stable vortex over the wing that *enhanced the lift*, and this effect is widely exploited in all designs. Flows with *coherent vortex separation* have led to a new pattern of stable separated flow, quite different from the desired streamline flow pattern for the high-*AR* wings, straight or swept. The discovery of vortex lift is quite an interesting story and the reader is referred to the full account of it by Luckring [20].

Nonlinear flow physics phenomena in the high-*AR* cases involve mainly the behavior of the attached boundary layer, along with its interaction with a shock wave. Hirschel et al. [12] present a book-length exposition. The delta-wing case, with coherent vortex separation, involves further interactions, including wing stall due to *vortex breakdown*. In supersonic flight, shock waves are ubiquitous, excited by sharp changes in the cross section of the configuration at rounded leading edges and nose, wing–body junctions, etc., leading to many variants of shock–vortex interactions and shock–vortex–boundary layer interactions.

Nevertheless, the maximum lift-to-drag ratio $(L/D)_{max}$ of the thin, low-*AR* wing is low at subsonic/transonic speeds, and becomes even lower at supersonic speeds, making flight in the low-supersonic speed range not very efficient, as we shall see in Chapter 9 for the TransCruiser (TCR) study. To regain efficiency, supersonic aircraft, such as the Concorde, cruise at considerably higher speed, $M_\infty \sim 2$.

2.1.4 Healthy Flow

For each of the above three flight conditions, the wing designer must consider effects of deviations in speed and attitude as well as of gusts and elastic deformations of the airframe, indicated by the region surrounded by the dashed circle in Figure 1.5. The aerodynamic forces must not change abruptly for any deviation (i.e. they must be *stable with respect to perturbations*, the property of Küchemann's "healthy" flows). Aerodynamic design aims to create a well-tempered aircraft with a *well-ordered*, "healthy" flow that provides stable forces in all flight conditions.

Some of the factors determining the health of a flow are discussed in Section 2.2, beginning with the lift and drag on an airfoil. Section 2.4.2 presents the classical Prandtl–Glauert model of vortical flow around high-*AR* straight wings at low speed. Then, in Section 2.5, we address the physics behind the nonlinear flow phenomena: vortices, shock waves, boundary layers, and stall; and how these combine into the favorable flow patterns leading to efficient cruise in Figure 1.13. Finally, Section 2.8 presents the mathematical models most commonly used today to predict these flows.

2.2 Shape Determines Performance

Forces on the airplane surface result from pressure acting normal to and shear stress (friction force) tangential to the surface. The net forces and moments acting on the body are the surface integrals of these forces over *flight shape*, the whole external surface "wetted" by the airflow. We first discuss drag and its associated flow physics, and then we turn to lift in Section 2.4.1.

A mental image of the flow around a body is most helpful to understand a given aerodynamic situation. The pressure distributions over the major lifting surface (i.e. the wing) are useful for the analysis of the flow. In a plane approximation, the flow streams past the free stream parallel wing section, the *airfoil*, which often looks similar to the one in the upper part of Figure 2.2. They can be computed by a mathematical flow model at every point of the surface. An alternative is to measure them by pressure taps in a number of well-chosen places on a model in a wind tunnel campaign. The forces and moments are obtained immediately by the wind tunnel balance.

This section will describe some of the systematics concerning the geometry of airfoils and wings, the pressures occurring on these bodies, their integration into forces and dimensionless coefficients, and finally the use of the coefficients in the estimation of airplane performance. We assume a basic familiarity with laminar and turbulent flows, and some details are given in Sections 2.3.1, 2.3.2, and 2.3.3.

A prime example of how shape influences flow and hence performance (see Figure 2.2) appears in the classic book *Applied Hydro- and Aero-Mechanics* from 1934 by Prandtl and Tietjens, now reprinted as [24]. The drag of the different objects in the figure are as follows.

Consider a circular cylinder whose diameter is the maximum thickness of the airfoil. You can then think of the streamlined airfoil as being a circular cylinder plus a slightly extended nose and a long, sharpened tail. The drag on the circular cylinder alone is about tenfold that on the airfoil.

Now reduce the diameter of the circular cylinder to 1% of the airfoil chord; the drag then is the same as that of the airfoil that is ten times thicker. The reason for this is that the turbulent wake of the cylinders carries away much more kinetic energy than that of the airfoil.

The scale for speed and size is the Reynolds number, the ratio of inertial forces to viscous forces. It was named for Osborne Reynolds (1883), who investigated transition from laminar to turbulent flow in pipes,

$$Re_\ell = \frac{\rho V \ell}{\mu} \tag{2.1}$$

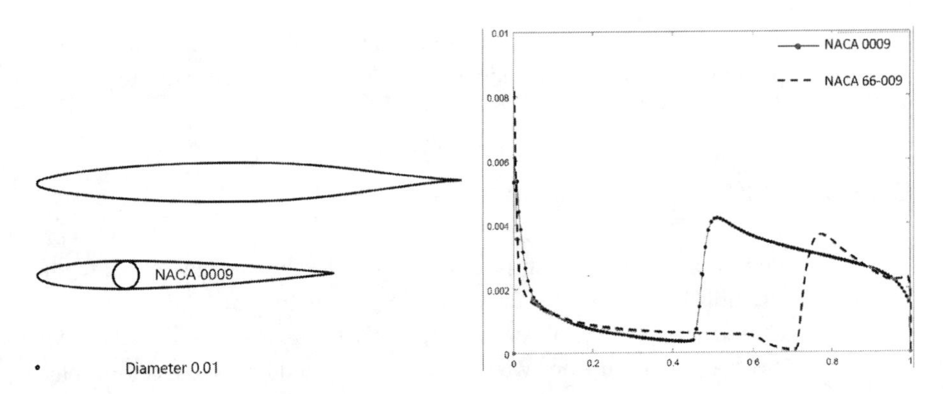

Figure 2.2 (Left) Circular cylinder, NACA0009 airfoil, and laminar airfoil NACA66-0009, all with the same drag; (right) friction coefficients for the two foils.

where V is the velocity, ℓ a characteristic length, and ρ and μ the density and dynamic viscosity of the fluid. If the choice of length scale is "obvious," the subscript ℓ is left out.

The flow next to the airfoil is laminar at the nose, and transitions to turbulent further along the surface. There is much to gain if the flow can be kept laminar for longer. The top image in Figure 2.2 is a NACA66-0009 airfoil designed for such laminar flow. When its surface is smooth enough and the oncoming flow has vanishing turbulence, it is about 40% longer than the NACA0009 foil for the same drag.

The *skin friction coefficient* C_f is the ratio between tangential force per unit area (the tangential component of the shear stress) and dynamic pressure of the stream just outside the boundary layer:

$$C_f = \tau/(1/2\rho V^2) \tag{2.2}$$

The plots in Figure 2.2 show the MSES computed friction coefficients along the chord, with free transition, at $Re_{chord} = 5M$. The laminar airfoil transition is at $x = 0.7$ instead of 0.5 for the classical one; since most of the drag comes from the turbulent part, the drag coefficients are approximately in the relation 3 to 5: 0.0030 to 0.0047.

2.2.1 Lift and Drag: Pressure Forces Acting on the Surface

The *pressure coefficient*, C_p, is defined as

$$C_p = \frac{p - p_\infty}{q} \tag{2.3}$$

where q is the free stream dynamic pressure. Figure 2.3 shows pressure coefficient distributions for a slightly cambered airfoil, the Eppler 387 airfoil at $\alpha = 2.04°$, $Re = 0.2M$ The pressure coefficient is illustrated as arrows normal to the surface. A vector pointing away from the contour represents *suction* $C_p < 0$ (i.e. $p < p_\infty$). The pressure coefficient is in general negative, except in small regions around the edges of the airfoil. This case is a little special: on the suction side, there is a large

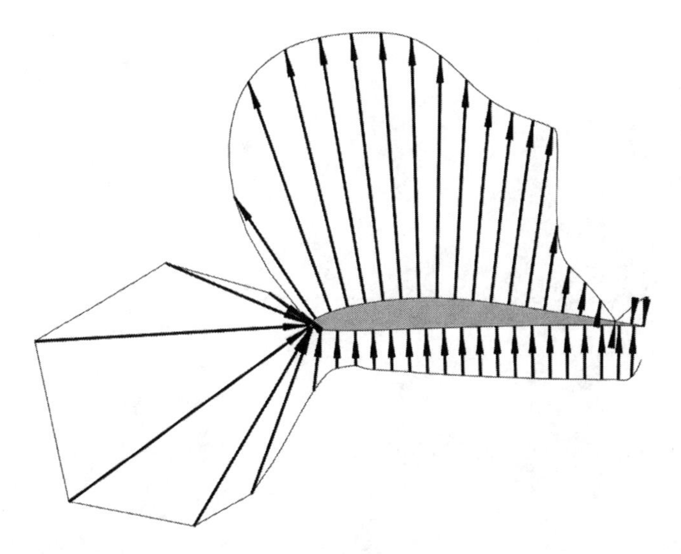

Figure 2.3 Pressure coefficient as arrows on the Eppler 387 airfoil.

laminar separation bubble, responsible for the pressure plateau, and the whole pressure side has $p > p_\infty$.

Airfoils are shaped to maximize lift and minimize drag. A visual inspection of the arrow lengths and directions in Figure 2.3 reveals that:

- The highest pressure appears at the leading-edge stagnation point where velocity vanishes. It exerts primarily a drag force.
- Past the nose, the arrows over the mainly horizontal upper surface point upward and are larger than those on the lower surface, producing a net lifting force. This case happens to have over-pressure on the whole lower side, giving lift.
- Near the trailing edge, the pressure increases again (arrows point into the surface), with a horizontal force counteracting the drag force at the nose (i.e. a thrust).
- The net horizontal force (drag at nose minus thrust at trailing edge) is the net drag. Depending on the *pressure recovery* toward the trailing edge, the net drag can be very small, even zero if the pressure recovers all the way to the stagnation value (d'Alembert paradox).

Note that the arrow spacing is much smaller at the nose. The stagnation pressure acts on a very small surface, so its drag is not so dominant as it appears here.

2.2.2 Pressure Distributions

The aerodynamic character of an airfoil is best studied in a graph with pressure coefficient plotted vs. chord coordinate x, with negative values upward in the graph, because then the upper surface of a conventional lifting airfoil corresponds to the upper curve (see Figure 2.4).

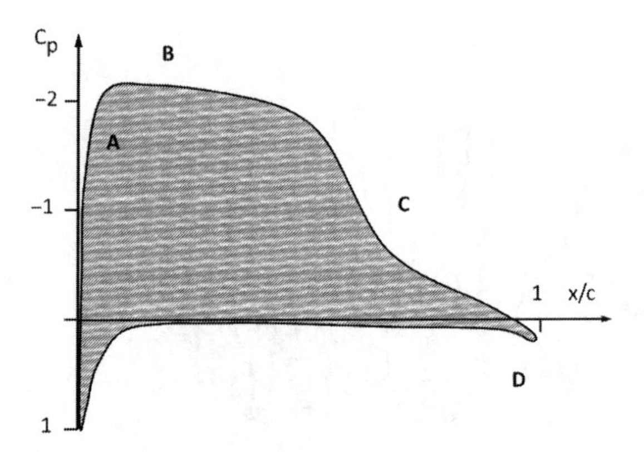

Figure 2.4 Features of airfoil pressure distribution.

The pressure changes very little through a thin boundary layer, so the velocity referred to is just outside it. C_p starts at the stagnation point where the velocity is zero on the leading edge. It is 1.0 in incompressible flow and slightly higher in compressible flow. The pressure decreases as the flow proceeds downstream, later to recover to a small positive value at the trailing edge.

- Trailing edge: the flow leaves the airfoil in the direction of the trailing-edge bisector.
- Upper surface: the upper surface flow makes a steep turn at the nose and the centrifugal forces are balanced with a low pressure at the surface.
- Lower surface: the lower surface pressure coefficient is both positive, which gives a lift, and negative, corresponding to a down-force.
- Pressure recovery: the pressure recovery region is a region of adverse pressure gradients. The pressure increases from a minimum value to the trailing-edge value. This represents a difficulty for the boundary layer fluid to move downstream, having lost some energy to friction.
- Trailing-edge pressure: the pressure depends on airfoil thickness and the shape close to the trailing edge. For thick airfoils, the pressure is slightly positive. For infinitely thin trailing-edge regions, $C_p = 0$.

Figure 2.4 [16] points out some elementary properties of desirable pressure profiles.

(A) Favorable pressure gradient maintains laminar flow; too small a gradient sacrifices lift.
(B) Minimum C_p determines maximal speed, indicating shock formation and start of pressure recovery region.
(C) Adverse pressure gradient determines transition to turbulence and separation. Avoid strong adverse pressure gradients close to the trailing edge.
(D) Trailing-edge pressure influences adverse gradient.

Means to Minimize Drag

The *pressure gradient*, a driving force for the flow along the wall, exerts an important influence on the transition from laminar to turbulent flow. It can be shown that laminar velocity profiles in adverse pressure gradients are less "stable" and more prone to transition than those in favorable pressure gradients. A more detailed discussion of laminar and turbulent boundary layers is found in Section 2.3.1.

Delayed transition to turbulence by a favorable pressure gradient is utilized as an underlying physical principle for laminar-flow airfoils which became known shortly before World War II. The result is achieved by displacing the section of maximum thickness downstream, which causes the point of minimum pressure also to move downstream.

In Figure 2.2, the "normal" NACA0009 foil has its maximum thickness at $x/c = 0.3$, whereas the NACA66-009 foil has its maximum thickness at $x/c = 0.45$. Both are symmetric and 9% thick. Owing to the downstream displacement of the point of maximum thickness, the point of minimum pressure has moved rearward by a considerable distance. The length of the laminar stretch of the boundary layer is considerably increased and this reduces the drag coefficient appreciably. The caveat is that it is very hard to keep the surfaces so smooth and clean that transition to turbulent flow is not triggered prematurely, voiding the drag improvement.

2.3 Boundary-Layer Development

Lift on the airfoil grows with increasing angle of attack, but only up to some limit, $C_{L,max}$, when it starts decreasing and the airfoil *stalls*. Figure 2.5 (top-left) shows the low-speed lift vs. angle of attack curve for the NACA 64-2-015 airfoil at $Re = 1M$, which stalls with decreasing lift from around 12°.

The flow adheres to the airfoil and, due to viscosity, the flow close to the body surface is retarded and the velocity decreases to zero on the surface. As a result, the airfoil experiences a wall shear-stress whose magnitude depends on the velocity gradient $\tau = \mu \frac{\partial V}{\partial y}$ at the wall, normal to the wall (Figure 2.5, bottom). The layer of fluid next to the boundary where the velocity $V(y)$ drops to zero is the *boundary layer*, outside which the effects of viscosity are very weak.

The cause of stall is the thickening of the boundary layer, which makes the exterior flow see a thicker trailing edge. This is known as "viscous de-cambering," shown by the edges of the boundary layer in the three images in the top-right of Figure 2.5. Note that the stall is very docile; lift decreases only slowly after the maximum.

Figure 2.5 (bottom) shows the boundary layer development with velocity profiles along the upper side of the airfoil. A is the stagnation point on the leading edge where the flow is divided into a stream following the upper side and one following the lower side. The flow accelerates around the nose. The boundary layer is laminar at the leading edge and transitions to turbulent at T. A fluid particle in the boundary

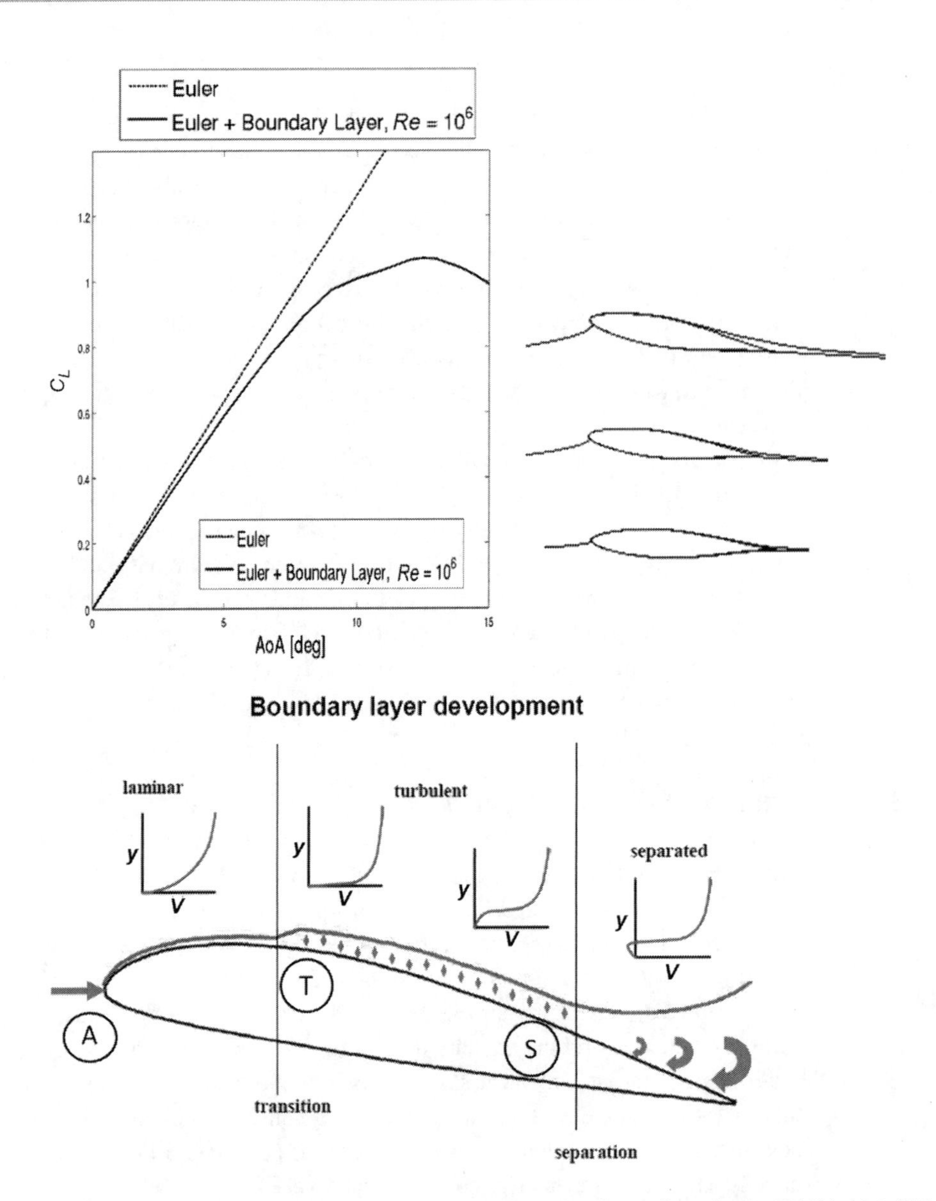

Figure 2.5 (Top) Lift vs. α curve and boundary-layer and wake outlines for the NACA 64-2-015 airfoil; (bottom) tangential velocity profiles $V(y)$ along the suction side of the airfoil; MSES computation.

layer subsequently is retarded due to the effect of friction until point S is reached. Upstream from S, the velocity profile in the boundary layer is as sketched to the left. At S, the wall-shear stress vanishes. Downstream of the separation point, to the right, the adverse pressure gradient reverses the flow direction and the shear stress. The velocity profiles show the sharper velocity gradient in the turbulent flow, responsible for the increased friction. The process can be observed by experimental visualization and computed by CFD.

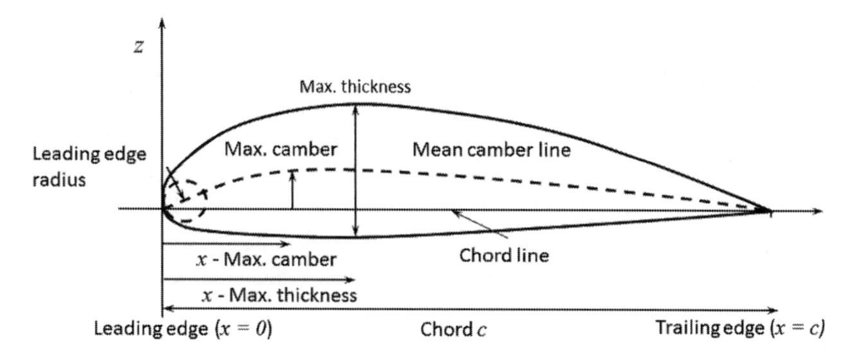

Figure 2.6 Main parameters defining airfoil shape.

The influence of the boundary layer is significant.

- Without the boundary layer (i.e. without viscosity), there is "ideal flow," hence:
  Lift increases indefinitely with increasing α.
  For subsonic flow, there is no drag, a peculiarity of 2D flow.
- With the boundary layer:
  Lift starts decreasing as the boundary layer thickens with increasing α.
  Airfoil *stalls* as $C_{L,max}$ is reached when the boundary-layer separation point has moved sufficiently upstream.

The aim of wing design is to match the useful operating range up to stall to the performance required for the aircraft.

Airfoil Geometry

Flow properties can be correlated to the airfoil geometry features shown in Figure 2.6. The nose radius of curvature may be of the order of a couple of percent of the chord, sufficiently large to avoid excessive suction that, for pressure recovery, would be followed by a strong adverse pressure gradient and could lead to flow separation. The airfoil thickness has a smooth maximum at a position along the chord of 25–60%. In addition, structural considerations enter into airfoil design, as a thicker wing makes a better structural beam, permitting the same load to be carried with less structural weight. Furthermore, the wing usually serves as the fuel tank, so maximum volume is desired. Counter to this, a large wing thickness induces higher velocities that, in turn, cause higher adverse pressure gradients on the aft portion, making thicker boundary layers and greater pressure drag. High-speed flight favors thinner airfoils because these higher local velocities due to thickness reach the speed of sound at a lower free stream Mach number, reducing the M_{dd} of the airfoil.

A mean or camber curve may be drawn midway between the airfoil's upper and lower contours, with equal distances along the camber curve normal to the upper and lower contours. The straight line connecting the most upstream point of the mean curve, the *leading edge*, with the most downstream one, the *trailing edge*, is the *chord line*, the length of which is the chord, c.

At the trailing edge, the airfoil is sharp, rounded, or even blunt (i.e. squarely cut off). A sharp or blunt trailing edge establishes the Kutta–Zhukovsky condition that the upper- and lower-side trailing pressures are equal so that there is no flow around the edge. A substantial radius at the trailing edge would allow air to flow part of the way from the lower surface to the upper surface without excessive velocities, thus reducing the lift. The leading edge is taken as the origin of a coordinate system, with x along the chord line and z (sometimes y) perpendicular and positive upward. The shapes of airfoil sections are usually given as tables. The angle of attack is measured to the chord line.

Camber

The maximum height of the camber curve above the chord line is denoted δ, which usually has a magnitude of a few percent. For some of the older airfoils, the camber line, $z_c(x)$, had a simple polynomial shape, or, as in the NACA four-digit airfoil family, was composed of two parabolic arcs. Modern airfoils do not have such simple camber geometry.

Thickness

The distances between the airfoil contour and the mean curve, measured normal to it, give the airfoil thickness distribution. The overall maximum thickness to chord ratio, $(t/c)_{max}$, may be of the order of 10–18% for subsonic airplanes, while for supersonic planes the maximum thickness may be as low as 3%. Older sections often had $(t/c)_{max}$ far forward, say around 30%, while modern sections may have the maximum at 60%.

Airfoil Composition

When the camber curve slope is small, half the thickness ordinate, $z_t(x)$, may be added to or subtracted from the camber ordinate, $z_c(x)$, to obtain the airfoil contour (see Figure 2.6). When maximal camber is significant and its position is close to the leading edge, the difference from the correct shape may become appreciable.

Can All Airfoils Be Generated in This Way?

Airfoils are defined from sets of points. An algorithm to extract camber curve and thickness is given on the book's website for the student to think about and experiment with. A closely related problem is to find the "medial axis transform", the locus of centers of circles inscribed in the foil contour. This recipe requires "extrapolation" at the rounded nose.

One may think of the flow field built from flow around the mean camber line without thickness and around the thickness with zero camber (see Figure 2.7). This connection can be made rigorous for linear flow models.

Airfoil Shapes Vary with Function

Figure 2.8 shows the lift-to-drag ratio (L/D) for some airfoil sections of lifting surfaces ranging from insects and birds to large subsonic aircraft. Keep in mind here that the L/D ratio for airfoils is necessarily higher than that for wings and complete aircraft

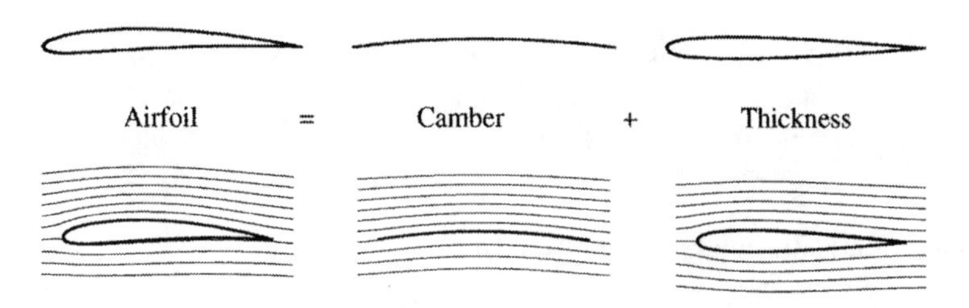

Figure 2.7 Wing-section contour built up of camber curve and thickness. Resulting flow-field also built up from solutions for camber curve and thickness.

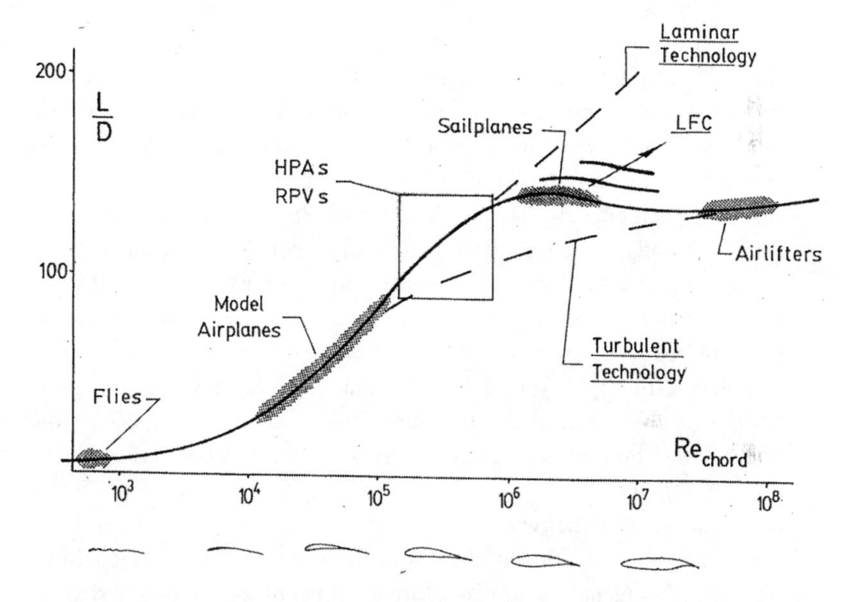

Figure 2.8 Flight "vehicles," airfoils, and speed regime. LFC = laminar flow control. (Courtesy of Mark Drela, MIT, private communication)

(cf. Figure 1.3). The flow is subsonic everywhere and the Mach number plays little role. Thus, the Reynolds number based on the chord characterizes the speed and size of the wing. Lift-to-drag increases with Re_{chord}. For wings of human-powered aircraft (HPA) and remotely piloted vehicles (RPVs), efficiency can be gained if profiles are designed for laminar flow. Sailplanes (optimized gliders) have very high-*AR* wings with low induced drag and the largest L/D, typically 50 and even 70. Airfoils for airlifters (military transport planes and commercial jet airliners) have a somewhat lower L/D than sailplane airfoils. Insects and birds can practise variable geometry. While in flight, they adjust the airfoil to give a suitable flow. Therefore, they do not need a big leading-edge radius to get a section tolerant to angle-of-attack changes. The early aircraft needed a larger nose radius to have some ability to change angle of attack in flight with a rigid airfoil. This trend is accentuated in the conventional airfoil design. Airfoils for high speed have less camber and smaller nose radii and the location of maximal thickness moved further downstream, as witnessed by the foil for

$Re = 10M$. In supersonic flow, the surface pressure depends primarily on the local surface inclination to the flight direction. To minimize drag, the inclination must be as small as possible, thus a sharp leading edge and t/c of only a few percent are required.

2.3.1 Laminar and Turbulent Boundary Layers

A discussion of aerodynamic drag would be incomplete without making the distinction between laminar and turbulent flows, which was touched on in Section 2.2. As a reminder, Figure 2.13 shows laminar and turbulent boundary layers. It brings out a new feature: the turbulent boundary layer stays attached longer than the laminar one. The mixing with the outside high-speed flow energizes it; the laminar layer does not mix. Figure 2.10 (left) quantifies the lower wall friction in laminar than in turbulent flow by plotting the skin-friction coefficient C_f vs. Re_x. We can think of the Re_x-axis as nondimensional streamline distance from the nose stagnation point, where the body first meets the flow. The flow is laminar until it abruptly transitions into turbulent flow at a position Re_{tr}. The actual value of Re_{tr} depends on factors such as surface roughness and turbulence intensity in the oncoming flow. Transition prediction models are as yet unreliable and outer pressure gradients have a strong influence, as we shall see in Section 9.5.3. The accuracy of the regions sketched on the aircraft below is quite limited.

Figure 2.10 (right) shows flat plate boundary-layer velocity profiles at $Re_{x,tr} = 0.5M$ with free stream velocity U and kinematic viscosity ν. The turbulent layer is thicker, yet with four times the wall stress. Note that the wall distance is measured in units of ν/U. Often such plots use velocity and length scales based on wall stress, but this is different for the two flow regimes.

The boundary layer is turbulent over most of the wetted surface. In Figure 2.9, the regions of turbulent flow are downstream of the black curves where $Re_x = Re_{tr}$, and the associated friction drag is larger there. For the airliner in Figure 2.9 flying at, say, 250 m/s, the transition on the nose is at 0.5 m, where $Re_x = 3M$. Figure 2.10 shows that the friction coefficient at that Re increases by a factor about 8 at transition. The jump is even more dramatic for transition at larger Re_{tr}. It follows that accurate drag prediction must include prediction of laminar–turbulent transition locations. This will

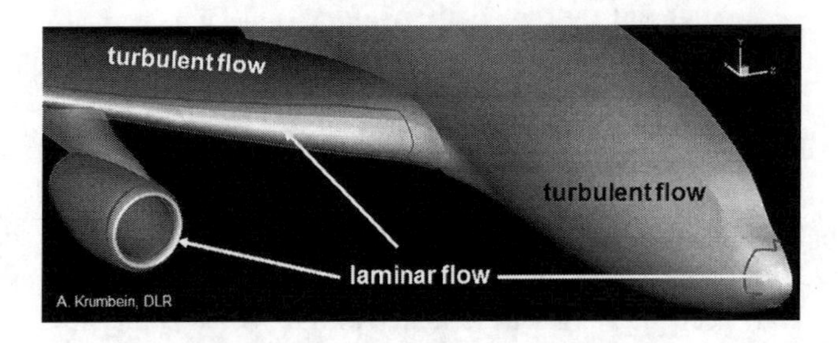

Figure 2.9 The high Reynolds number airflow over an aircraft is almost entirely turbulent. Only the limited regions indicated have laminar flow. (Courtesy of Andreas Krumbein, DLR Göttingen, personal communication)

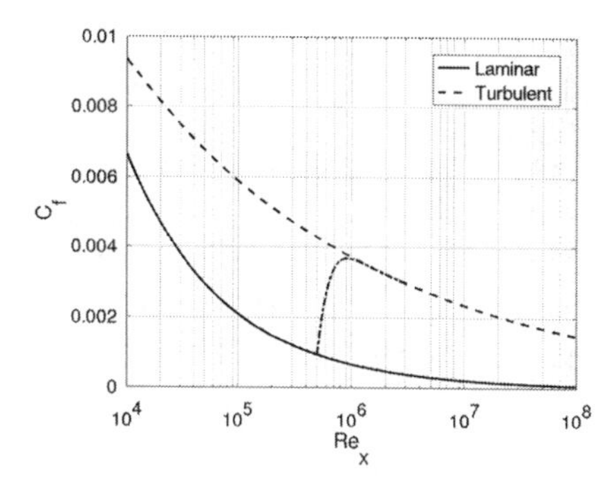

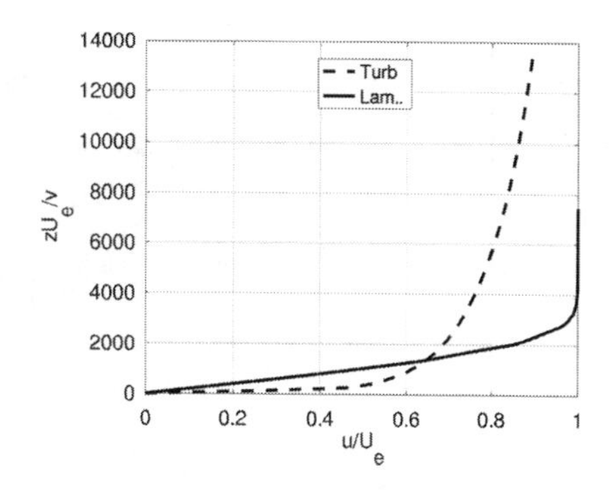

Figure 2.10 (Left) Flat plate skin friction coefficient in laminar flow transitioning to turbulent at $Re_x \approx 5 \times 10^5$; (right) velocity profiles at that Re_x. The wall distance is measured in units of ν/U_e, ν is the kinematic viscosity, and U_e is the free stream velocity.

be touched on in the discussion of turbulence models and in Chapter 7 on airfoil design.

2.3.2 Turbulence

In experiments on fluid systems, it is observed that at values of Re below the critical Reynolds number Re_{crit}, the flow is smooth and adjacent layers of fluid slide past each other in an orderly fashion. If the applied boundary conditions do not change with time, the flow is steady. This regime is called *laminar* flow.

At values of the Reynolds number above Re_{crit}, a complicated series of events takes place, which eventually leads to a radical change of the flow character. The motion becomes intrinsically unsteady even with constant imposed boundary conditions, and velocity and all other flow properties vary in a random and chaotic way. This regime is called *turbulent* flow. An important characteristic is the richness of scales of the eddy motion in such flows. In a fully developed turbulent motion, all scales appear to be fully occupied or saturated in a sense, from the largest ones that can fit within the size of the flow region down to the smallest scale allowed by dissipative processes, the Kolmogorov dissipation scale. In-between, there are the integral scales and the inertial subrange, which follows the −5/3 Kolmogorov law. See Figure 2.11 for the turbulent kinetic energy spectrum as a function of wave number $k = 2\pi/\lambda$. λ is the length scale of the flow features, the size of eddies. These scales and features of the spectrum are determined in terms of the dissipation rate of the turbulent kinetic energy by dimensional and symmetry arguments. See Wilcox [35] for a derivation. Figure 2.11 also shows the ranges of length scales resolved by three numerical models: direct numerical simulation (DNS)), large/detached eddy simulation (LES/DES) and Reynolds-averaged Navier–Stokes (RANS) models, to be discussed in Section 2.9.3 and in Chapter 6.

Turbulence is so prevalent because steady laminar flows tend to become unstable at high Reynolds numbers and therefore cannot be maintained indefinitely as steady

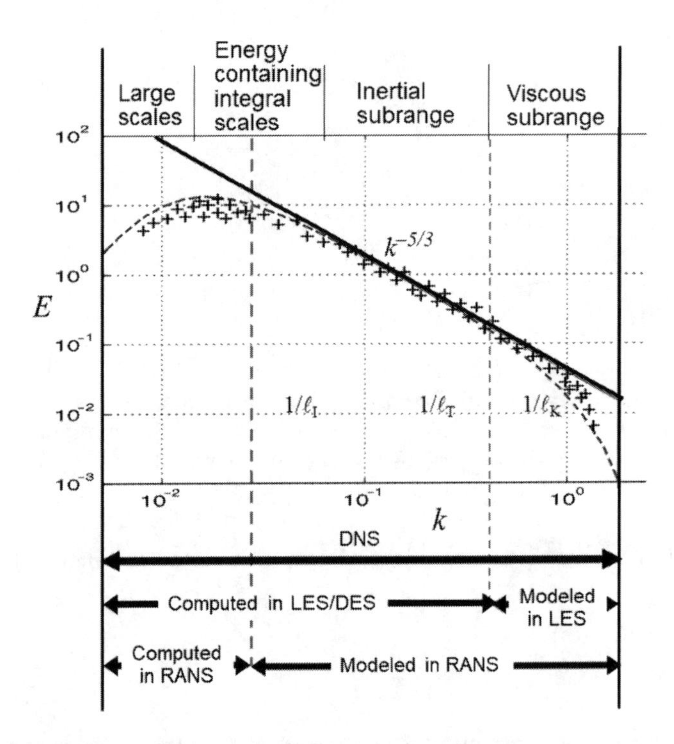

Figure 2.11 Energy spectrum of turbulence as a function of wave number k. (Courtesy of Christer Fureby, Lund University, personal communication)

laminar flows. Instability to small disturbances is an initial step in the process whereby a laminar flow goes through transition to turbulence.

2.3.3 Instability and Transition

Consider a laminar flow that is slightly disturbed locally in space and time. If the flow returns to its original state after the disturbance, it is called stable, whereas if the disturbance grows, the flow is called unstable. Instability often leads to turbulence, but may also take the flow to a different laminar, usually more complicated state.

Flows of fluids of low viscosity may become unstable when large gradients of kinetic and/or potential energy are present. The flow field set up by the instability generally tends to smooth out the velocity and temperature differences causing it. The available kinetic or potential energy released by the instability may be so large that transition to a fully developed turbulent flow occurs. Figure 2.12 shows how instability mechanisms acting on a boundary layer over a flat plate lead to turbulent flow.

Transition is influenced by many parameters. An important one is the level of pre-existing disturbances in the fluid; a high level would generally cause early transition. Another cause of early transition in the case of wall-bounded shear flows is surface roughness. The manner in which transition occurs may also be very sensitive to the detailed flow properties.

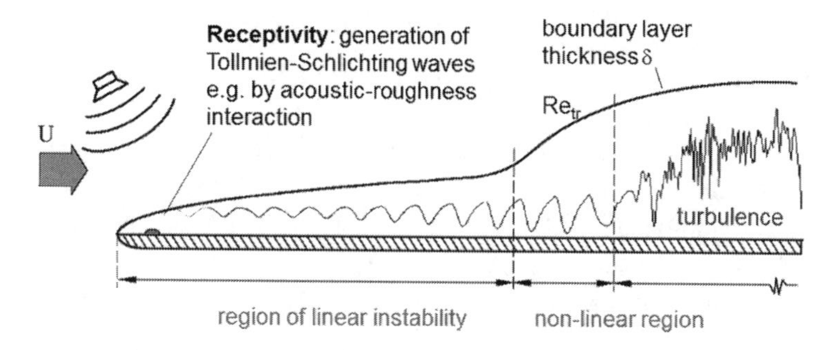

Figure 2.12 Instability mechanisms acting on a boundary layer over a flat plate. (Courtesy of Ulrich Rist, Stuttgart University, personal communication.)

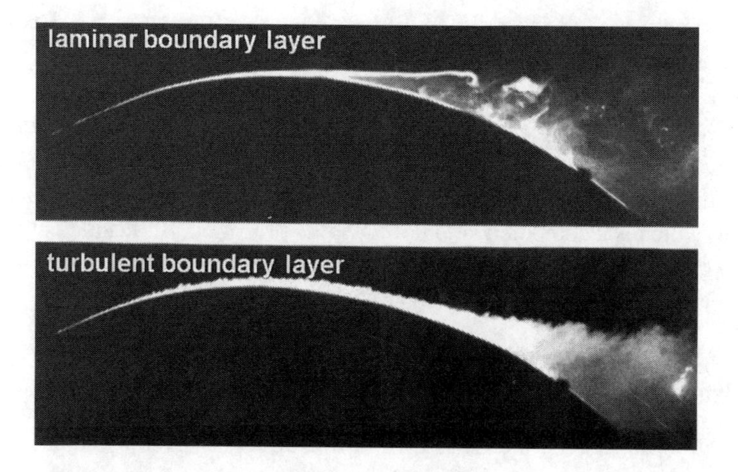

Figure 2.13 Wind tunnel visualizations of laminar (top) and turbulent (bottom) boundary layers over a circular cylinder. (From M. R. Head [9], reprinted with permission)

Figure 2.13 visualizes laminar (top) and turbulent (bottom) boundary layers over a circular cylinder.

2.4 Physics of Wing-Lift Creation

The creation of lift for carrying a load is the primary purpose of the airplane. The cost of doing so is the drag, which has to be overcome by the thrust. For efficiency reasons, as high a value of the lift-to-drag ratio as possible is wanted, what we call *aerodynamic quality*. This chapter explains how the flight shape determines the surface pressures that produce forces and moments yielding good aerodynamic quality. The flow around the aircraft mixes together nonlinear phenomena such as vortices, vortical shear layers, viscous boundary layers, and shock waves, all embedded in an overall field of essentially nonviscous flow. The flow can be modeled accurately by the compressible Navier–Stokes PDEs, and CFD represents the computerized solution of the equations.

CFD provides the relationship between the shape of the aircraft and the forces and moments acting upon it by the air flowing past it: lift, drag, and pitching moment.

However, CFD methods currently must use simplifications of the unsteady Navier–Stokes equations for compressible flow. There are many variants with different ranges of application – too many to present here. The full spectrum ranges from DNS, with resolution of all length and time scales, to heuristic handbook methods that combine formulas derived as curve fits to wind tunnel measurements. We limit ourselves to three levels of approximation, in order of computational resources necessary.

- Steady RANS, our L3 tool: the steady, Reynolds, and Favre averaged Navier–Stokes equations. This is what most CFD analyses attempt to solve, for external aerodynamics as well as internal flows in turbines, mixers, heat exchangers, etc. An introduction to a modern CFD package such as EDGE and SU2 is presented in Chapter 6.
- Steady Euler equations, our L2 tool: capture the compressibility effects at all Mach numbers. They can be augmented by a separate simulation of the boundary layers, the subject of Chapter 7.
- Steady vortex lattice approximation, our L1 tool: for incompressible potential flow, which is a computerization of Prandtl's flow models. It exploits the importance of concentrated vorticity to produce surprisingly accurate predictions of forces and moments (see Chapter 3).

2.4.1 Circulation Theory of Lift

The need to understand the remarkable lifting effect achieved by cambered surfaces led to the birth of modern aerodynamics. Experimental results for such surfaces, provided by Lilienthal (1889), among others, indicated enhanced lifting effects compared to those achieved with the flat surfaces hitherto investigated, and this needed to be explained and understood in order to be applied in design work. The close and fundamental connection between *vortical* flow and *lift* force on a wing was discovered at the beginning of the twentieth century and published by Lanchester [18], Kutta [17], and independently by Zhukovsky. The works by Kutta and Prandtl [22] from 1904 are the origins of modern lift and drag theories.

Flow of Ideal Fluids

At this stage, it may be pertinent to introduce the simplifications of the Navier–Stokes equations for a compressible gas, which enabled the mathematization of aerodynamics. For our present purpose, we need only the most simplified version for inviscid, incompressible, irrotational flow on the extreme right in Figure 2.14. When vorticity $\nabla \times \mathbf{u}$ is neglected, the steady velocity field $\mathbf{u}$ is the gradient of a velocity potential Φ. Since density variations are negligible for slow flows, the mass balance $\nabla \cdot (\rho \mathbf{u}) = 0$ translates to

$$\nabla \cdot (\nabla \Phi) = \Delta \Phi = 0,$$

or the Laplace equation, almost everywhere.

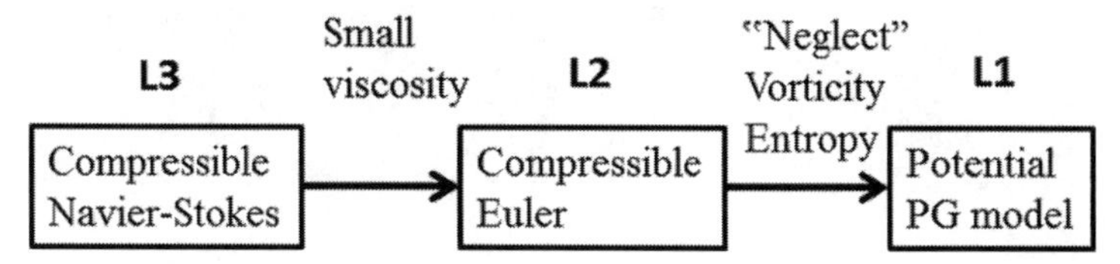

Figure 2.14 Hierarchy of mathematical flow models and our computational tools: L1, L2, and L3.

For a still body in an infinite fluid at a free stream velocity, the boundary conditions are that through solid surfaces there is no flow, $\mathbf{u} \cdot \mathbf{n} = \partial\Phi/\partial n = 0$, and that $\mathbf{u}$ tends to the free stream velocity far away.

Satisfaction of these conditions everywhere lead to the d'Alembert paradox; the resulting flow exerts no net force, but only a moment on the body. This was a result of studying "ideal fluids." James Lighthill in [19] credits Sir Cyril Hinshelwood, Nobel Laureate in Chemistry 1956, with the observation that

> fluid dynamicists in the nineteenth century were divided into hydraulic engineers, who observed what could not be explained, and mathematicians, who explained things that could not be observed.

In developing a circulation theory for a finite wing, Prandtl's work reconciled these two differing viewpoints with the concept of the boundary layer.

At the time, the practical tool was 2D potential flow theory, and it turned out that the class of solutions must be extended by *vortices*. Ideal vortices make the flow irrotational everywhere *except* (possibly) on curves, the vortex filaments, and *vortex sheets*, where the vorticity may be unbounded. The potential function is not differentiable everywhere; for an airfoil solution, it has a jump that equals the lift force. Such solutions are constructed by superposing particular solutions to the Laplace equation whose velocities are well behaved everywhere except at the source, filament, sheet, etc.

Trailing Edge Sheds Starting Vortex

Consider now the airfoil (i.e. an infinite wing) shown in the middle row of Figure 2.15 a short instance after an impulsive start to the left, visualized by aluminum particles floating on a water surface. The "starting vortex" is clearly seen to rotate counterclockwise and cancels the not very obvious clockwise circulation around the airfoil, which can be modeled by the so-called *bound vortex*. Figure 2.15 tells a story about lift generation (cf. Prandtl's [23] circulation theory of lift). The confluence of fluid streams from the upper and lower surfaces of a wing generates vorticity confined to a sheet at the "aerodynamically sharp" trailing edge, and the flow at the trailing edge does *not* turn around the edge. This is an established experimental fact. A potential flow without concentrated vorticity would turn, producing very large velocities requiring extreme pressure differences. The potential flow model is insensitive to such unphysical behavior and must be constrained by the extra condition of *smooth flow at the trailing edge* (Figure 2.15, bottom-left), named the

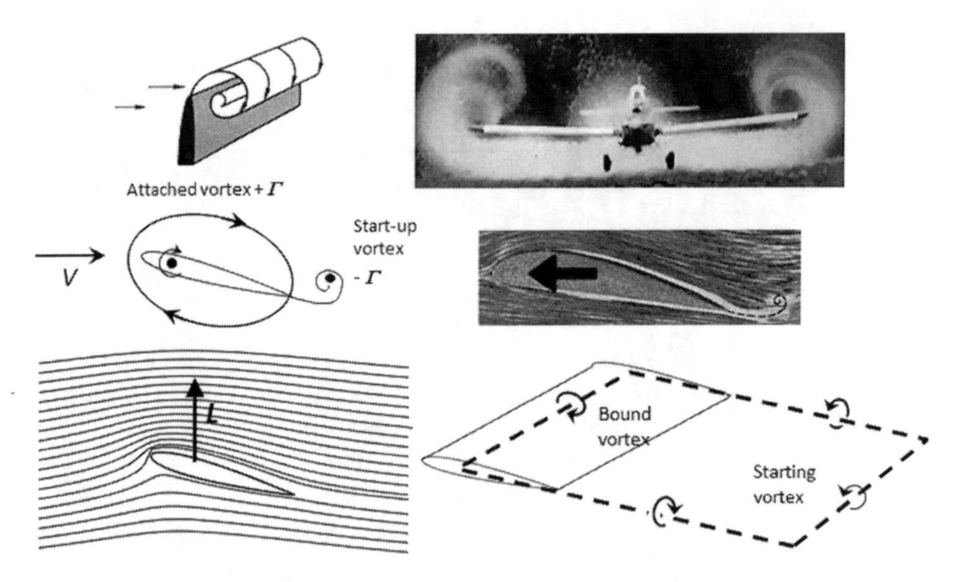

Figure 2.15 Impulsive start of finite wing creating circulation and resulting vortex system. (Embedded photographs reprinted from Campbell and Chambers [3], public domain; and from Prandtl [23] with permission)

Kutta condition after M. Kutta. In a plane accelerated flow, this produces the circulation revealed by the Prandtl experiment (Figure 2.15, center-left). For steady plane motion, when the airfoil no longer accelerates, the Kutta condition makes the bound vortex exactly cancel the vorticity shed during acceleration and left far downstream.

Now consider a finite-wing flow (right column in Figure 2.15). An ideal vortex cannot simply end somewhere in the fluid, so the start-up and bound vortices must be joined together by two trailing tip vortices to form a ring vortex (Figure 2.15, lower-right). This was the initial notion to Prandtl's lifting-line theory of lift, later refined to a number of trailing vortices. In steady motion of the wing, they represent a significant sheet of vorticity shed from the trailing edge. This sheet is invisible, but each edge of the sheet rolls up into a vortex trailing from the wing tips that can be seen when they pass through partly cloudy skies (Figure 2.15, upper-right) or in humid conditions as tip condensation trails.

The L1 airfoil computational model becomes one of superposing the following.

- A flow derived from the Laplace equation with slip, no-flow-through boundary (i.e. zero normal gradient) at the airfoil surface
- A flow associated with a point vortex inside the airfoil

The Kutta–Zhukovsky law (Eq. (2.4)), follows from the construction and the Kutta condition: the force **L** per unit length on a vortex filament is the cross product of mass flow $\rho \mathbf{u}$ with vortex strength $\mathbf{\Gamma}$,

$$\mathbf{L} = \rho \mathbf{u} \times \mathbf{\Gamma} \tag{2.4}$$

for any form of section.

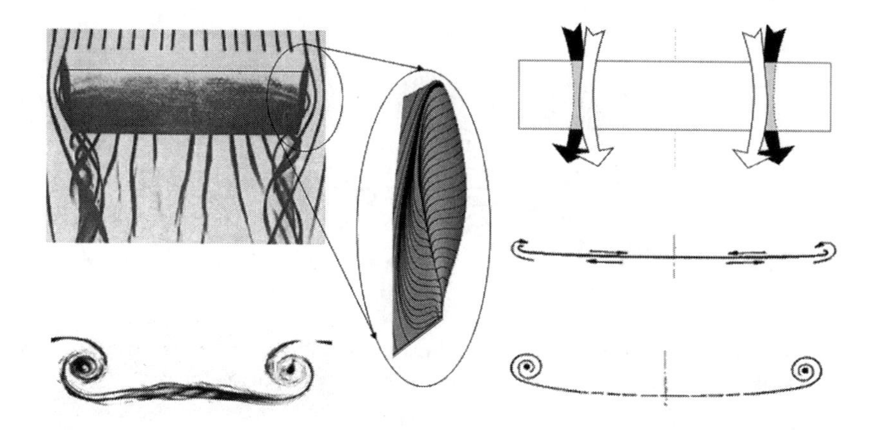

Figure 2.16 Vortices around a lifting wing; generation of trailing tip vortices and wake due to spanwise pressure distribution. (Left) Experiment with smoke trails (photographs from M. R. Head [9], reprinted with permission). (Right) Detail sketches along with skin-friction lines computed on a NASA Common Research Model wingtip that coalesce, indicating separation.

The lift computed by the Kutta–Zhukovsky law agrees well with experimental data, being proportional to the angle of incidence of the oncoming stream on the airfoil, but for small such angles only. When the angle is increased past a critical value, the lift force decreases quickly and the airfoil stalls (see Figure 2.5).

The disagreement is the consequence of flow separation, in turn caused by the neglected viscosity.

The shedding of vorticity from the trailing edge when the airfoil accelerates from rest is the cause of lift. To work as expected, airfoil surfaces should impose boundary-layer separation at their tails. Understanding the boundary layer helps us control its behaviour so as to maximize lift while keeping drag limited. The recipe has been to adopt a smooth, carefully contoured, or streamlined, shape tapering gently to a sharp tail, at which point boundary-layer separation is encouraged. This boundary-layer behavior controls the inviscid flow exterior to it and the top-to-bottom pressure imbalance that produces lift.

2.4.2 Physical Description of a Lifting Finite Wing

One might start out simply by assuming that each section of a finite wing could behave as described by our 2D analysis. However, things do not work this way, because the pressure difference between the upper and lower surfaces causes the air to leak around the tips, reducing the pressure difference in the tip regions and shedding the vortex sheet that produces tip vortices, as suggested in Figure 2.16. In fact, the lift must go to zero at the tips because of this effect.

Figure 2.16 suggests how the upper- and lower-surface wing flows lead to a shear layer shed at the trailing edge. Experimental observations find distinct concentrations of vorticity close behind the tips. Skin-friction patterns on wing surfaces are visualized

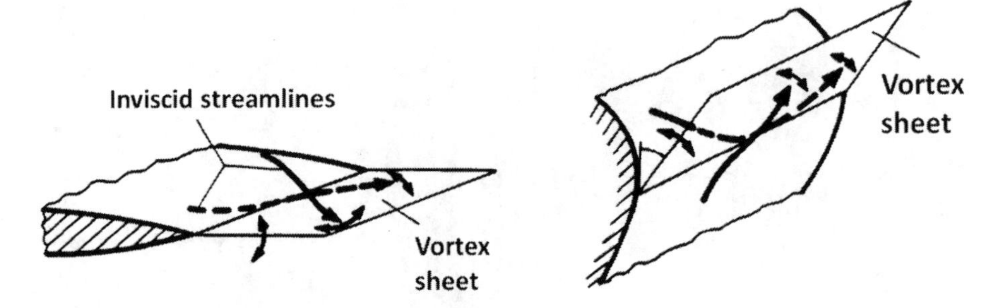

Figure 2.17 (Left) Vortex sheet shed from a sharp trailing edge and (right) from a smooth surface. (Courtesy of Ernst Hirschel [11], reprinted with permission)

by *skin-friction lines* everywhere parallel to the tangential component of the friction forces. A thin layer of fluid smeared on the surface flows along the skin-friction lines to form visible streaks in oil flow visualization.

2.4.3 Vortex Filaments and Sheets

As a first step toward a quantitative vortex theory, we need a few considerations about how vortices behave. A useful fundamental flow idealization is that of a *vortex filament*. Consider a thin tube in which the flow has nonvanishing vorticity, $\omega = \nabla \times \mathbf{u}$, directed along the tube axis, whereas outside, the flow is irrotational. The *circulation* Γ is the integral around a small curve C surrounding the tube of the velocity component parallel to the curve,

$$\Gamma = \int_C \mathbf{u} \cdot d\mathbf{l} = \int_S \omega \cdot \hat{\mathbf{n}} dS$$

where S is a surface bounded by C. The equality is Stokes' theorem, or the definition of the $\nabla \times$ *curl* operator. In the limit, as the diameter of the tube is made small, but Γ is held fixed, this region of vorticity is called a vortex filament.

A *vortex sheet* is a discontinuity in the velocity component tangential to each side of the sheet, illustrated in Figure 2.17. This tangential discontinuity is a weak solution of the Euler equations, as is the shock wave, a discontinuity in the normal velocity (see Chapter 4). In potential theory, a sheet is created by stacking vortex filaments close to one another to make up a surface, thus concentrating the vorticity completely to the surface.

Vortex Properties

Drela [5] derives the Helmholtz transport equation for the behavior of vorticity by formally taking the curl of the momentum equation, using vector identities, holding ρ and μ constant, and combining and rearranging terms:

$$\frac{D\boldsymbol{\omega}}{Dt} = (\boldsymbol{\omega} \cdot \nabla) v + \nu \nabla^2 \boldsymbol{\omega} \tag{2.5}$$

The term $(\omega \cdot \nabla)v$ on the right-hand side represents *vortex tilting* and *vortex stretching*. A rotating fluid's vorticity intensifies when stretched by the components of the velocity gradient matrix ∇v, which are parallel to ω itself: think of a pirouetting ice-skater stretching his or her arms up. However, if $\omega = 0$ to begin with, as in flight into still air, then this term is disabled, since there is no initial vorticity to stretch or tilt. The baroclinic source term $\nabla\rho \times \nabla \text{p}$ appears in the compressible version of the Helmholtz transport equation. It causes vorticity to appear wherever density and pressure gradients are present. However, in homentropic flow, a $f(p, \rho) = 0$ relation holds, the ρ and p gradients are parallel, and the baroclinic term vanishes.

The strength of a vortex must be constant along the vortex line. Thus, a vortex cannot end in the fluid. It can make a closed loop as in a smoke ring, end on a boundary, or extend to infinity. Of course, in a real, viscous fluid, vorticity is diffused through the action of viscosity, the last term of Eq. (2.5). This widens the vortex tube until it is barely recognizable as a filament. A tornado is an interesting example. One end of the twister is on a boundary, but at the other end, the vortex diffuses over a large area with vorticity.

2.5 Behavior and Interaction of Flow Phenomena

We explained earlier in this chapter the familiar 2D case of flow past an airfoil: how the boundary layer in a healthy flow separates close to the trailing edge. As a section of a straight high-*AR* wing, it creates lift. The details of flow separation from a wing, however, are complex, with lifting surface shape, Reynolds number, and speed regime M_{cruise} playing important roles. Nevertheless, a significant fraction of the separation structures falls into a set of well-organized basic separation forms that can be recognized and identified. Two important examples are the *bubble* separation and the *free vortex-layer* or *open* separation.

The discussion in Section 2.3 gives reasons for why the chosen pattern of streamlined flow gives the best performance for high-*AR* straight or swept wings designed to maintain attached flow. If the boundary layer separates before reaching the trailing edge, lift is reduced and drag increases: wing performance degrades until ultimately it stalls.

German aerodynamicist Alexander Lippisch designed tailless and delta-winged aircraft in the 1920s and 1930s. Their slender configurations performed better at high speed and led the way for the development of high-speed jet and rocket airplanes. The slenderness is measured as s/ℓ, where s is half span-width and ℓ is configuration length from nose to tail. Since $\ell \geq c_{root}$, a slender configuration requires a low-*AR* wing.

The flow pattern on thin, slender wings is quite different from that on a high-*AR* wing. It is one of coherent leading-edge vortex separation, and the design strives to hold the vortex stable to provide a reliable lift increment. If it bursts, the flow becomes incoherent. Depending on the location of the burst, there is loss of lift and stall, or

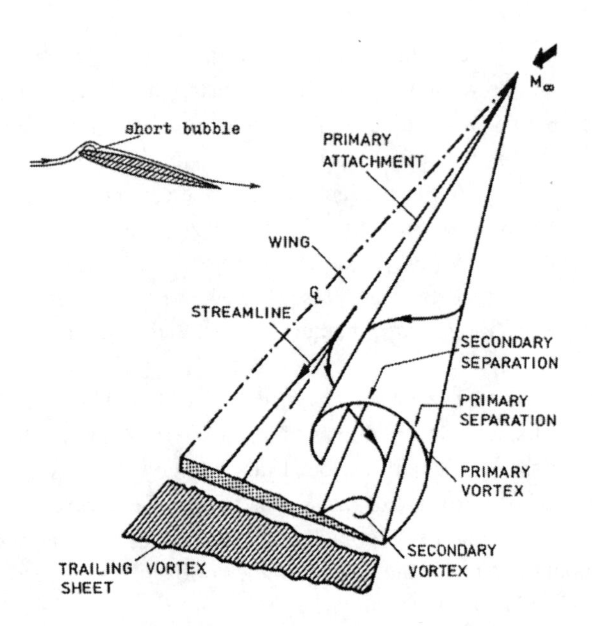

Figure 2.18 (Left) Airfoil with leading-edge separation bubble; (right) delta-wing leading-edge separation and secondary separation from a smooth surface.

deleterious effects on lifting surfaces downstream. The phenomenon is, in the words of Bram Elsenaar [7], both "the beauty and the beast." Let us examine the situation more closely.

2.5.1 Low-*AR* Wing Vortex Separation and Lift

For thin wings with high sweep, separation likely takes place at the leading edge. The 2D bubble-type separation from the leading edge of a thin airfoil comes into play, as shown in Figure 2.18 (left). The boundary layer lifts off the surface at the leading edge and reattaches a short distance downstream, forming a bubble. When this occurs in 3D on a swept-wing leading edge, the surface streamlines detach from the leading edge and are swept downstream into the flow field as a vortex layer that, under self-induction, rolls up into the primary vortex, drawn in Figure 2.18 (right).

The vortex sheet leaves the leading edge along the primary separation line, and the streamlines spiraling around the vortex rejoin the wing surface along the primary attachment line (Figure 2.18, right). In fact, this is the same flow mechanism that acts on the round tip of the high-*AR* wing shown by the skin-friction lines in Figure 2.16 (right column), except now it occurs over most of the wingspan instead of being localized at the tip.

The essential parameters governing the stability of this flow pattern are the angle of incidence α and the leading-edge sweep angle of the delta. If the leading edge of the wing is sharp enough, this geometric feature fixes the location where the vortex sheet leaves the surface. The flow could be handled as inviscid separating from an

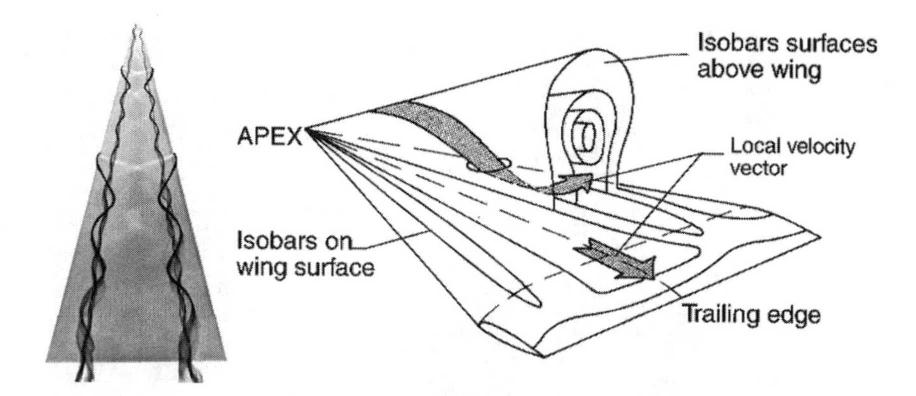

Figure 2.19 (Left) Symmetrical pair of vortices shed from aerodynamically sharp swept leading edge (photograph by Henri Werlé showing dye streaks in water, courtesy of ONERA, reprinted with permission). (Right) Schematic surfaces of iso-pressure around the vortex and its low-pressure footprint on the upper surface. (Courtesy of Bram Elsenaar [6], reprinted with permission)

aerodynamically sharp edge. But when the leading-edge radius is larger, the location where the vortex sheet forms is far from obvious and 3D boundary-layer separation must be resolved.

Vortex Effect on Surface Pressure

Let us look at the pressure field associated with the streamlines spiraling around the primary vortex drawn in Figure 2.18 (right) and shown as spirals in Figure 2.19 (left). Characteristic of any vortex, the circumferential velocity around its core increases as the distance to the core decreases. Higher velocity translates into lower pressure. The resulting iso-pressure surfaces centered on the vortex core are shown in Figure 2.19 (right). The footprint of the core on the wing surface is the low-pressure trough drawn in Figure 2.19 (right), which enhances the lift. Such vortex lift is exploited on many configurations (see Figure 9.23 for the Aerospatiale-BAE Concorde in landing attitude).

Vortex–Boundary Layer Interaction and Secondary Vortex

The pressure field induced by the primary vortex has further effects. In Figure 2.19, the velocity vector spirals under the vortex and through the low-pressure trough from left to right. As it underpasses the vortex core from the left, it experiences a decreasing pressure that is beneficial to holding the boundary layer attached. However, past the core, the pressure gradient becomes adverse and causes the boundary layer to separate as a vortex sheet rolling up into the secondary vortex drawn in Figure 2.18 (right).

Vortex Burst and Stall

The ultimate downstream fate of a leading-edge vortex is to dissipate and disappear just like wingtip vortices. But it may also burst into chaotic motion prior to expending

Figure 2.20 Vortex breakdown visualized by dye streaks in water. (Photograph by Henri Werlé, courtesy of ONERA, reprinted with permission)

most of its rotational kinetic energy. As the angle of attack is increased, such a dramatic event can take place at some position along the axis of the vortex where its ordered structure breaks down.

Figure 2.20 shows *vortex burst*. From a tightly wound spiral motion upstream, the flow suddenly decelerates along the axis. The core itself then begins to spiral with lower velocity in expanding loops, and it eventually decays into large-scale turbulence. If this happens over lifting surfaces, they are subjected to unsteady forces with potentially deleterious effects.

For instance, vortex burst over the aft part of a delta wing is accompanied by loss of lift and pitch-up, much like tip stall on a high-*AR* swept wing. Leading-edge vortices hitting lifting surfaces further aft are often a recipe for disaster. The chaotic air motion gives buffeting loads on the surfaces that are hit, and the undesirable buffet decreases their fatigue life. The F-18 Hornet is a documented example, where the main wing LEX generates a vortex, which serves its purpose to enhance lift, but also hits the twin vertical tails.

Vortex bursting is modeled by the unsteady Navier–Stokes equations and can be predicted. An open issue is how well it can be predicted by RANS instead of unsteady-flow models. Two issues are important for numerical modeling. The first is the vortex separation: with a sharp geometric feature, the inviscid flow model may be accurate enough. The second is numerical dissipation: a fine computational grid is required, lest the vortex dies before it travels very far.

2.5.2 Shock Waves

For flight speeds close to the speed of sound, the simple linear models break down due to effects created by the compressibility of air. When the airplane travels close to Mach 1, the air it encounters has very little time to get out of the way because the air

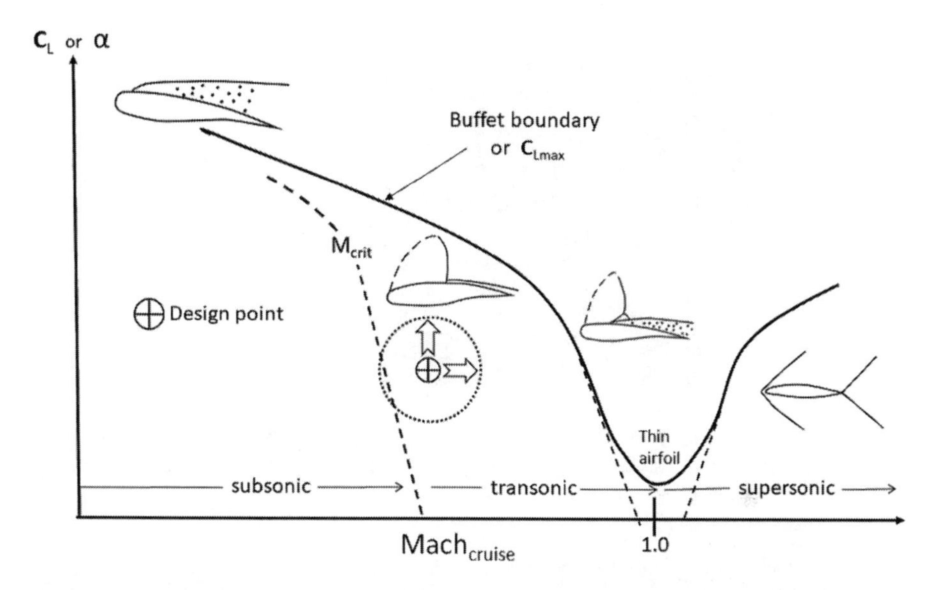

Figure 2.21 Buffet boundary, indicated by $C_{L,max}$ vs. Mach dipping down severely at near-sonic speed because of shock–boundary layer-induced separation. (Adapted from Thomas [31], reprinted with permission)

particles are alerted to the impending collision by sound waves. This mental picture was popularized by the notion of the "sound barrier," which seemed very hard to cross. However, rifle bullets are supersonic, so it could not be impossible. Indeed, in October of 1947, Col. "Chuck" Yeager piloted the rocket plane Bell X-1 to Mach 1.06. It had a bullet-shaped fuselage and thin, unswept wings. Wing sweep was introduced on the DH-5882-II Skyrocket to investigate its effects on handling, buffet, etc., and it was the first piloted plane to achieve Mach 2 in 1953. Compare that to Figure 2.21 for the transonic dip. See also Section 1.1.4 regarding the high-speed "buffet boundary," which limits the available lift.

Shock–Boundary Layer Interaction

On swept high-*AR* wings, weak shocks are present at and around the design condition. For military combat aircraft, massive separation due to strong shock wave–boundary layer interaction is a very critical phenomenon that determines wing buffet and hence the limits of the flight envelope. It can also give rise to unsteady phenomena such as self-sustained "limit cycle oscillation" (LCO) of the shock and the boundary layer/wake.

Figure 2.22 displays shock waves over an airfoil in transonic flow, with flow from the left. The photograph shows smoke in the boundary layer and Schlieren-visualized density gradients. These appear at the λ-shock, which is a compression fan close to the boundary layer, converging into a normal shock further out. Note the separation of the upper-side flow at the foot of the shock, a typical case of shock–boundary layer interaction. On the pressure side, the flow does not separate so easily, since the

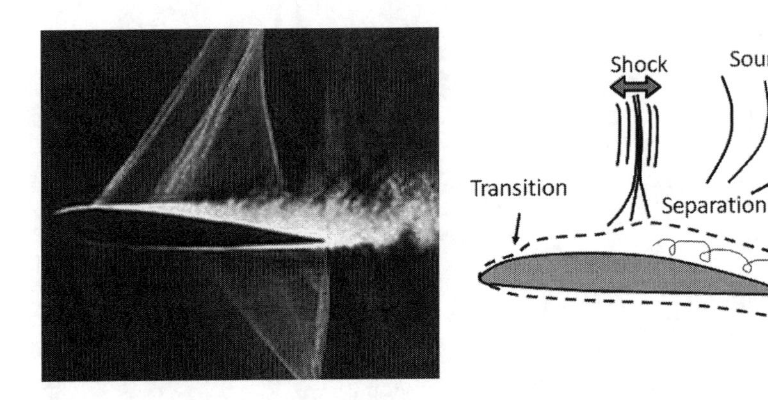

Figure 2.22 Shock wave–boundary layer interaction. (Left) Schlieren photograph of experiment with smoke (NASA website [14], public domain). (Right) Schematic drawing of the mechanism in unsteady interaction.

pressure gradient is more favorable there. These boundary-layer features are sensitive to the Reynolds number.

Figure 2.22 (right) indicates how these small-scale features can have substantial large-scale flow effects. For example, the combined effect of the shock wave–boundary layer interaction and the subsequent downstream boundary-layer development determines the conditions at the trailing edge. These, in turn, affect the overall circulation and hence the shock wave strength. Shock wave–boundary layer interaction and wave drag are very sensitive to shock strength, so all of the ingredients are present for a complicated Reynolds number-sensitive viscous–inviscid interaction that can be either steady or unsteady.

The sketch in Figure 2.22 (right) shows self-sustained shock oscillation during transonic buffeting with fully separated flow where the shock wave oscillates on the upper surface of the airfoil about a mean position. The shock movement sends pressure waves propagating downstream in the separated flow region. On reaching the trailing edge, the disturbances generate upstream moving waves, either from the wake fluctuation or from the trailing-edge boundary layer. These waves then interact with the shock and impart energy to it that maintains its oscillation. A more thorough treatment of shock/boundary layer flow physics can be found in Vos and Farokhi [34].

The flow variations resulting from incipient separation are generally gradual and continuous, until the point is reached when massive separation occurs to involve a topological change in the overall flow field (e.g. Figure 2.22, left). If the shock is sufficiently weak, a little bit of separation at the trailing edge, causing some extra drag, is not necessarily bad. This is somewhat similar to the situation at the optimal lift-to-drag ratio of a transonic wing, where a weak shock is normally present and the resulting wave drag is acceptable.

Buffet

The above has illustrated the feedback loops in shocked flow over an airfoil. Separation patterns in 3D flows are more complex, so we limit the text below to a phenomenological presentation of buffet and buffeting.

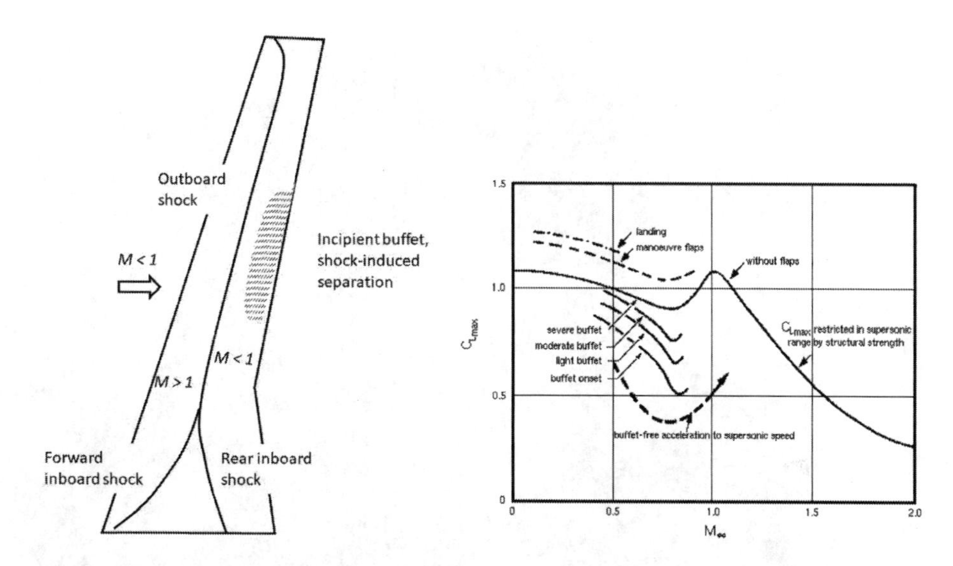

Figure 2.23 (Left) Flow and separation patterns on an airliner wing; (right) typical buffet boundaries in a Mach/$C_{L,max}$ diagram. (Courtesy of Klaus Huenecke [15], reprinted with permission)

Downstream of separations, whether induced from leading edges or from shock waves, the pressure becomes fluctuating, as for example in Figure 2.22 (right). It acts on the flexible structure of the wing, which will oscillate. The pilot senses this vibration, or buffet, as an oscillating normal g-load. The flight regimes in which buffet will be encountered are described by buffet boundaries in diagrams with Mach number or speed vs. lift or load factor or angle of attack, and can be shown in a $V - n$ diagram. The dominating aerodynamic performance issue next to ML/D is the *buffet boundary*, which is closely associated with the wing design. This boundary can at times be nearly as important a performance parameter as drag, since the maximum cruising lift coefficient is limited by the certification requirement to maintain a $1.3g$ maneuver margin to buffet onset. The normal encounter with buffeting in transonic cruise occurs when an aircraft hits a strong gust. This will increase the incidence to the point where upper-surface flow separation occurs, typically along the aft part of the wing. A fighter typically cruises well below the buffet-onset boundary, but vigorous maneuvering can take it into regions of moderate to heavy buffeting (compare Figure 2.23). Note that the buffet boundary is not precisely defined, since continued light buffet may be tolerable, but serious buffet is acceptable only occasionally (Figure 2.23, right).

Shock–Vortex–Boundary Layer Interactions

A shock wave encountering a vortex can trigger its bursting. Figure 2.24 shows an example of this. A delta wing meets an $M = 1.1$ supersonic stream in a wind tunnel at 14° incidence. The vortex core over the wing is tightly wound until it interacts with the oblique shock on the upper surface, where it bursts, grows larger in diameter, and becomes turbulent. B marks the bow shock; the wide shock angle is testimony to the low supersonic free stream.

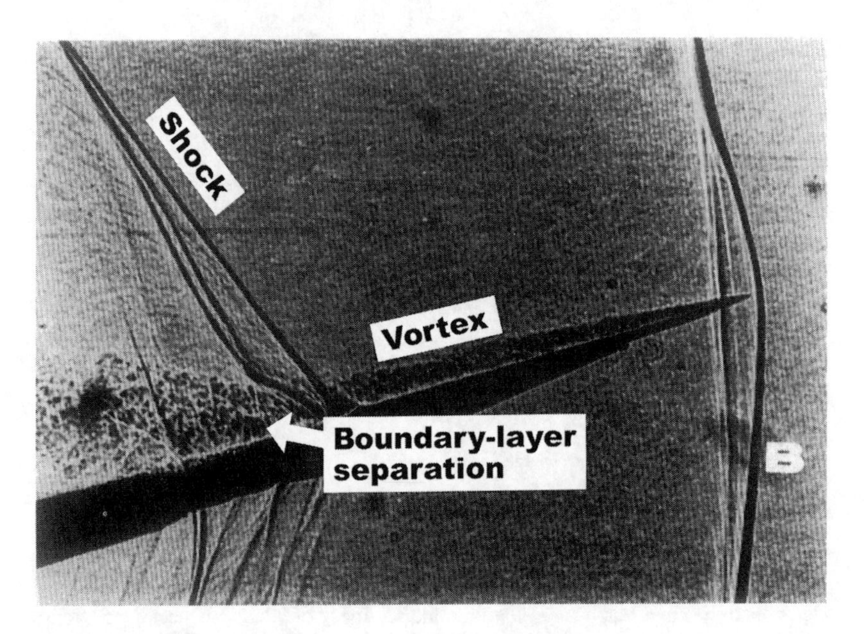

Figure 2.24 Shock–vortex–boundary layer interaction at supersonic speed. (Unpublished, courtesy of Anders Gustafson, Swedish Aeronautics Research Establishment FFA)

2.6 Drag Taxonomy

Lift and drag are the two most important forces entering into aerodynamic design considerations. Drag warrants a special discussion because it has so many different components and builds up in so many different ways.

The terminology often creates confusion. Several schemes are possible for classifying the drag of an airplane; for example, breaking it down according to the physics responsible for the drag forces. They are manifested in near-field physics by normal and tangential forces on the surfaces and in far-field flow phenomena such as wake, vortices, and waves, as in Figure 2.25. An applied aerodynamicist laying out an aircraft configuration shape finds it practical to look at drag contributions from different components, as well as to separate lift-*induced drag* from the *form drag* created by undesirable flow separations and boundary-layer momentum deficit. The lift distribution (span loading) exerts a major influence on the induced drag, as we shall see in Chapter 3, and so does the total wetted area on the friction drag.

An aerodynamicist concerned with the flow phenomenological origins of drag would look at it from the bottom up, considering *vortex, wave,* and *wake* contributions. Another viewpoint considers the physical effects on the airplane surfaces and splits the total into *pressure* and *friction* drag. Note that both form drag and skin friction are the result of viscous effects acting in different ways. Wave drag is thermodynamic in nature and also appears in inviscid flow models.

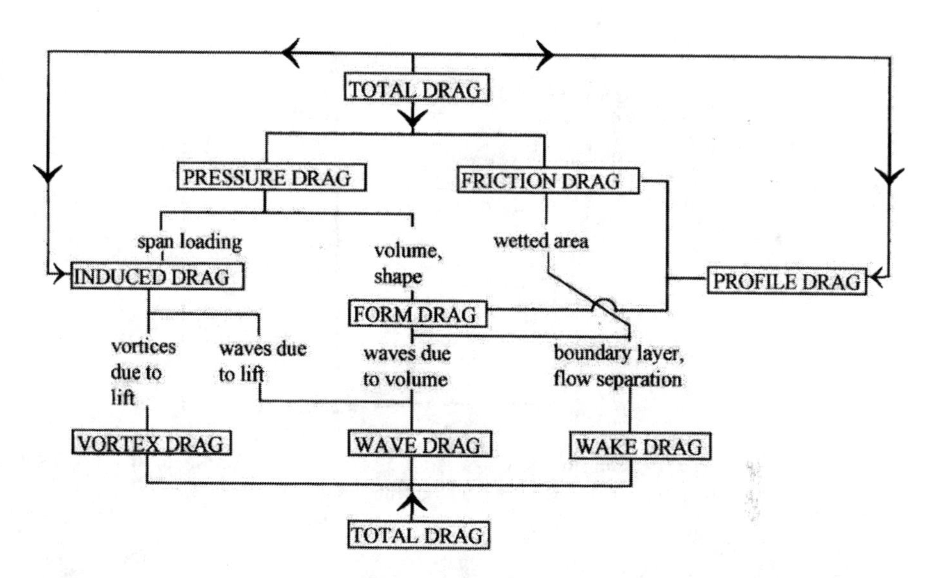

Figure 2.25 Drag breakdown of a body without internal flow. (From Torenbeek [33], reprinted with permission)

Near-Field Analysis

- *Skin-friction drag* results from the viscous tangential stress on the wetted surface. It is larger for turbulent boundary layers than laminar, and the designer may try to reduce it by delaying the laminar–turbulent transition by proper shaping of the pressure profile.
- *Pressure drag* is the pressure force component in the free stream direction and is the consequence of incomplete pressure recovery on aft portions of the configurations, such as by flow separation.

Far-Field Analysis

- *Wave drag* is due to the creation of entropy in shock waves and is responsible for the rapid growth of drag as the velocity approaches the speed of sound.
- *Wake drag* is the momentum deficit in the wake, which contains the shed boundary layer with its lower velocity.
- *Vortex drag* is the force that powers the lifting surface trailing vortices discussed in Section 2.4.1.

Figure 2.26 shows the relative magnitudes of the various types of drag for low transonic speed, broken down according to physics for the whole plane. Friction drag is listed per component, and its strong relation to wetted surface area is evident. Just over 50% of the total drag is due to friction, and this is the major impetus of current research to maintain more laminar flow over the surface. Drag due to lift is the next largest category that motivates much effort to produce lift at lower induced drag by, for example, wing tip devices.

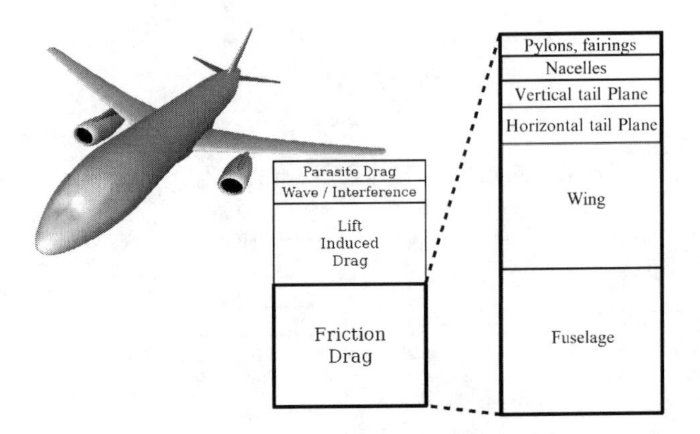

Figure 2.26 Drag contributions on aircraft broken down according to their origins. (Courtesy of Maximilian Tomac, PhD thesis [32], reprinted with permission)

Induced drag is the sum of vortex-induced drag and wave drag due to lift. For a wind tunnel model, wake drag can be derived from static and total pressure distribution measurements in the wake of the model. The induced drag is then found by subtracting the wake drag from the total drag.

2.6.1 Skin-Friction Drag and Pressure Drag

The total drag that a fluid exerts on a body can be considered as the result of the elementary forces on all points of the surface, both normally and tangentially. Consider the drag on a closed body (i.e. one with no internal flow due to power plant installation or internal systems): the *skin friction* is generated by tangential forces, while *pressure drag* or *form drag* is caused by the pressures acting normal to solid surfaces. The lift can be subdivided in a similar manner, but the contribution of friction is negligible in practice. The pressure distribution is also affected by the boundary layer and regions of separated flow. In wind tunnel experiments, the pressure forces can be obtained by measuring the distribution of pressure at small holes in the surface of the body, while friction drag is obtained by subtracting the pressure drag from the total drag. The skin friction drag is

$$D_F = \bar{C}_f q S_{wet}.$$

$\bar{C}_f$ is the mean friction drag coefficient based on the wetted areas of the various aircraft parts, typically in the order of 0.003–0.005 for most subsonic aircraft. The wetted area of a body or wing is therefore the main geometric parameter determining the skin-friction drag. Friction drag is lower for laminar flow than for turbulent, and it is strongly dependent on the Reynolds number. Wind tunnel testing at a Reynolds number lower than the flight value may be in error due to incorrect transition location. CFD can potentially do better. Accurate computation of drag requires a transition model to predict where the flow transitions to turbulent.

Form drag or *pressure drag* is determined only by the pressure distribution. When the flow is attached and follows the surface without breaking away, the form drag is small. It is strongly related to how the separation occurs. For instance, the pressure in the wake behind a bluff body is lower than at the upstream stagnation region at the nose, whereas on a streamlined profile, the pressure recovers almost to the stagnation pressure and gives small form drag.

2.6.2 Vortex-Induced Drag, Wake Drag, and Wave Drag

The work produced in overcoming the aerodynamic drag of a body moving at constant speed is equal to the energy increase in the surrounding fluid. The forms of energy transmitted to the fluid are vortex-induced energy, wake energy, and shock wave energy. The aerodynamic drag can be subdivided accordingly (Figure 2.25).

Vortex-Induced Drag

This is part of the pressure drag, corresponding to the kinetic energy that is distributed throughout the fluid and is lost with the trailing vortices shed by a lifting wing. For high-*AR* wings with relatively thick sections, the vortex-induced drag can be computed accurately with lifting-line or lifting-surface theories such as the vortex lattice method of Chapter 3. When leading edges are sharp, angles of sweepback are large or aspect ratios are low, the leading-edge suction force begins to break down at some critical lift and the vortex-induced drag increases considerably. Lifting-line theory is accurate for high-*AR* wings in subcritical flow and predicts that the lift-induced drag is

$$C_{D,i} = k' \cdot \frac{C_L^2}{\pi AR}$$

where the efficiency factor $k' \geq 1$ and the minimum obtains when the spanwise lift force distribution is elliptic. This translates into

$$\frac{D_i}{W} = \frac{k'n^2}{\pi q} W/b^2 \tag{2.6}$$

where n is the load factor and $n = 1$ for straight and level flight. *AR* plays a major part in determining the induced drag coefficient, but it is drag, not the drag coefficient, that must be overcome, so it is *span loading* W/b^2 and not *AR* that determines how efficiently a wing performs its lifting function. Note the relation to *wing loading* $W/S = AR \cdot W/b^2$.

Wake Drag

This is caused by the boundary layer and regions of separated flow. The main source of drag generated by the boundary layer is shear action, resulting in skin-friction drag. For a well-streamlined body at small angles of attack, the skin-friction drag is the dominant part of the wake drag. *Separation drag* increases sharply when stall of the lifting surface is approached. But it is also manifest in areas near ill-shaped wing–body junctions, blunt bases, sharp corners, etc. Only in a few simple cases can flow separation be predicted and analyzed theoretically, and drag prediction therefore requires high-accuracy CFD and cannot be well predicted until the configuration details have been settled.

Wave Drag

This is another part of the pressure drag, associated with the work produced by compression of the fluid at high (local) flow velocities, which manifests itself in the form of shock waves. A complicating factor is that strong shock waves may induce flow separation, resulting in an increase in both the wake drag and the vortex-induced drag. At supersonic speeds, assuming linear theory, wave drag can be subdivided into wave drag due to the body volume and wave drag due to lift, but at transonic speeds, when mixed flow is present, this subdivision is less evident.

It will be clear from the foregoing that the subdivision into vortex-induced drag, wake drag, and wave drag components is not very well defined in cases where appreciable interactions between the various flow fields occur. These interference effects result in drag increments, frequently referred to as *interference drag*.

Wave drag is strongly dependent on the slenderness (half wingspan to total length) of the configuration, as well as on its cross-section area. Figure 3.22 shows the relation between slenderness and speed for different classes of aircraft. For real supersonic flight, slenderness is required.

2.7 Example: Swept-Wing Flow Physics

Let us conclude this chapter by making what we have been discussing more concrete with some examples of shock wave phenomena in high-speed flight. We discuss the *drag divergence* in transonic flight and three possible means to shift it to higher Mach number, namely:

(1) The *swept wing* in Chapter 9
(2) The *supercritical airfoil* in Chapter 8
(3) The *area rule* discussed in Chapter 3

We could illustrate flow phenomena with observations in flight and/or with photographs taken in wind tunnel experiments, but instead we choose in this book to illustrate them with numerical simulations. One can interrogate the solution with visualization software to display all of the phenomena that are captured in the numerical formulation. Figure 2.27 presents one such example of this for a RANS simulation of steady transonic flow around the NASA Common Research Model (CRM) wing–body configuration defined by the AIAA Aerodynamic Design Optimization Discussion Group (ADODG) for CFD validation and shape optimization experiments. The CRM airfoil will appear again in Chapter 8 and the wing (alone) in Chapter 9. Turbulence modeling is used, so only the largest scales are resolved. This is sufficient for our purpose, because the emphasis here is on how the shape influences the flow and how the Reynolds number influences boundary-layer properties.

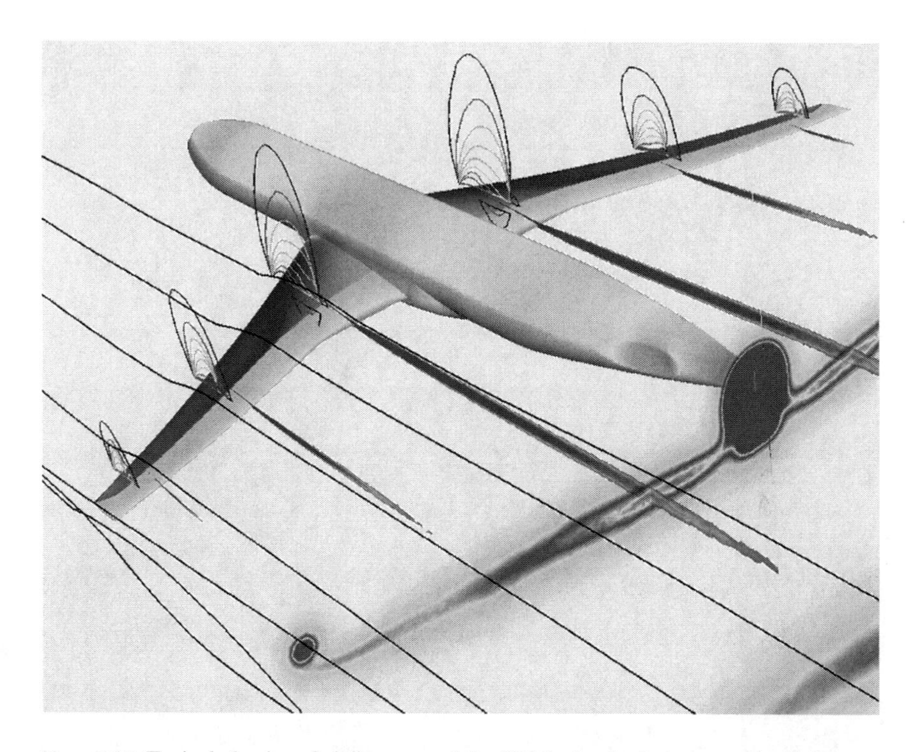

Figure 2.27 Typical physics of airflow around the CRM wing-body in transonic flight; $Re = 20M$ per reference chord, $M_\infty = 0.85$, $C_L = 0.47$, Spalart–Allmaras turbulence model. (Courtesy of Mengmeng Zhang, personal communication)

Figure 2.27 shows the following.

(1) Pressure coefficient as color on the surfaces
(2) Mach number as colored iso-curves in cuts parallel to the symmetry plane
(3) Vorticity as color in cuts orthogonal to flight
(4) Streamlines

On the left wing tip, the streamlines show the vortex that results from the high-pressure flow on the lower surface flow spilling over to the lower-pressure region on the upper surface.

The oncoming airstream accelerates rapidly to supersonic speed as it progresses around the curved leading edge of the wing and advances until the isobars coalesce into a shock wave, with almost instantaneous deceleration to subsonic speed. In this case, the shock interacts with the boundary layer, but does not separate it, at least not over the inboard region of the wing.

At its trailing edge, the wing sheds the flow in a shear layer that forms the wake. At this high Reynolds number, the boundary layer is thin, and so the shear layer grows only slowly as it travels downstream. Notice also that the wake is nearly but not completely planar. In the wake, one also sees the boundary layer shed from the fuselage. At the wing–body juncture, the two boundary layers interact and can shed a *juncture vortex* that leads to *interference drag*.

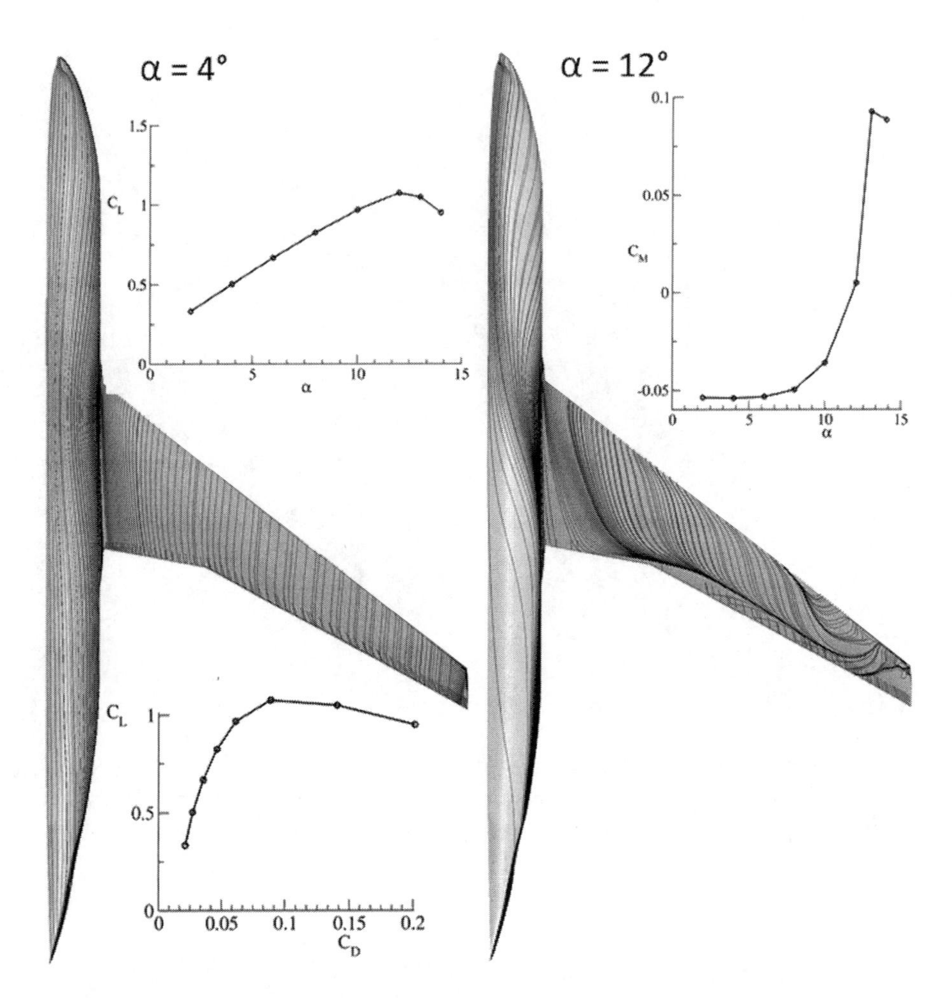

Figure 2.28 Computed skin friction lines on the CRM wing for angle of attack 4° (left) and 12° (right); the flow condition is $Re \approx 1.2M$ per reference chord, $M_\infty = 0.2$, Spalart–Allmaras turbulence model. (Courtesy of Peter Eliasson, personal communication)

Spanwise Boundary-Layer Flow

The boundary-layer flow is more clearly seen in the computed oil-flow visualization in Figure 2.28 in steady transonic flow around a swept wing–body configuration. At 4°, the spanwise flow is insignificant. The tip shows traces of the vortex separation. At 12°, the spanwise flow is accentuated, and the whole outer wing panel shows patterns of vortical flow, possibly with impending tip stall. The flow separates upstream of the trailing edge, all the way from tip to kink.

Figures 2.27 and 2.28 introduce for us a number of important phenomena of flow over a wing. Among the phenomena to be studied in the coming chapters are flow separation at or before the wing trailing edge, compressibility effects, boundary layers, shock waves, vortices, and turbulence.

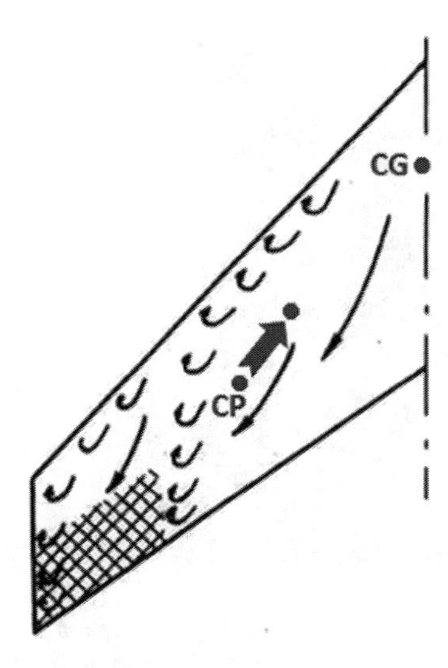

Figure 2.29 Tip stall on a swept wing leads to CP movement and pitch-up.

Tip Stall and Pitch-Up of a High-AR Wing

Stall implies loss of lift, and also a change in the pitching moment. If the moment around the aircraft center of gravity (CG) changes to more nose-up rapidly, the plane tends to pitch up, which increases the angle of attack and the stalled area with further loss of lift. Such pitch-up is an issue for swept-back wings if their tips stall first, since that means lift loss on the aft part of the wing. Figure 2.29 shows the aircraft CG and how the wing center of pressure (CP) moves when the flow on the tip separates (cross-hatch) so lift there is lost. CG ahead of CP is necessary for pitch stability, so the indicated CP movement does not make the plane unstable, but its associated nose-up moment must be rapidly countered by the pilot to avoid runaway pitch-up. The pitching moment plot in Figure 2.28 shows just how rapid this nose-up effect can be at $\alpha = 10^\circ$ and above.

Tip stall is also problematic for straight wings because it makes the ailerons less effective. If the stall is asymmetric, significant rolling moments appear and make control of the plane difficult. Tip stall is mitigated in design by reducing the tip lift by wash-out (i.e. twisting it nose-down) and by installing vortex generators to energize the boundary layer. Boundary-layer fences with outer-wing panel chord extensions have also been used. Chapter 9 shows examples from Saab aircraft, the J-29, and the SF340. In addition, there is some discussion in Chapter 3 related to mitigation of low-speed tip stall.

Skin-Friction Lines

The flow in the boundary layer on a 3D surface can be visualized in experiments as streaks forming in a coating of oil on the surface. The oil flow is driven by the viscous stresses in the boundary layer. The streaks form the skin-friction lines along

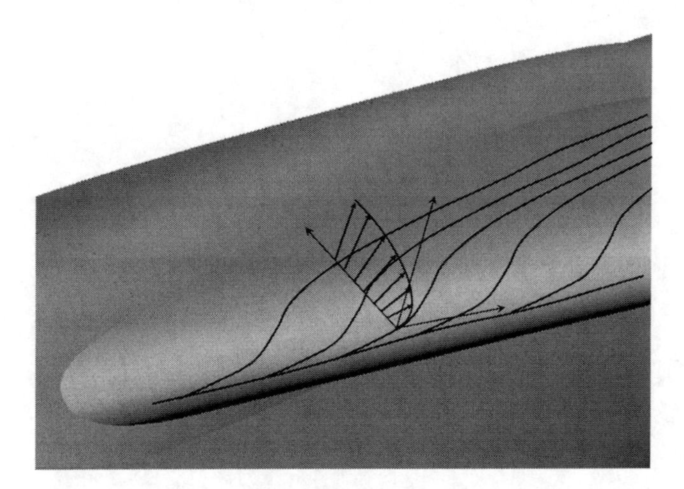

Figure 2.30 Rotation of velocity vector through boundary layer close to separation line on leading edge of swept wing. Skin friction lines emanate from the stagnation line, below the apex, and curve outward to follow the stream lines, almost parallel to the free stream, further aft.

the tangential component of the wall stress. Thus, oil particle movement forms a 2D dynamic system on the surface, so concepts of dynamic systems on manifolds, such as critical points and their type, confluence of path lines, etc., can help with our understanding and classification of separation types. Skin-friction lines can be computed from CFD (RANS) results (see Figure 2.28).

The 4° case shows attached flow, with skin-friction lines almost parallel to free the stream and outer flow. However, the boundary-layer flow deviates more from the free stream direction toward the separation at the trailing edge. Similar deviation happens at the leading edge around the stagnation line, which is on the pressure side and not visible here. Note that the pressure gradient on the wing is not in the free stream direction; rather, it is normal to the wing sweep. This may seem confusing since we think of pressure gradient as the driver for the flow, but a constant flow along the wing requires no pressure gradient and can be superposed. The 12° with separated flow regions on the suction side is clear evidence that pressure gradients are misleading indicators of boundary-layer flow on the wing near separation lines. The full Navier–Stokes (RANS) CFD models these phenomena accurately.

Close to separation lines, the velocity vector must rotate through the boundary layer as shown in Figure 2.30. The details of velocity variation are influenced by wing planform through pressure gradients and by the surface curvatures. Figure 2.30 shows a case with attached flow on the leading edge. There is no laminar separation bubble, since the wing leading edge is rather blunt and the angle of attack is small. The boundary-layer spanwise drift is small except at the leading and trailing edges, as in Figure 2.28 (left).

2.8 Physics Models: The Equations

The multidisciplinary optimization paradigm encompasses three levels of aerodynamic modeling: L1, L2, and L3. The sections below describe the mathematical bases for these models.

2.8.1 Navier–Stokes Equations: The Complete Model L3

The Navier–Stokes equations are, as far as we understand today, a complete continuum model for the flow of fluids, compressible or incompressible, laminar or turbulent, with ideal or real gas thermodynamics. Only for flight regimes with very low air density must the model be amended. They are presented and derived in many books on fluid mechanics, and we refer the reader to, for example, C. Hirsch [10]. They are written here in nonconservative form for the flow of an ideal gas with ratio $\gamma = \frac{C_p}{C_v}$ of specific heats at constant pressure and volume.

$$
\begin{aligned}
\frac{\partial \rho}{\partial t} + \mathbf{u} \cdot \nabla \rho + \rho \nabla \cdot \mathbf{u} &= 0, \\
\frac{\partial \mathbf{u}}{\partial t} + (\mathbf{u} \cdot \nabla)\mathbf{u} + \frac{1}{\rho} \nabla p &= \frac{1}{\rho} \nabla \cdot \boldsymbol{\tau}. \\
\frac{\partial p}{\partial t} + (\mathbf{u} \cdot \nabla) p + \gamma p \nabla \cdot \mathbf{u} &= (\gamma - 1)[(\boldsymbol{\tau} \cdot \nabla) \cdot \mathbf{u} + \nabla \cdot (\kappa \nabla T)],
\end{aligned}
\tag{2.7}
$$

for the fields of density ρ, Cartesian velocities $\mathbf{u} = (u, v, w)^T$, and pressure p. The heat conductivity is κ, and by the Stokes hypothesis that the bulk viscosity vanishes for a Newtonian fluid, the dynamic viscosity coefficient μ makes the shear stress tensor $\boldsymbol{\tau}$

$$
\boldsymbol{\tau} = \mu[\nabla \mathbf{u} + (\nabla \mathbf{u})^T] - \frac{2}{3}\mu(\nabla \cdot \mathbf{u})\mathbf{I} \tag{2.8}
$$

To formulate the above equations as conservation laws, we remind the reader of the thermodynamic quantities (per unit mass); *internal energy* $e = c_V T$, *enthalpy* $h = e + p/\rho$, *total energy* $E = e + 1/2|\mathbf{u}|^2$, and *total enthalpy* $H = E + p/\rho$.

It is convenient sometimes to use the vector notation above and sometimes the scalar-plus-index notation, which is similar to the formalism of tensor calculus. In the discussion of spatial discretization in Chapter 6, subscripts will also be needed for cells and edges, etc., so vector notation hides one set of indices. In the derivation of averaged equations and turbulence models below, it is more convenient to use the scalar-plus-index notation. We shall always work in Cartesian coordinates, so there is no need to distinguish between co- and contra-variant quantities. Thus, we use the following.

- $(u, v, w) = v_i, i = 1, 2, 3$
- The convention of summation over repeated indices

For example, $\nabla \cdot \mathbf{u} = \frac{\partial v_i}{\partial x_i}$ and $\tau_{ij} = \mu \left(\frac{\partial v_i}{\partial x_j} + \frac{\partial v_j}{\partial x_i} - 2/3 \frac{\partial u_k}{\partial x_k} \delta_{ij} \right)$.

With body forces $f_i(N/m^3)$ and heat source $Q(W/m^3)$ added, we have the Navier–Stokes equations in conservation law form.

$$\frac{\partial \rho}{\partial t} + \frac{\partial}{\partial x_i}(\rho v_i) = 0$$

$$\frac{\partial}{\partial t}(\rho v_i) + \frac{\partial}{\partial x_j}\left(\rho v_j v_i + p\delta_i j - \tau_{ij}\right) = f_i \tag{2.9}$$

$$\frac{\partial}{\partial t}(\rho E) + \frac{\partial}{\partial x_j}\left(\rho v_j H - v_i \tau_{ij} - \kappa \frac{\partial T}{\partial x_j}\right) = Q$$

The components of the *strain-rate* S_{ij} and *laminar stress* tensors τ_{ij} are as follows.

$$S_{ij} = \frac{1}{2}\left(\frac{\partial v_i}{\partial x_j} + \frac{\partial v_j}{\partial x_i}\right), \tau_{ij} = 2\mu(S_{ij} - 1/3 S_{kk}\delta_{ij}) \tag{2.10}$$

For slow flows with $M \ll 1$, a further simplification is to neglect density variations, and such flows are called incompressible. For incompressible flows, we can reduce the equations in Section 2.8.1 to the following form.

$$\frac{\partial v_i}{\partial x_i} = 0$$

$$\frac{\partial v_i}{\partial t} + v_j \frac{\partial v_i}{\partial x_j} = -\frac{1}{\rho}\frac{\partial p}{\partial x_i} + \nu \nabla^2 v_i + f_i/\rho \tag{2.11}$$

$$\frac{\partial T}{\partial t} + v_j \frac{\partial T}{\partial x_j} = \alpha \nabla^2 T + Q/(C_p \rho)$$

with $\nu = \mu/\rho$ being the kinematic viscosity coefficient, assumed constant, just like $\alpha = \frac{\kappa}{C_p \rho}$, denoting the heat diffusivity, and $\nabla^2 = \frac{\partial^2}{\partial x_i \partial x_i}$, denoting the Laplace operator. In the absence of buoyancy effects, the internal energy equation expressed in temperature T becomes decoupled from the mass conservation and momentum equations.

DNS and Its Computational Requirements

Numerical solution of the unsteady Navier–Stokes equations is called DNS. It computes the evolution of all scales of motion. They range from the largest, imposed by the boundaries or geometry, to the smallest scales given by the Kolmogorov dissipation scale (see Figure 2.11) that indicates the turbulent energy spectrum as a function of wave number $k = 2\pi/\lambda$, and the limits of the range of application of computational large-eddy simulation (LES) and RANS models (to be discussed later). The wave number has been nondimensionalized by multiplication with the Kolmogorov dissipation scale. Available computational resources limit DNS to low Reynolds numbers. An estimate of the mesh requirements can be made by considering the Kolmogorov dissipation scale $\eta = (\nu^3/\epsilon)^{1/4}$ (e.g. [36]) and the largest scale l that must be resolved. The length scale l can be related to the characteristic velocity scale v of the large-scale energy-containing eddies by $v \sim (\epsilon l)^{1/3}$, where $\epsilon = dk/dt$ is the rate of dissipation of turbulent kinetic energy k. Then the ratio of the largest to the smallest scales is given as follows.

$$\frac{l}{\eta} \sim \left(\frac{vl}{\nu}\right)^{3/4} = Re_l^{3/4} \tag{2.12}$$

where Re_l is the large-scale turbulent Reynolds number. For example, for homogeneous isotropic turbulence, the mesh size Δ must be smaller than $\eta/2$, and the size L of the computational domain using periodic boundary conditions must be larger than l. Hence, the number of mesh points for a 3D domain scales with Re_l as follows.

$$N_{xyz} = \left(\frac{L}{\Delta}\right)^3 \sim Re_l^{9/4} \tag{2.13}$$

This means that tripling the turbulent Reynolds number requires increasing the number of mesh points by more than a factor of 10. Spalart [29] performed a DNS of a flat plate boundary layer at $Re_x = U_\infty x/\nu \approx 0.5M$, using a mesh with approximately $10M$ grid points. An estimate for a similar boundary-layer simulation can then be given as follows.

$$N_{xyz} \approx 10^7 \times (Re_x/5 \cdot 10^5)^{2.2} \tag{2.14}$$

However, the Reynolds numbers encountered in flows of practical aeronautical interest are, in general, orders of magnitude larger than those given above, and the corresponding grid sizes and computation times are far beyond the capacity of present supercomputers. Therefore, DNS is, and will be at least in the near future, used mainly to study turbulence physics and to develop or improve turbulence models.

The Steady RANS Equations: L3 Workhorse of CFD

This model looks for the time average of the unsteady flow. The RANS equations, as the most used model in CFD, merit a more detailed discussion (see Section 2.9). They look much like the Navier–Stokes equations with additional terms – the Reynolds stresses – resulting from the averaging operation. These terms, or their approximated evolution PDEs, are modeled again with contributions from the mean flow quantities. The simplest such models replace the heat conductivity and the viscosity coefficients with expressions in mean flow quantities; more advanced models formulate and solve the evolution equations. A single such turbulence model typically can be made to work accurately for a specific class of flows, for which components of the model have been fitted to experimental data. It follows that much experience may be needed in the choice of turbulence model: the EDGE software offers more than 10 different such models. Turbulence models represent a vast subject, so we will use the turbulence models available in the CFD code with parameters that have been found to work well for transonic external aerodynamics. An overview is provided in Section 2.9.3.

Unsteady-Flow Models: LES and Hybrids with RANS

As we shall see below, the RANS models are developed by time-averaging of the unsteady equations. If the averaging period is finite, it can be considered a temporal low-pass filter to produce the unsteady RANS (URANS) model.

The option to average in space (only) leads to the LES model proposed by the meteorologist Joseph Smagorinsky in 1963 [27] and implemented by James Deardorff in 1970 [4]. The idea is to apply a spatial filter to the equations, producing a model for the filtered quantities that resolves all length scales down to the numerical grid size. The set of equations must be closed by a model for the effect of the subgrid-scale eddies, the *subgrid-scale* model.

Recently, several hybrid RANS–LES models such as that presented in [30] have been proposed to reduce further the requirement on grid resolution in order to make time-accurate flow simulations more acceptable for engineering purposes. The HYB0 model in EDGE is discussed in Chapter 6.

2.8.2 Convection Phenomena: The Euler Equations L2

As we shall see, apart from the resolution issues, the convection part of the equations is the most nonlinear and causes the most difficulty numerically. Therefore, we initially neglect viscosity and heat conduction to produce the Euler equations, formulated here in the primitive variables $\boldsymbol{u}$, ρ, and p as follows.

$$\begin{aligned} \frac{\partial \rho}{\partial t} + \mathbf{u} \cdot \nabla \rho + \rho \nabla \cdot \mathbf{u} &= 0 \\ \frac{\partial \mathbf{u}}{\partial t} + (\mathbf{u} \cdot \nabla)\mathbf{u} + \frac{1}{\rho} \nabla p &= 0 \\ \frac{\partial p}{\partial t} + (\mathbf{u} \cdot \nabla) p + \gamma p \nabla \cdot \mathbf{u} &= 0 \end{aligned} \tag{2.15}$$

This form of the equations is called quasi-linear since the differentiated variables appear linearly. Euler equations admit solutions with shocks and vorticity layers. Indeed, such discontinuities appear spontaneously even in cases with smooth initial data. Chapter 4 is devoted to a description of the solution properties and how the discontinuities are handled by numerical methods.

2.8.3 Linear-type Flow Models for L1

We come now to the classical model of *linearized subsonic and supersonic wing flow* built on Prandtl's concept of a thin layer where viscosity is important and outside of which it is neglected. The basic linearized PDE governing the flow field is derived under the assumption that the wing is thin and perturbs the oncoming stream only slightly such that perturbation velocity components are small relative to the free stream velocity V_∞. The perturbation velocity potential $\phi(z, y, z)$ for entirely subsonic or supersonic flow satisfies the Prandtl–Glauert (P–G) equation.

$$(1 - M_\infty^2)\phi_{xx} + \phi_{yy} + \phi_{zz} = 0 \tag{2.16}$$

For $M_\infty < 1$, the P–G transformation of introducing a new x-coordinate, $x' = \beta \cdot x$, with

$$\beta = \frac{1}{\sqrt{|1 - M_\infty^2|}},$$

turns the equation into the canonical elliptic Laplace equation for flow of an ideal fluid, introduced in Section 2.4.1.

For $M_\infty > 1$, the P–G model is a wave equation with the free stream direction x a time-like variable and "wave speed" β, so we write it as follows.

$$\phi_{xx} = \beta^2 \left(\phi_{yy} + \phi_{zz}\right) \tag{2.17}$$

The adjective *linear* in this specific context refers to the technique of superposing simple solutions of linear equations to make new ones, and this applies to both equations, as further discussed in Chapter 3. It works for the subsonic as well with a few twists to the supersonic model. A more detailed description is given in Drela's textbook [5] (pp. 171–193). The *vorticity* ω of the flow is as follows.

$$\omega = \nabla \times \mathbf{u}.$$

Similarly, let us define the source strength σ.

$$\sigma = \nabla \cdot \mathbf{u}$$

Any velocity field $\mathbf{u}(\mathbf{x}, t)$ may be decomposed into a sum of a solenoidal field $\nabla \times \mathbf{A}$, which is obviously divergence free, and a gradient of a scalar field $\nabla\Phi$, which is irrotational: the Helmholtz decomposition. We shall gloss over issues of unicity. The velocity field constructed from a given source strength and vorticity is as follows.

$$\mathbf{v}(\mathbf{x}) = \int_\Omega \left([\sigma(\xi) \cdot \; + \omega(\xi) \times] \frac{\mathbf{x} - \xi}{4\pi R^3} \right) d\Omega \tag{2.18}$$

where $R = |(\mathbf{x} - \xi)|$ is the distance between field point $\mathbf{x}$ and source point ξ. The latter part that produces the solenoidal part is called the Biot–Savart law, and the former can be recognized as the gradient of "the" solution to the Poisson equation with source term $\Delta\phi = \sigma$. We write "the" because a unique solution requires stipulation of boundary conditions, such as that $\mathbf{u}$ tends to a free stream velocity $\mathbf{U}_\infty$ far away. The formula cannot create a velocity field both divergence and rotation free everywhere, such as the free stream constant, because that would make both source functions vanish and produce a vanishing velocity, so only in special cases is $\mathbf{u} = \mathbf{v}$. Linear aerodynamics seeks to find the σ and ω to give the velocity field around an aircraft. σ must then vanish everywhere except on and inside solid surfaces, whereas we allow vorticity concentrated into sheets and curves to appear anywhere.

For fluids idealized as inviscid, motions initially irrotational will stay so forever unless driven by vortical forces such as those that appear in boundary layers and wakes.

Further development of vortex-based computational models for unsteady flow continues to this day. Steady-flow software is exemplified by the TORNADO vortex lattice flow solver, described in more detail in Chapter 3.

2.8.4 Viscous Effects: The Boundary-Layer Equations – L3

Complete neglect of viscosity has serious consequences. The order of the differential equations is reduced by one, so some physically required boundary conditions must be dropped. The result is an inviscid analysis that cannot enforce the *no-slip condition* on the wall and that predicts zero frictional drag. This analysis is unacceptable in most circumstances. We can use Prandtl's boundary-layer hypotheses and derive a model for the flow in a thin layer close to walls. This model will be used together with a model for inviscid flow outside the boundary layer. The combination is described in Section 7.5.1.

To this end, consider the 2D, incompressible, steady Navier–Stokes equations for the simplest case (constant density, constant property flow), $\rho = \rho_{ref}$, $\mu = \mu_{ref}$. We wish to assess the magnitudes of the different terms to see which can be dropped when the equations shall model the flow in a thin boundary layer. Thus, we expect streamwise gradients to be much smaller than normal gradients. The idea is to introduce nondimensional variables by choosing representative scales for each, and then to judge magnitudes from the coefficients appearing in the scaled equations. Note that several different scalings may work to produce different simplified models for different flow situations. Introduce scaled nondimensional variables and stretch the normal coordinate z and the velocity in that direction, w, to expand the thin region for cases with high Re. Let the small parameter that will accomplish the stretching be represented by ϵ and remain unspecified for now: $u^* = u/U_\infty$, $w^* = w/(U_\infty \epsilon)$, $x^* = x/L$, $z^* = z/(L\epsilon)$, $P^* = p/(\rho_{ref} U_\infty^2)$, $\rho' = \rho/\rho_{ref}$, and $\mu' = \mu/\mu_{ref}$ turns the steady, incompressible 2D Navier–Stokes equations into the following.

$$\frac{\partial u^*}{\partial x^*} + \frac{\partial w^*}{\partial z^*} = 0$$

$$u^* \frac{\partial u^*}{\partial x^*} + w^* \frac{\partial u^*}{\partial z^*} = -\frac{\partial P^*}{\partial x^*} + \frac{1}{Re}\left(\frac{\partial^2 u^*}{\partial x^{*2}} + \frac{1}{\epsilon^2}\frac{\partial^2 u^*}{\partial z^{*2}}\right) \tag{2.19}$$

$$u^* \frac{\partial w^*}{\partial x^*} + w^* \frac{\partial w^*}{\partial z^*} = -\frac{1}{\epsilon^2}\frac{\partial P^*}{\partial z^*} + \frac{1}{Re}\left(\frac{\partial^2 w^*}{\partial x^{*2}} + \frac{1}{\epsilon^2}\frac{\partial^2 w^*}{\partial z^{*2}}\right) \tag{2.20}$$

where $Re = \rho_{ref} U_\infty L/\mu_{ref}$. We see that the streamwise second derivative terms are small compared to the normal second derivatives. This is expected from the assumed thinness of the boundary layer. Dropping just those terms produces the "parabolized Navier–Stokes equations" with x as the time-like variable, but we wish to simplify further. Then, choosing $\epsilon^2 = 1/Re$ removes the dependence on Re. Thus, to order $\mathcal{O}(Re^{-1/2})$, the untransformed system of equations reduces to the following.

$$\frac{\partial u}{\partial x} + \frac{\partial w}{\partial z} = 0 \tag{2.21}$$

$$u\frac{\partial u}{\partial x} + w\frac{\partial u}{\partial z} = -\frac{1}{\rho}\frac{\partial p}{\partial x} + \nu\frac{\partial^2 u}{\partial z^2} \tag{2.22}$$

$$\frac{\partial p}{\partial z} \approx 0 \tag{2.23}$$

Two boundary conditions can and must be imposed for u, $u = 0$ at $z = 0$ on the wall and $u \longrightarrow U_e$ as $z \longrightarrow \infty$. The continuity equation needs one boundary condition $w(0) = 0$. It is natural to think of an initial-boundary value problem for Eq. (2.23) on the domain $x > 0, 0 < z < H$ with some large H with a prescribed value for $\partial p/\partial x$. It is, however, a differential-algebraic system of one ordinary differential equation and an algebraic equation, because $\partial w/\partial x$ does not appear. The marching in the positive x-direction is stable as long as $u > 0$. But in adverse pressure gradients, u will decrease and may vanish. The flow may separate, and the boundary-layer equations break down as a model for the flow. This is obvious from the pictures of oil flow (Figure 2.28). The driving force for the boundary layer equations is the exterior pressure gradient, orthogonal to the isobars. But oil flow close to separation lines is almost parallel to the isobars, as shown by the separation line close to the trailing edge in the right part of Figure 2.28. It follows that simulations of separation phenomena need the full RANS or even LES.

Section 7.5.1 also presents a model for the boundary layer integrated across its height, valid for laminar and turbulent flow. That model can incorporate transition to turbulence and deal with reversed flow in separated recirculation bubbles, but not with significant separated flow areas.

2.9 Averaging for Turbulent Flows

We indicated above that DNS needs to resolve all length and time scales, which is very expensive. Averaging procedures can alleviate the problem by modeling only large scales. The first approach for the approximate treatment of turbulent flows was presented by O. Reynolds in 1895 [25]. The methodology is based on the decomposition of the flow variables into a mean and a fluctuating part. A set of governing equations derived from Eq. (2.7) and a *turbulence closure model*, to be discussed in Section 2.9.3, is then solved for the mean values, which are the most interesting for engineering applications. Thus, considering first incompressible flows, the velocity components and the pressure are split up as follows.

$$v_i = \overline{v}_i + v_i', \; p = \overline{p} + p' \tag{2.24}$$

The mean value is denoted by an overline and the turbulent fluctuations by a prime. The mean values are obtained by an averaging procedure in time or space.

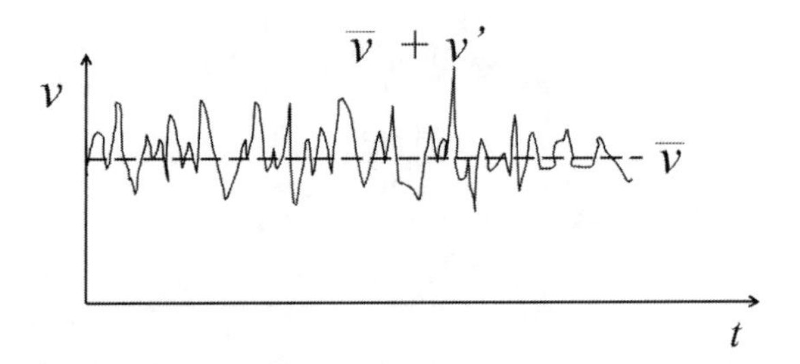

Figure 2.31 Reynolds averaging: turbulent velocity fluctuations v' and statistical mean value $\overline{v}$.

Time or Reynolds Averaging

$$\overline{v}_i = \frac{1}{T}\int_t^{t+T} v_i dt. \tag{2.25}$$

The situation is sketched in Figure 2.31. The time interval T should be large compared to the typical timescale of the turbulent fluctuations. For $T \to \infty$, the steady RANS equations appear. A finite (but unspecified) T allows unsteady URANS simulations.

Spatial Averaging

$$\overline{v}_i = \frac{1}{|\Omega|}\int_\Omega v_i d\Omega \tag{2.26}$$

with Ω a control volume. In CFD, Ω is thought of as a control volume associated with the computational grid.

For both approaches, the average of the fluctuating part is zero (i.e., $\overline{v_i'} = 0$). However, it can be easily seen that $\overline{v_i'v_i'} \neq 0$. The same is true for $\overline{v_i'v_j'}$, if the fluctuations in velocity components are correlated.

2.9.1 Favre (Mass) Averaging

In cases where the density is not constant, it is advisable to apply the *density (mass) weighted* or Favre decomposition to certain quantities in Eq. (2.7) instead of time averaging. Otherwise, the averaged governing equations would become considerably more complicated due to additional correlations involving density fluctuations. The most convenient approach is to employ Reynolds averaging for density and pressure and Favre averaging for other variables such as velocity, internal energy, enthalpy, and temperature.

$$\widetilde{v}_i = \frac{1}{\overline{\rho}}\lim_{T\to\infty}\frac{1}{T}\int_t^{t+T} \rho v_i dt, \tag{2.27}$$

where $\overline{\rho}$ denotes the Reynolds-averaged density. Hence, the Favre decomposition reads as follows.

$$v_i = \widetilde{v}_i + v_i'', \tag{2.28}$$

where $\widetilde{v}_i$ represents the mean value and v_i'' the fluctuating part of the velocity v_i. Again, the average of the fluctuating part is zero (i.e., $\widetilde{v_i''} = 0$, etc.).

The following relationships can be derived for a mix between Favre and Reynolds averaging.

$$\widetilde{\rho v_i} = \bar{\rho}\tilde{v}_i, \quad \overline{\rho v_i''} = 0, \quad \text{but} \quad \overline{v_i''} \neq 0 \tag{2.29}$$

2.9.2 Favre- and Reynolds-Averaged Navier–Stokes Equations

Application of the Reynolds-averaging Eq. (2.25) to density and pressure and of the Favre-averaging Eq. (2.27) to the remaining flow variables in the compressible Navier–Stokes equations (Section 2.8.1) yields the following.

$$\frac{\partial \bar{\rho}}{\partial t} + \frac{\partial}{\partial x_i}(\bar{\rho}\tilde{v}_i) = 0 \tag{2.30}$$

$$\frac{\partial}{\partial t}(\bar{\rho}\tilde{v}_i) + \frac{\partial}{\partial x_j}\left(\bar{\rho}\tilde{v}_j\tilde{v}_i + \bar{p}\delta_{ij} - \tilde{\tau}_{ij}\right) = -\frac{\partial}{\partial x_j}(\bar{\rho}\widetilde{v_i''v_j''}) \tag{2.31}$$

$$\frac{\partial}{\partial t}(\bar{\rho}\tilde{E}) + \frac{\partial}{\partial x_j}\left(\bar{\rho}\tilde{v}_j\tilde{H} - \kappa\frac{\partial \tilde{T}}{\partial x_j} - \tilde{v}_i\tilde{\tau}_{ij}\right) = \frac{\partial}{\partial x_j}\left(-\bar{\rho}\widetilde{v_j''h''} - \bar{\rho}\tilde{v}_i\widetilde{v_i''v_j''}\right). \tag{2.32}$$

These are the *Favre- and Reynolds-averaged Navier–Stokes* equations. Some terms have been neglected, and the terms on the right-hand side must be modeled somehow. Similarly to the Reynolds averaging, the viscous stress tensor in the momentum (and energy) equation is extended by the *Favre-averaged Reynolds-stress* tensor as follows.

$$\tau_{ij}^F = -\bar{\rho}\widetilde{v_i''v_j''} \tag{2.33}$$

The form is similar to that for incompressible flow but with Favre instead of Reynolds averaging. The components of the laminar (molecular) viscous stress tensor $\tilde{\tau}_{ij}$ are evaluated using Favre-averaged velocity components.

If we employ the definition of the Favre-averaged turbulent kinetic energy,

$$\bar{\rho}\tilde{k} = \frac{1}{2}\bar{\rho}\widetilde{v_i''v_i''}, \tag{2.34}$$

we can express the total energy in Eq. (2.31) as follows.

$$\bar{\rho}\tilde{E} = \bar{\rho}\tilde{e} + \frac{1}{2}\bar{\rho}\tilde{v}_i\tilde{v}_i + \frac{1}{2}\bar{\rho}\widetilde{v_i''v_i''} = \bar{\rho}\tilde{e} + \frac{1}{2}\bar{\rho}\tilde{v}_i\tilde{v}_i + \bar{\rho}\tilde{k} \tag{2.35}$$

The total enthalpy is defined as

$$\bar{\rho}\tilde{H} = \bar{\rho}\tilde{h} + \frac{1}{2}\bar{\rho}\tilde{v}_i\tilde{v}_i + \frac{1}{2}\bar{\rho}\widetilde{v_i''v_i''} = \bar{\rho}\tilde{h} + \frac{1}{2}\bar{\rho}\tilde{v}_i\tilde{v}_i + \bar{\rho}\tilde{k} \tag{2.36}$$

The individual parts of the Favre- and Reynolds-averaged Navier–Stokes equations (Eq. (2.31)) that must be modeled to close the system have the following physical meaning:

$$\tau_{ij}^{F} - \text{Favre} - \text{Reynolds stresses}$$

$$\frac{\partial}{\partial x_j}\left(\bar{\rho}\widetilde{v_j'' h''}\right) - \text{turbulent transport of heat}$$

$$\frac{\partial}{\partial x_j}\left(\tilde{v}_i \tau_{ij}^{F}\right) - \text{work done by the Favre} - \text{averaged Reynolds stresses}$$

2.9.3 Turbulence Closure Models

The Favre/Reynolds-averaged transport equations contain unknown variables coming from averaged products, where ϕ is a fluctuation

$$\overline{\phi v_i'}$$

like the Reynolds stresses in incompressible flow $\tau_{ij}^{R} = -\rho\overline{v_i' v_j'}$. These variables need to be supplemented before any solution can be obtained. A set of mathematical equations that provide the unknown variables is called a *turbulence closure model.* Note that these variables are always indexed as higher order (in the tensor sense) than the basic variables (i.e. if ϕ is scalar, $\overline{\phi v_i'}$ is a vector; if ϕ is a vector (first-order tensor), $\overline{\phi v_i'}$ is a second-order tensor). We need to solve more equations than for a laminar flow. The type (algebraic or differential) and the number of auxiliary equations defines the closure level. For most engineering purposes, only the effects of the turbulence on the mean flow are sought: the Reynolds stresses and the turbulent scalar transport terms.

The *classical models* use the Favre/Reynolds equations developed in this section and form the basis of turbulence calculations in currently available commercial CFD codes. *LESs* are turbulence models where the time-dependent, spatially averaged flow equations are solved for the mean flow and the largest eddies and where the effects of the smaller eddies are modeled. The largest eddies interact strongly with the mean flow and contain most of the energy, so this approach results in a good model of the main effects of turbulence. Smaller flow features that cannot be resolved accurately on the computational grid are dissipated by a subgrid-scale filter.

Two basic levels of classical models are as follows.

- *Eddy viscosity/diffusivity models* are known also as first-order models, since the quantities modeled are scalars, such as turbulent kinetic energy k and its dissipation rate ε.
- *Second-moment closure models*, also known as *Reynolds stress models* (RSMs), model the individual Favre/Reynolds stress components.

Each category has a number of variants. The first-order models are categorized according to the number of equations they solve. In the second-order models, if the equations for the stresses are algebraic, the models are known as algebraic RSMs (ARSMs).

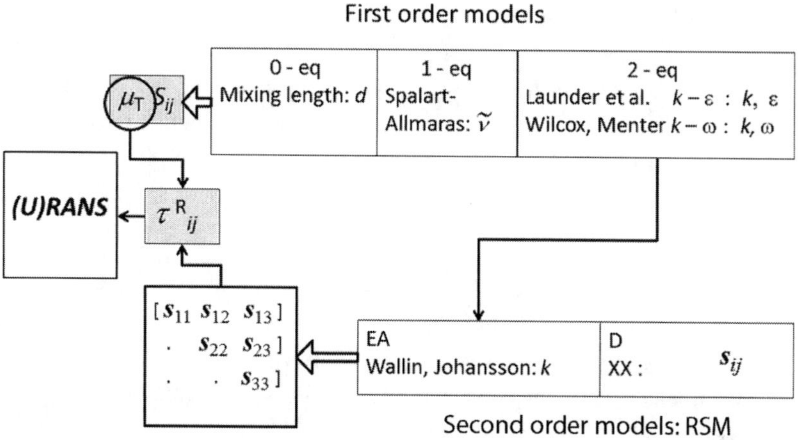

Figure 2.32 Taxonomy of a few turbulence models.

The most advanced models solve differential transport equations for the second moment, hence the acronym DRSM. Some common models are illustrated in Figure 2.32.

LESs are time-accurate and require 3D geometry even for essentially plane flows. The computational grids need to be isotropic and resolve all eddies carrying significant energy, which is a stringent requirement in boundary layers. (U)RANS models use anisotropic boundary-layer grids to save very significantly on grid points. Combinations of (U)RANS models close to walls with LESs in the outer regions are known as "hybrid RANS–LES" and "detached eddy simulations." Current efforts to model separated flows, as required for maneuver simulations, for example, have produced a plethora of hybrids, such as that given in [28] and the Peng Hyb0 model in EDGE.

The different turbulence models described above are represented in most CFD codes. For details of implementation, applicability, and grid requirements, we refer to the documentation of the CFD code in use.

2.9.4 First-Order Closures: Eddy-Viscosity Models

Eddy-Viscosity Models

The first attempt to model the Reynolds stresses was made by Boussinesq [1], who introduced an eddy viscosity μ_T in complete analogy with the molecular viscosity μ for a Newtonian fluid. The eddy viscosity may be interpreted as a diffusivity coefficient (of momentum) related to the velocity scale V and length scale L of the large energetic turbulent eddies, $\mu_T \propto \rho\ V\ L$. Under the eddy-viscosity hypothesis, turbulence modeling reduces to modeling the eddy viscosity μ_T in terms of the mean quantities. The Reynolds stress tensor is then related to the mean flow field through the following.

$$\tau_{ij}^F = -\bar{\rho}\widetilde{v_i''v_j''} = 2\mu_T\tilde{S}_{ij} - \left(\frac{2\mu_T}{3}\right)\frac{\partial \tilde{v}_k}{\partial x_k}\delta_{ij} - \frac{2}{3}\bar{\rho}\tilde{k}\delta_{ij}, \tag{2.37}$$

where $\tilde{S}_{ij}$ and $\tilde{k}$ are the Favre-averaged strain rates and turbulent kinetic energy, respectively. Note the similarity to Eq. (2.10).

The turbulent heat flux vector will be modeled by defining a turbulent thermal conductivity coefficient κ_T through a turbulent Prandtl number from the turbulent viscosity as follows.

$$\kappa_T = \frac{C_p \mu_T}{Pr_T}$$

First-order models can be classified according to the number of additional transport equations for the turbulent quantities they require. The algebraic or zero-equation models are therefore the simplest. Methods using one additional differential equation (the Spalart–Allmaras model) or two additional differential equations (e.g. the $k - \omega$ models) are common.

2.9.5 Second-Moment Closures: RSMs

Eddy-viscosity models perform reasonably well in attached boundary-layer flows as long as *only one component* of the Reynolds stress tensor is significant. In these cases, one could consider the eddy viscosity to represent that significant Reynolds stress component. But for more complicated flows where this is not the case, the validity of the eddy-viscosity approach is questionable.

Two-equation turbulence models, such as $k - \varepsilon$ and $k - \omega$, are incapable of capturing the effects of anisotropic normal stresses. They also fail to represent correctly the effects on turbulence of extra strains and body forces. The RSM incorporates these effects exactly, but several unknown turbulence processes (pressure–strain correlations, turbulent diffusion of Reynolds stresses, dissipation) need to be modeled.

Taking the second moment of the Navier–Stokes equations leads to the transport equation for the Reynolds stress tensor.

$$\frac{\mathrm{D}\overline{v_i v_j}}{\mathrm{D}t} = \underbrace{\mathcal{P}_{ij}}_{production} - \underbrace{\varepsilon_{ij}}_{dissipation} + \underbrace{\Pi_{ij}}_{pressure\ strain\ rate} + \underbrace{\mathcal{D}_{ij}}_{diffusion} \tag{2.38}$$

where $\mathcal{D}_{ij}$ represents both molecular and turbulent diffusion. The dissipation ε_{ij}, pressure–strain rate Π_{ij}, and diffusion terms must be modeled.

The advection and diffusion terms account for transport effects for the individual Reynolds stress components, whereas eddy-viscosity two-equation models only consider such effects for the trace of the Reynolds stress tensor (i.e. the turbulent kinetic energy k).

Most important are the different local source terms. No modeling is needed for the production term, $\mathcal{P}_{ij}$, which is a significant improvement compared to the modeling of the trace of the production in eddy-viscosity models. The inter-component energy transfer, represented by $\mathcal{D}_{ij}$ needs to be modeled, but already the simplest possible models represent improvements compared to eddy-viscosity models in which this effect is not present at all.

The most used models in the EDGE CFD code are presented in some detail in Chapter 6.

- The one-equation Spalart–Allmaras model, solving a single transport equation for the eddy viscosity $\tilde{\nu}$.
- The two-equation Hellsten $k - \omega$ model, with transport of turbulent kinetic energy k and $\omega = k/\nu_T$, a frequency scale for the turbulent fluctuations.
- The Wallin–Johansson explicit ARSM that computes the Reynolds stress tensor from Hellsten's k and ω.
- The Peng LES model Hyb0.

With an appreciation of airflow physics that is beneficial for performance and with the mathematical models now defined, the next chapter combines the models and the physics to arrive at suitable L1 computational procedures suited for Cycle 1 wing design.

2.10 Learn More by Computing

Gain hands-on experience of the computational tools for the topics in this chapter by working with the on-line resources. Exercises, tutorials, and project suggestions are found on the book website www.cambridge.org/rizzi. The "Rotor Plane Kite" example is a very concrete demonstration of lift and circulation. Software used to compute many of the examples shown is available from http://airinnova.se/education/aero-dynamic-design-of-aircraft

References

[1] J. Boussinesq. Essai sur la théorie des eaux courantes. *Mémoires présentés par divers savants à l'Académie des Sciences*, 23(1), 1877.

[2] G. E. P. Box, W. G. Hunter, and J. S. Hunter. *Statistics for Experimenters: Design, Innovation, and Discovery*, 2nd edition. Wiley, 2005.

[3] J. F. Campbell and J. R. Chambers. *Patterns in the Sky: Natural Visualization of Aircraft Flow Fields*. NASA Langley Research Center, 1994.

[4] J. Deardorff. A numerical study of three-dimensional turbulent channel flow at large reynolds numbers. *Journal of Fluid Mechanics*, 41(2): 453–480, 1970.

[5] M. Drela. Newton solution of coupled viscous/inviscid multielement airfoil flows. Presented at AIAA Aerospace Sciences Meeting, 90-1470, 1990.

[6] A. Elsenaar. Separation in transonic flow: a shocking experience. Technical report NLR TP 97151, National Aerospace Laboratory, 1997.

[7] A. Elsenaar. Vortex formation and flow separation: the beauty and the beast in aerodynamics. *Aeronautical Journal*, 104(1042): 615–633, 2000.

[8] A. B. Haines. Computers and wind tunnels: complementary aids to aircraft design. *Aeronautical Journal*, 81(799): 306–321, 1977.

[9] M. R. Head. Flow visualization at cambridge university engineering department. Presented at Second International Symposium on Flow Visualization, September, Bochum, West Germany, 1980.

[10] C. Hirsch. *Numerical Computations of Internal and External Flows*. Wiley, 1990.

[11] E. H. Hirschel. On the creation of vorticity and entropy in the solution of the euler equations for lifting wings. Technical report, Ottobrunn, Germany, 1985. MBB-LKE122-AERO-MT-716.

[12] E. H. Hirschel, J. Cousteix, and W. Kordulla. *Three-Dimensional Attached Viscous Flow: Basic Principles and Theoretical Foundations*. Springer-Verlag, 2014.

[13] E. H. Hirschel, A. Rizzi, and C. Breitsamter. *Separated and Vortical Flow in Aircraft Aerodynamisc*. Springer, 2020.

[14] NASA history website. Figure [105], schlieren photograph of transonic flow. Available from https://history.nasa.gov/SP-440/ch7-2.htm.

[15] K. Huenecke. *Modern Combat Aircraft Design*. Naval Institute Press, 1987.

[16] I. Kroo. *Applied Aerodynamics: A Digital Textbook*. Desktop Aeronautics, 1997.

[17] M. W. Kutta. Auftriebskräfte in strömenden flüssigkeiten. *Illustrierte Aeronautische Mitteilungen*, 6: 133–135, 1902.

[18] F. W. Lanchester. *Aerodynamics, Constituting the First Volume of a Complete Work on Aerial Flight*. A. Constable and Co., Ltd., 1907.

[19] J. Lighthill. *An Informal Introduction to Theoretical Fluid Mechanics*. Clarendon Press, 1986.

[20] J. M. Luckring. The discovery and prediction of vortex flow aerodynamics. *Aeronautical Journal*, 123(1264): 729–804, 2019.

[21] B. Melvil Jones. The streamline aeroplane. *Proceedings of the Royal Aeronautical Society*, January: 357–385, 1929.

[22] L. Prandtl. Über flüssigkeitsbewegung bei sehr kleiner reibung. *Verhandlungen des dritten internationalen Mathematiker-Kongresses*, pp. 489–491, 1904.

[23] L. Prandtl. The generation of vortices in fluids of small viscosity. *Journal of the Royal Aeronautical Society*, 31(200): 718–741, 1927.

[24] L. Prandtl and O. G. Tietjens. *Applied Hydro- and Aeromechanics*. Dover Books on Aeronautical Engineering. Dover Publications, 2012.

[25] O. Reynolds. On the dynamical theory of incompressible viscous fluids and the determination of the criterion. *Philosophical Transactions of the Royal Society of London A* 186: 123–164, 1895.

[26] A. Rizzi and J. M. Luckring. Evolution and use of cfd for separated flow simulations relevant to military aircraft. Presented at Symposium on Separated Flow: Prediction, Measurement and Assessment for Air and Sea Vehicles, paper 11. Neuilly-sur-Seine, October 2019.

[27] J. Smagorinsky. General circulation experiments with the primitive equations. *Monthly Weather Review*, 91(3): 99–164, 1963.

[28] P. R. Spalart et al. A new version of detached-eddy simulation, resistant to ambiguous grid densities. In *Theoretical and Computational Fluid Dynamics*. Springer Science and Business Media LLC, 2006, pp. 181–195.

[29] P. R. Spalart. Young-person's guide to detached eddy simulation grids. Technical report CR-2001-211032, NASA, July 2001.

[30] P. R. Spalart, W.-H. Jou, M. Strelets, and S. R. Allmaras. Comments on the feasibility of LES for wings and on a hybrid RANS/LESapproach. In C. Liu and Z. Liu, editors, *Advances in DNS: Proceedings of the First AFOSR International Conference on DNS/LES*. Greyden Press, 1997.

[31] P. Thomas. Entwurfsgerechte tragflugelaerodynamik. Presented at Bericht über die Sitzung des WGLR-Fachausschusses für Aerodynamik, number 67-24 in DLR Mitteilung, Darmstadt, November 1967.

[32] M. Tomac. *Towards Automated CFD for Enginerring Methods in Airfcraft Design*. Trita ave 2014:11. KTH School of Engineering Sciences, 2014.

[33] E. Torenbeek. *Synthesis of Subsonic Airplane Design*. Springer Netherlands, 1982.

[34] R. Vos and S. Farokhi. *Introduction to Transonic Aerodynamics*. Fluid Mechanics amd its Applications. Springer, 2014.

[35] D. C. Wilcox. Reassessment of the scale determining equation for advanced turbulence models. *AIAA Journal*, 26(11): 1299–1310, 1988.

[36] D. C. Wilcox. *Turbulence Modeling for CFD*. DCW Industries, Inc., 1998.

3 Concepts and Computational Models in Wing Design

All human knowledge begins with intuitions [or sensibility], proceeds from there to concepts [or understanding the relation of concepts], and ends with ideas [or reason/knowledge achieved through that understanding] ...

Immanuel Kant, 1781 (translated from the German)

There's nothing so practical as good theory.

Kurt Lewin, German–American psychologist

Preamble

The previous chapter showed us that the aerodynamic characteristics of a wing change markedly with planform shape and the other primary design parameters, discussed the flow physics responsible for these changes, and spelled out the partial differential equations governing such changes. The design task before us is to shape the wing so as to realize aerodynamic characteristics that are suitable to the mission. Doing this requires a *prediction* method, of either L1, L2, or L3 genus, that maps the given geometry to its pressure field and ultimately its performance.

For the range of flight missions, the subject of this chapter then is to survey the various wing design considerations, explain which design method is suitable, and describe the L1 model implementation into a practical computational tool – the vortex lattice method for subsonic missions. This method is applied to practical planform studies that illustrate the design task with primary geometry parameters.

The design paradigm assumes that the flow is attached over the whole wing surface. A straight high aspect ratio wing has flow essentially parallel to the free stream. The geometry of the section parallel to the free stream determines the behavior of the boundary layer, friction drag, separation, and onset of compressibility effects. The design of such a classical pre-jet-age wing could successfully be decomposed into planform design, treated in this chapter, and section design or airfoil design, taken up in Chapter 8.

With high-speed swept and delta wings, the decomposition loses prediction power and the design task becomes more involved, as the cases studied in Chapter 9 show. The discussion of the design task continues for high-speed flight missions and indicates where L2 and L3 tools must be used.

3.1 Introduction: Mapping Planform to Lift and Drag

This section briefly discusses how the central geometric parameters that define the gross shape of the wing influence its aerodynamic properties. The discussion addresses an isolated *clean* wing devoid of engine nacelles, high-lift devices, and control surfaces and not mounted on the fuselage.

As discussed in Chapter 1 (Table 1.1), the primary wing geometry parameters affecting its aerodynamic characteristics and used for sizing the wing to carry the aircraft weight are: aspect ratio AR, span b, wing area S, average section thickness ratio $(t/c)_{ave}$, taper ratio λ, sweep angle Λ, and cruise speed M_{cruise}. Figure 2.1 in Chapter 2 maps three different planforms – straight, swept, and slender delta wings – to three distinct flight regimes: subsonic, transonic, and supersonic. This mapping achieves the high aerodynamic efficiency offered by the respective planform's *flow pattern* – stable and predictable – that results in low drag. The prescription of a particular type of flow pattern is in itself a powerful factor in defining the shape of a practical airplane. The actual design and sizing of any specific wing requires a predictive tool, based on the flow models outlined in Chapter 2, which relates the wing geometry to its resulting pressure distribution that determines the wing's aerodynamic characteristics. For example, Figure 3.1 indicates large differences in the lift generated by, as well as the stall behavior of, high-AR, low-Λ wings compared with low-AR, high-Λ delta wings. Our predictive aerodynamic tools must accurately capture these characteristics. Let us look at this more closely. Chapter 2 showed us how low-speed aerodynamics phenomena differ markedly from those encountered at high speed. Thus, our discussion begins with wings in low-speed flight, $M_{cruise} \lessapprox 0.7$, and continues in Section 3.4 for high-speed flight, $M_{cruise} \gtrapprox 0.7$.

3.1.1 Classical Wing Design by Prandtl–Glauert Models

Until near-sonic flight speeds became a practical possibility, the broad outline of the airplane changed very little. Unswept wings of relatively high AR, mounted on a slender, essentially nonlifting fuselage, were almost universal, and established the *classical aircraft genus* with engines in separate nacelles. In this flight condition with wings primarily straight, the Cayley paradigm provides components without strong interference. For example, the drag of the wing, fuselage, and tail, measured separately, could be added together with only minimum consideration of interference effects to obtain the drag of the entire aircraft.

Chapter 2 (see Figure 2.1) presents the dominant pattern of *streamlined flow* attached everywhere on the lifting surfaces. Flow separation takes place only along the trailing edge, so that extensive regions of separated flow on the airfoil surface itself are avoided and viscous effects are confined to a thin boundary layer. Furthermore, the flow is smooth everywhere and there are no discontinuities such as shock waves.

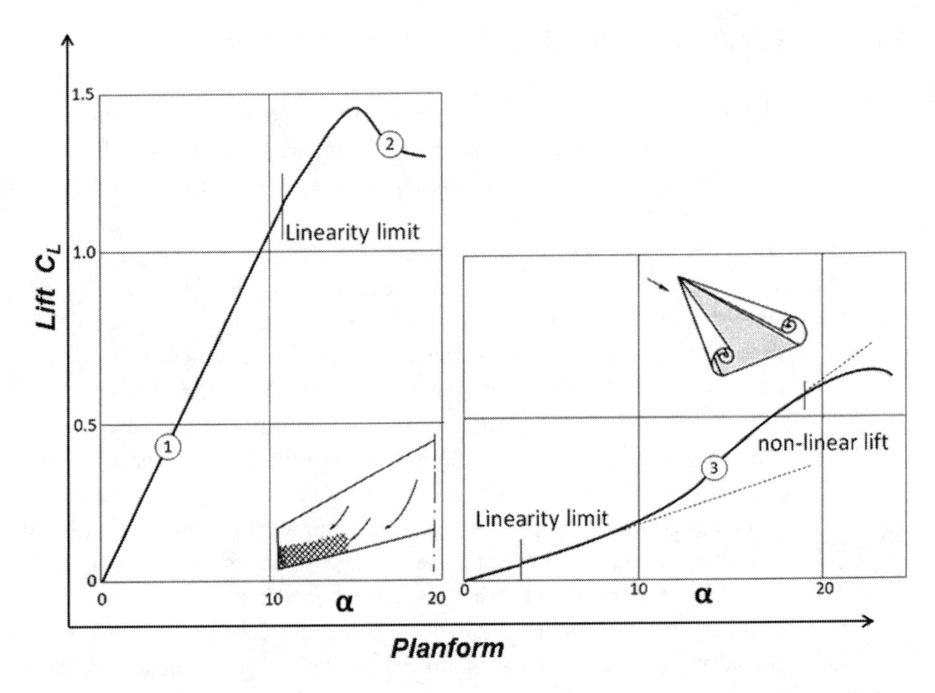

Figure 3.1 High-*AR* airliner wing and low-aspect ratio delta wing, optimal for different flight regimes, with lift curves. (Courtesy of K. Huenecke [10], reprinted with permission)

The Prandtl lifting-line theory predicts sufficiently well the effects of the finite span of such a wing in low subsonic speed. The induced drag and sizing of the wing load-bearing structures could be managed by design of the spanwise lift distribution (e.g. for minimum induced drag). Thus, the main task was to determine the chordwise flow over the wing section so that the stall limits and force distribution, important for structural design and flight dynamics via the moments, could be predicted. Practical design needs were largely met by 2D theory, either inviscid or coupled with boundary-layer analysis. Lifting-line theory and improvements such as the Weissinger model accounted for finite-wing effects by the Prandtl–Glauert model presented in Section 2.8.3.

This state of affairs lasted for nearly 50 years, during which a vast amount of systematic data was amassed, and the theory and design of the 2D airfoil reached an extremely advanced stage of development.

Wing design begins with airfoil selection. The chordwise shaping of the wing is achieved primarily by selecting airfoils with desirable aerodynamic characteristics, one for the root section and another for the tip section, and then ruling the surface between them. Next comes planform shaping to acquire the desired span loading and bending moments for low induced drag. Finally, once all of the components have been assembled into the overall configuration, streamline the design to maintain attached flow and minimize pressure drag. By the late 1930s, aircraft showed very clean designs with engine cowlings, cantilever monoplane wings, retractable undercarriages, smooth surfaces, and wing flaps for higher wing loadings. Airliners at the dawn of the jet age had become three times cleaner than those in the early 1920s. These designs were achieved with basically analytical formulations of the Prandtl-Glauert (PG) model (e.g. lifting-line theory).

3.1.2 Lift Slope Predictions

The wing *AR* affects the slope of the lift curve, $dC_L/d\alpha$.

High-AR Lifting-Line Predictions

For high-*AR* wings, lifting-line theory predicts the lift-curve slope to be as follows.

$$C_{L,\alpha} = \frac{c_{\ell,\alpha} AR}{AR + 2}$$

Helmbold refined the model with an elliptic single lifting vortex at quarter-chord, satisfying the boundary condition at the 3/4-quarter chord point.

$$C_{L,\alpha} = \frac{c_{\ell,\alpha} AR}{\sqrt{AR^2 + 4} + 2} \tag{3.1}$$

The effect of *AR* on lift and drag coefficients – the drag polar – can be seen in Figure 3.2. The plane flow over an airfoil ($AR = \infty$) produces no induced drag. Still, drag grows somewhat with lift due to thickening of the aft part of the boundary layer. The shape of the wing ($AR < \infty$) drag polar is close to parabolic up to $C_{L,max}$ when stall sets in and drag continues to rise but lift drops off.

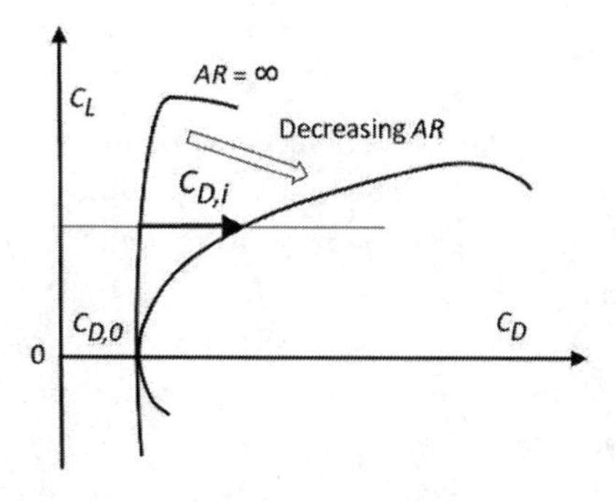

Figure 3.2 Effect of *AR* on drag polar.

Low-AR and Delta Wings

Interest rose in Germany during the 1920s and 1930s for alternative planforms to the classical high-*AR* straight wing when Alexander Lippisch and the Horten brothers, among others, were flying prototype delta wings.

Wind tunnel experiments of low-*AR* planforms, however, brought consternation when the measured lift differed widely from lifting-line theory, which, for small angles of attack and *AR*, gives twice the observed lift. This lasted until Robert Jones [12] devised a theory in 1945 that accurately predicts the lift-curve slope for high-sweep delta wings.

$$C_{L,\alpha} = \frac{\pi}{2} AR$$

Vortex Lift

Looking now at the lift curve for the delta wing in Figure 3.1 (right), we see that after the linear range, lift *increases* nonlinearly with further increase in α due to the suction effect of the vortex over the wing's upper surface. The vortex lift is predicted neither by lifting-line theory nor by the Jones theory. Luckring [15] explains how this nonlinear vortex effect on lift was discovered through wind tunnel testing of the Lippisch DM-1 delta model.

Polhamus subsequently devised an ad hoc correction to PG models that turns the leading-edge forward suction peak of the vortex into lift and drag. We refrain from further development of the leading-edge suction model and advocate moving up the ladder of fidelity to L2 and L3 models over heuristics and empiricisms.

As we shall see in Chapter 9, vortex lift can often be modeled by inviscid flow – the Euler equations (L2) – when the leading edge is sharp, so the separation position is well determined by the geometry. However, delta wing lee-side flow at higher angles of attack may show complicated separation and vortical features, such as bursting, which require Reynolds-averaged Navier–Stokes (RANS) or large-eddy simulation (LES) models.

3.1.3 Numerical Solution to the PG Model

Just as jet propulsion opened up new vistas in high-speed aerodynamics, the arrival of the programmable electronic computer at about the same time revolutionized aerodynamic theory by making algorithms – too lengthy for computation by hand – accessible by machine computation.

The vortex lattice method (VLM) is one of the first of many examples of this. All PG methods seek a solution for the Laplace equation as a superposition of elementary horseshoe vortices. Lifting-line theory uses a single vortex, as does the Weissinger method, putting it at 1/4 chord with the collocation point at the 3/4-chord point. The calculations become cumbersome as the number of vortices or the resolution along a single vortex grows. Nevertheless, vortex superposition methods were in use in many places, and we point to Ref. [8] from 1947 as an early accessible description.

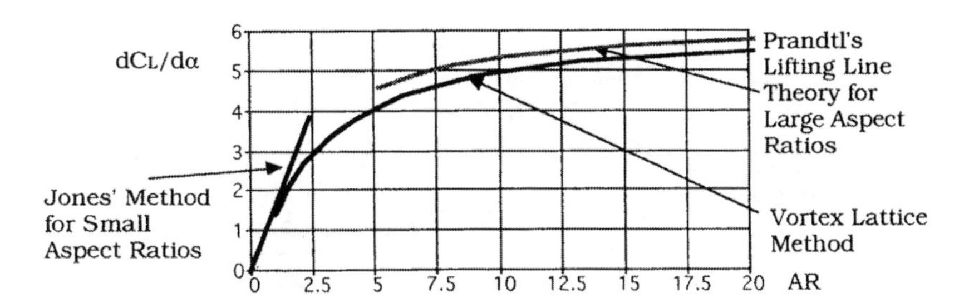

Figure 3.3 Lift-curve slopes for wings: limited lifting-line and Jones analytical models compared with the computational VLM model.

But machine computation allows representation of the lifting surface as a *lattice* of *many* horseshoe vortices, and this improves the accuracy.

Numerical computer solution of (2D) partial differential equations (PDEs) quickly became popular even on early computers with their small memories, since finite-difference relaxation methods could be programmed with very few instructions and needed no grid data structures.

In contrast, the VLM (as originally conceived) needs a full matrix and a Gaussian elimination program. That may be why VLM became widely computerized only as late as the 1960s, in the IBM 360 and CDC 6600 era. This was the start of the computation age of fluid dynamics. The birth of computational fluid dynamics (CFD) in its different shapes is usually dated to around 1965, when larger machine memories and advanced operating systems made computers much more capable and easy to program. A NASA report from 1976 [6] points precisely to 1965 as the start of the VLM wave of computerization of aerodynamics, although computations had then been made for 20 years: "In the mid 1960's four independent papers appeared on vortex-lattice methods, respectively by Rubbert, Dulmovits, Hedman, and Belotserkovskii."

Vortex Lattice Prediction: Lift versus AR

Figure 3.3 shows how the VLM prediction of the lift-curve slope covers both lifting-line results at high *AR* and Jones' method at low *AR*: the factor of two error in the lifting-line result appears clearly only for $AR < 1$, so is not very visible in the figure.

In conclusion, the VLM that solves the PG model by machine computation, with the Prandtl–Glauert transformation accounting for linear compressibility effects in the range $M_{cruise} \lessapprox 0.7$, is the preferred L1 method in this book. The following two sections show that it yields accurate predictions for wing lift curve segment ① and induced drag (Figure 3.1, left). As an inviscid attached-flow model, it does not predict segments ② or ③.

Section 3.4 considers high-speed flow when $M_{cruise} \gtrapprox 0.7$, VLM is no longer valid, and the L2 Euler equations are the model of choice. Chapter 4 then continues with the development of that model.

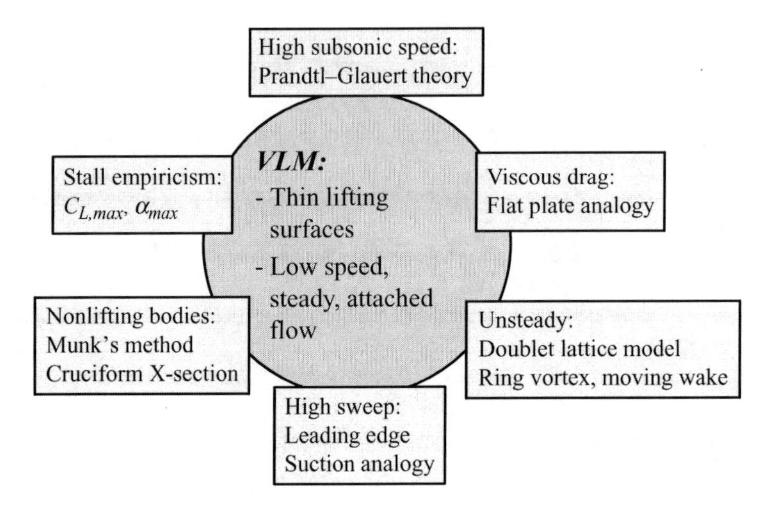

Figure 3.4 Computational VLM and its add-ons.

3.2 Computational VLMs

This section deals with several aspects of thin-wing theory – the PG model in Section 2.8.2 – from the development of a theoretical vortex-sheet model of the finite-wing flow, to its implementation in the VLM, a numerical solution procedure for the PG model.

3.2.1 Introduction to VLM

The real 3D flow around a finite wing will be modeled as inviscid, essentially irrotational, and slow enough to be incompressible or within the Prandtl–Glauert approximation. Boundary layers and shocks are neglected at this stage of the discussion. Additionally, we shall assume that only lifting components of the configuration are important and that they are thin enough to be idealized as surfaces. This will lead the way toward the VLM as implemented in the TORNADO code. Vortex lattice codes have been around for more than 50 years, and developers have come up with ingenious add-ons to extend the flow effects modeled. Six such issues are shown in Figure 3.4.

- High Mach-number correction to pressure coefficient (and hence forces and moments): the Prandtl–Glauert theory.
- Empirical relation for high-angle-of-attack stall, prediction of the maximal lift coefficient $C_{L,max}$ and its associated angle of attack α_{max}.
- Drag coefficient at zero lift, C_{D0} is the skin-friction drag, by analogy with zero pressure gradient flow over a flat plate.

- Treatment of nonlifting bodies, such as fuselage, tail booms, and nacelles.
- Treatment of unsteady flow by ring vortex lattice models with moving wake and the doublet lattice method for harmonic motion. The latter is the preferred flow model for subsonic flutter analysis. We shall give a brief account in Chapter 11 and refer the interested student to texts devoted entirely to that subject, for instance [14].
- Treatment of leading-edge vortex lift by the suction analogy.

The PG correction has sound theoretical backing. However, it is inaccurate once the flow develops supersonic pockets. But that may be hard to diagnose from the vortex lattice results, and experience is required to judge how close to $M_\infty = 1$ the PG correction remains valid.

The α_{max} empiricism is a quantitative warning not to trust the vortex lattice model at high angles of attack.

The C_{D0} estimate is an estimate based on flow over a flat plate, and its accuracy is hard to assess.

VLM without add-ons can model a fuselage as a "cruciform" body, replacing the volume with two thin sheets at right angles. An airliner model with a cruciform body is shown in Figure 3.5. The engine nacelles are cylinders; horseshoe vortices can only model surfaces created by extrusion of a polygon along an axis. But the "normals" can be taken as those of the real doubly curved nacelles.

A slender body such as an aircraft fuselage can also be modeled by superposition of sources and sinks along its axis. The simplest such models assume that the body flies in undisturbed free stream, and they model the influence of the body on the wings, but not the converse. An educated guess of the "base area" is needed to produce a net force in addition to the moment.

More information and a discussion of the unsteady ring vortex model and the doublet lattice model are found on the book's website.

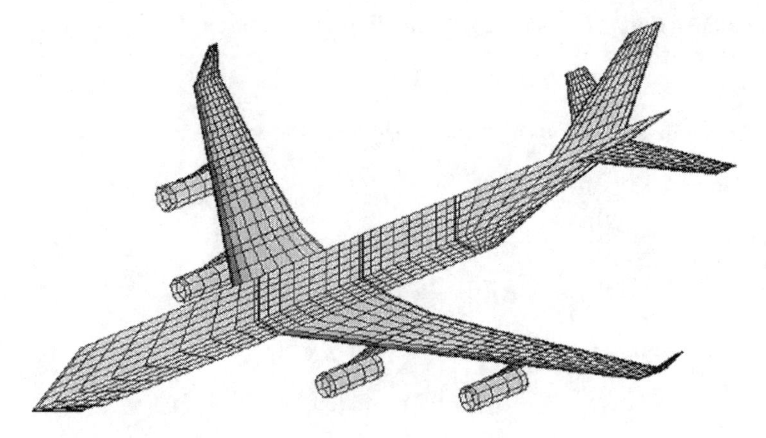

Figure 3.5 Computational VLM model of airliner using a cruciform fuselage model.

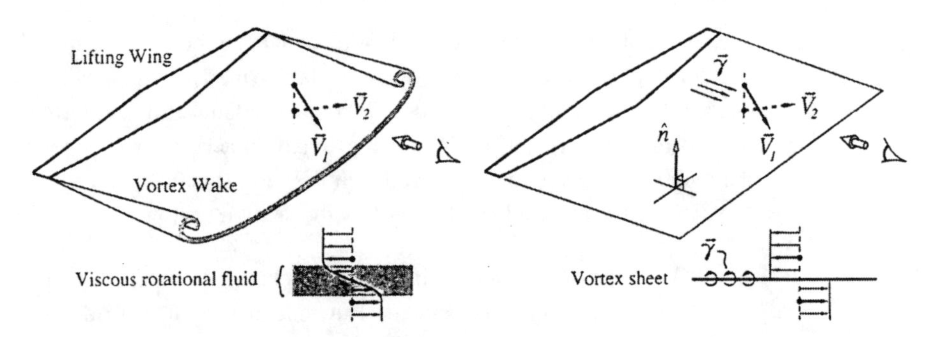

Figure 3.6 Vortex-sheet model for wing flow: velocity components in plane normal to flight direction. "Reality" (left) is discussed in Chapter 2. The vortex-sheet model (right) is the basis for VLM. (Courtesy of Mark Drela, private communication)

3.2.2 Vortex-Sheet Model for Lifting Surface

As shown in Figure 3.6, one possible model for wing flow is a lifting wing surface modeled by a vortex sheet with a vortex wake (i.e. a shear layer) that rolls up, which can then be idealized by a *rigid* vortex-sheet model in inviscid flow that is not allowed to roll up. This is our model for wing flow.

Let us develop the vortex-sheet flow model further. First, recall its velocity field (Equation (2.18)) here with vorticity concentrated to a sheet S.

$$\mathbf{v}(\mathbf{x}) = \int_S \gamma(\xi) \times \frac{\mathbf{x} - \xi}{4\pi|\mathbf{x} - \xi|^3} dS \tag{3.2}$$

$\mathbf{v}$ is discontinuous across the sheet, and the jump $\Delta\mathbf{v}$ is as follows.

$$\Delta\mathbf{v} = \gamma \times \hat{\mathbf{n}} \tag{3.3}$$

(This can be proved somewhat laboriously by actual computation.)

This jump relation is useful, for instance, in computing the force on a vortex sheet modeling a thin lifting surface. By the Bernoulli relation,

$$\Delta p/\rho = 1/2(|\mathbf{v}_+|^2 - |\mathbf{v}_-|^2) = \mathbf{v}_A \cdot \Delta\mathbf{v}$$

where the mean velocity in the sheet is

$$\mathbf{v}_A = 1/2(\mathbf{v}_+ + \mathbf{v}_-)$$

so that

$$\Delta p/\rho = \mathbf{v}_A \cdot \gamma \times \hat{\mathbf{n}} = \hat{\mathbf{n}} \cdot \mathbf{v}_A \times \gamma$$

The pressure force is normal to S so we may write

$$\Delta p = \hat{\mathbf{n}} \cdot \mathbf{F} \text{ where } \mathbf{F} = \rho\mathbf{v}_A \times \gamma$$

where the force per unit area is $\mathbf{F}$. If the vorticity is now further concentrated from a quadrilateral of spanwise extent Δy and chordwise Δx to a line vortex in the sheet, aligned of course with γ, there obtains the force per unit span

$$\Delta x \rho \mathbf{v}_A \times \gamma$$

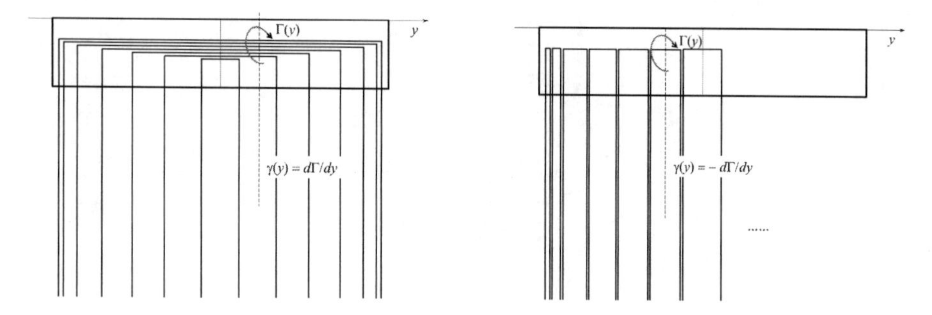

Figure 3.7 Horse-shoe vortices (left) in a vortex lattice (right) to model a wing.

which is the Kutta–Zhukovsky "law." It was derived first in 2D where the total circulation around the airfoil equals the strength of the vortex Γ needed to fulfill the Kutta condition at the trailing edge. Note for the wake pressure jump to vanish, as it should for a steady wake, $\mathbf{v}_A$ must be parallel to γ (i.e. the vorticity is aligned with the downstream flow).

Lifting-Line Theory and Horseshoe Vortices

The thin lifting surface in steady flow can now be represented by a system of *horseshoe vortices* where the trailing vortices represent the wake vortex sheet (see Figure 3.7). The left-hand image in Figure 3.7 is the common illustration motivating the "horseshoe" shape. The right-hand image in Figure 3.7 is the arrangement used in the VLM with shapes that might more aptly be called "hairpins." In the limit, as the number of horseshoe vortices goes to infinity, the trailing wake becomes a sheet of vorticity. The trailing vortex strength per unit length in the y-direction (i.e. the vorticity strength) is the spanwise derivative $\gamma = -\frac{d\Gamma}{dy}$ of the total circulation on the wing at that station. From this model, we can derive the basic relations for finite wings. Since the wing circulation changes most quickly near the tips, the trailing vorticity is strongest in this region. This is why we see tip vortices.

The flow field w_i induced by the trailing vortex sheet has several very significant consequences.

- It creates downwash which *changes the inflow angle* from the geometric angle of attack of the airfoil section, α, to an effective angle, $\alpha_{eff} = \alpha + \Delta\alpha$ (see Figure 3.8). This changes the lift-curve slope and has many implications.
- It exerts *induced drag*. In 2D steady flow, there is no trailing vorticity, by the Kutta condition. The aerodynamic force is normal to the free stream, so there is no drag. However, in 3D, the flow at the position of the bound vortex is deflected downward due to the trailing vortex sheet, and the total wing force is tilted aft, causing the induced drag, as shown in Figure 3.8. The figure shows an elliptic lift distribution with the same effective angle of attack across the span. The magnitude of the downwash can be estimated using the Biot–Savart law and the formula in Section 3.3.3. This drag is significant enough to merit consideration.

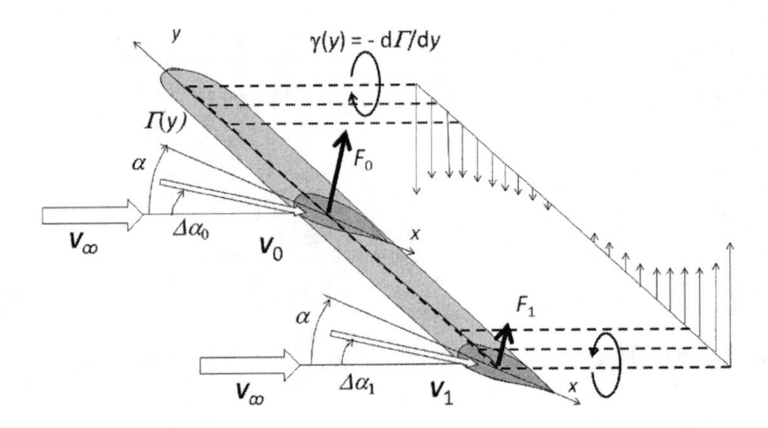

Figure 3.8 Influence of downwash on wing velocities and forces and tilting of lift vector.

3.2.3 Vortex Lattice Model

Prandtl's lifting-line theory uses a single vortex along the wingspan (e.g. presented in [2]) and applies only to high-*AR* wings. Jones' method [11] applies only to wings of small *AR*. Neither theory could predict the chordwise and spanwise distributions of the load over the surface, nor the effects of control surface deflections. A method is needed that can be used for all *AR*s, and that can describe the load distribution. This can be accomplished by covering the thin lifting surfaces, idealized as curved lamina, by discrete vortices. The VLM does this, and it applies to combinations of lifting surfaces, such as a wing with flaps or a wing and horizontal and vertical tail.

(0) The lifting surfaces are replaced by surfaces, such as the camber curves for airfoils.
(1) The surfaces are divided into elements or panels.
(2) The influence on the flow field from the lift of a panel is simulated with a bound vortex: (1) a ring around the panel; or (2) a hairpin with the short leg somewhere on the panel. The hairpin's long leg need not be straight.
(3) For every panel, the tangential flow condition is satisfied at one point: the collocation point.
(4) The assembly of flow tangency conditions gives a system of equations, describing the relation (1) between the camber and the vortex strengths, hence (2) between the camber and aerodynamic pressure forces.

Ring vortices can be put around each panel, on the camber surface, and follow control surfaces as they are deflected. But most codes use hairpin vortices, whose long legs should follow the wing camber surface to simulate attached flow. To do this, the long legs must be broken into segments. The resulting bookkeeping is complicated, so most codes put the hairpin vortices on a flattened surface, endowed with the normals of the real camber surface. Using this method, control surface deflections are also modeled (Figure 3.9). We shall return to this issue in Chapter 11, where the deformation of surfaces under aerodynamic loads is considered. Downstream of the wing, the free

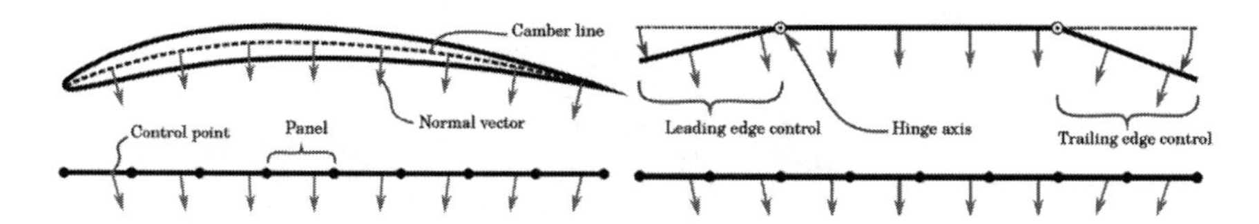

Figure 3.9 Camber and control surface deflection modeled by normal rotation.

vortices should really be allowed to develop according to their own dynamics to "relax" the wake. For a steady-flow model, they are usually assumed to be straight and fixed to follow the free stream, or placed in the wing reference plane.

It remains to place the vortex on the panel and the collocation point. We can start by considering the vortex rings for which the natural collocation point is the center of the panel, giving "second-order" accuracy of the approximation to the continuous problem. If the panel edges line up with the free stream, the set of ring vortices can be replaced *exactly* by the horseshoe vortices we have been discussing. Clearly, the Kutta condition states that there shall be no vortex on the trailing edge. Glauert's thin airfoil theory places the transversal leg on the 1/4-chord line of the panel. It follows that the shift of 1/4 panel puts the mid-ring collocation point at the 3/4-chord line, and that, too, comes out of the Glauert theory, as shown below.

3.2.4 Vortex and Collocation Point Placement

Glauert's thin airfoil theory in section 4.8 in Ref. [2] describes the chordwise vortex distribution γ over a cambered thin airfoil.

$$\gamma(\theta) = 2V_\infty \left(A_0 \frac{1 + \cos\theta}{\sin\theta} + \sum_{n=1}^{\infty} A_n \sin n\theta \right) \tag{3.4}$$

$$x = \frac{c}{2}(1 - \cos\theta), 0 \le \theta \le \pi$$

$$A_0 = \alpha - \frac{1}{\pi}\int_0^\pi \frac{dz}{dx} d\theta, A_n = \frac{2}{\pi}\int_0^\pi \frac{dz}{dx}\cos n\theta d\theta$$

where V_∞ is the free stream velocity and $x, z(x)$ are the airfoil camber-curve coordinates.

For a flat plate, $dz/dx = 0$, so $A_0 = \alpha$ and $A_n = 0, n > 0$. The center of gravity for this distribution lies in the point $x = c/4$. Replacing $\gamma(\theta)$ with the resultant $\Gamma = V_\infty \pi c \alpha$ at $x = c/4$, lift force per unit span and moment remain the same (Figure 3.10, left). The flow tangency condition is satisfied in one point, on the distance kc from the leading edge.

$$w(x = kc) = -\frac{\Gamma}{2\pi(kc - c/4)} + V_\infty \alpha = 0 \Rightarrow \quad k = 3/4.$$

If the airfoil chord were to be divided into several parts – N panels – and on each panel i is put a vortex Γ_i, it can be shown that the boundary condition is satisfied at

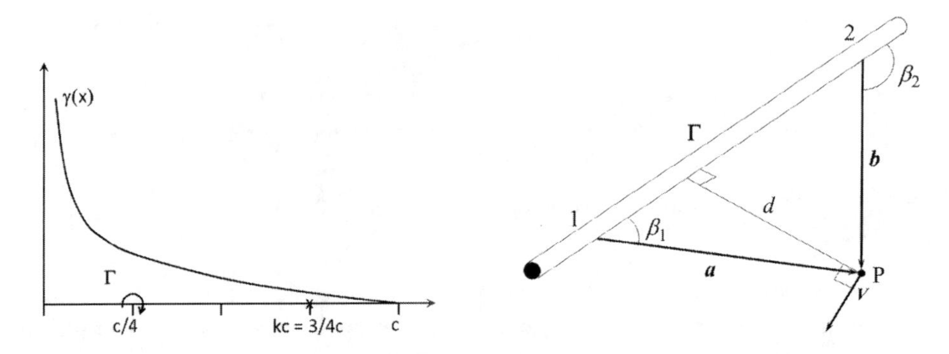

Figure 3.10 Substitution of the distributed vorticity by a point vortex (left). Definition of variables according to the Biot–Savart law, (right).

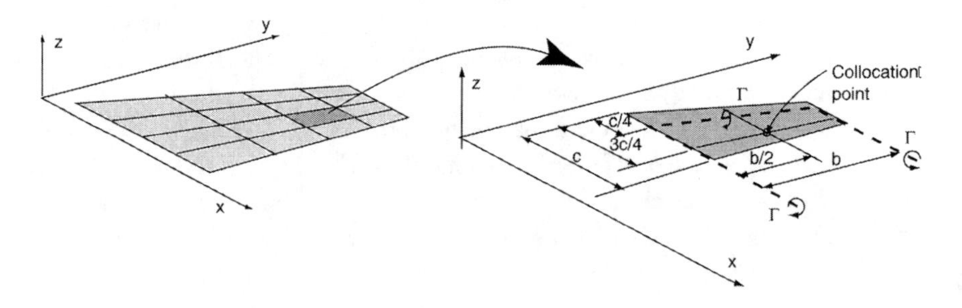

Figure 3.11 Wing divided into vortex panels; bound vortex and collocation point.

the 3/4 local chord on each panel. It will be assumed that this argument, which is valid for a section, can be extended to apply to a wing. This hypothesis cannot be proved formally, but its success in a large number of applications provides strong support for it. A wing divided into a system of panels and the effect from each panel simulated by a horseshoe vortex is shown in Figure 3.11.

3.2.5 Biot–Savart Law

The Biot–Savart law provides the increment $d\mathbf{u}$ to the velocity at $\mathbf{r}$ from a part $d\mathbf{l}$ at $\mathbf{r}'$ of a vortex filament of strength Γ.

$$d\mathbf{u} = \frac{\Gamma}{4\pi} \cdot \frac{d\mathbf{l} \times (\mathbf{r} - \mathbf{r}')}{|\mathbf{r} - \mathbf{r}'|^3}$$

which is a specialization of Eq. (2.18). The relation admits easy integration to give the circumferential velocity, V_θ, due to a portion of a vortex line.

$$V_\theta = \frac{\Gamma}{4\pi d}(\cos\beta_1 - \cos\beta_2)$$

where d, β_1, and β_2 are defined in Figure 3.10, right. An infinite straight vortex gives the 2D result, $V_\theta = \frac{\Gamma}{2\pi d}$.

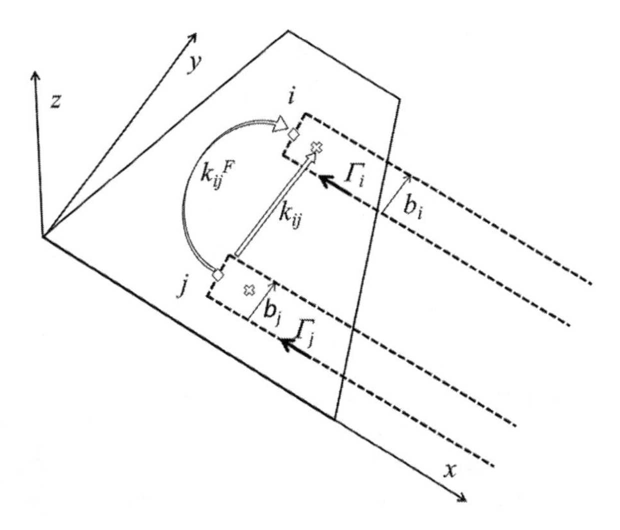

Figure 3.12 Vortices i and j, collocation points x, on wing camber surface in (x, y) plane.

In vector notation, one obtains for the induced velocity $\mathbf{v}$, using $a = |\mathbf{a}|$ etc.

$$\mathbf{v} = \frac{\Gamma}{4\pi} \cdot \frac{\mathbf{a} \times \mathbf{b}}{ab + \mathbf{a} \cdot \mathbf{b}} \left(\frac{1}{a} + \frac{1}{b} \right) \tag{3.5}$$

3.2.6 System of Equations

The horseshoe vortex with index j induces a velocity w_{ij} in the collocation point i (see Figure 3.12).

$$w_{ij} = k_{ij} \Gamma_j$$

where $K = \{k_{ij}\}$ is the aerodynamic influence coefficient matrix that can be evaluated from Eq. (3.5). All of the vortices together give the following.

$$w_i = \sum_{j=1}^{N} w_{ij} = \sum_{j=1}^{N} k_{ij} \; \Gamma_j$$

In each collocation point, the no-flow-through boundary condition is applied with free stream $\mathbf{U}_\infty = V_\infty(\cos\alpha, 0, \sin\alpha)$.

$$(V_\infty \cos\alpha, 0, V_\infty \sin\alpha + w_i) \cdot \hat{\mathbf{n}}_i = 0$$

This is simplified to neglect second-order terms in α.

$$V_\infty (dz/dx)_i - V_\infty \alpha = \sum_{j=1}^{N} k_{ij} \; \Gamma_j .$$

The force on the bound vortex i is computed by the Kutta–Zhukovsky formula from the free stream and the induced velocity at its midpoint, $w_i^F = \sum_{j=1}^{N} k_{ij}^F \Gamma_j$, the "F" superscript reminding us that the induced downwash is different on the bound vortex than on its associated collocation point where the no-flow-through condition is applied. The lift force on panel i is, to first order in α, $L_i = b_i \rho V_\infty \Gamma_i$, where b_i is the span of the panel. In dimensionless form, it is $(\Delta C_p)_i = L_i/(q\ b_i\ c_i)$, where c_i is the panel chord and $q = \rho V_\infty^2/2$, so for all of the panels, a linear system of equations appears

$$\{(dz/dx) - \alpha\} = 1/2K\{\mathbf{c}\Delta\mathbf{C_p}\} \tag{3.6}$$

- The *design task* is to compute (by matrix multiplication) the slope of the wing camber surface, dz/dx, for a given load distribution, $\Delta\mathbf{C}_p$.
- The *analysis task* is to solve the system of equations for the vortex strength/pressure differences for a given wing camber surface shape $z(x, y)$ and then to compute forces and moments.

  The lift is $L = \sum L_i$ and the y-aerodynamic moment around $(x_{ref}, 0, 0)$ is $M = \sum L_i(x_i - x_{ref})$.

  It was found in Section 3.2.2 that the wake causes induced drag. For a flat plate at angle of attack, the resultant force from the pressure difference between the suction and compression sides is perpendicular to the surface. Yet, $D/L = \tan\alpha$ is wrong. Still, the force on a panel can be computed with VLM from the Kutta–Zhukovsky force on the bound vortex, as indicated in Section 3.2.2. VLM gives correct induced drag as a result of proper approximation of the pressure singularity appearing at the leading edge; The trailing edge has no singularity because of the Kutta condition. Ref. [7] discusses these issues in more detail. The lift-induced drag D_i is actually a quadratic function of all of the Γ_i

  The drag can also be computed from the momentum deficit in the flow far downstream. The Trefftz plane analysis is described in Section 3.2.8.

  The wing bending moment along the span is $M(y) = \sum_{y_i \geq y} L_i(y_i - y)$ and is used to incorporate structural aspects in wing-shape optimization. A thin wing needs heavier structural elements to absorb the bending load.

3.2.7 VLM Software

Many implementations have been made of variants of the VLM described above. A short list is given here, from Public Domain Aeronautical Software: www.pdas.com/index.html.

- Free, or GNU public license
  - AVL, initially developed in the MIT Athena project in the 1980s by M. Drela and H. Youngren,

http://web.mit.edu/drela/Public/web/avl/. The code is still under development.
  - TORNADO, http://tornado.redhammer.se/index.php, developed by T.Melin at KTH in the 1990s.
  - LAMDES, J. Lamar's code [13] for minimal induced drag (Fortran source and possibly Windows executable).
- Commercial
  - VLAERO+, Analytical Methods, Inc.,
  - www.flightlevelengineering.com/surfaces.
  - ZAERO from ZonaTech, www.zonatech.com/index.html.

This text uses the open (GNU public license) TORNADO code. The documentation available from the code website is somewhat outdated; the book's website has an updated manual for running TORNADO, practical hints on how to model a given geometry in VLM, and how the results converge as the number of panels is increased.

3.2.8 Far-Field Analysis of Drag

The forces can be computed by integration over the wetted surface of *pressures* and stress tangential components, computed by CFD or measured. This is accurate for lift force but more dubious for drag, where the final result is a serious cancellation of the positive and negative contributions. Total forces can also be computed from the flow on a control surface enclosing the aircraft – the Trefftz analysis discussed below.

Put a control surface around the airplane, Figure 3.13. First, indent it so that all concentrated vorticity is excluded from its enclosed volume V. Integration by parts (Gauss's law) of the conservative form of the Euler equations of momentum balance over V produces the following.

$$\int_S \left(\rho(\mathbf{u} - \mathbf{U}_\infty)(\mathbf{u} \cdot \hat{\mathbf{n}}) + (p - p_\infty)\hat{\mathbf{n}}\right) dS = 0$$

The $\mathbf{U}_\infty$ and p_∞ terms have been added; their integrals vanish. The integral over the lifting surfaces becomes the force on the wing. The integral over the wake vanishes because $\Delta p = 0$ and there is no mass flow through it, and the result is that the force on the lifting surfaces matches the integral over the far boundary, no matter if it is close to the aircraft. Actually, viscous forces $\nabla \cdot \tau$ could have been included, assuming that they vanish quickly enough to make no contribution on the far boundary. Now $\mathbf{u} = \mathbf{U}_\infty + \mathbf{v}$, where $\mathbf{v}$ is the velocity perturbation created by the lifting

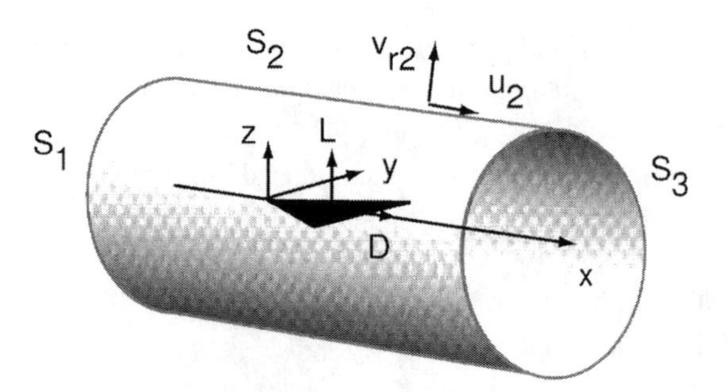

Figure 3.13 Airplane inside a control volume.

surfaces and wake. This is so small that we set $\rho = \rho_\infty$ everywhere. The energy equation states $p_\infty - p = \frac{1}{2}\rho(|\mathbf{v} + \mathbf{U}_\infty|^2 - U_\infty^2) = \frac{1}{2}\rho(v^2 + 2\mathbf{v} \cdot \mathbf{U}_\infty)$ So, finally,

$$-\mathbf{F}/\rho = \int_S \left(\mathbf{v}(\mathbf{U}_\infty + \mathbf{v}) \cdot \hat{\mathbf{n}} - \left(\frac{1}{2}(v^2 + 2\mathbf{v} \cdot \mathbf{U}_\infty) \right) \hat{\mathbf{n}} \right) dS \tag{3.7}$$

The minus sign appears because we look for the aerodynamic force on the surface, not the converse. Choose $S = S_1 \cup S_2 \cup S_3$ in the manner of Figure 3.13, remembering to cut out the wake as it cuts the downstream plane, S_3. This, in the vocabulary of aerodynamicists, is the *Trefftz* plane. The cylindrical surface S_2 is far away from the aircraft and its trailing wake and has its axis parallel to $\mathbf{U}_\infty$, so we choose that as the x-direction.

Low-Speed Induced Drag

As the radius of S_2 is increased and S_1 is sent to $x = -\infty$, on $S_1 \cup S_2$, the integrand vanishes quickly enough that the integral also vanishes. There remains only the contribution from S_3 with normal $\hat{\mathbf{x}}$.

On S_3, the velocity components are $(U_\infty + u_3, v_3, w_3)$. We obtain, since $\mathbf{F} = (D, Y, L)$ with Y the sideforce,

$$\frac{D}{1/2\rho} = \int_{S_3} (v_3^2 + w_3^2 - u_3^2) dS$$

and since $u^2 \ll v^2 + w^2$ on S_3,

$$D/q = \frac{1}{U_\infty^2} \int_S (v^2 + w^2) dS$$

q is dynamic pressure. This formula shows that drag is non-negative, which was hard to see from the VLM Kutta–Zhukovsky expression.

Ref. [7] shows how, assuming the wake is flat, this surface integral equals a line integral over a cross section of the wake, with contributions from both sides.

$$D/q = -\frac{1}{U_\infty^2} \int_{-b/2}^{b/2} \Gamma(y) w(y) dy$$

where b is the span of the wing and of the wake, since the trailing vortices are supposed to be parallel. Γ is the total circulation at y such that $\gamma(y) = -d\Gamma/dy$, and w is the vertical velocity at downstream (half) infinity. Since w is linear in γ, we see again induced drag as a quadratic function of the vortex strengths.

3.3 Planform Design Studies with VLM

The key to the success of any commercial transport aircraft is its ability to carry the design payload over the required range with maximum efficiency and safety at a competitive speed. A major contribution comes from the quality of the wing design.

It is important to realize that the commercial transport is not a single operating point configuration. Due to the assignment of cruise altitudes and the limited number of step altitude changes allowed by air traffic control during a given flight, and the reduction in aircraft weight due to fuel burn, the cruise lift coefficient throughout a given flight typically varies by as much as ± 0.1. Admittedly, this is a much smaller variation than a combat aircraft, but nevertheless it cannot be ignored.

Some of the performance criteria that must be met in the design of an efficient transonic transport wing include the following.

(1) Good drag characteristics (parasite, induced, wave/compressibility) over a range of lift coefficients at cruise (i.e. $C_{L,design} \pm \delta$)
(2) Buffet boundary high enough (1.3g margin required) to permit cruising at design lift coefficients
(3) No pitch-up tendencies near the stall or buffet boundary
(4) No unsatisfactory off-design performance
(5) Must be structurally efficient to minimize weight and provide sufficient space to house main landing gear, fuel, etc.

3.3.1 Span Loading

Before we move on, consider the relation (discussed in Section 3.3.3) for lift-induced drag Eq. (2.6), repeated here for convenience.

$$\frac{D_i}{W} = \frac{k'}{\pi}\frac{L^2}{b^2} = \frac{n^2 k'}{\pi q} W/b^2$$

n is the load factor, 1 for straight and level flight. The efficiency factor $k' \geq 1$ accounts for deviation from elliptic loading and varies little with planforms of conventional sweep and taper ratio. Hence, the induced drag-to-weight ratio is primarily proportional to the span loading W/b^2.

The cost in terms of drag for producing lift becomes less as the wingspan is increased. Unfortunately, large span means increased weight, which calls for greater lifting power. For a wing of a given planform, the lower the span loading and the faster and higher the aircraft flies, the lower the induced drag. This had been predicted by Lanchester and Prandtl, but it was Max Munk, who left Germany for the USA in the 1920s, who gave the name "induced drag" to the drag caused by trailing vortices [16]. Drag due to lift constitutes approximately 40% of the total drag of typical transport aircraft at cruise conditions, and as much as 80% of the drag at takeoff.

Simple theoretical results for induced drag have been available since the earliest days of aircraft development. Achieving minimal vortex drag is a major concern of aircraft designers, and estimating the theoretical minimum, subject to a variety of constraints and simplifying approximations, has been an important topic in aerodynamics since its inception.

3.3.2 *AR*, Sweep, and Taper Effects at Low Speed

The wing planform influences the lift distribution in other systematic ways. The *section lift coefficient distribution* $c_\ell(y)$ indicates wing stall properties, whereas the *span loading parameter* (*SLP*) indicates the section lift. It is linked to the distribution of the wing bending moment, which is important for structural design.

Section Lift Coefficient and SLP

The section lift coefficient $c_\ell(y)$ is the lift per unit span at span coordinate y, normalized by the chord $c(y)$ and the dynamic pressure q.

$$c_\ell(y)c(y) = \int_C \frac{p(x, y) - p_\infty}{q} n_L ds = \int_C C_p(x, y) n_L ds \tag{3.8}$$

where C is the closed section contour for constant y, n_L is the wing normal component along the lift axis, and ds is the arc length differential along C. The wing lift coefficient becomes, with S = wing area,

$$C_L S = \int_{-b/2}^{b/2} c_\ell(y)c(y)dy$$

$SLP(y)$ is the section lift nondimensionalized by the product of C_L and the mean aerodynamic chord MAC.

$$SLP(y) = \frac{c_\ell(y)c(y)}{C_L MAC} \tag{3.9}$$

Sweep and Taper

Figure 3.14 shows SLP for quadrilateral wings of *AR* 5 with 1/4-chord sweep 0°, 30°, and 60° and taper ratios 0.2, 0.6, and 1.0. The section lift coefficient peak moves outboard with increased sweep and *reduced* taper ratio, and its overall level is influenced by the sweep. Wing loading SLP is a smoother distribution that moves outboard with increased sweep *and* taper ratio. Note that SLP is normalized to be insensitive to total lift, which is clearly seen for taper ratio 1 to the right in Figure 3.14.

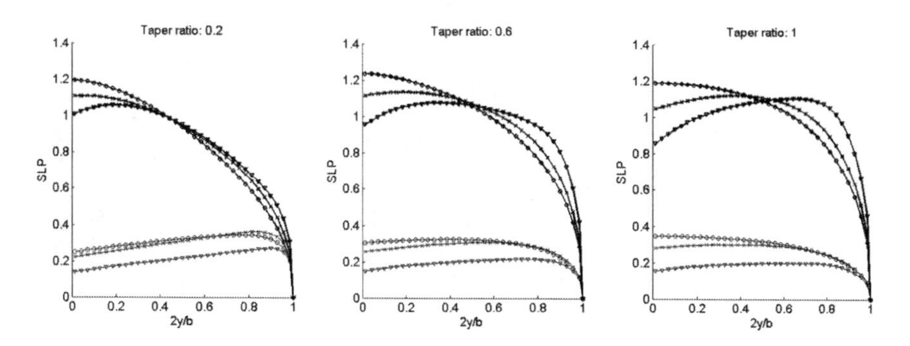

Figure 3.14 $SLP(y)$ (upper curves) and local section lift coefficient $c_\ell(y)$ (lower curves) for different sweep and taper ratios. Sweep 0°: o, 30°: x, and 60°: ∇.

Heuristic Stall Prediction

The following discussion assumes that wing sections work like airfoils. We know that this model is flawed; nevertheless, it gives useful information. The idea is that stall is initiated where c_ℓ first exceeds its section $c_{\ell,max}$. A wing that stalls first at the tip is considered undesirable, as discussed in Section 2.7. In this sense, a rectangular wing is safer than a tapered wing. It also has higher induced drag, since its lift distribution is farther from elliptic, but the differences are quite small.

A plane elliptical wing has a constant local lift coefficient, and all wing sections stall simultaneously. A tapered wing stalls first at the wing tips, which is bad from an airplane-control point of view. A rectangular wing is most loaded at the root and would stall there first. This is better because airflow disturbances from the beginning of the stall will shake the tail and warn the pilot. The taper ratio also effects the drag. High-*AR* rectangular wings have about 7% more induced drag than the optimal elliptic wing. A taper ratio of 0.45 gives only about 1% more drag than the elliptic wing. Lower taper ratios lead to a lower wing weight, to an internal volume that is more useful for fuel, and to a larger root chord for accommodating the landing gear. As a compromise between induced drag and weight, a taper ratio smaller than 0.45 may be selected – the SAAB SF340 studied in Chapter 9 has a taper ratio of 0.375.

The actual planforms for the Spitfire, Me-109, and B-58 are shown in Figure 3.15 with section lift coefficient $c_\ell(y)$, an assumed stall limit $c_{\ell,max}$ indicated by a dashed line, the $SLP(y)$ distribution, and an elliptic distribution (thin dashed line) with the same total lift. The elliptic wing (Figure 3.15, left) has a constant section lift coefficient and elliptic lift distribution. The polygonal approximation to the wing tips is responsible for the small wiggles of c_ℓ at the tips. The Me-109 wing (Figure 3.15, middle) has slightly larger induced drag and a section lift coefficient growing to a maximum at about 70% span. The B-58 (Figure 3.15, right) a Mach 2 strategic bomber of the early 1960s, had a 60° swept delta wing to keep the leading edge subsonic. Its section lift coefficient rises to a strong peak at the tips, which signals problematic stall behavior. This was indeed the case: landing attitude was around 14° and stall set in at 17°. There is also vortex lift, and the question is when the vortex will burst and the wing stall. VLM does not model the vortex lift and cannot predict vortex burst.

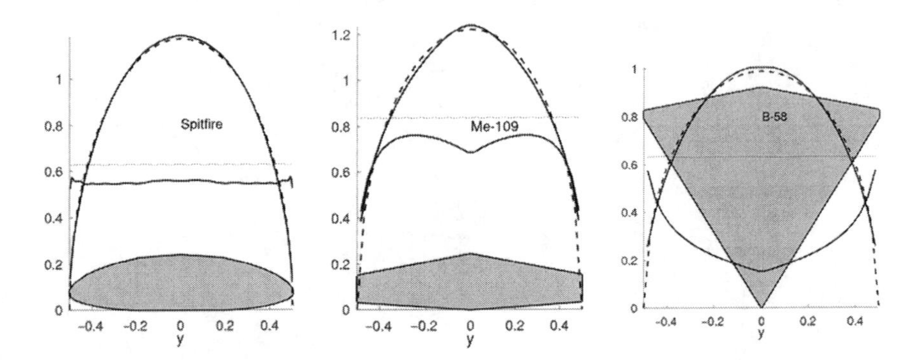

Figure 3.15 Section lift coefficient $c_\ell(y)$ and $SLP(y)$ (full line), elliptic lift distribution (dashed line), $c_{\ell,max}(y)$ (thin dashed line), and planform shape for the Spitfire, Me-109, and B-58 (left to right).

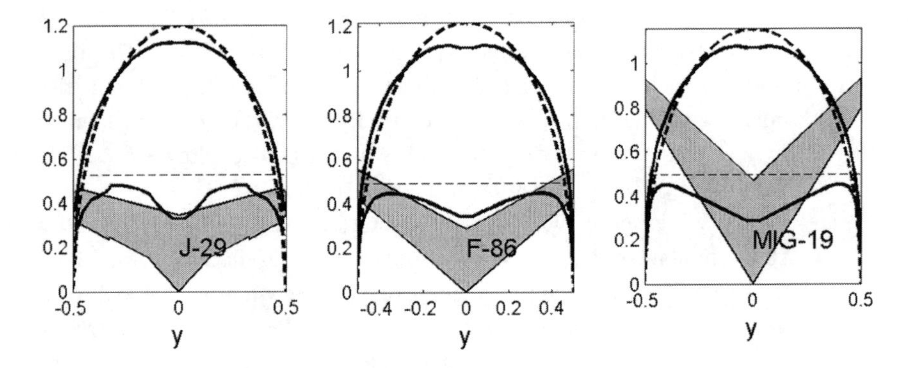

Figure 3.16 Section lift coefficient $c_\ell(y)$ and $SLP(y)$ (full line), elliptic lift distribution (dashed line), $c_{\ell,max}(y)$ (thin dashed line), and planform for J-29, F-86, and MiG-19 wings (left to right).

The effect of sweep (and taper ratio) on the lift distribution and $SLP(y)$ is shown for the wings of the SAAB J-29, North American Aviation F-86, and the Mikoyan/Gurevich MiG-19 in Figure 3.16, with sweep increasing from left to right. The trend in loading and section lift coefficient for increasing sweep is obvious. Note also the jump in section lift coefficient at the sawtooth on the J-29 wing, while SLP looks smooth. The section lift coefficient curve slope has a break at the junction with the leading edge root extension.

Wing Design for Maneuvering

Span loading is the determining factor for drag in cruise lift condition. If the goal is to obtain efficient lift at high lift coefficients using attached flow design, the emphasis switches from low induced drag to a span loading that pushes each section lift coefficient to its limit. Thus, if the planform is a simple plane trapezoidal planform with a single airfoil section, the goal is to attain a constant section c_ℓ across the wing. The penalty for the nonelliptic spanload is small compared to the additional profile drag for airfoils operating past stall on portions of the wing.

Two other considerations need to be addressed. Wings designed to operate over a wide range of conditions can use automatic scheduled deflection of the leading- and trailing-edge devices to approximate the optimum wing shape. Although research has been done on smooth elastic surfaces to achieve this, in most cases the devices are rigid flaps and slats. With proper arrangement of the link mechanism, the slats can be deployed automatically through the aerodynamic forces.

The second consideration is airfoil–planform integration. If the airfoil is designed to be heavily loaded, there is likely to be a fairly strong shock well aft on the wing at transonic speed. To obtain low drag, this shock should be highly swept. This means that the trailing edge of the wing should be highly swept. This can be achieved through inverse taper or a forward-swept wing, which is one reason to consider forward sweep.

Finally, when the airfoils operate close to stall, planform kinks are a very poor idea. The tendency of the spanload to remain smooth means that the local lift coefficients change rapidly in the kink region, often becoming excessively large. The sawtooth on the J-29 wing in Figure 3.16 is an example.

3.3.3 Span Loading and Induced Drag

Once the role of wing tip vortices in creating "lift-induced" drag was understood, the question of how to minimize it for a given lift could be addressed. Prandtl's lifting-line theory gave an answer: *for a given wingspan, the induced drag at given lift is minimized when the wing is shaped so that the spanwise distribution of the load is elliptic.*

Fuller descriptions are given in Drela [7] and Bertin [4]. If the total lift is L, its spanwise distribution – lift force per unit span – is dL/dy. Consider now the wing to be a single vortex of strength $\Gamma(y)$ along the span, shedding vorticity $d\Gamma/dy$ in the wake and from the tips to comply with the Helmholtz rules of conduct. It follows by the Kutta–Zhukovsky formula that $dL/dy = V_\infty \rho \Gamma(y)$. Determination of $\Gamma(y)$ requires solution of an integral equation that relates the local angle of attack induced by the wing itself and the wake to Γ. Prandtl found an analytical solution for elliptical $\Gamma(y)$ because it gave a constant induced angle of attack along the span. Now, with a Fourier ansatz,

$$\Gamma(y) = 2bV_\infty \sum_{n=1}^{\infty} A_n \sin n\theta, \ \cos\theta = y/s, \ 0 \le \theta \le \pi$$

it turns out that only A_1 contributes to lift. Drag becomes a sum involving all A_i^2, so the minimal drag appears when $A_i = 0, i > 1$, and then

$$C_{Di} = \frac{C_L^2}{\pi AR} \tag{3.10}$$

and the lift distribution is elliptic. The result seems to imply that we should expect vanishing drag at zero lift. This is not so; it is quite possible to cancel lift by twisting the wing, but drag contributions do not cancel in the same way. It is not clear that the optimum can be realized by some reasonable wing shape, but it was found that an elliptical planform would do so. This result also illustrates how optimal C_{Di} depends on the *AR* with the correct limit 0 as $AR \to \infty$. Lifting-line theory is not accurate for low *AR*. The span efficiency, *e*, for a real wing appears in

$$C_D = C_{D0} + \frac{C_L^2}{e\pi AR} \tag{3.11}$$

where C_{D0} is the zero-lift friction drag and $e \leq 1$. The optimum is quite flat. The World War II Spitfire had an elliptical wing planform, but its opposite number, the Me-109, had a quadrilateral wing, demonstrating only slightly increased induced drag, but being much easier to manufacture.

Minimal Induced Drag with Constraints

The celebrated Prandtl lifting-line result in Eq. (3.10) shows how induced drag decreases with increased *AR*. Gliders do have very long wings, as do aircraft designed for extreme endurance, such as the Rutan Voyager, which flew nonstop around the world. Transport aircraft prototypes have been built with *AR* around 20 (Chapter 1). However, current airliners have *AR*s around 10, and fighters have *AR*s of no more that 5. Obviously, a very long span produces strong wing bending moments, requiring strong and heavy structures, so there is a limit to the gains obtainable by increasing span.

Much work and many papers are devoted to realistic conceptual wing optimization with simple aerodynamic models. Ref. [20] considers the following:

1. **Aerodynamic model for lift**
   The "Weissinger model" (i.e. VLM with only one chordwise panel) and Prandtl–Glauert Mach scaling.
2. **Drag**
   Drag is estimated as the sum of the computed induced drag and empirical correlations for wave and viscous drag.
3. **Weight**
   Weight is estimated from wing bending moment and assumptions on wing box size, stringer, etc. Both stresses in the sizing load case, a 2.5g pull-up, and local buckling of the skins are checked.
4. $C_{L,max}$
   $C_{L,max}$ assessment is done by considering each section to behave like its airfoil at the local angle of attack. Experimental data for airfoil $c_{\ell,max}$ is used.
5. **Fuel inertia relief**
   Fuel inertia relief also works for steady flight. The gravity force on the fuel weight decreases the wing bending moment.

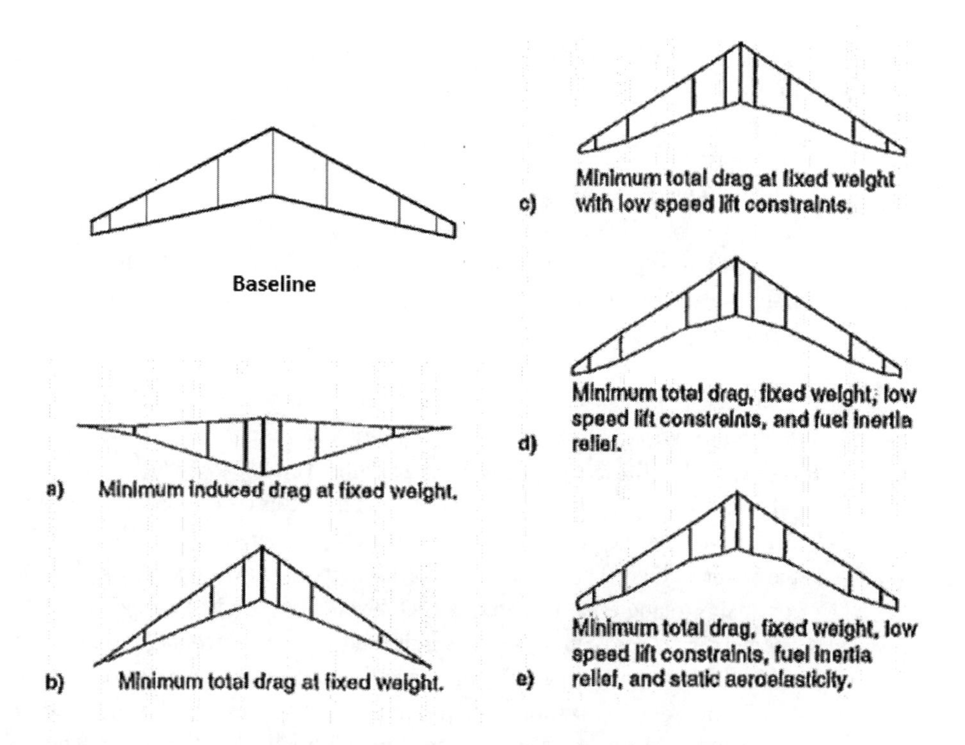

Figure 3.17 Baseline planform and effects of including the constraints successively. (Courtesy of Wakayama and Kroo [20])

6. **Static aero-elastic effect**
   The wing distorts aero-elastically under aerodynamic forces: a structural beam model of the wing is used to compute the flight shape.

Figure 3.17 shows the baseline and the optimal planforms for a fixed-weight wing when the constraints are applied successively. The point here is that simplistic constraints produce unfamiliar shapes. The fuselage is not modeled, nor are the constraints to housing the landing gear, so the shape next to the centerline is irrelevant.

3.3.4 VLM Application: Induced Drag Optimization

Lifting-line theory successfully models induced drag for high-*AR* unswept wings at low speed and low angle of attack. But delta wings, combinations with close-coupled canards, and low-*AR* wings in general, such as those seen on modern fighters, are beyond its scope. VLM also works well for low-*AR* wings and interacting lifting surfaces as long as the wake from a surface does not cut another lifting surface, which should be avoided for a number of reasons. Note also that highly swept leading edges produce strong vortices over the wing, which give lift and drag increments not modeled by VLM. VLM can be used in different ways to optimize wing shape.

Inverse Design

VLM is a classical tool for the creation of a lifting surface with given, desirable pressure distribution – the "design task" (see Section 3.2.6). We shall return to the challenging problem of finding a desirable pressure (or pressure difference) distribution in Chapter 8.

$\boldsymbol{\Gamma}$ *Optimization: Optimization of Vortex Strengths*

A VLM-specific technique is to regard the vortex strengths Γ_i as parameters to optimize with a given planform. Only the streamwise slope dz/dx will then be determined, so the geometry must be suitably constrained spanwise (e.g by freezing the leading edge). The expressions for lift and drag have appeared in Section 3.2.6, so it remains to add other constraints such as on wing bending moment, and possibly on geometry $z(x, y)$. With vortex strengths and α so obtained, dz/dx can be computed, and then the shape $z(x, y)$. The drag itself is a quadratic function of Γ_i. Thus, the task is a quadratic minimization. Lift and moment constraints are linear functions of $\boldsymbol{\Gamma}$, as is $z(x, y)$, but the geometric constraints may introduce nonlinearities. Thus, if the geometric constraints are simple, this is a convex problem with a single global minimum. J. Lamar [13] describes such an algorithm for minimizing low-speed induced drag by optimizing twist and camber for nonplanar configurations of lifting surfaces.

Multidisciplinary Optimization: Loosely Coupled Shape Optimization

Induced drag can be minimized, under suitable constraints on geometry, etc., by varying the planform, camber, and twist in a process outlined in Section 1.3.3. A formulation for cruise might be as follows.

- Minimize C_D subject to given C_L, with limits on root bending moment, C_m, wing thickness, etc., by varying:
    - Angle of attack
    - Planform (sweep and taper)
    - Camber surface (twist and camber)

Neglecting the effect of thickness, VLM can provide the forces and moments – the analysis task of Section 3.2.6 – for a given candidate shape whose parameters are the ones to optimize: its planform, camber, and twist. Trim by control surfaces and the associated drag can be included in the constraints. However, trim analysis requires knowledge of the position of the center of mass; it is customary, therefore, to do the optimization for a range of center of gravity positions. The dihedral plays little, if any, role here. It influences roll–yaw coupling, and hence the adverse yaw tendency, but these properties can be assessed only when the moments of inertia can also be estimated.

Induced Drag Planform Optimization with VLM

Induced drag contributes a significant fraction of the total wing drag. VLM describes the wing planform and camber surface and computes induced drag for quite general systems of lifting surfaces. Planform geometry variation for a single planar wing gives a quite flat optimum. We have seen above that a simple tapered straight wing comes within 1% of the minimal induced drag for a given span. Nonplanarity and combinations of several lifting surfaces offer other geometric handles and may yield more substantial gains. The American Institute of Aeronautics and Astronautics (AIAA) Aerodynamic Design Optimization Discussion Group (ADODG) Test Case 3 [1] treats twist optimization of a rectangular wing of $AR = 6.12$ in inviscid flow. The multidisciplinary optimization-type, loosely coupled induced drag minimization is formulated as follows.

$$\min C_{D,i}(\theta) \tag{3.12}$$

$$\text{Subject to } C_L \geq C_{L0}$$

$$\theta \in [-5^o, 5^o] \tag{3.13}$$

The twist angle is $\theta(y), 0 < y < b/2$. The design parameters are the spline control points for the twist profile from root to tip. The angle of attack is taken to be the twist angle of the root section at $y = 0$. A rectangular wing has too high tip loading. This can be decreased by washout to approximate elliptic loading, and the optimization problem allows for a detailed variation of the twist along the span. The gain is, however, only of the order of one drag count, from 82 to 81.

3.4 Wings for High Speed

The potential for high speed offered by the revolutionary new jet propulsion systems that had their beginnings in the 1940s, however, could in no way be realized by typical low-speed propeller-driven aircraft designs, usually characterized by straight wings having thickness ratios in the range of 14–18%. For flight in the high-subsonic, transonic, and supersonic flight regimes, however, linear component-by-component addition of drag and other time-honored tenets of airplane design needed to undergo radical change. The classic shape of the airplane of the 1940s had to be fundamentally altered to permit efficient and safe operation in these new speed ranges because compressibility effects become strong enough to provoke flow separations on the wing surface, leading to large increases in drag, as well as to stability problems and frequently to unsteady flows.

Planform designs became highly swept, as in, for example, the F-86, MiG-15 and SAAB J-29, all of which were flying by 1948. As flight speeds increased further toward sonic values and above, *AR*s became even lower and thickness decreased to offset the increased drag, evident in virtually all supersonic designs from the 1950s on.

3.4.1 Transonic Cruise: $M_{cruise} \approx 0.85$

If for moderate increases in cruise Mach number M_{cruise} there occurs no supersonic region in the whole flow, then the flow properties can be easily estimated by the Prandtl–Glauert model. But as the cruise Mach number M_{cruise} increases further, the flow past the classical airplane will become supersonic in a limited region near the surface of the wing and will eventually become supersonic everywhere, except perhaps in a small region near a rounded nose. The M_{cruise}, at which the local flow velocity at some point on the wing first reaches the local speed of sound and discontinuous flows become possible, is designated the critical Mach number, M_{crit}.

The value of M_{crit} on a wing clearly depends on both the wing geometry and the angle of attack. The appearance of shock waves is accompanied by a pressure drag force on the body as a manifestation of the change in entropy through the shock. This is rapidly followed in many cases by flow separation behind the shock wave, and this in turn frequently leads to unsteady flows due to interactions between the shock-induced separation and the trailing-edge separation. Such phenomena require higher-fidelity models, and we turn to L2 or L3 simulations for their prediction. Examples will be given in Chapters 6–9.

To some extent, these difficulties could be countered by using thinner wings, but this solution could not be taken too far for reasons of wing strength/weight and fuel volume. The great advance came from sweepback, first postulated in Germany in the 1930s and used very successfully since 1945 to improve both transonic and supersonic characteristics.

Swept Wings

The critical Mach number is dependent on the excess speed (relative to the free stream velocity), which is roughly proportional to the ratio of the body thickness to the body length d/L. For a cylinder in cross flow $d/L = 1$, a free stream Mach number of 0.5 is critical. The more slender the body, the higher its M_{crit}. Hence, one is able to shift, by means of a very thin wing, M_{crit} (and consequently the drag-divergence Mach number M_{dd}) to higher flight Mach numbers M_{∞}. But because a wing needs a certain thickness for strength reasons and to provide the required volume to accommodate fuel and control surface actuators, this approach can only be pursued in exceptional cases. The unswept supersonic Lockheed F-104 Starfighter's "thin" wing of 3% thickness ratio is such a case.

If a certain thickness is to be kept, an approach other than thinning the wing must be found to raise M_{crit} and M_{dd}. At the Volta conference in Rome in 1935, Adolf

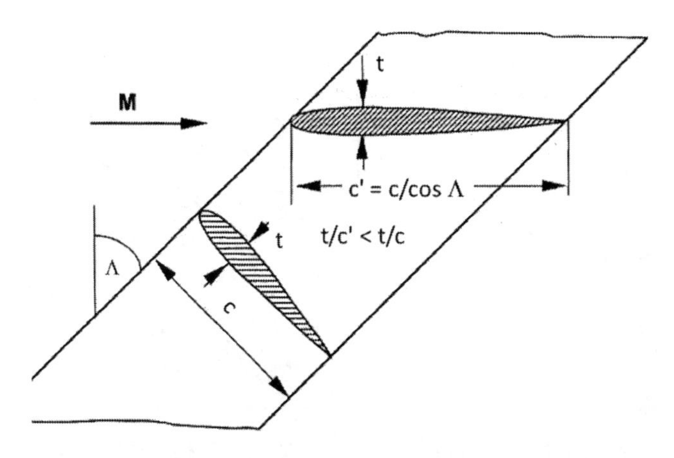

Figure 3.18 The principle of wing sweep.

Busemann theorized the significant reduction in transonic drag achieved by "sweeping" the wing (i.e, rotating it as a rigid body through the sweep angle Λ). The flow then "sees" a wing with a thickness-to-chord ratio reduced by cos Λ, and M_{crit} and drag divergence at M_{dd} occur at a higher flight Mach number M_∞ than in the case of the unswept wing (Figure 3.18).

Another way to understand the effect of sweep is to consider an *infinite* wing of constant section. Then the velocity component along the wing creates no pressure gradients since its path is flat, and thus the isobars are swept parallel to the wing leading edge. This can be "proved" by considering the Euler equations in a suitably rotated coordinate system. The Mach number $M_\perp$ normal to the leading edge is reduced, $M_\perp = M_\infty \cos \Lambda$, and hence $M_{crit,\Lambda} = M_{crit,\Lambda=0}/\cos \Lambda$.

This simple principle is obviously only true (if at all) on an infinite-span wing of constant section in inviscid flow. Nevertheless, the initial suggestions of Busemann and Betz led to wind tunnel tests that substantiated the essence of the theory. Results of the wind tunnel measurements performed at Göttingen in 1939 by H. Ludwieg are reproduced in Figure 3.19. These data show clearly that the effects of the shock waves on the wing at high subsonic speeds will appear only at higher Mach numbers on swept-back wings. Wing-end effects at the tip and root clearly invalidate the premise for the parallel isobars of the infinite-span wing. Chapter 9 discusses the design steps taken to counteract this effect and applications to example aircraft.

Prone to Tip Stall

Despite the benefits of sweep in delaying drag rise, there are penalties associated with swept wings, so that the aerodynamicist will want to use as little sweep as possible. Sweepback alters the lift distribution in much the same way as decreasing taper ratio. All outboard sections of the wing are affected by the upwash produced by preceding inboard sections. Thus sweepback – as well as taper, with which it is always combined – leads to higher local lift coefficients c_L toward the wing

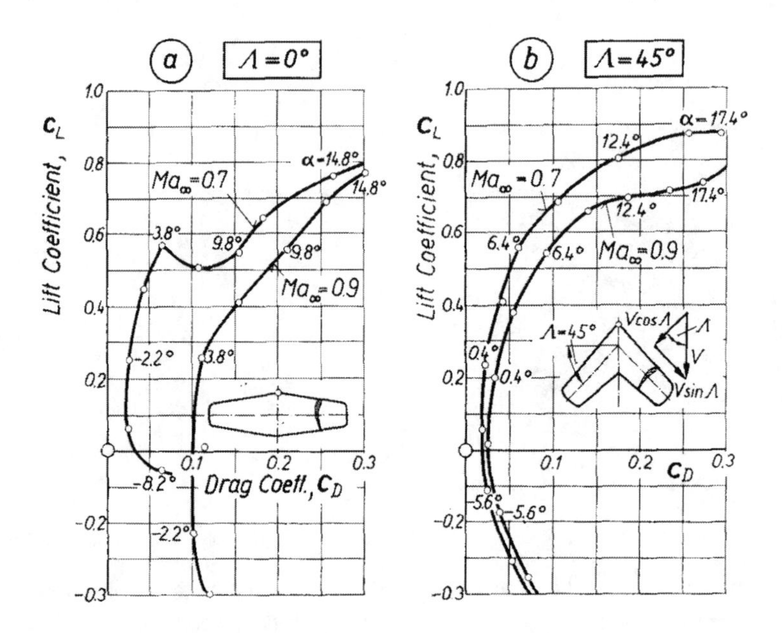

Figure 3.19 Polar curves of the unswept and swept wing at the transonic Mach numbers $M_\infty = 0.7, 0.9$. Measurements of H. Ludwieg, Göttingen, 1939. (From Schlichting [17], reprinted with permission)

tips and the possibility of outboard stall, accompanied by pitch-up, as indicated in Figures 3.15 and 3.16.

Twisting the wing tip nose down (washout) to unload the tip is one way to mitigate pitch-up. Another is to add small vortex generators to the outer wing that effectively delay flow separation by creating vortices in the boundary layer.

Example: B-47

Cook [5] explains how the Boeing B-47 is a case in point. With a wing *AR* of 6.0, a 1/4-chord sweepback angle of 35°, and a taper ratio of 0.23, the B-47 wing would be expected to suffer pitch-up, and indeed it did. However, vortex generators added to the outer wing panel reduce the pitch-up instability by delaying tip stall. Vortex-generating wing leading-edge notches/cutouts (such as a dog-tooth), the pylons carrying the underslung engines, and so-called stall-fences have similar effects.

Sweep brings in further repercussions. The lift curve slope also decreases. In addition, for a given span, the wing is longer, and hence it requires stronger structural elements to obtain sufficient stiffness. High-lift devices are less effective when the trailing edge is swept, and finally, swept wings are prone to flutter.

Thus, the total system design must be considered when selecting the wing sweep. One of the benefits of advanced airfoils is that, with less sweep, they can achieve the same performance as a wing with a less capable airfoil. This explains the general trend that modern transports have less sweep than earlier transports.

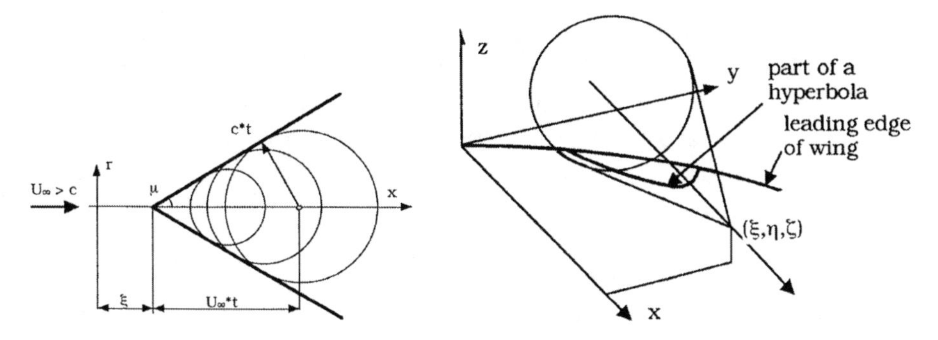

Figure 3.20 The Mach cone and region of influence.

3.4.2 Wing in Supersonic Flight: $M_{cruise} > 1$

Sweeping the wing lowers the lift at $M_{cruise} \approx 0.85$, but at $M_{cruise} \gtrapprox 1$, the maximum lift-to-drag ratio drops dramatically, primarily because of the steep drag rise. Flight in the low-supersonic speed range is not very efficient, and the design approach must contend with thin and slender configurations. This section sketches some of the issues that this involves. It begins with a brief description of the expected behavior of supersonic flow around a thin wing modeled by the supersonic PG model in Eq. (2.17), the linear wave equation.

The Mach Cone

In an airstream of velocity U_∞, where $M_\infty > 1$, consider a fluid particle following the stream. At time $t = 0$, it passes a point $x = \xi$ (see Figure 3.20, left). At that point and time, a pressure disturbance is emitted. Relative to the stream, the disturbance travels radially outward at the speed of sound to form the surface of a growing sphere. The envelope to the moving and expanding sphere is a cone with the tip in $x = \xi$ and with μ as half the tip angle, $\sin \mu = c_\infty / U_\infty = \frac{1}{M_\infty}$. In mathematical terms, this envelope is called the *characteristics* cone, which is further discussed in Chapter 4.

The equation for the Mach cone from the point $\mathbf{P} = (\xi, \eta, \zeta)$ (Figure 3.20, right), assuming free stream in the x-direction, is as follows.

$$[(y-\eta)^2 + (z-\zeta)^2]^{1/2} = (x-\xi)\tan\mu, x \geq \xi \tag{3.14}$$

where $\tan\mu = \beta = 1/\sqrt{M_\infty^2 - 1}$.

Only points (x, y, z) that are situated downstream of this conical surface can "hear" disturbances at $\mathbf{P}$, so von Kármán called the region outside of the cone "the zone of silence." The set of points whose disturbances can be heard at $\mathbf{P}$ is its domain of dependance, with the interior of the cone pointing in the opposite direction.

$$[(y-\eta)^2 + (z-\zeta)^2]^{1/2} = (\xi - x)\tan\mu, x \leq \xi \tag{3.15}$$

When the flow is supersonic, the domain of dependance is only the upstream Mach cone. A wing in the plane $z = 0$ has an intersection with the upstream Mach cone from **P**, which is a hyperbola. Only the part of the wing between the hyperbola and the wing leading edge has an effect in **P** (see Figure 3.20). This is quite different from subsonic flow, governed by an elliptic equation, where *all* of the wing is "felt" *everywhere* in the flow field. The basis for this statement to be exactly true is the "small disturbance" assumption that the speed of sound is constant and that the longitudinal velocity is everywhere U_∞.

Subsonic and Supersonic Edges

The zone of silence determines how the edges of the wing influence the relation between wing loading and incidence. The influence depends on the following

- Whether the edge is a leading or a trailing edge.
- Whether the Mach number vector component $M_\perp$ normal to the edge is less or larger than 1. Note that $M_\perp \leq 1 \Leftrightarrow \Lambda \geq \pi/2 - \mu$.

The sketches in Figure 3.21 illustrate the different flow cases for a wing with straight edges.

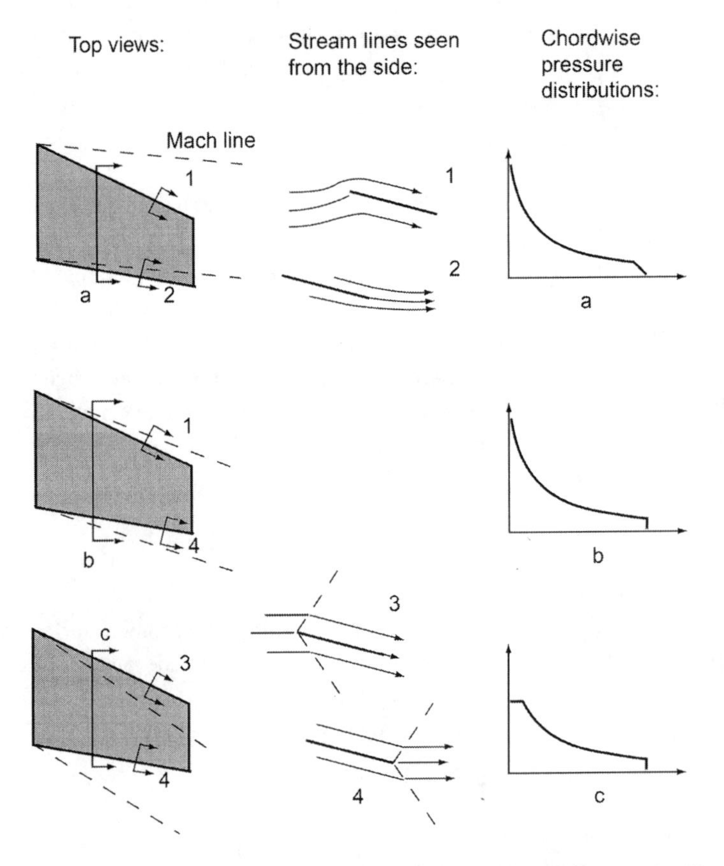

Figure 3.21 Supersonic flow over a wing: (top) subsonic leading and trailing edges; (middle) subsonic leading and supersonic trailing edges; (bottom) supersonic leading and trailing edges.

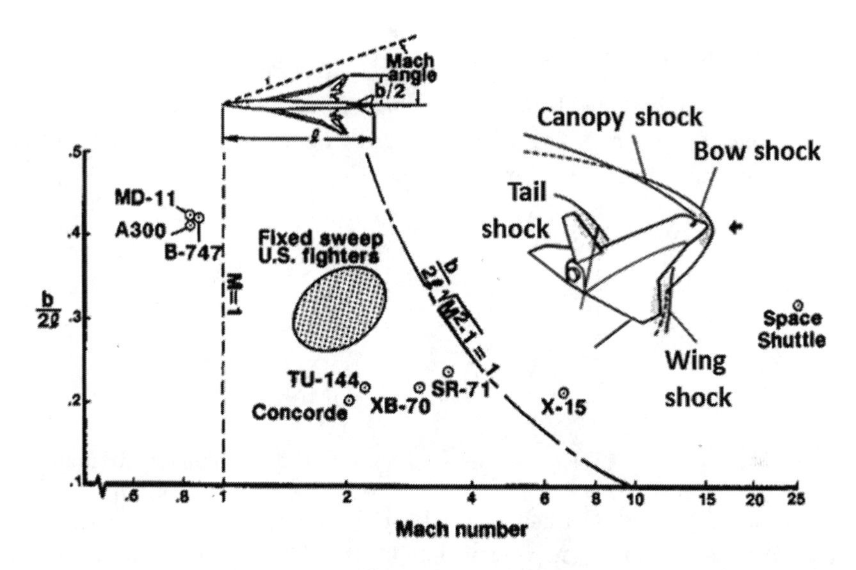

Figure 3.22 Supersonic aircraft have small slenderness ratios and subsonic leading edges. The Space Shuttle has supersonic edges with more complex shock interactions and embedded subsonic flow regions that increase wave drag. (After Harris [9], NASA public domain)

When the leading edge is swept inside the wing apex downstream Mach cone, the normal flow is subsonic and "sees" the wing coming, adapting with upwash (situation 1). With less sweep, the normal flow is supersonic, and the domain of dependance of a point on the leading edge includes only the free stream, so there is no upwash, and discontinuous slopes of streamlines (situation 3). The pressure peak becomes flatter.

Flow at a trailing edge swept inside the Mach cone sees the free stream and adapts to a smooth separation (situation 2). But with less sweep, the domain of dependance of a trailing-edge point includes no free stream, so the flow hits the trailing edge with discontinuous streamlines and a pressure jump occurs (situation 4). Chapter 6 presents flow simulations for an airfoil at different Mach numbers illustrating situations 3 and 4.

Swept arrow or delta wings are the norm for supersonic flight. A very important parameter is the slenderness ratio s/ℓ of wing half-span to total configuration length and its relation to the Mach cone angle μ, which are discussed further in Chapter 9.

Most supersonic aircraft have subsonic leading edges (see Figure 3.22).

3.4.3 Wing Shape Influence on Supersonic Drag

Chapter 8 will present computational results showing the large effect of airfoil thickness on drag in supersonic speed. Chapter 9 will continue this computational investigation by indicating how the slenderness ratio s/ℓ and planform shape have marked effects on supersonic drag, especially for a wing with a supersonic leading edge.

3.4.4 Supersonic Wave Drag

In supersonic flight, there are more drag sources than the subsonic friction, form, and induced drag terms. The minimal drag for a body with given lift L, span b, length l, and volume Vol is as follows.

$$
\begin{aligned}
D &= q\bar{C}_f S && \text{Friction drag} \\
&+ \frac{L^2}{\pi q b^2} && \text{Induced drag} \\
&+ (M^2 - 1)\frac{L^2}{2\pi q l^2} && \text{Vortex wave drag} \\
&+ q\frac{128 \cdot Vol^2}{\pi l^4} && \text{Volume wave drag}
\end{aligned}
\tag{3.16}
$$

where S is wetted area and, as usual, q is dynamic pressure. The approximation is valid for low-*AR* bodies. The volume drag looks independent of Mach-number but appears only for M > 1.

Consider again the airplane inside a cylindrical control volume to estimate the drag force (Figure 3.13). In low-speed flight, the S_2 contribution is negligible because the governing equation is elliptic and disturbances die off with distance. But the wave equation governs supersonic flight and disturbances travel along the characteristic cones forever and decrease in amplitude only slowly.

Thus, for supersonic flight S_2 contributes to the total drag, and it is this contribution D_w that is the *wave drag*. A brief discussion of it follows below.

$$
\begin{aligned}
D_w/q &= \int_0^{2\pi} d_w(\theta)d\theta, \text{ where} \\
d_w(\theta) &= \frac{-2R}{U_\infty^2}\int_{-\infty}^{\infty} v_{r2}(x,\theta)u_2(x,\theta)dx
\end{aligned}
\tag{3.17}
$$

$d_w(\theta)$ is the wave drag contribution at the peripheral angle θ along a strip on the far-field cylinder of radius R.

In the remainder of the text, we consider only that part of the wave drag that exists when lift is zero. The manipulations required to turn Eq. (3.17) into the von Kármán slender-body formula of Eq. (3.18) published in [18] are too technical for our purposes here, and instead we refer the interested reader to the exposition by Drela [7].

A source distribution on the surface of the body is required for the fulfilment of tangential flow. At a point (x, y, z) on the surface of the control cylinder, contributions from all of the sources to the perturbation velocity at (ξ, η, ζ) on the configuration inside the upstream Mach cone are summed together. The influence function for the supersonic Prandtl–Glauert (Eq. (2.17)) is proportional to the following.

$$1/\{\beta^2(x-\xi)^2 - [(y-\eta)^2 + (z-\zeta)^2]\}^{1/2},$$

$\beta = 1/\sqrt{M_\infty{}^2 - 1}$ (e.g. [7]). The dominating contributions come from the intersection of body and Mach cone, where the denominator vanishes. As the control cylinder is far

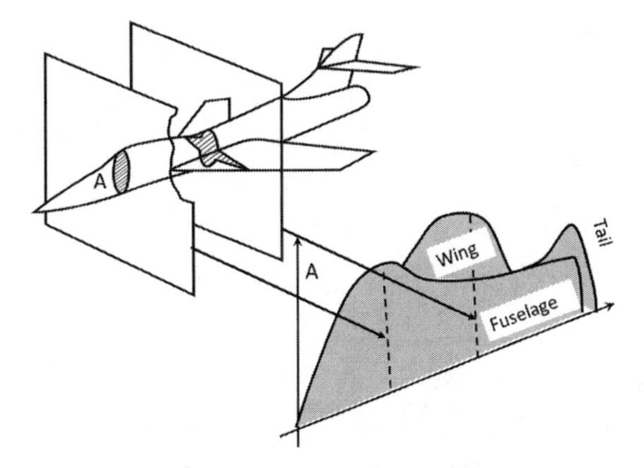

Figure 3.23 Intersection between configuration and $M_\infty = 1$ planes (i.e. yz-plane cuts through the airplane). (After Baals et al. [3], NASA public domain)

away from the airplane, the conical surface may be approximated by a plane tangential to the cone at the intersection with the body axis (Figure 3.20), making the Mach cone angle μ along with it. The sources along the intersection of the Mach plane and the body surface can be replaced by their sum on the body centerline with an error $O(d/R)^2$ with d body diameter. This works well for a slender configuration that, in effect, is replaced by an axisymmetric body, dependent on M_∞.

Sweeping for each value θ in the x-direction from $-\infty$ to $+\infty$ gives a longitudinal distribution of the projections of the intersection surface $S(x, M_\infty, \theta)$. The total wave drag becomes the foundation for the supersonic area rule.

$$D_w/q = -\frac{1}{2\pi}\int_0^{2\pi} d_W(\theta)d\theta \tag{3.18}$$

$$d_W(\theta) = \frac{1}{2\pi}\int_0^L\int_0^L S''(x,\theta,M_\infty)S''(\xi,\theta,M_\infty)\ln|x-\xi|d\xi dx$$

At Mach number 1, the intersecting planes are normal to the x-axis (Figure 3.23). The drag depends on the distribution of the cross sections of the configuration. The von Kármán formula is the basis for various optimal drag configurations, with constraints on area, volume, length, etc. Chapter 9 gives examples of designs using the principle of area ruling (if not the formula) to make the cross-section variation smooth – note the second derivative on the cross-section area – from nose to tail.

3.4.5 Area Rule: Surrogate Model for Wave Drag

Theoretical analysis and wind tunnel tests have shown that slender, pointed bodies of revolution (Eq. (3.19)) with an area distribution as in Figure 3.24 have minimum wave drag for their length and volume by Eq. (3.18). These shapes are known as Sears–Haack bodies after the engineers who initially studied them.

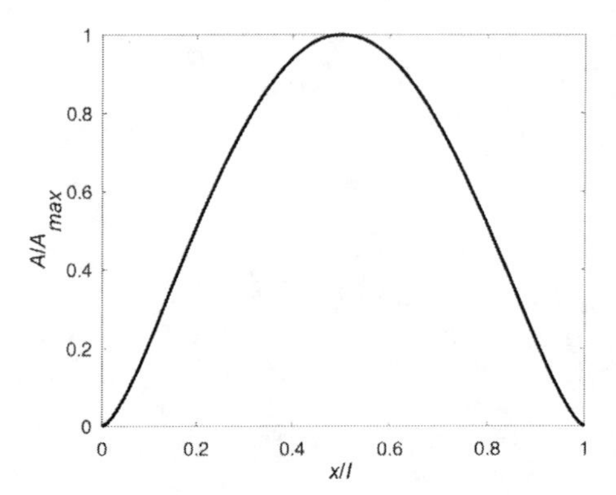

Figure 3.24 Cross-sectional area distribution of a Sears–Haack body.

$$r/r_{max} = (1 - (1 - 2x/l)^2)^{3/4}, \; C_{D_{\text{wave}}} = \frac{4.5\pi}{S}\left(\frac{A_{max}}{l}\right)^2 \tag{3.19}$$

S is the planform reference area, $A_{max} = \pi r_{max}^2$ is the maximum cross-sectional area of the body and l is its overall length. Figure 9.34 shows the shape for $r_{max}/l = 1/18$. To achieve minimum wave drag for supersonic aircraft, designers strive to make the cross-sectional areas of their designs vary smoothly, inspired by Figure 3.24. The process is called applying the area rule or just *area ruling*.

The rule was first discovered by Heinrich Hertel and Otto Frenzl working in a transonic wind tunnel at Junkers Aircraft between 1943 and 1945. It was then rediscovered independently by Richard Whitcomb nearly a decade later, in 1952, at the NACA Langley research center. He pointed out that area ruling should apply to the aircraft as a whole rather than just the fuselage, which means that the extra cross-sectional area of the wing and tail had to be accounted for in the overall shaping.

Since then, the area rule has been applied to aircraft that have to fly at transonic or supersonic speeds. The wing must perform well across the speed and lift range. To maintain the wing shape in order to avoid a bump in the area plot, the area of the fuselage must be reduced where protrusions such as the wing and pod-mounted engines are. The result is a fuselage with a *wasp waist* or *Coke bottle* shape, as is seen on early area-ruled aircraft such as the Convair F-106 Delta Dart and the Convair F-102 Delta Dagger.

The engines in today's supersonic fighters are much more powerful than were available in the 1950s. While the area rule is not as essential as it used to be, it is still used for combat aircraft to reduce cruise drag, as discussed in Section 9.5.1 on the Cycle 3 downselection of candidate configurations for the Swedish multi-role fighter JAS-39. The rule has also found great application in subsonic aircraft, particularly commercial airliners, since they cruise at the lower end of the transonic regime. A good example is the Boeing 747, known for its distinctive hump that houses the cockpit and upper passenger deck. The hump increases the cross-sectional area of the forward fuselage and has the effect of evening the volume distribution over the length of the aircraft.

It should be noted, however, as discussed in Ref. [19], that the supersonic area rule is not an exact theory. In addition to the slender body theory assumptions, the supersonic area rule assumes that an aircraft, which usually departs considerably from a body of revolution, can be represented by a series of equivalent bodies of revolution. The theory, therefore, does not account for wave reflections that may occur due to the presence of the fuselage, wing, or tail surfaces. In addition, the theory does not account for the induced lift of configurations with highly twisted and cambered lifting surfaces. At supersonic speeds, the fuselage itself generates shock waves that cause wave drag, and the shaping of the wing–body junction requires special attention.

Nevertheless, for most configurations, the supersonic area rule does account for the major part of the wave drag and provides a useful procedure for the analysis of aircraft wave drag.

This chapter developed computational methods describing *linear* aerodynamics, in particular linear compressibility. More realistically, we must be able to compute flows with shock waves. Classical numerical methods could not compute solutions containing discontinuities such as shocks, and new methods had to be invented. The next chapter gives us some insights into these methods and what the user today needs to know in order to apply them to practical problems.

3.5 Learn More by Computing

Gain hands-on experience of the computational tools for the topics in this chapter by working with the on-line resources. Exercises, tutorials, and project suggestions are found on the book website www.cambridge.org/rizzi. Go through the VLM applications which demonstrate planform design. Software used to compute many of the examples shown is available from http://airinnova.se/education/aero-dynamic-design-of-aircraft

References

[1] AIAA ADODG. Available from http://mdolab.engin.umich.edu/content/aerodynamic-design-optimization-workshop.

[2] J. D. Jr. Anderson. *Fundamentals of Aerodynamics*, 5th edition. McGraw-Hill, 1995.

[3] R. A. Warner, D. D. Baals, and R.V. Harris. Aerodynamic design integration of supersonic aircraft. Technical report. October 1968. AIAA Paper 68-1018.

[4] J. J. Bertin. *Aerodynamics for Engineers*, 4th edition. Prentice-Hall, 2002.

[5] W. H. Cook. *The Road to the 707: The Inside Story of Designing the 707*. TYC Publishing Company, 1991.

[6] J. deYoung. Historial evolition of vortex-lattice methods. In *Vortex-Lattice Utilization*, National Aeronautics and Space Administration, 1976, pp. 1–9.

[7] M. Drela. *Flight Vehicle Aerodynamics*. MIT Press, 2014.

[8] V. I. Falkner. The solution of lifting plane problems by vortex lattice theory. Technical report. R. & M. 2591 2591, British A.R.C., September 1947.

[9] R. V. Harris. On the threshold – The outlook for supersonic and hypersonic aircraft. Technical report. August 1989. AIAA Paper 89-2071.

[10] K. Huenecke. *Modern Combat Aircraft Design*. Naval Institute Press, 1987.

[11] R. T. Jones. Properties of low-aspect-ratio pointed wings at speeds below and above the speed of sound. Technical report, NACA, 1946. NACA R-835.

[12] R. T. Jones and D. Cohen. *High Speed Wing Theory*. Princeton University Press, 1960.

[13] J. E. Lamar. A vortex-lattice method for the mean camber shapes of trimmed noncoplanar planforms with minimum vortex drag. Technical Note D 8090, 1976.

[14] M. Landahl. Kernel function for non-planar oscillating surfaces in a subsonic flow. *AIAA Journal*, 5(5): 1045–1046, 1967.

[15] J. M. Luckring. The discovery and prediction of vortex flow aerodynamics. *Aeronautical Journal*, 123(1264): 729–804, 2019.

[16] M. Munk. The minimum induced drag of aerofoils. Technical report. NACA-TR 121, National Advisory Committee for Aeronautics, 1923. NACA-TR-121.

[17] H. Schlichting. Some developments in boundary layer research in the past thirty years. *Aeronautical Journal*, 64(590): 63–80, 1960.

[18] T. von Karman. The problems of resistance in compressible fluids. In *Proceedings of the Fifth Volta Congress*. R. Accad. D'Italia (Rome), 1936, pp. 222–283.

[19] R. Vos and S. Farokhi. *Introduction to Transonic Aerodynamics*. Fluid Mechanics amd its Applications. Springer, 2014.

[20] S. Wakayama and I. Kroo. Subsonic wing planform design using multidisciplinary optimization. *Journal of Aircraft*, 32(4): 746–753, 1995.

4 Finite-Volume Schemes for the Euler Equations

Shock problems are a striking example of a class of mathematical problems where the practical use of computational methods plays a significant role well ahead of the theoretical mathematical/numerical analysis.

Bertil Gustafsson[1]

Preamble

In this chapter, we show the numerical schemes used to treat flows containing shock waves as modeled by the Euler equations of gas dynamics. These developments will be described for flow in one space dimension. The following topics are covered:

- Conservation law form of governing equations
- Conservative discretization
- Constructing the numerical flux
- Artificial dissipation schemes
- Approximate Riemann schemes
- High-resolution schemes

Most of the material can also be found in more detail in Refs. [1, 5, 9, 15].

4.1 Introduction to Computing Flow with Shock Waves

The overall goal of computational fluid dynamics (CFD) in aeronautics is to solve the Navier–Stokes equations for the flow around the configuration. In order to reach this goal, we need to obtain some fundamental understanding of the mathematical behavior of these partial differential equations (PDEs) and we need to analyze their properties along with the properties of the numerical methods used to solve them. As we saw in Chapter 2, numerical solution of the nonsimplified Navier–Stokes equations is extremely laborious and beyond the reach of the computing power of the machines of even the near future. Therefore, we shall first discuss the simplifications that led the way to mathematization of aerodynamics and to successful engineering calculations.

[1] With his permission, personal communication.

Prandtl kicked off this development in 1904 when he introduced boundary-layer theory to describe the flow behavior of a viscous fluid near a solid boundary.

Whichever simplification (i.e. flow model) and whichever discretization scheme (i.e. CFD code) we might select, due diligence is of the essence, as pointed out in Section 1.4.2. It is important to ensure that the numerical solution converges when the grid is sufficiently refined. If so, there is convergence to the solution of a PDE, and it is the *right* PDE if the scheme is consistent.

The properties of schemes are analyzed by application to model problems. They are designed simplifications, often linearizations, of the Navier–Stokes equations to highlight some specific feature, yet simple enough to allow analysis. Such studies illustrate difficult phenomena and provide insight that results in suitable physical and numerical modeling techniques. The following developments assume that the reader is familiar with basic nomenclature, mathematics, and numerical finite-difference methods for PDEs.

Numerical schemes essentially approximate derivatives by differences, assuming the differenced functions are smooth enough. But high-speed flows develop discontinuities, so special treatment becomes necessary. How to compute shocks is the subject of this section. Here, the following topics are covered.

- Governing equations
- Model problems and classification
- Analysis of discretized equations
- Exploring numerical behavior with DemoSolv

The demo program DemoSolv is available for download from the book's website. There you also find an associated tutorial with suggested experiments to concretize the theoretical development in this text.

Historical Factors Driving Technical Developments

The aerodynamics of airplanes designed before and during World War II springs from linear potential theory (discussed in Chapter 3), data from wind tunnel experiments, empirical correlations derived from such experiments and experience, together with lessons learned from previous airplane designs. Two new developments in the 1940s, the jet engine and rocket propulsion, enabled vehicles to fly much faster. This brought forth hitherto unknown high-speed aerodynamic phenomena that were difficult to understand with the then-available tools, and therefore were hard to control, namely shock waves.

Race to Space and the Blunt-Body Problem

The first US human spaceflight program, running from 1958 through 1963, was Project Mercury. Its goal was to put a man into Earth orbit and return him safely. The space capsule reentering the Earth's atmosphere at $M_\infty = 25$ posed enormous technical challenges. They were met by a blunt shape with an ablative heat shield protecting it from the intense heat of atmospheric reentry: *the blunt-body problem.*

In addition to the heating problem, the stable flight of such a vehicle was also a technical challenge, and Figure 4.1 shows a Schlieren photo of a Mercury capsule-like

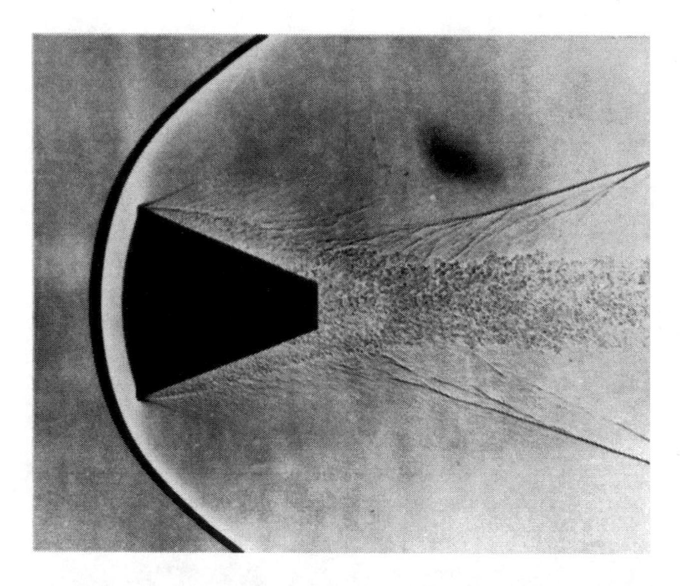

Figure 4.1 Shadowgraph photo of experiment showing bow shock ahead of Mercury reentry space capsule. (From NASA [12], public domain)

object at $M1.4$. Higher Mach numbers push the bow shock wave closer to the heat shield with higher temperatures.

Swept-Wing Jets and the Transonic Flow Problem

Jet engines provide thrust at high speed where propellers are inefficient. Design of high-speed wings called for swept shapes. These delay the appearance of shocks with accompanying drag rise so that higher speeds can be reached. The design challenge was to understand and predict the *transonic flow problem.*

The Blunt-Body Problem

Pressure disturbances travel at the speed of sound, about 330 m/s at ground level. When a body moves faster than the speed of sound (supersonic speed, $M_\infty > 1$), the air cannot adjust gradually, but will only react abruptly in front of the body to form a shock wave. Dependent on the shape of this *bow shock* wave, a new Mach number distribution evolves behind it where the flow speed is subsonic $M_\infty < 1$, as drawn at the bottom of Figure 4.2. The flow rapidly accelerates to supersonic values again just a short distance downstream of the nose. The distinguishing feature of the blunt-body problem is a small pocket of subsonic flow embedded in a supersonic stream. Across a shock wave, density, temperature, and pressure increase, so it is called a *compression shock wave*. The flow velocity decreases abruptly. The shock wave has a thickness of only a few mean-free paths and therefore appears in reality, and is handled in mathematical models, usually as a discontinuity in the flow.

Transonic Airfoil Problem

From a body in subsonic flight, pressure disturbances can precede upstream to "announce" that it is coming, as in the top of Figure 4.2. The air ahead of it can

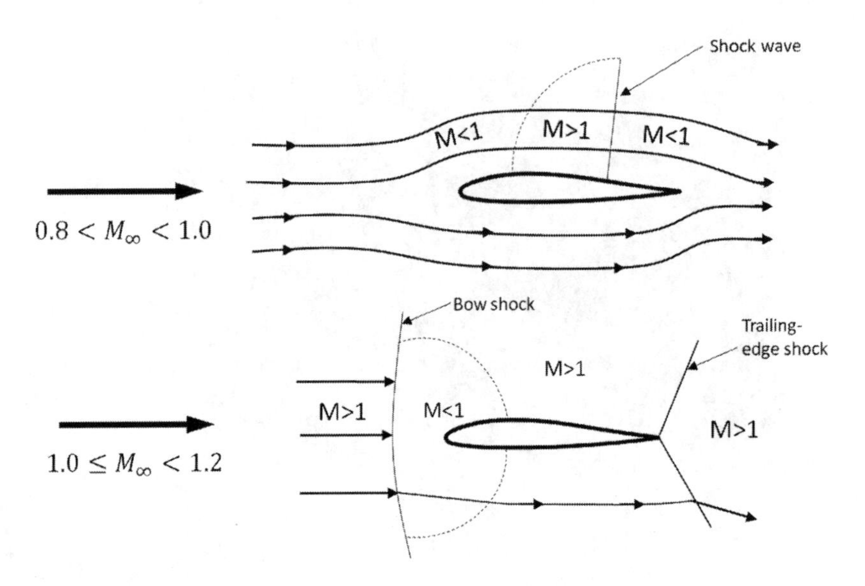

Figure 4.2 An airfoil in subsonic (top) and in supersonic (bottom) flow.

gradually adjust, with streamlines smoothly bending around it, while the flow accelerates downstream from the nose. If the subsonic flight speed, M_∞, is sufficiently high, supersonic velocities appear over the front part of the airfoil. The air must decelerate again over the rear part of the airfoil to return to the subsonic speed far downstream. This deceleration takes place through a shock wave. The *transonic airfoil* flow field is one of subsonic flow with an embedded pocket of supersonic flow. It begins at that smallest $M_\infty < 1$ for which any supersonic velocity appears, the *critical speed* M_{crit}, marking the start of *supercritical flow*. The entropy of the flow rises across shock waves. The consequence is the *wave drag*, which is to be minimized. But despite such drawbacks in drag, supercritical flight offers the possibility of better flight performance with a proper trade-off between drag rise and higher speed, the *supercritical airfoil design* problem.

4.1.1 The Challenge to Compute Shock Waves: Mixed Elliptic–Hyperbolic Equation Model

When the steady flow field is entirely subsonic, the equations are elliptic. This means roughly that perturbations are felt everywhere, decaying rapidly with distance from their sources. Classical relaxation procedures are suitable solution methods. An entirely supersonic steady flow field makes the equations hyperbolic. Localized disturbances are washed away so as to be felt only in limited regions, and the method of characteristics is a suitable solution procedure. Numerical schemes using relaxation and the method of characteristics were well known in the late 1930s and 1940s.

Both the blunt-body and the transonic airfoil problems have a mixture of subsonic and supersonic flow, and finding the boundary between them is part of the solution. The model for steady flow, the steady Euler equations, is then mixed-type elliptic–hyperbolic. But solving equations of mixed type stymied the best mathematical minds at the time, and it was only in the late 1950s that the blunt-body problem yielded to a specialized computational marching procedure. The transonic airfoil problem had to wait until 1970 when a relaxation technique produced a solution. Since only numerical methods could solve either of these problems, the two problems became the drivers for the development of CFD from the 1960s and into the 1980s. In the 1940s, even studying flows with shock waves in wind tunnels was difficult because of tunnel choking and shock wave reflections. Only early in the 1950s did slotted wind tunnel walls relieve these difficulties and yield reliable measurements. All of these efforts have paid off. Today, it is possible, with at least reasonable confidence, to compute a numerical solution of the flow around a complete aircraft configuration. All major physical effects, such as shock waves, turbulent boundary layers, and vortices, are resolved. The discussion in Section 2.7 illustrates the current state of computation.

Mathematical and Numerical Difficulties

Discontinuous solutions of the type shown above clearly do not satisfy the PDEs in the classical sense at all points, since the derivatives are not defined at discontinuities. To define what we mean by a solution to the PDEs, we must first understand the derivation of conservation laws from physical principles. This leads first to an integral form of the conservation law. The "pointwise" differential equation is derived from this by imposing additional smoothness assumptions. The crucial fact is that the integral form continues to be valid even for discontinuous solutions. Lax introduced the *weak form* of the differential equations, which will be fundamental in the development and analysis of numerical methods.

When we attempt to calculate these solutions, we face a new set of problems: a finite-difference discretization of the PDE is expected to be inappropriate near discontinuities, where the PDE does not hold. Standard methods, developed under the assumption of smooth solutions, typically produce strongly dissipative numerical results that smear steep gradients excessively, or strongly dispersive solutions that exhibit Gibbs phenomenon of wiggly profiles.

Shock Capturing

A numerical method should produce sharp approximations to discontinuous solutions automatically, without explicit tracking and use of jump conditions. Methods that attempt to do this are said to *capture* the shock. Over the past four decades, great progress has been made in this direction, and a variety of such methods is available. A method should provide the following.

1. At least second-order accuracy in smooth regions of a solution, even when discontinuities are present elsewhere
2. Sharp resolution of discontinuities without excessive smearing
3. Absence of spurious oscillations in the computed solution
4. Appropriate consistency with the weak form of the conservation law, required for convergence to weak solutions

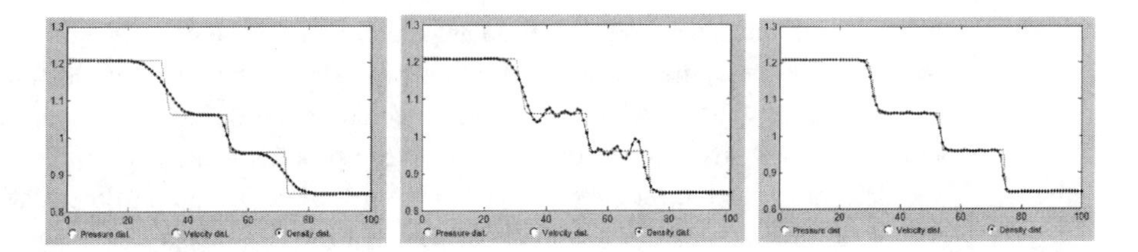

Figure 4.3 Solution of a shock tube problem. The true density profile is shown along with the numerical solution computed using three different methods: first-order Roe (left); Jameson scheme (middle); Hi-Res Roe scheme (right).

5. Solution bounds that, together with consistency, guarantee convergence as the grid is refined
6. A discrete form of the entropy condition, allowing us to make the approximations converge to the physically correct weak solution

The methods with the accuracy and resolution properties indicated above are often referred to as *high-resolution* (Hi-Res). Our goal is to study the development and analysis of these methods. Figure 4.3 shows the results obtained on a moving shock in a tube – the shock tube problem – with methods developed in this chapter.

Understanding these methods requires a good understanding of the mathematical theory of conservation laws, as well as some physical intuition about the behavior of solutions. Consequently, we will cover a considerable amount of material to introduce these methods.

For linear hyperbolic systems, characteristics play a major role. For nonlinear problems, the generalization of this theory, which is most frequently used in developing numerical methods, is the solution of a Riemann problem (Section 4.1.3). This is a pure initial value problem for the conservation law together with particular initial data consisting of two constant states separated by a single discontinuity.

For the Euler equations, this is the *shock tube problem*, detailed in Section 4.1.3. Its solution has a relatively simple structure and can be explicitly calculated in many cases.

Research at the Courant Institute of New York University on hyperbolic systems of conservation laws led to the development in 1960 of the first second-order accurate dissipative methods for these problems by Peter Lax and Burton Wendroff [8]. This class of methods forms the basis for the efficient explicit method developed in 1969 for the Navier–Stokes equations by Robert W. MacCormack [10].

Historical Evolution of Shock Computation

A dissipative scheme smears all steep gradients unphysically, and a dispersive scheme turns steep fronts into wave trains, often with over- and under-shoots. First-order schemes are very dissipative and standard second-order methods are dispersive. Figure 4.3 shows the results of schemes in DEMOFLOW applied to a shock tube problem. The exact density profile is given as a thin line. Left is the first-order Roe scheme, middle is the Jameson scheme (which is second order but adds first-order artificial viscosity at pressure jumps), and right is a Hi-Res scheme. These results are typical of first-order, second-order, and Hi-Res methods.

A breakthrough on the transonic airfoil problem was accomplished by Earll Murman and Julian Cole [11] in 1970. They obtained stable solutions using a relaxation method and switching from central differencing in the subsonic zone to upwind differencing in the supersonic zone. Their discovery demonstrated that solutions for steady transonic flows could be computed, and it provided major impetus for the further development of CFD. With growing computer speeds in the 1970s, an alternative became practical: to solve the time-accurate equations by time marching. The equations constitute a well-posed initial-boundary value problem, even when shock waves appear. Of course, the solver must deal with solutions that are discontinuous in space, and the dimensionality of the problem was increased by one, requiring great development efforts to speed up convergence to the steady state.

Initial work by Magnus and Yoshihara in 1970 met with little appreciation because of the large computer resources needed, but this was about to change.

With five additional years of algorithm development and growth in computer hardware speed and size, at Symposium Transsonicum II in Göttingen 1975, Rizzi demonstrated practical solutions to the transonic problems illustrated in Figure 4.2 by time marching. This was the only paper on Euler time marching at the symposium. All other methods solved the transonic potential equation by relaxation procedures, such as Jameson's fully conservative scheme.

Focus on the Euler Equations

The solution of the Euler equations became a central focus of CFD research in the 1980s. Most of the early solvers tended to exhibit post-shock oscillations. In addition, in a workshop held in Stockholm in 1979, it was apparent that none of the existing schemes truly converged to a steady state.

Antony Jameson decided to address the full unsteady Euler equations. The Jameson–Schmidt–Turkel (JST) scheme (e.g. see [6]) used Runge–Kutta time stepping and a blend of second and fourth differences (both to control oscillations and to provide background dissipation). It consistently demonstrated convergence to a steady state, and consequently it became one of the most widely used methods. It is hard to overestimate Jameson's influence on the mathematization and computerization of aerodynamic analysis and design. Thus, the case for using the unsteady equations was made. If time stepping produces a steady state, it is improbable that the flow is unstable. The review by Rizzi and Engquist [13] explains the development in theory and software of compressible inviscid flow simulation up through the mid-1980s. Today, a similar capability for prediction of turbulent flows around complete configurations is at hand, and CFD software and its application is a multibillion dollar business run by a time-stepping industry.

Canon of Schemes: Explicit or Implicit in Time, Centered or Upwind in Space

Evolution has created today's fauna of CFD schemes for compressible flows. Most solvers provide variants of the Jameson scheme, and Hi-Res upwind schemes have

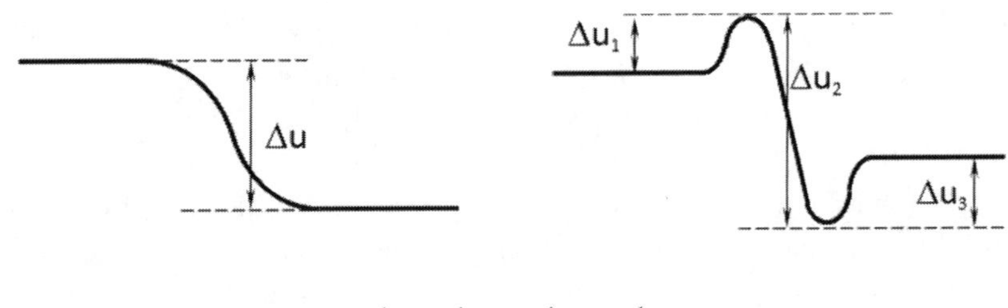

Figure 4.4 Monotone initial data that cease to be monotone; total variation increases.

been constructed to avoid the overshoots and wiggles in standard second-order schemes. Time-stepping methods include explicit Runge–Kutta methods and a greater variety of implicit methods.

Monotone Schemes

For computing a solution with sharp gradients, and even discontinuities, we would like a scheme that creates no new extrema, because the flow model does not. This will guarantee that, with monotone initial data, no over- or under-shoots or "wiggles" are produced. Figure 4.4 shows a solution that, after some time steps, ceases to be monotone, although the initial data are.

The total variation of a function $f(.)$ over an interval $[a,b]$ is $TV(f) = \int_a^b |\frac{df}{dx}|dx$ and for a discrete solution $\mathbf{U} = \{u_i\}$, $TV(\mathbf{U}) = \sum |u_i - u_{i-1}|$.

A time-stepping scheme for which $TV(\mathbf{U}(t + \Delta t)) \leq TV(\mathbf{U}(t))$ is total variation nonincreasing, but is usually called TV diminishing (TVD). The issues of oscillation control and positivity were addressed by Sergei Godunov in his pioneering work [2]. Monotonicity-preserving algorithms are called *monotone schemes*. This condition is very restrictive. Godunov showed that for schemes that form the solution at the next time step as a linear combination of the values and derivatives at current time, *no* monotone scheme exists that is better than first-order accurate. TVD schemes are monotonicity preserving. The challenge is to develop a scheme that is higher-order accurate in smooth parts of the solution, yet TVD, so it must be first-order accurate at extrema. Such Hi-Res schemes must be nonlinear. Note that the TVD property, and thus convergence, has only been shown for 1D scalar hyperbolic conservation laws. However, TVD schemes have been proven to work well also for multidimensional nonlinear hyperbolic systems such as the Euler equations.

4.1.2 Conservation-Law Form of Euler Equations

We want to treat compressible inviscid flows as governed by the Euler equations Equation (2.15). As long as there are no shock waves or other discontinuities, this non-conservative form is valid and equivalent to the conservation-law form, given for the Navier–Stokes equations in Eq. (2.7). However, if a shock wave is present, then some of the derivatives in the quasi-linear form break down. The equations are no longer equivalent to the conservation laws, which are still valid across a shock wave. Thus, numerical methods to treat flows with shocks must satisfy the conservation laws in integral form.

In integral form, the Euler equations are as follows.

$$\int_{\Omega} \frac{\partial \mathbf{U}}{\partial t} dV + \int_{\partial\Omega} \mathbf{F} \cdot \mathbf{n} \, dA = 0, \tag{4.1}$$

where the conserved variables $\mathbf{U}$ and the inviscid flux tensor $\mathbf{F}$ are as follows.

$$\mathbf{U} = \begin{pmatrix} \rho \\ \rho\mathbf{u} \\ \rho E \end{pmatrix}, \quad \mathbf{F}(\mathbf{U}) = \begin{pmatrix} \rho\mathbf{u} \\ \rho\mathbf{u}\,\mathbf{u} + p\mathbf{I} \\ \rho H\,\mathbf{u} \end{pmatrix} = \mathbf{U}\,\mathbf{u}^T + \mathbf{P}, \quad \mathbf{P} = p\begin{pmatrix} 0 \\ \mathbf{I} \\ \mathbf{u}^T \end{pmatrix}$$

and the specific total energy E is the sum of internal energy e and kinetic energy as follows.

$$E = e + \frac{1}{2}u^2$$

and $H = E + \frac{p}{\rho}$ is the total enthalpy. For an ideal gas with specific heats c_p at constant pressure and c_v at constant volume,

$$p = \rho e\left(\gamma - 1\right), \qquad \gamma = \frac{c_p}{c_v}.$$

The Euler differential equations read as follows.

$$\frac{\partial \mathbf{U}}{\partial t} + \nabla \cdot \mathbf{F} = 0. \tag{4.2}$$

In Cartesian coordinates $(x_i) = (x, y, z)^T$:

$$\frac{\partial \mathbf{U}}{\partial t} + \frac{\partial \mathbf{F}_1}{\partial x_1} + \frac{\partial \mathbf{F}_2}{\partial x_2} + \frac{\partial \mathbf{F}_3}{\partial x_3} = 0,$$

$$\mathbf{F}_1 = \begin{pmatrix} \rho u \\ \rho uu + p \\ \rho vu \\ \rho wu \\ \rho Hu \end{pmatrix}, \quad \mathbf{F}_2 = \begin{pmatrix} \rho v \\ \rho uv \\ \rho vv + p \\ \rho wv \\ \rho Hv \end{pmatrix}, \quad \mathbf{F}_3 = \begin{pmatrix} \rho w \\ \rho uw \\ \rho vw \\ \rho ww + p \\ \rho Hw \end{pmatrix}.$$

Rankine–Hugoniot Relations

The Euler equations can produce shocks and contact discontinuities in the solution. Initially, smooth waves grow and turn into moving shocks. Suppose a flow discontinuity is propagating with the constant velocity $\mathbf{v}$ as illustrated in Figure 4.5. Consider a control volume Ω containing the discontinuity and moving with that velocity $\mathbf{v}$. The Euler Eq. (4.1) become the following.

$$\int_{\Omega} \frac{\partial \mathbf{U}}{\partial t} dV + \int_{\partial\Omega} (\mathbf{U}\,(\mathbf{u} - \mathbf{v}) + \mathbf{P}) \cdot \mathbf{n} \, dA = 0 \tag{4.3}$$

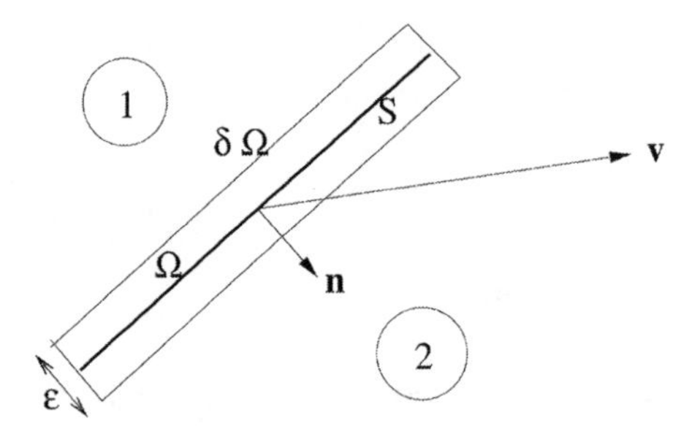

Figure 4.5 Control volume Ω moving with velocity $\mathbf{v}$ of flow discontinuity S.

Let the side faces shrink to zero (i.e. $\varepsilon \longrightarrow 0$). Then, the volume integrals in Eq. (4.3) vanish, and the boundary $\partial\Omega$ becomes the upstream (state (1)) and downstream (state (2)) sides of the discontinuity S.

Since $\mathbf{n}_1 = -\mathbf{n}_2$, the Euler equations (Eq. (4.3)) reduce to the following.

$$\int_S [\mathbf{U}\,(\mathbf{u} - \mathbf{v}) + \mathbf{P}] \cdot \mathbf{n}\, dA \;= 0,$$

where $[\mathbf{u}] = \mathbf{u}_2 - \mathbf{u}_1$ denotes the jump of $\mathbf{u}$ across the discontinuity, etc. As that relation holds for any surface S along the discontinuity, we obtain the Rankine–Hugoniot relations, published in 1870 (W. J. M. Rankine) and 1887 (H. Hugoniot),

$$[\mathbf{U}\,\mathbf{u} + \mathbf{P}] \cdot \mathbf{n} = [\mathbf{U}]\,\mathbf{v} \cdot \mathbf{n}. \tag{4.4}$$

For a stationary discontinuity, $\mathbf{v} = 0$ and the Rankine–Hugoniot relations (Eq. (4.4)) simplify to the following.

$$[\mathbf{U}\,\mathbf{u} + \mathbf{P}] \cdot \mathbf{n} = 0. \tag{4.5}$$

If the discontinuity is moving with the velocity $\mathbf{v}$ in the inertial frame of reference, the discontinuity is stationary in the frame of reference moving with $\mathbf{v}$. The velocities in the moving and inertial frames are related by

$$\mathbf{u}_{movingframe} = \mathbf{u}_{inertialframe} - \mathbf{v}.$$

The Rankine–Hugoniot relations for a stationary discontinuity (Eq. (4.5)) can be expressed as follows.

$$\begin{aligned} [\rho\,\mathbf{u} \cdot \mathbf{n}] &= 0 \\ [\mathbf{u}]\rho\,\mathbf{u} \cdot \mathbf{n} + [p]\mathbf{n} &= 0 \\ [H]\rho\,\mathbf{u} \cdot \mathbf{n} &= 0. \end{aligned} \tag{4.6}$$

Tangential Discontinuity or Vortex Sheet

If there is no mass flow through the discontinuity (i.e. $\mathbf{u} \cdot \mathbf{n} = 0$), it is a tangential discontinuity (sometimes called a contact discontinuity):

$$u_n = \mathbf{u}_1 \cdot \mathbf{n} = \mathbf{u}_2 \cdot \mathbf{n} = 0, \;\; [p] = 0, \qquad \text{but in general } [\rho] \neq 0, \;\; [\mathbf{u}_t] \neq 0, \;\; [H] \neq 0,$$

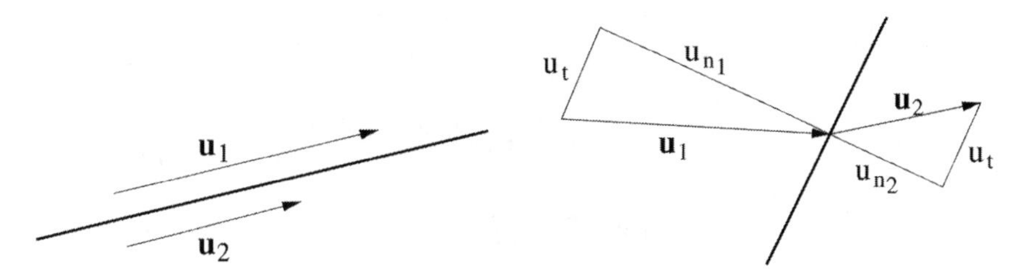

Figure 4.6 Vortex sheet–tangential velocity discontinuity (left); shock wave–normal velocity discontinuity (right).

where $\mathbf{u}_{t,k} = \mathbf{u}_k - u_n\mathbf{n}$, $k = 1, 2$ are the tangential velocities. Figure 4.6 shows the velocities on each side of the discontinuity. Left, a vortex sheet with velocities parallel to it, and right, a shock wave with mass flow through it. The non-vanishing jumps in density and total enthalpy imply a non-vanishing jump in entropy. For inviscid flows with $M \ll 1$, which we classify as incompressible, the energy equation is decoupled from the momentum and mass balances, and in the absence of heat sources, H becomes constant along streamlines to become the Bernoulli relation,

$$p + 1/2\rho|\mathbf{u}|^2 = \text{const. along streamlines.} \tag{4.7}$$

So in this case $|\mathbf{u}_1| = |\mathbf{u}_2|$, and the velocity jump must be orthogonal to the sheet normal,

$$\mathbf{u}_1 - \mathbf{u}_2 = \gamma \times \mathbf{n}$$

We encountered such a discontinuity in Chapter 3 as a sheet of concentrated vorticity with γ being the vorticity strength vector. A vortex sheet is a tangential discontinuity.

Compression Shock Wave

When there is mass flow $\rho\,\mathbf{u} \cdot \mathbf{n} \neq 0$ through it, the discontinuity is a shock wave:

$$[\mathbf{u}_t] = 0, \quad [H] = 0, \quad [u_n] < 0, \quad [\rho] > 0, \quad [p] > 0, \quad [s] > 0,$$

where s is entropy. This type of discontinuity does not appear in slow flows. The Mach number upstream must exceed 1. The conservation equations do not require growth of entropy through the shock, but the second law of thermodynamics does: shocks are always compressive. The weak solutions mentioned above are in general not unique; the entropy condition is needed to single out the physically admissible ones.

4.1.3 Model Problems: 1D Euler Equations

The Euler equations in differential conservation-law form in the (x, t) plane become

$$\mathbf{U}_t + \mathbf{F}_x = 0, \tag{4.8}$$

where

$$\mathbf{U} = \begin{pmatrix} \rho \\ \rho u \\ \rho E \end{pmatrix}, \qquad \mathbf{F}(\mathbf{U}) = \begin{pmatrix} \rho u \\ \rho u^2 + p \\ (\rho E + p)\, u \end{pmatrix}$$

Applying the chain rule $A = \frac{\partial \mathbf{F}}{\partial \mathbf{U}}$ yields the quasi-linear form

$$\frac{\partial \mathbf{U}}{\partial t} + A \frac{\partial \mathbf{U}}{\partial x} = 0 \tag{4.9}$$

The "flux Jacobian" matrix A has a complete set of *real* eigenvalues and eigenvectors and can be diagonalized as follows.

$$A = R \Lambda R^{-1}, \Lambda = \begin{pmatrix} u & 0 & 0 \\ 0 & u + c & 0 \\ 0 & 0 & u - c \end{pmatrix} \tag{4.10}$$

where R is the matrix of right eigenvectors and c is the (local) speed of sound, $c = \sqrt{\frac{\gamma p}{\rho}}$. Conservation laws (Eq. (4.8)), whose flux Jacobian A matrices admit such diagonalization, are called *hyperbolic*.

The eigenvalues are wave speeds for the PDE. One may say that three different pieces of information – the Riemann variables discussed below – travel with the different wave speeds along curves in the (x,t) plane to build the solution. For instance, if $u > c$, all waves move to the right, and there is no "information" traveling to the left. The flow is "unaware" of what lies downstream, leading to the formation of a discontinuous solution – a shock wave. Numerical methods should take the direction of the wave speeds into account, and such schemes are called upwind or upstream.

Riemann Variables

We can now transform Eq. (4.9) into a set of three uncoupled scalar equations as follows.

$$\begin{aligned} R^{-1} \frac{\partial \mathbf{U}}{\partial t} + R^{-1} A R \; R^{-1} \frac{\partial \mathbf{U}}{\partial x} &= 0 \\ R^{-1} \frac{\partial \mathbf{U}}{\partial t} + \Lambda \; R^{-1} \frac{\partial \mathbf{U}}{\partial x} &= 0 \end{aligned} \tag{4.11}$$

The elements of $R^{-1} \partial \mathbf{U}/\partial t$ times proper integrating factors can be written as $\partial \mathbf{W}/\partial t$, where $\mathbf{W} = (w_1, w_2, w_3)$ are called the *Riemann variables* or *characteristic variables*. The differentials can be integrated to the following.

$$w_1 = \frac{p}{\rho^\gamma}, \; w_2 = u + \frac{2c}{\gamma - 1}, \; w_3 = u - \frac{2c}{\gamma - 1} \tag{4.12}$$

assuming constant entropy for the latter two. This is the set of variables that best reflects the physics of the flow, and it will be used to model the properties at the boundaries. Note that not all conservation laws with more than two equations admit such integration in closed form.

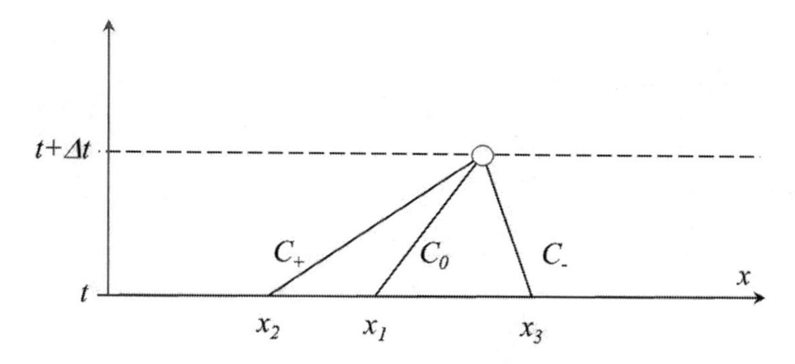

Figure 4.7 Construction of solution by characteristics.

In terms of the Riemann variables, Eq. (4.9) becomes a set of uncoupled wave equations as follows

$$\partial \mathbf{W}/\partial t + \Lambda \partial \mathbf{W}/\partial x = 0 \tag{4.13}$$

Each Riemann variable follows its own first-order wave or advection equation, of the type $\partial u/\partial t + a\partial u/\partial x = 0$.

Thus, each quantity w_j propagates along a *characteristic curve*, one of C_0, C_+, C_-, given by the following.

$$C_0 : \frac{dx}{dt} = u, \quad \partial w_1/\partial t + u\partial w_1/\partial x = 0$$

$$C_+ : \frac{dx}{dt} = u + c, \quad \partial w_2/\partial t + (u + c)\partial w_2/\partial x = 0$$

$$C_- : \frac{dx}{dt} = u - c, \quad \partial w_3/\partial t + (u - c)\partial w_3/\partial x = 0$$

The Riemann variables for Eq. (4.9) are actually constant along the characteristic curves since the right-hand side is zero, so they are known as *Riemann invariants*, but the velocities change over the flow field so that the characteristic curves are not in general straight. A solution can be built over a short time interval by the *method of characteristics* (Figure 4.7) as follows.

Given the $w_j(x)$ at time t and a point $(x, t + \Delta t)$, find the points x_i from which the $C_.$-characteristics meet at $(x, t + \Delta t)$. The solution there has

$$w_j(x, t + \Delta t) = w_j(x_j, t), \quad j = 1, 2, 3$$

The slopes of the characteristics are as follows.

$$u = (w_2 + w_3)/2, \quad u \pm c = u \pm (\gamma - 1)/4 \cdot (w_2 - w_3)$$

so there is no need to compute all of the conserved or primitive variables. This looks attractive, but characteristic curves can intersect at $t + \Delta t$ from more than one x_2, say. This signals a shock formation and the need to use the Rankine–Hugoniot relations (i.e. the conserved variables) to propagate the shock. The method outlined here can be used for steady isentropic flows in 2D, for example, but becomes quite involved in 3D and for shocked flows.

But as it turns out, multidimensional finite-volume schemes need only 1D fluxes across interfaces between computational cells, as shown in Chapter 6. Thus, it makes sense to continue the development here of 1D (or quasi-1D) flows with shocks and contact discontinuities.

The model of the quasi-1D nozzle considered in Section 4.3 has a source term. The Riemann variables still evolve along the characteristic curves, but they are not constant. For w_1, the source term vanishes; physically, this expresses the transport of constant entropy at speed u in the flow.

The Shock Tube and Riemann's Problems

Consider a long pipe (tube) divided by a diaphragm at $x = 0$ into two chambers (see Figure 4.8). Initially, the chamber to the left of the diaphragm contains gas at rest at a high pressure p_4, and in the chamber to the right the gas at rest is at a low pressure p_1. When the diaphragm is ruptured at $t = t_0$, the high pressure accelerates the gas toward the right with the formation of a shock wave, a contact discontinuity, and an expansion fan or rarefaction wave, as shown in Figure 4.8 at t_1. Although there is neither viscosity nor heat conduction in the model, dissipation is still present: the shock

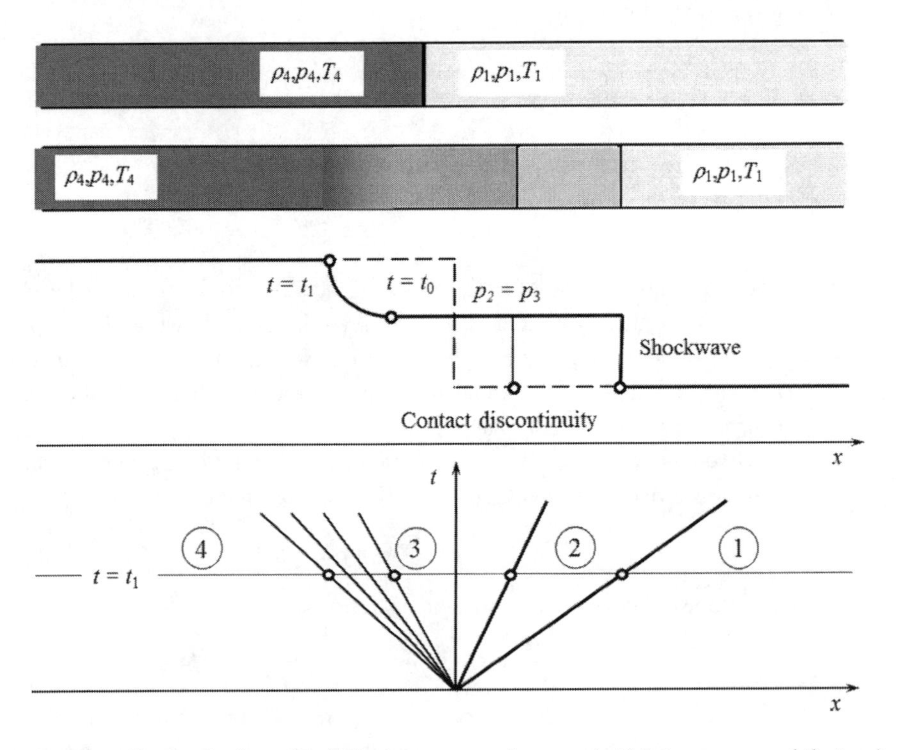

Figure 4.8 Shock tube flow. (Top) Flow at $t = t_0$ and $t = t_1$; (middle) pressure; and (bottom) $x - t$ diagram with jump traces.

wave converts kinetic energy into heat and increases entropy. The jump traces show that the solution is a function of $\eta = x/t$. The velocities of the jumps are related to the characteristics and must be computed from the Rankine–Hugoniot jump relations together with the states that they separate. The leftmost point of the rarefaction wave or expansion fan is at $x = -a_4(t_1 - t_0)$, where a_4 is the speed of sound to the left of the membrane. Entropy is piecewise constant with jumps between regions 3 and 2 and between 2 and 1. Pressure varies continuously across the fan and jumps only between 2 and 1.

The Euler equations can be solved exactly for Riemann's problem, and analytical expressions can be derived that relate the conditions between the various states from 1 to 4 in Figure 4.8.

The good news is that there is a unique solution if an entropy condition (i.e. presence of physically correct dissipation at discontinuities) is satisfied. The bad news is that the relations are nonlinear, requiring iterative solution, which makes them unattractive for use in computations. Nevertheless, Sergei Godunov in 1959 [2] based his scheme on this solution, and flows with very strong shocks may require an exact Riemann solver.

Quasi-1D Nozzle Flow

A steady flow through a convergent–divergent nozzle of specified cross-section distribution is driven by a fixed pressure ratio between the exit and inlet; this models a physical situation where the inlet is connected to a large reservoir with known pressure and temperature p_0, T_0, and the outlet exits in a large domain with given back pressure $p_1 < p_0$ (Figure 4.9). The resulting steady flow depends on the pressure ratio p_0/p_1,

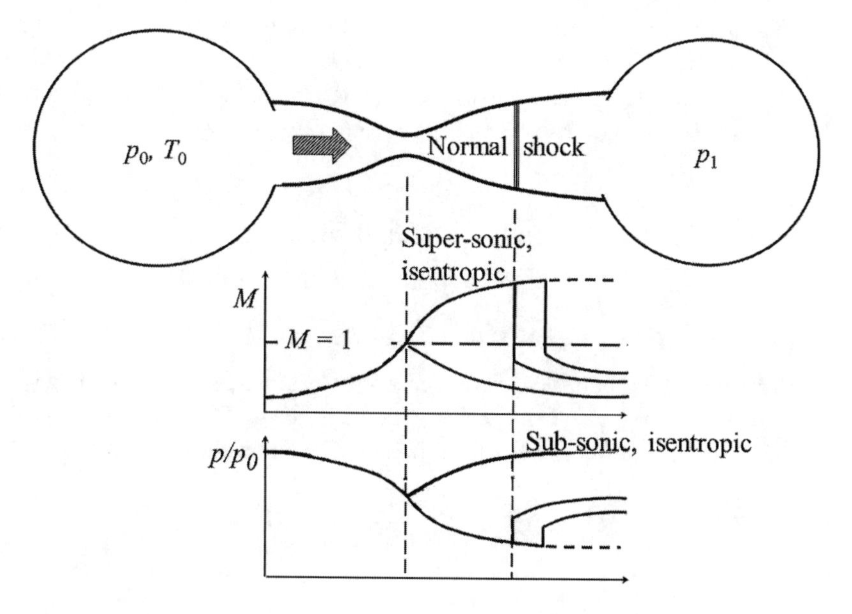

Figure 4.9 Quasi-1D nozzle setup and solutions for different back pressures p_1.

and we can control p_1. When p_1 is large, the flow accelerates to the throat, the minimal cross-section area, and then decelerates to the exit, and decreasing p_1 increases the mass flow. With p_1 small enough, the flow at the throat becomes sonic and returns to subsonic speed in the divergent part through a shock wave. The flow is choked, and the flow profiles from inlet to throat do not change with further decrease of p_1. Lowering p_1 further makes the shock move downstream until the exit is also supersonic. This is the lowest attainable pressure at the exit. p_1 lower than this limit makes the flow expand further downstream of the nozzle, so a shock/expansion wave pattern appears in the reservoir. Pressure and Mach number profiles for a few cases are given in Figure 4.9.

The model neglects viscosity and assumes that the nozzle is so slender that the flow is *quasi-1D*: flow properties vary only with the coordinate x along the nozzle, and velocity is assumed parallel to the nozzle axis. This illustrates the technique of capturing a *steady shock* within a numerical solution by time marching of the unsteady equations.

One must ask the question of how well the model predicts reality. The answer is that no a priori check is available. Obviously, the boundary conditions replace the rest of the system, and they can do so only in certain cases. A look at a jet engine or rocket nozzle in operation confirms that the flow also shows systematic patterns outside the nozzle. The neglect of viscosity means that the flow does not separate at the nozzle walls, as it should if the area grows too quickly downstream. Such separations entail total pressure loss and are to be avoided.

Quasi-1D Euler Equations

The derivation of the governing equations for unsteady flow in the nozzle begins with the control volume Ω, a dx slice perpendicular to the nozzle axis, fixed in time and space, as shown in Figure 4.10, and the conservation laws for mass, momentum, and energy. The nozzle cross-section area is $A(x), 0 \le x \le L$.

$$\begin{aligned}
Adx\partial\rho/\partial t - A\rho u + A_+\rho_+u_+ &= O(dx^2)\\
Adx\partial(\rho u)/\partial t - A\ \rho u^2 + A_+\rho_+u_+^2 + p_+A_+ - pA\ - p(A_+ - A) &= O(dx^2)\\
Adx\partial(\rho E)/\partial t - A\ \rho H u + A_+\rho_+H_+u_+ &= O(dx^2)\\
p = (\gamma - 1)\rho(E - u^2/2)&
\end{aligned}$$

To improve legibility, the dependence on t has been suppressed, as on x when the argument is just x; x_+ means $x + dx$, A_+ means $A(x_+)$, etc. Division by dx and sending dx to zero produces the differential equations in Eq. (4.14).

$$\frac{\partial \mathbf{U}}{\partial t} + \frac{\partial \mathbf{F}}{\partial x} = \mathbf{G} \tag{4.14}$$

where

$$\mathbf{U} = A\begin{pmatrix}\rho\\ \rho u\\ \rho(e + u^2/2)\end{pmatrix}, \quad \mathbf{F} = A\begin{pmatrix}\rho u\\ \rho u^2 + p\\ \rho(e + u^2/2)u + pu\end{pmatrix} \quad \mathbf{G} = \begin{pmatrix}0\\ p\frac{\partial A}{\partial x}\\ 0\end{pmatrix} \tag{4.15}$$

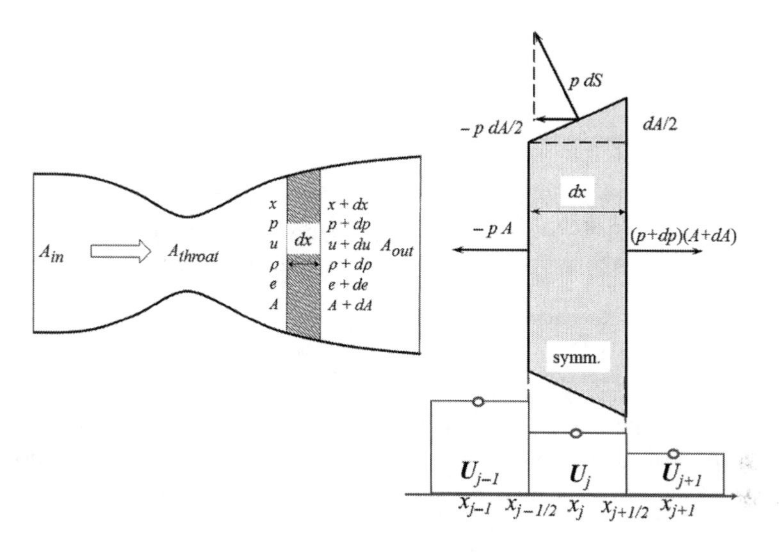

Figure 4.10 Discretization of a quasi-1D nozzle flow problem.

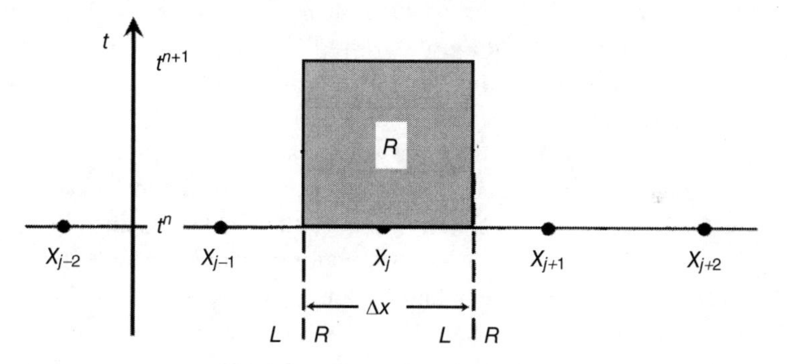

Figure 4.11 Control volume R in 1D, or grid cell in space–time mesh. L/R stands for left/right cell face.

These equations are equivalent to the 1D Euler equations studied in Section 4.1.3 with an added source term $\mathbf{G}$. This term has no space derivative of the solution and plays a minor role in the analysis of solution character, etc. We also need relations from thermodynamics for air to treat it as a perfect gas; $p = \rho RT$ and $e = RT/(\gamma - 1)$ where $R = 288.7$ J/kg/K is the gas constant for air, $\gamma = 1.4$, and the speed of sound (a in the preceding chapter) is $c = \sqrt{\gamma \frac{p}{\rho}}$.

4.2 Finite-Volume Methodology

We now repeat the process to develop a numerical method for the "abstract" conservation law expressed in Eq. (4.8). In order to handle a shock wave correctly, the numerical method must maintain the conservation property (i.e. a balance of mass, momentum, and energy in an integral sense in the (x, t) plane).

Begin by integrating Eq. (4.8) over a *cell R* in the space–time mesh (Figure 4.11) (i.e. from $x_{j-\frac{1}{2}}$ to $x_{j+\frac{1}{2}}$ and over one time step from t^n to t^{n+1}).

$$\int_{x_{j-\frac{1}{2}}}^{x_{j+\frac{1}{2}}} \left[\mathbf{U}(x,t^{n+1}) - \mathbf{U}(x,t^n)\right] dx + \int_{t^n}^{t^{n+1}} \left[\mathbf{F}\left(\mathbf{U}\left(x_{j+\frac{1}{2}},t\right)\right) - \mathbf{F}\left(\mathbf{U}\left(x_{j-\frac{1}{2}},t\right)\right)\right] dt = 0 \tag{4.16}$$

Equation (4.16) is valid in the presence of shock waves even when $\mathbf{U}$ is not differentiable and can satisfy the differential equations only in a *weak* sense.

Define the *cell average* of $\mathbf{U}_j$ in cell with midpoint x_j as follows.

$$\mathbf{U}_j(t) = \frac{1}{\Delta x} \int_{x_{j-\frac{1}{2}}}^{x_{j+\frac{1}{2}}} \mathbf{U}(x,t)\, dx \tag{4.17}$$

The discrete approximate solution becomes piecewise constant, with jumps at $x_{j-1/2}$. This leads to a grid of control volumes or cells along the x-axis, as shown in Figure 4.11. Figure 4.10 right shows the spatial control volumes and grid points for the quasi-1D nozzle model.

At this point, anticipating application to multidimensional flow domains, it makes sense to distinguish between the *primal* grid, here the set of x_j, and the computational cells C_i, which form the *dual* grid, $C_i = \left[x_{i-1/2}, x_{i+1/2}\right]$ (see Figure 4.10). This constitutes a *cell-center* scheme. Now, only if the cells are the same size will the cell centers be x_i, the points of the primal grid.

The variables solved for in the finite-volume approach here are the cell averages. To consider fields defined by point values, it is natural to define the value at the midpoint of C_i as the "best" point value. That is exact if the field is assumed linear in the cell, the cell average $\mathbf{U}_j(t) = \mathbf{U}(x_j,t) + O(\Delta x^2)$ for smooth $\mathbf{U}(x,t)$.

In a *cell-vertex* scheme, the unknowns are associated with the points of the primal grid, and the cells are defined so cell faces are midway between the neighbor points. This makes the divided difference between neighbor values a second-order accurate approximation to the derivative. In the current 1D case, the differences are minor, but in multidimensional flow models, there are significant differences between solver data structures and discretization details. The EDGE code is a cell-vertex scheme, with details reviewed in Chapter 6.

Semi-Discrete Formulation

The *semi-discrete* approach discretizes first in space, leaving the time variable continuous. The advantage to this approach is that it postpones the choice of time integration scheme and its accuracy. In addition, any steady-state solution will be independent of the time discretization procedure and therefore independent of the time step.

$$\frac{d}{dt}\mathbf{U}_j(t) = -\frac{1}{\Delta x}\left[\mathbf{F}(\mathbf{U}(x_{j+1/2},t)) - \mathbf{F}(\mathbf{U}(x_{j-1/2},t))\right] \tag{4.18}$$

Define $\hat{\mathbf{F}}_{j+1/2}(t)$ to be the *numerical flux* approximation to the right-hand side as follows.

$$\hat{\mathbf{F}}_{j+1/2}(t) \approx \mathbf{F}\left(\mathbf{U}\left(x_{j+\frac{1}{2}},t\right)\right) \tag{4.19}$$

A semi-discrete scheme to advance the cell average $\mathbf{U}_j(t)$ forward in time is as follows.

$$\frac{d}{dt}\mathbf{U}_j(t) = -\frac{1}{\Delta x}\left(\hat{\mathbf{F}}_{j+1/2}(t) - \hat{\mathbf{F}}_{j-1/2}(t)\right) \tag{4.20}$$

Equation (4.20) is said to be in *conservation* form because it is a numerical approximation of the integral conservation law. Lax and Wendroff [8] have proved that discontinuous solutions to conservation laws can be computed without special treatment of the discontinuity if the numerical approximation is consistent, stable, and conservative.

Example

Equation (4.20) may be discretized by a forward time difference, also known as the explicit Euler time stepping scheme,

$$\mathbf{U}_j^{n+1} - \mathbf{U}_j^n = -\frac{\Delta t}{\Delta x}\left(\hat{\mathbf{F}}_{j+1/2}^n - \hat{\mathbf{F}}_{j-1/2}^n\right) \tag{4.21}$$

to become a *fully discrete* difference approximation to the PDE. EDGE and DEMOFLOW use explicit Runge–Kutta methods as described in Section 4.3.

4.2.1 Finite-Volume Discretization of Quasi-1D Nozzle Flow

The governing in Eq. (4.14) are semi-discretized by the dual control–volume method, meaning that cell boundaries are midway between neighbor gridpoints, to produce following.

$$\frac{d}{dt}\mathbf{U}_j(t) = \mathbf{R}_j \equiv -\frac{1}{\Delta x}\left(\hat{\mathbf{F}}_{j+1/2} - \hat{\mathbf{F}}_{j-1/2}\right) + \mathbf{G}_j$$

with the *residual* $\mathbf{R}_j$ as the right-hand side. The numerical flux function for a central difference scheme is as follows.

$$\hat{\mathbf{F}}_{j+\frac{1}{2}} = \frac{1}{2}\left(\boldsymbol{F}(\mathbf{U}_{j+1}) + \boldsymbol{F}(\mathbf{U}_j)\right) - \hat{\mathbf{D}}_{j+\frac{1}{2}} \tag{4.22}$$

where the *generic dissipation flux function* $\hat{\mathbf{D}}_{j+\frac{1}{2}}$ is specified by the choice of a particular spatial discretization scheme. $\mathbf{G}$ is not involved in the numerical flux function. This means that our development above of space discretizations carries over unchanged.

4.2.2 Space-Centered Scheme with Added Dissipation

The numerical flux for the space-centered scheme is as follows.

$$\hat{\mathbf{F}}_{j+1/2}(t) = \frac{1}{2}\left(\mathbf{F}(\mathbf{U}_{j+1}(t)) + \mathbf{F}(\mathbf{U}_j(t))\right)$$

This makes a second-order truncation error in space, but it is unstable when used in Eq. (4.21). It can be stabilized by adding a second-difference *dissipation term*, which makes the solution smooth [17]. We are then solving the following.

$$\mathbf{U}_t + \mathbf{F}(\mathbf{U})_x = (\mu \Delta x\, \mathbf{U}_x)_x$$

If we approximate spatial derivatives by second-order accurate central differences, the dissipative numerical flux in the semi-discrete scheme becomes the following.

$$\hat{\mathbf{D}}_{j+\frac{1}{2}} = \mu_{j+1/2}(\mathbf{U}_{j+1}(t) - \mathbf{U}_j(t))$$

to produce a first-order accurate conservative scheme. If the artificial dissipation is activated only at discontinuities, the scheme is second-order accurate except at jumps. The Jameson scheme in the DEMOFLOW nozzle simulator is of this kind.

Jameson Flux Approximation

The dissipative flux $\hat{\mathbf{D}}^n_{j+\frac{1}{2}}$ is constructed artificially in an ad hoc manner. The Jameson approach activates the dissipative term near discontinuities where the scheme becomes first-order accurate. Everywhere else, it is switched off by a sensor that turns on a fourth-order dissipation term. The switch function is designed to detect a shock and toggle between the dissipative modes. The dissipation is a blend of second and fourth differences scaled by the spectral radius (the maximal absolute eigenvalue) of the convective flux Jacobian matrix A.

Switch

Shocks are located by monitoring the variation of pressure, using the second-order central difference approximation to the nondimensional quantity $\frac{d^2p/dx^2 \Delta x^2}{4p}$.

$$\mu_j = \frac{|p_{j+1} - 2p_j + p_{j-1}|}{p_{j+1} + 2p_j + p_{j-1}} \tag{4.23}$$

This quantity is small in smooth flow domains and large in regions of strong pressure gradients. The span is then extended over the neighboring nodes, and the actual switch is found as follows.

$$\upsilon_j = (\mu_{j-1} + 2\mu_j + \mu_{j+1})/4 \tag{4.24}$$

Second- and Fourth-Order Coefficients

The sensor υ_j can now be used to toggle between dissipative modes such that the fourth-order dissipation is automatically switched off in the vicinity of a discontinuity as follows:

$$\varepsilon_j^{(2)} = k^{(2)}\upsilon_j, \varepsilon_j^{(4)} = \max(0, k^{(4)} - \varepsilon_j^{(2)}) \tag{4.25}$$

where the two constants $k^{(2)}$ and $k^{(4)}$ are specified by the user and called $Vis2$ and $Vis4$ in DEMOFLOW.

Scaling

A mathematical analysis of upwind schemes shows that the numerical viscosity is proportional in cell j to the spectral radius $\lambda_j = |u_j| + c_j$ of the flux Jacobian matrix $A = \partial \mathbf{F}/\partial \mathbf{U}$, where c_j is the local speed of sound. The final scaling factor is obtained by averaging.

$$r_{j+1/2} = \frac{1}{2}(\lambda_j + \lambda_{j+1}) \tag{4.26}$$

Dissipation Flux

All of the tools are now in place to finally form the expression for the artificial dissipation flux $\hat{\mathbf{D}}_{j+1/2}$. Since the flux *difference* comprises second- and fourth-order terms, the flux itself must be formed with first- and third-order differences. Compounding with the coefficients and scaling factors yields the following.

$$\hat{\mathbf{D}}_{j+1/2} = r_{j+\frac{1}{2}}\left[\, \varepsilon_j^{(2)}(\mathbf{U}_{j+1} - \mathbf{U}_j) \; - \; \varepsilon_j^{(4)}(\mathbf{U}_{j+2} - 3\mathbf{U}_{j+1} + 3\mathbf{U}_j - \mathbf{U}_{j-1})\,\right] \tag{4.27}$$

4.2.3 Space Schemes using Approximate Riemann Solvers

In many cases, *central schemes* draw numerical information from points without respecting the slopes of the characteristic lines, and this can compromise the accuracy of the solution.

• *For flow fields that involve smooth, continuous variations of the flow field variables*, this does not appear to cause a major problem.

• *When discontinuities exist in the flow* such as shock waves, central schemes do not work very well. They produce oscillations around the shock wave if no explicit artificial viscosity is added. It is this type of problem that convinced people to develop upwind schemes in modern CFD.

Upwind schemes are designed to simulate more accurately the propagation of information along the characteristic curves. With proper upwinding, the calculation of discontinuities spread over only two grid cells with no oscillations is possible.

To reduce or eliminate the spurious diffusive property, while at the same time keeping the advantages of an upwind scheme, some new algorithms have been developed over the past decades. These modern algorithms have introduced terminology such as the following.

- TVD schemes
- Flux splitting
- Flux and slope limiters
- Godunov schemes
- Approximate Riemann solvers

Godunov's Scheme and the Roe Linearization

Godunov [2] based his scheme on the exact solution to the Riemann problem, defined below. The state variables are the cell averages $\mathbf{U}_j$, and the steps in a Godunov-type scheme are as follows.

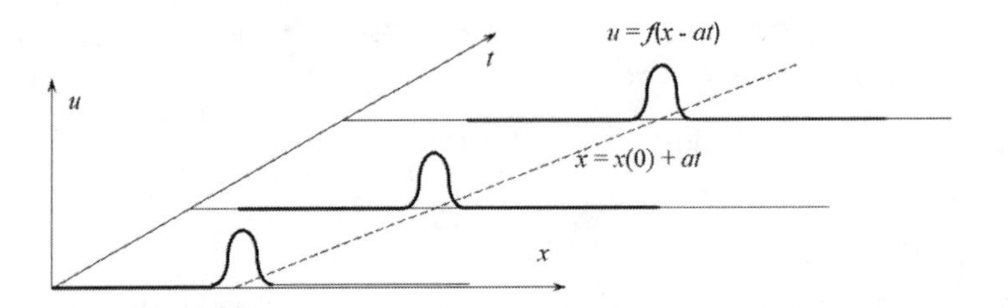

Figure 4.12 Propagation of wave form $f(x)$ along the characteristic line $x(t) = x(0) + a \cdot t$ of Eq. (4.29).

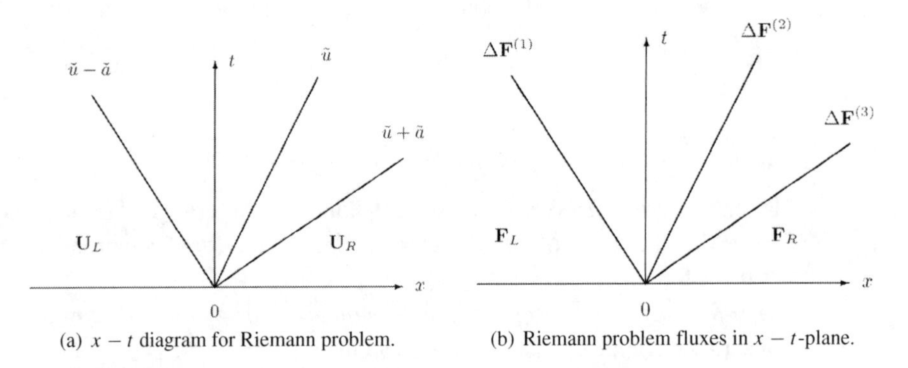

(a) $x - t$ diagram for Riemann problem.

(b) Riemann problem fluxes in $x - t$-plane.

Figure 4.13 Jumps in: (a) the solution and (b) flux across the characteristic lines.

1. Reconstruct cell distributions (constant, linear, ...) from cell averages. From this, find values at cell interfaces, which are different on the left and right sides.
2. Solve the Riemann problem at each interface. This determines the interface fluxes.
3. Use interface fluxes to evolve the cell averages over a time step.

The major disadvantage with the Godunov scheme is the expensive exact solution of the Riemann problem. Philip Roe [14] improved on this in 1981 by devising an approximate *linearized* solution satisfying the exact Rankine–Hugoniot relations.

Linear Riemann Problems for Systems

The $x - t$ diagrams in Figures 4.12 and 4.13 define the various quantities used in the formulation for the solution to a hyperbolic linear problem $\mathbf{U}_t + A\mathbf{U}_x = 0$ with constant matrix A. A has eigenvalues λ_k, the characteristic velocities, and eigenvectors sorted so that $\lambda_{k-1} < \lambda_k$.

$$A\mathbf{r}_k = \lambda_k \, \mathbf{r}_k, \; k = 1, \ldots, N$$

Then we have $AR = R\Lambda, R = [\mathbf{r}_1, \mathbf{r}_2, \ldots, \mathbf{r}_N]$, $\Lambda = diag(\lambda_k)$
(see also Eq. (4.10)). For the characteristic variables $\mathbf{W} = \{w_k\} = R^{-1}\mathbf{U}$, there obtains the set of N uncoupled first-order wave (also known as advection) equations.

$$\frac{\partial w_j}{\partial t} + \lambda_j \frac{\partial w_j}{\partial x} = 0, \ j = 1, \ldots, N$$

Solution of Advection Equations

We study a pure initial value problem for one of the set, with initial data on the whole real line – a *Cauchy problem.*

$$u_t + au_x = 0, \quad a \in \mathbb{R} > 0, \quad t > 0 \tag{4.28}$$

$$u(x,0) = f(x), \quad -\infty < x < \infty$$

Look at how the solution $u(x,t)$ behaves along a curve $x = X(t)$ in the $x - t$ plane. The time derivative of $u(X(t),t)$ along the specific curve $x = X(t)$ becomes as follows.

$$\frac{d}{dt}u(X(t),t) = \{\text{chain rule}\} = u_t(X(t),t) + \frac{dX}{dt}u_x(X(t),t) = 0 \quad \text{if} \quad \frac{dX}{dt} = a$$

This means that the solution, u, is constant along the *characteristic* curve $(t, X(t))$ defined by $dX/dt = a$. Thus, the solution to the initial value problem is as follows.

$$u(x,t) = f(x - at)$$

In other words, the initial profile of u is advected with constant velocity a. “Upwind differencing” is an attempt to mimic this one-sided information propagation along the characteristics. Note that even if $f(x)$ is not differentiable, the above form of the solution solves the PDE in a weak sense.

All we have to do is to compose the solutions to the characteristic equations as follows.

$$w_j(x,t) = w_j^L + (w_j^R - w_j^L)\, H(x - \lambda_j t) \tag{4.29}$$

with the Heaviside function

$$H(s) = 1, \text{ if } s > 0, = 0 \text{ otherwise}$$

and

$$\mathbf{W}^L = R^{-1}\,\mathbf{U}_L, \ \mathbf{W}^R = R^{-1}\,\mathbf{U}_R$$

The solution gives the state along $x/t = 0, t > 0$, the cell interface, $w_j = w_j^L + (w_j^R - w_j^L)\, H(0 - \lambda_j)$ or

$$w_j(x/t = 0) = w_j^L \text{ for } \lambda_j > 0, = w_j^R \text{ for } \lambda_j < 0$$

so the flux through the interface is $\hat{\mathbf{F}} = AR\mathbf{W}(x/t = 0)$, which can be used in the time stepping of Eq. (4.20). It can be written as follows.

$$\hat{\mathbf{F}} = \frac{1}{2}(A\mathbf{U}_L + A\mathbf{U}_R) - \frac{1}{2}|A|(\mathbf{U}_R - \mathbf{U}_L)$$

where the matrix $|A| = R\, diag(|\lambda_i|)\, R^{-1}$. Equation (4.20) with this numerical flux is a spatially first-order generalization of the scalar upwind scheme of a system.

Figure 4.13a displays the characteristic lines $x/t = \lambda_k$ of the Riemann problem and the jumps $\Delta\mathbf{U}^k$ across them. We have

$$\mathbf{U}_R - \mathbf{U}_L = \sum_{k=1}^{N} \Delta\mathbf{U}^k, \quad \Delta\mathbf{U}^k = \alpha_k \mathbf{r}_k, \quad \alpha_k = w_k^R - w_k^L,$$

so

$$\mathbf{F}_R - \mathbf{F}_L = \sum_{k=1}^{N} \Delta\mathbf{F}^k, \quad \Delta\mathbf{F}^k = \lambda_k \alpha_k \mathbf{r}_k.$$

4.2.4 Roe Linearization

For the Euler equations of gas dynamics, Roe managed to find a mean value Jacobian matrix $\tilde{\mathbf{A}}(\mathbf{U}_L, \mathbf{U}_R)$ with eigenvalues and right eigenvectors $\tilde{\lambda}_k, \tilde{\mathbf{r}}_k$ such that

$$\mathbf{F}(\mathbf{U}_R) - \mathbf{F}(\mathbf{U}_L) = \tilde{\mathbf{A}}\,(\mathbf{U}_R - \mathbf{U}_L)$$

Then, the formulas for the linear Riemann problem can use the $\tilde{A}$ matrix, with Rankine–Hugoniot jumps $\Delta\mathbf{U}^k = \tilde{\alpha}_k \tilde{\mathbf{r}}_k$ across a discontinuity at $x/t = \tilde{\lambda}_k$, etc.

The numerical flux at t^n is $\mathbf{F}(x/t = 0)$, and

$$\hat{\mathbf{F}}^{Roe} = \frac{1}{2}\,(\mathbf{F}(\mathbf{U}_R) + \mathbf{F}(\mathbf{U}_L)) - \frac{1}{2}|\tilde{\mathbf{A}}|(\mathbf{U}_R - \mathbf{U}_L) \tag{4.30}$$

which is a generalization to a nonlinear system like the Euler equations of the upstream flux for the linear Riemann problem.

The exact Riemann problem has a rarefaction fan, but the linearized problem has only discontinuities propagating at the characteristic speeds of the Roe matrix.

4.2.5 Hi-Res Schemes

In a Godunov scheme, $\mathbf{U}_L$ can be taken as $\mathbf{U}_j$ in the flux at $x_{j+1/2}$, etc. This makes the scheme first-order accurate everywhere. A flux-difference splitting (FDS) scheme, such as Roe's, works with the two fluxes on either side of the cell interfaces to update the conservative variables. It is necessary to interpolate/extrapolate the flow properties from cell averages to the cell interfaces by reconstructing the distributions. Bram van Leer developed Monotone Upstream-centered Schemes for Conservation Laws (MUSCL) [16] variable interpolation/extrapolation. With degree-one or even higher-degree polynomials, MUSCL can achieve second-order accuracy. The idea in a Hi-Res scheme is to limit the dissipative term(s) to vanish in smooth portions of the flow and turn them on at discontinuities and extrema. This is achieved by the limiters described next.

Slope Limiters $\phi(.)$

Consider the reconstruction of a linear distribution in cell j for the left flux at $x_{j+1/2}$. Using only the nearest cells,

$$u_j(x) = U_j + (x - x_j)(U_{j+1} - U_j)/\Delta x$$

or

$$u_j(x) = U_j + (x - x_j)(U_j - U_{j-1})/\Delta x$$

to give

$$u^L_{j+1/2} = U_j + 1/2(U_{j+1} - U_j) \text{ or } U_j + 1/2(U_j - U_{j-1})$$

which is equivalent to

$$u^L_{j+1/2} = U_j + \frac{1}{2}\phi(r_{j+1/2})(U_j - U_{j-1}), \quad r_{j+1/2} = \frac{U_{j+1} - U_j}{U_j - U_{j-1}}$$

with the *slope limiter function* $\phi(.)$ to define how the choice is made. Limiting is defined for scalars. ϕ is applied component-wise to its vector argument of component-wise ratios, and we write

$$limit(\boldsymbol{a}, \boldsymbol{b}) = \phi(\boldsymbol{a}/\boldsymbol{b})\boldsymbol{b} \text{ for } \{\phi(a_i/b_i)b_i\}$$

to define the $limit$ function.

Choosing the smallest slope, if the two have equal signs, and zero, if the signs are different, makes a first-order scheme at extrema. This defines the *minmod* function as follows.

$$minmod(r) = \max(0, \min(r, 1))$$

It is clear that ϕ must vanish for $r \leq 0$. Likewise, $\phi(1) = 1$ is necessary for second-order accuracy when the two slopes are equal. Another natural property enjoyed by most (but not all) limiters is symmetry, $\phi(\frac{1}{r}) = \frac{\phi(r)}{r}$. Limiter functions must also satisfy a number of TVD technical requirements as presented in, for example Hirsch [5]. The allowed region for a limiter to give a Hi-Res scheme is shown in Figure 4.14. The heavy curve in Figure 4.14 is the van Albada limiter $\phi(r) = \frac{r+r^2}{1+r^2}$, $r \geq 0$ used in DemoFlow.

Roe Flux Approximation

Roe's FDS scheme, by means of an eigen-analysis, decomposes the flux difference over a control surface into a sum of left- and right-running wave contributions that propagate with the speed of the eigenvalues. These wave contributions then determine the last, and distinguishing, term of the numerical flux approximation as follows.

$$\hat{\mathbf{D}}^n_{j+\frac{1}{2}} = -\frac{1}{2}\mathbf{R}_{j+\frac{1}{2}}\Psi_{j+\frac{1}{2}}\mathbf{R}^{-1}_{j+\frac{1}{2}}(\mathbf{U}^n_{j+1} - \mathbf{U}^n_j) \tag{4.31}$$

where $\boldsymbol{\Psi} = diag(|\lambda_i|\,\psi_i)$ with ψ_i, which are functions of the slope limiters ϕ_i. Note that they operate on the column vector of differences to their right. λ_i is the ith eigenvalue and $\boldsymbol{R}$ is the matrix of right-hand eigenvectors of the matrix $\widetilde{\mathbf{A}}$

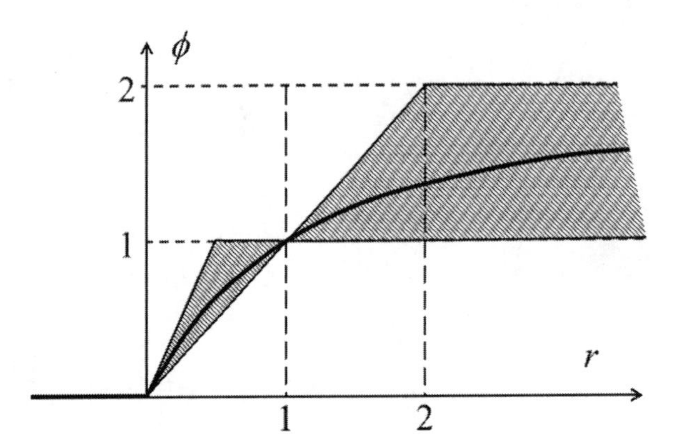

Figure 4.14 A Hi-Res slope limiter $\phi(r)$ must stay in the shaded area. The *minmod* limiter is on the lower bound.

(see Section 4.2.3). Limiting the MUSCL interpolation in this way ensures that the correct amount of dissipation is given to each eigen-component.

Flux-Difference Splitting

From Roe's expression for the flux, we see where the term FDS arises. Compare Roe's scheme to the first-order upwind scheme for linear systems of equations: the result is a second-order spatial discrete approximation of

$$\mathbf{U}_t + \mathbf{F}_x = \frac{\Delta x}{2} \left(|\mathbf{A}| \, \mathbf{U}_x \right)_x$$

where

$$\mathbf{A} \equiv \frac{\partial \mathbf{F}}{\partial \mathbf{U}}, \qquad |\mathbf{A}| = \mathbf{S} \, |\mathbf{\Lambda}| \, \mathbf{S}^{-1}$$

Thus, Roe's scheme is a conservative generalization of the first-order upwind scheme for linear systems. It is useful to think of *all* FDS schemes as discretizations of the above equation. This is perfectly true for smooth solutions. When, additionally, the Roe linearization is applied to produce the **A** matrix, shock solutions can be computed that satisfy the correct Rankine–Hugoniot relations.

It is necessary to avoid expansion shocks where velocity is sonic and one eigenvalue vanishes, and with it the dissipation on its field. The "entropy fix" proposed by Harten [4] uses a smooth function, say φ, instead of the absolute value of an eigenvalue, to guarantee that it never drops below $\delta/2 > 0$, a tuning parameter, as follows.

$$\varphi(\lambda) = \begin{cases} |\lambda| & \text{if } |\lambda| \geq \delta, \\ \frac{\lambda^2+\delta^2}{2\delta} & \text{if } |\lambda| \leq \delta \end{cases} \tag{4.32}$$

4.2.6 MUSCL Extension to Hi-Res Scheme

We shall illustrate the limited MUSCL extension of Roe's approximate Riemann solver to a Hi-Res scheme. This is an option in DEMOFLOW. The approach can be used with any Riemann solver. With constant distribution of the flow variables cell by cell, we obtain the first-order spatial discretization derived in Section 4.2.4. With reconstruction using adjacent cell averages, we can reconstruct the flow variables on either side of a cell interface more accurately. A quadratic ansatz for the distribution yields for the extrapolated primitive variables **V** at the interface $i + 1/2$ – left and right sides – for an equidistant grid $\{x_i\}$ as follows.

$$\mathbf{V}^L_{i+1/2} = \mathbf{V}_i + \frac{1}{4}[(1+\kappa)\Delta\mathbf{V}_{i+1/2} + (1-\kappa)\Delta\mathbf{V}_{i-1/2}] \tag{4.33}$$

$$\mathbf{V}^R_{i+1/2} = \mathbf{V}_{i+1} - \frac{1}{4}[(1+\kappa)\Delta\mathbf{V}_{i+1/2} + (1-\kappa)\Delta\mathbf{V}_{i+3/2}] \tag{4.34}$$

where $\Delta\mathbf{V}_{i+1/2} = \mathbf{V}_{i+1} - \mathbf{V}_i$.

$\kappa = 1$ corresponds to central differencing, which leads to the wiggles known from Fourier analysis as the Gibbs phenomenon. But even the upwind schemes (e.g. the second-order pure upwind scheme with $\kappa = -1$, Fromm's second-order scheme with $\kappa = 0$, and the third-order upwind-biased scheme with $\kappa = \frac{1}{3}$) lead to oscillations at discontinuities. Limiting the slopes in Eq. (4.33) and Eq. (4.34) yields the values of the primitive variables **V**, which has been found to work better than the conserved variables **U**.

$$\mathbf{V}^L_{i+1/2} = \mathbf{V}_i + \frac{1}{4}\left[(1+\kappa)\phi\left(\frac{\Delta\mathbf{V}_{i-1/2}}{\Delta\mathbf{V}_{i+1/2}}\right)\Delta\mathbf{V}_{i+1/2} + (1-\kappa)\phi\left(\frac{\Delta\mathbf{V}_{i+1/2}}{\Delta\mathbf{V}_{i-1/2}}\right)\Delta\mathbf{V}_{i-1/2}\right] \tag{4.35}$$

$$\mathbf{V}^R_{i+1/2} = \mathbf{V}_{i+1} - \frac{1}{4}\left[(1+\kappa)\phi\left(\frac{\Delta\mathbf{V}_{i+3/2}}{\Delta\mathbf{V}_{i+1/2}}\right)\Delta\mathbf{V}_{i+1/2} + (1-\kappa)\phi\left(\frac{\Delta\mathbf{V}_{i+1/2}}{\Delta\mathbf{V}_{i+3/2}}\right)\Delta\mathbf{V}_{i+3/2}\right] \tag{4.36}$$

If the limiter is symmetric, the two limited slopes in an interface variable become identical, the result is independent of κ, and

$$\mathbf{V}^L_{i+1/2} = \mathbf{V}_i + \frac{1}{2} limit(\Delta\mathbf{V}_{i-1/2}, \Delta\mathbf{V}_{i+1/2}) \tag{4.37}$$

$$\mathbf{V}^R_{i+1/2} = \mathbf{V}_{i+1} - \frac{1}{2} limit(\Delta\mathbf{V}_{i+1/2}, \Delta\mathbf{V}_{i+3/2}). \tag{4.38}$$

The construction of the left and right states is illustrated in Figure 4.15.

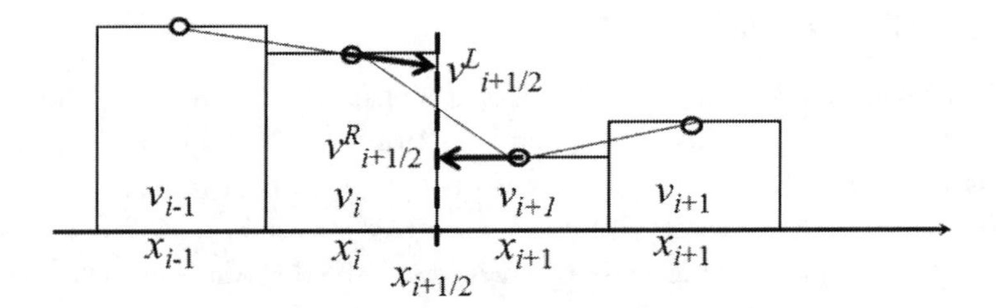

Figure 4.15 Limited MUSCL for v at $x_{i+1/2}$ with *minmod* limiter.

Knowing $\mathbf{V}^{L,R}_{i+1/2}$, we can compute $\mathbf{U}^{L,R}_{i+1/2}$ if needed. We obtain a second-order spatial discretization (except at extrema and discontinuities) with

$$\hat{\mathbf{F}}_{i+1/2} = \mathbf{F}^{Roe}(\mathbf{U}^L_{i+1/2}, \mathbf{U}^R_{i+1/2}) \tag{4.39}$$

(i.e. using Roe's approximate Riemann solver with $\mathbf{U}_L = \mathbf{U}^L_{i+1/2}$ given by Eq. (4.35)) and $\mathbf{U}_R = \mathbf{U}^R_{i+1/2}$ given by Eq. (4.36)). The scheme becomes a five-point scheme and this must be considered in the design of boundary conditions.

Connection to the Artificial Dissipation Approach

Consider the dissipative term in Eq. (4.31), which limits all components of the solution individually. A number of simplified dissipation models with a reduced number of limiters are considered to reduce the computational effort. The simplest methods use only *one* limiter. The dissipation model by Jameson belongs to this class. The switch in the Jameson–Schmidt–Turkel scheme is neither a flux nor a slope limiter that produces a TVD scheme, but it efficiently curbs overshoots and undershoots. It can actually be reformulated so that the scheme diminishes local extrema, and it may be interpreted as an example of a "symmetric limited positive scheme" [6, 7].

The centered scheme with added dissipation then can be formulated by choosing a *scalar* dissipation factor as follows.

$$\boldsymbol{\Psi} = \rho(\boldsymbol{\Lambda})\psi\,\mathbf{I}$$

where $\boldsymbol{I}$ is the identity and the spectral radius of the flux Jacobian is $\rho(\boldsymbol{\Lambda}) = \max_i(|\lambda_i|)$. Then

$$\mathbf{R}_{j+\frac{1}{2}}\Psi_{j+\frac{1}{2}}\mathrm{R}^{-1}_{j+\frac{1}{2}} = \rho(\boldsymbol{\Lambda}_{j+\frac{1}{2}})\psi_{j+\frac{1}{2}}\boldsymbol{I},$$

and the numerical flux is

$$\hat{\mathbf{F}}^n_{j+\frac{1}{2}} = \frac{1}{2}(\mathbf{F}^n_{j+1} + \mathbf{F}^n_j) - \frac{1}{2}\rho(\boldsymbol{\Lambda}_{j+\frac{1}{2}})\psi_{j+\frac{1}{2}}(\mathbf{U}^n_{j+1} - \mathbf{U}^n_j).$$

If we add the fourth order dissipation and choose

$$\psi_j = \kappa^{(2)}\frac{|p_{j+1} - 2p_j + p_{j-1}|}{|p_{j+1} + 2p_j + p_{j-1}|}$$

we obtain the centered Jameson scheme. The choice of $\kappa^{(2)} > 0$ will be discussed later, as used in DEMOFLOW.

4.3 Time Integration Schemes

Before detailing the time-stepping schemes, we need to consider the accuracy and stability of the numerical schemes. The order of accuracy measures the rate at which the numerical solution $\mathbf{u}(\Delta t)$ converges to the exact solution $\mathbf{U}$. If there is a constant K such that $||\mathbf{u}(\Delta t) - \mathbf{U}|| < K \cdot \Delta t^p$, we say the order of accuracy is p.

Instability is a much used term that, in the context of time stepping, means unacceptable error growth with time. Since the differential equation is solved inexactly over each time step, the discretization errors accumulate over time. This is so even for numerical quadrature, where the final error is simply the sum of the local errors in each subinterval. However, differential equations add dynamic behavior to this picture. The error in a step can be seen as a deposit into a global error account, the balance of which develops at a rate – positive or negative – determined by the dynamics. There is often a sharp limit on the time step: if exceeded, the numerical solution grows unacceptably fast compared to the exact solution. Consider, for example, the initial value problem with velocity $a > 0$ for the advection equation, $u_t + au_x = 0, u(x,0) = f(x)$. Let it be discretized by the simplest differences on an equidistant grid with step Δx and time step Δt,u_j^n to approximate $u(j\Delta x, n\Delta t)$ as follows.

$$u_j^{n+1} = u_j^n - \frac{a\Delta t}{\Delta x}\left(u_j^n - u_{j-1}^n\right)$$

This is the upwind scheme, which is first-order accurate in time as well as space. The exact solution exhibits no growth, but only if

$$\frac{a\Delta t}{\Delta x} \leq 1$$

will the numerical solution stay bounded for all times. The computational grid laid out over the x,t plane shows which initial data influence u_j^n: only the interval $[x_j - n\Delta x, x_j]$. But the exact solution at that point is $f(x_j - at^n)$ (see Figure 4.16). Unless $x_j - an\Delta t > x_j - n\Delta x$ (i.e. Eq. (4.3)), there cannot be convergence, since the numerical scheme does not pick up data that include the right point. This is the Courant–Friedrichs–Lewy (CFL) argument formulated in 1926. The time-step limit is called the CFL limit and $\frac{a\Delta t}{\Delta x}$ is the Courant or CFL number. This argument says nothing about growth. But another famous theorem, the Lax equivalence principle of Peter Lax, says that, for consistent schemes, stability and convergence are equivalent. "Consistent" means that the discretization formally converges to the correct PDE when step sizes tend to zero. The upstream scheme is consistent and conditionally stable.

With the direction of the space difference changed to $u_{j+1}^n - u_j^n$ (the dashed cells to the right of Figure 4.16), the scheme is still consistent, but the numerical solution grows exponentially, no matter how small the CFL number: the downstream scheme is unconditionally unstable. The CFL condition can be interpreted to say that a wave must travel less than a grid cell in a time step. But only for three-cell, one-time-step schemes

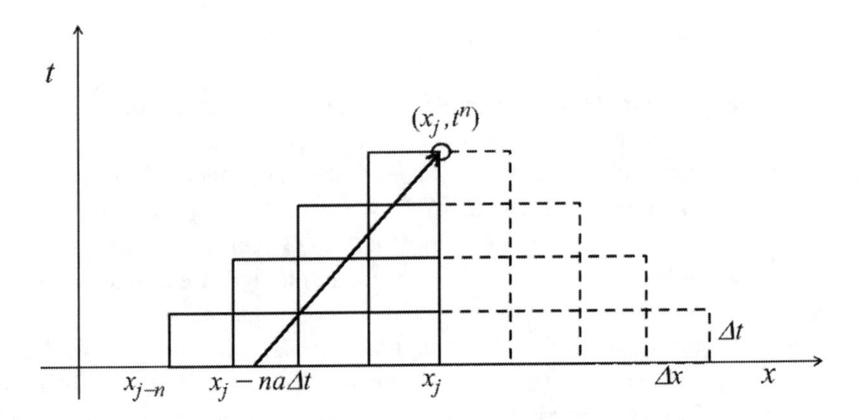

Figure 4.16 Courant–Friedrichs–Lewy convergence condition.

is this so. The explicit Runge–Kutta and implicit time-stepping schemes treated next do better.

Runge–Kutta and Implicit Time-Stepping Schemes

A time step, which is defined in terms of the CFL number as follows.

$$\Delta t = CFL \frac{\Delta x_{min}}{|\lambda_{max}|}$$

where the minimum and maximum are taken over all cells. The allowable CFL number depends on the spatial discretization and the time-stepping scheme. Not only the cell size, but also the local wave velocities are involved, so it is not possible to choose the time step in advance of the calculation. The maximum allowable time step Δt can become very small as the mesh becomes finer.

The motivation behind the development of implicit schemes is to relieve the severe CFL condition limitation above on the permissible time step of explicit schemes.

Implicit schemes can have unbounded CFL limits for constant-coefficient model problems and very large limits also for Euler problems.

Explicit Runge–Kutta Schemes

Runge–Kutta schemes are popular for time stepping to steady state when designed to provide appropriate dissipation. Consider again the semi-discrete system in Eq. (4.3), which introduces the residual $\boldsymbol{R}(t)$ as shorthand for the right-hand side in Eq. (4.18).

$$\frac{d}{dt}\mathbf{U}_j(t) = \mathbf{R}_j(t) \equiv -\frac{1}{\Delta x}\left(\hat{\mathbf{F}}_{j+1/2} - \hat{\mathbf{F}}_{j-1/2}\right)$$

A class of Runge–Kutta schemes, often referred to as “Gary-type” schemes, evaluates the fluxes in each stage only for the last solution update. This makes them memory efficient and suitable for time stepping of PDEs. An m-stage scheme is defined as follows.

$$
\begin{aligned}
&\mathbf{W}_j^0 = \mathbf{U}_j^n \\
&\textbf{For } k = 1 \textbf{ to } m \\
&\mathbf{W}_j^k = \mathbf{W}_j^0 + \alpha_k \Delta t\, \mathbf{R}_j^{k-1} \\
&\textbf{end } k \\
&\mathbf{U}_j^{n+1} = \mathbf{W}_j^m
\end{aligned}
\tag{4.40}
$$

A five-stage, second-order accurate Runge–Kutta scheme is used in DEMOFLOW and a three-stage, first-order scheme is used in EDGE. Their stage coefficients, α_i, given in Table 4.1, can be chosen to control the order of time accuracy and the stability margin of the scheme (i.e. the CFL limit). The stability properties are further discussed in Section 4.3.

Stability of Runge–Kutta schemes

An m-stage Runge–Kutta scheme applied to an ordinary differential equation (ODE) $dq/dt = \lambda q$ with time step Δt produces the recursion $q_{n+1} = P(z)q_n$ with $z = \lambda\Delta t$ and P a degree m polynomial. For this time stepping to be at least first-order accurate, which seems a very desirable property, it is necessary that $\alpha_m = 1$. The *stability region* is the subset S of the complex z-plane where $|P(z)| \leq 1$. When $\lambda\Delta t \in S$, the numerical solution exhibits no exponential growth. The stability regions for EDGE and DEMOFLOW schemes are shown in Figure 4.17.

The Jacobian matrix of the large system of ODEs resulting from semi-discretization can be diagonalized and growth of solutions analyzed by considering each of the scalar advection equations. Central differences give purely imaginary eigenvalues, and upstream differences give negative real ones. The Roe Hi-Res and the Jameson schemes give eigenvalues in the left-hand plane, many with large imaginary parts and

Table 4.1 Stage coefficients for the DEMOFLOW and EDGE Runge–Kutta schemes.

Method	α_1	α_2	α_3	α_4	α_5
DEMOFLOW	0.0695	0.1602	0.2898	0.5060	1.0000
EDGE	0.6667	0.6667	1.0000	–	–

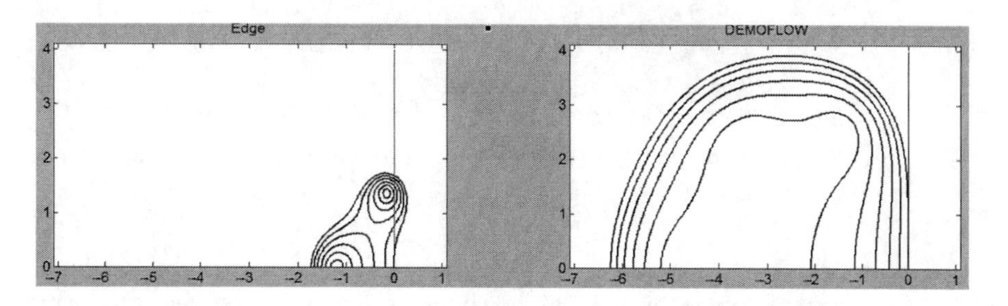

Figure 4.17 Time-stepper stability regions: EDGE (left) and DEMOFLOW (right).

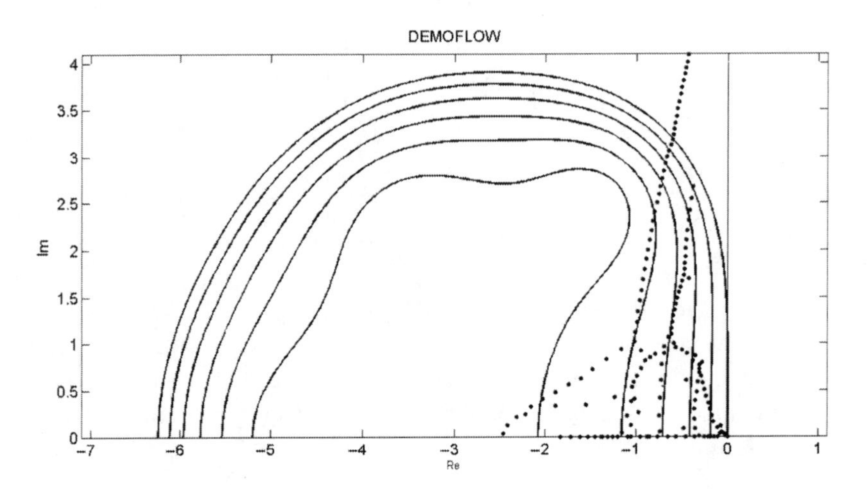

Figure 4.18 Stability region of the DEMOFLOW Runge–Kutta scheme and the $\Delta t\times$ Jameson spectrum.

small real parts. A rough and ready assessment therefore looks at purely imaginary eigenvalues for the CFL limit. The five-stage scheme has a CFL limit of about 1.0 and the three-stage scheme obtains 1.7. On the basis of stability considerations for hyperbolic PDEs, the five-stage scheme obtains a time step of 0.2 per stage (flux evaluation) and the three-stage scheme obtains about 0.6 per stage, a factor of three better. However, the dissipation terms push the poles from the imaginary axis, where they sit for central difference discretizations of hyperbolic equations, into the left half of the plane (see Figure 4.18). The five-stage scheme gives stronger damping there, illustrated by the level curves of $|P|$ in Figure 4.17. Numerical experiments confirm that $CFL > 1$ can be used for the five-stage scheme.

The semi-discretized PDE is a large, nonlinear set of ODEs. It can be linearized around some notional solution and its eigenvalues μ_i can be computed. These are not to be confused with the eigenvalues – the characteristic speeds – of the Euler equation Jacobian flux, to which they are related in a complicated fashion. The point is that the time step should be chosen to make all of these $\mu_i \Delta t \in S$. Figure 4.18 shows the $\Delta t \mu_i$ of the Jameson spatial scheme overlaid on the stability region of the DEMOFLOW Runge–Kutta scheme. There are 100 grid points, so there are about 150 eigenvalues in $\Im(z) \geq 0$, as is shown in Figure 4.18. Clearly, all dots lie in $\Re(z) \leq 0$, so the ODE system is stable. A few transgress the stability boundary for the timestep chosen.

Implicit Methods: Implicit Euler

When $\frac{d}{dt}\mathbf{U}_j(t)$ is approximated by the implicit Euler scheme, the following obtains.

$$\frac{\mathbf{U}_j^{n+1} - \mathbf{U}_j^n}{\Delta t} = -\frac{1}{\Delta x}\left(\hat{\mathbf{F}}_{j+1/2}^{n+1} - \hat{\mathbf{F}}_{j-1/2}^{n+1}\right) \tag{4.41}$$

Time stepping by Eq. (4.41) requires solution of a system of equations because values at the unknown time level $n + 1$ appear in the nonlinear flux functions.

The numerical flux term $\hat{\mathbf{F}}_{j+1/2}$ depends for the central scheme on $\mathbf{U}_j$ and $\mathbf{U}_{j+1}$. The dissipative terms in the Jameson scheme and the limiters in the MUSCL/Roe scheme also introduce dependence on $\mathbf{U}_k, j-2 \leq k \leq j+2$. To solve the system, we linearize the numerical flux terms about the time level n by a Taylor series expansion that leads to the following expression for $\hat{\mathbf{F}}^{n+1}_{j+1/2}$.

$$\hat{\mathbf{F}}^{n+1}_{j+1/2} = \hat{\mathbf{F}}^{n}_{j+1/2} + \sum_{k=-2}^{2} A_j^{+k} \Delta \mathbf{U}_{j+k} \text{ where} \tag{4.42}$$

$$A_j^{+k} = \frac{\partial \hat{\mathbf{F}}^{n}_{j+1/2}}{\partial \mathbf{U}_{j+k}}$$

and a similar expression for A_j^{-k}, where $\Delta \mathbf{U}_j = \mathbf{U}_j^{n+1} - \mathbf{U}_j^n$.

Insert Eq. (4.42) into Eq. (4.41) and rearrange as follows.

$$\frac{1}{\Delta t}\left(\sum_{k=-2}^{2} (A_j^{+k} - A_j^{-k})\Delta \mathbf{U}_{j+k} + \Delta \mathbf{U}_j\right) = -\frac{1}{\Delta x}\left(\hat{\mathbf{F}}^{n}_{j+1/2} - \hat{\mathbf{F}}^{n}_{j-1/2}\right) \tag{4.43}$$

Equation (4.43) represents a large, sparse, linear system of equations to be solved at each time step $\rho, \rho u$, and ρE for every value of j. The coefficient matrix can be ordered into *block penta-diagonal* form with 3×3 (i, j) blocks and zero blocks for $|i - j| > 2$. DEMOFLOW uses a direct method to solve the linear system. However, computational cost and storage requirements for Gaussian elimination are prohibitive in 3D models. Instead, iterative methods are applied and/or approximations are made to the linear system itself. Chapter 6 discusses this issue with reference to the EDGE CFD code.

The implicit Euler conservative discretization of Eq. (4.14) for the nozzle flow adds the source term $\mathbf{G}(\mathbf{U}_j)$, which contributes to the diagonal blocks only. All in all, exact analytical differentiation is cumbersome. DEMOFLOW computes the coefficient matrix by differencing. It is block-banded, and the number of flux evaluations to make a finite-difference approximation does not grow with N.

If Eq. (4.18) were linear with constant coefficients, a stability analysis would show Eq. (4.41) to be unconditionally stable (i.e. Δt could be arbitrarily large). In practice, the nonlinearity reduces the time-step limit to some large but bounded value.

4.4 CFD Workflow for Nozzle Problems

Section 1.4 contains a *recipe for CFD* pictured in Figure 1.19, to be used as a guide on how to approach a physical problem and analyze it with computational tools. In order to illustrate the method, we apply the recipe to a practical problem, starting from the description of the case, up to the implementation of the theory in a MATLAB code, namely DEMOFLOW. The book's website provides DEMOFLOW as well as associated tutorial material on actually running the code, analyzing the results, and drawing conclusions.

Recall the first three items in the *recipe for CFD*:

(1) Study the physical flow
(2) Construct the mathematical problem
- Analyze PDEs
- Choose boundary conditions

(3) Formulate the numerical problem
- Construct a mesh
- Time differencing
- Space differencing
- Initial conditions
- Boundary conditions
- Solve difference equations, determine stability

Section 4.1.3 analyzed the physical flow and identified the main flow phenomena. Items (2) and (3) then, are the next steps, to be illustrated here for the DEMOFLOW nozzle flow simulator.

So what does a quasi-1D solver such as DEMOFLOW contribute to the learning process? An important point is that 3D CFD schemes also work with fluxes across cell boundaries in perfect analogy with the 1D formulas developed above. We have been able to solve quasi-1D nozzle problems as far back as the middle of the nineteenth century. The theory has remained the same, but the ways to implement it numerically have changed. Today, there exists a variety of commercial solvers based on different types of solution methods and discretization schemes. Some of those schemes are represented in DEMOFLOW. For reasons of complexity and the computer resources required, it is nearly impossible for a student to work with a full-blown commercial CFD code by reading about it and then gain understanding of how it works and its characteristics. Keep in mind that best clinical experience is what guides the medical profession, just as best engineering experience should guide the CFD practitioner. DEMOFLOW aims to be a solver with which the user can interact and so explore the characteristics of the different solution schemes to reflect on their pros and cons. The graphical user interface (GUI) will help by keeping the inputs within reasonable bounds and restricting some inputs that may not be valid at a certain instance. A *n*ozzle tutorial on the book's website has been created to help run the code.

DEMOFLOW supports the following four representative types of solvers.

- Explicit Runge–Kutta time marching using Roe's approximate Riemann solver in space
- Explicit Runge–Kutta time marching using Jameson's centered space scheme with added scalar dissipation
- Implicit time marching using Roe's approximate Riemann solver in space
- Implicit time marching using Jameson's centered space scheme with added scalar dissipation

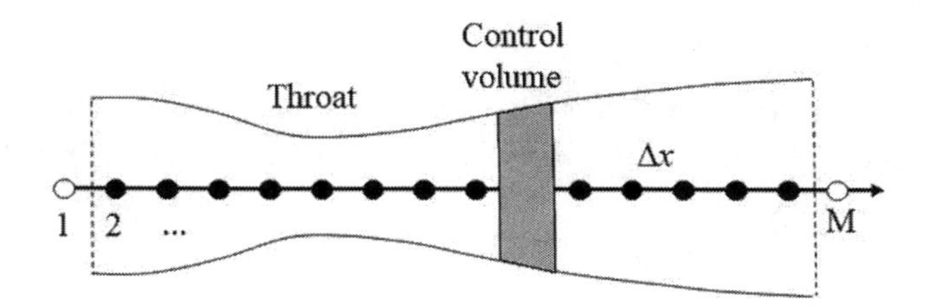

Figure 4.19 Grid and control volume for the 1D Euler solution.

The objective of the exercise then is to compute numerical solutions using these four different solvers with different parameter settings, to compare the computed results with the exact solution, and to draw some conclusions about the effect of varying these parameters. In particular, the objective is to:

- Test the philosophy of time marching to obtain steady-flow solutions
- Gain practical experience with time-integration methods
- Illustrate the effects of mesh refinement and artificial viscosity
- Understand better the stability limits of the various schemes.

Parameters such as the upstream and downstream reservoir conditions, number of nodes, number of time steps, the CFL number, and viscosity coefficients can be changed. The program shows the exact steady solution and the numerical approximation every Nth step. With the DEMOFLOW code at hand, the student can experiment freely and do CFD due diligence. Sections 4.4.5 ff. provide an introduction to the hands-on exercises.

4.4.1 Construct Grid

DEMOFLOW uses an equi-spaced mesh $x_j = j\Delta x$, where $\Delta x = L/(M-2)$. The first step is to generate the x-grid of points, as illustrated in Figure 4.19. Cells with centers numbered $i = 1$ and $i = M$ are the *ghost cells* used to hold the boundary conditions. The states (ρ, u, p) in the ghost cells are constructed from given reservoir conditions and the computed flow in neighbor interior cells as described in Section 4.4.4. This allows the fluxes to be computed by the standard differences for all cells $2, \ldots, M-1$.

4.4.2 Definition of Exact Solution

An exact solution can be determined by the isentropic flow relations and the shock jump conditions. One way to define this exact solution is to specify the location of the shock wave and then determine the solution upstream and downstream from the isentropic relations. This option in DEMOFLOW sets the pressure at the exit boundary, which sets the shock at the desired position. The alternative is to give the back pressure.

4.4.3 Initialization of the Flow Field

The code initializes the flow field by linearly interpolating between the exact values of pressure and temperature at the inlet and at the exit. There is no informed guess for velocity, so it is set to zero.

4.4.4 Boundary Conditions

The time-marching approach requires stepping the solution forward in time until all disturbances have been expelled through the boundaries and the flow becomes steady. Thus, it is important to be able to reach the steady-state flow in a *small* number of time steps. Conditions at open boundaries should be set to allow waves to pass out from the domain without reflections. Such conditions are called nonreflective or absorbing boundary conditions (ABCs).

The method of characteristics determines the number of boundary conditions required at each boundary to uniquely define the solution from the boundary and initial conditions (i.e. to make the initial–boundary value *well posed*). Well-posedness also implies limited sensitivity to the data prescribed, a necessary condition for conclusions about the physical setup to be learned from solution of the mathematical problem. These conditions, called *physical* conditions, are too few to determine the complete states in the ghost cells, so information from the flow must be used. These are called *numerical* boundary conditions. The Hi-Res finite-volume schemes described above are five-point schemes, requiring data from two cells on either side, so there must be two *ghost* cells at each end of the computational domain with values provided by the physical and numerical conditions.

Characteristic Lines and Physical Boundary Conditions

The considerations of Riemann variables have a strong bearing on the boundary conditions to be imposed at the inlet and outlet of the monodimensional flow. At the inlet, the lines C_0 and C_+ have slopes u and $c+u$, which are always in the positive x direction. Therefore, they carry the boundary information into the computational domain.

The third characteristic has a slope $u-c$, whose sign depends on the inlet Mach number. When the flow is supersonic, the information is transported into the flow domain and the whole state is specified. This does not happen for a convergent–divergent nozzle. When the flow is subsonic, the value of w_3 is carried from the computational domain. So extrapolating w_3 from inside and combining with isentropic expansion from p and T in the upstream reservoir does the trick. Actually, the theory (linearized) gives only the rank of the boundary conditions to be specified for well-posedness, not the exact variable combinations. But some combinations are more effective than others for letting waves out, which influences the rate of approach to steady state in the time stepping.

This analysis can be repeated at the outlet, and the results are summarized in Table 4.2 and illustrated in Figure 4.20.

Table 4.2 Physical and numerical boundary conditions for 1D inviscid flows.

	Subsonic	Supersonic
Inlet	Physical conditions: p, T Numerical conditions: w_3	Physical conditions: all Numerical conditions: none
Outlet	Physical conditions: p Numerical conditions: w_1, w_2	Physical conditions: none Numerical conditions: all

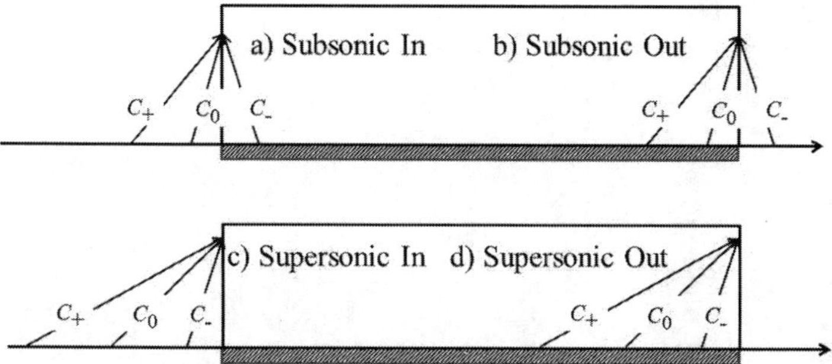

Figure 4.20 Number of boundary conditions imposed by characteristics: (a) subsonic inflow – two, (b) subsonic outflow – one, (c) supersonic inflow – three, (d) supersonic outflow – none.

Inflow Boundary Conditions

The inflow in these simulations will always be subsonic (i.e. there are two ingoing and one outgoing characteristics). One characteristic variable is extrapolated from inside to form a complete state with pressure and temperature equal to those in the reservoir.

Outflow Boundary Conditions

The outflow boundary conditions will depend on the flow case.

- For *supersonic* outflow, all characteristics are going out from the domain and physical boundary conditions are neither required nor allowed. Values of the flow variables at the boundary (i.e. numerical boundary conditions) are obtained by extrapolation of all variables from the interior field points.
- For *subsonic* outflow, one characteristic is ingoing, so one variable, the pressure, is set to the exit pressure $p(L,t) = p_{exit}$. The remaining necessary numerical boundary conditions are obtained by linear extrapolation from the interior field points.

Physical and Numerical Boundary Conditions

As outlined in the previous section, the *physical boundary conditions* have to be complemented by extrapolation of information – the *numerical boundary conditions* from the inner computational domain. Linear extrapolation of the Riemann invariants from interior points determines the values for the numerical boundary conditions. This leads to the completed boundary conditions table for the nozzle model (Table 4.2).

First-Order Extrapolation

In order to guarantee a second-order accuracy in the inside of the flow calculation, it is necessary that the numerical boundary conditions be evaluated with a first-order accurate estimate. This is obtained by extrapolating the Riemann variables linearly from two interior nodes. Note that only one time level is involved: we are looking for a steady flow. The extrapolation for time-accurate simulation needs to involve at least two time levels.

Example: Subsonic Outlet

At time n, the extrapolation of a variable v is as follows.

$$v_M^n = 2v_{M-1}^n - v_{M-2}^n$$

The Riemann variables

$$w_{1,M} \text{ and } w_{2,M} \tag{4.44}$$

are extrapolated. The pressure is known from the reference (back) pressure, $p_M = p_{ref}$, so there are three equations to solve for the complete U vector.

4.4.5 Selecting Schemes and Parameters

The DemoFlow GUI is shown in Figure 4.21. The nozzle, grid, and case is defined in the upper-left panel. The throat area is 1 at $x = 0.35$. Its inlet and outlet areas (1.5 and 2.5), number of grid points (150), as well as reservoir temperature and pressure ($t01, p01$, 288 K, and 100 kPa) are input, together with the desired shock position (0.7). The exact solution is computed and that provides the back pressure $p2$ required. It is an option to give $p2$ as an input instead of the shock position.

The numerical scheme is controlled in the lower panel in Figure 4.21. At the bottom, "Explicit Jameson" has been selected, and the relevant parameters can then be given: number of time steps (400), artificial dissipation parameters ($Vis2$, $Vis4$ at 0.25 and 0.02), and the CFL number (1). The solution is plotted every 10 steps. Plotting and other window and user interface management may actually take more time than all of the computations! To the right, we see the numerical solution (line with markers) that virtually overplots the exact solution (black dashed line), except for the over-/under-shoots, and the plot of $\log_{10}$ density RMS residual versus time-step number. A total of 2000 have brought the residual to 7×10^{-3} of its initial value. We have clicked the solve button five times.

The shock is in the correct position, smeared over a couple of cells, with some over- and under-shoots upstream and downstream. This is typical of the central difference Jameson scheme. Some tuning of $Vis2, Vis4$ may reduce the wiggles at the cost of further smearing of the shock.

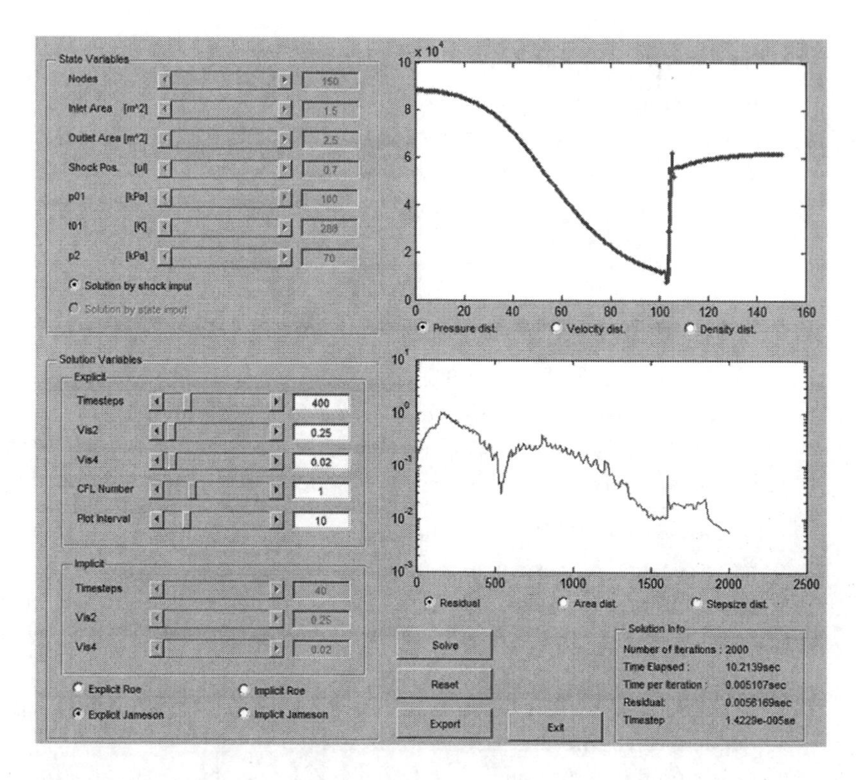

Figure 4.21 User interface for DemoFlow after solution.

Convergence in Time to Steady State

In a *time-marching solution*, the flow variables change from one time step to the next: we solve the set of ODEs

$$\frac{d\mathbf{U}}{dt} = \mathbf{R}(\mathbf{U})$$

and as the steady state is approached, the *residuals* **R** should approach zero. The algorithm that gives the fastest decay, in terms of computing time, of the residuals to arbitrarily small values – to "machine zero" – is usually looked upon most favorably.

One generally uses *a norm of the residual vector* $||\mathbf{R}^n||$ at time iteration n as a measure of the evolution toward convergence. A practical convergence criterion can be set by requiring that the residual drops by a predetermined number of orders of magnitude.

The residuals are generally first *nondimensionalized* by introducing reference quantities $\mathbf{U}_{ref}$. For external flows, the free stream properties can provide these, often via the initial guess for the flow field. The nozzle case has no free stream as such. A crude estimate follows from assumption of isentropic, isenthalpic expansion. Now, $\mathbf{R}^n$ is an array $R_{i,j}, i = 1, \ldots, M, j = 1, \ldots, N$ where M is the number of grid points and N is the number of variables (3) for DemoFlow. Measures generally used to indicate residual magnitude for each variable are as follows.

– The RMS, or $||L_2||$-norm residuals $RMS_k^n = \sqrt{\sum_i (R_{i,k}^n)^2}/\sqrt{M}$.
– the L_∞ norm, max-residuals, $||R_k||_\infty^n = \max_i |R_{i,k}^n|$. The initial values of the residuals are somewhat irrelevant; the focus instead is on how quickly they decay with n. The lower-right plot in Figure 4.21 shows how the RMS density residuals decrease during 2000 iterations for the Runge–Kutta Jameson scheme with $CFL = 1$. The convergence rate is not impressive.

4.4.6 Active Learning with DEMOFLOW

The schemes employed in DEMOFLOW are representative of current compressible flow CFD codes, among them the Swedish contender EDGE, the German Aerospace Center (DLR) code TAU, and the open source code SU2 from Stanford University. The schemes used in DEMOFLOW have been described above and are repeated here.

- Jameson is a centered-plus-artificial-scalar-dissipation space semi-discretization, being second order except at shocks.
- Roe relies on a linearized Riemann solver, with limited linear reconstruction at cell faces, which gives no over- or under-shoots at shocks. It is second order except at shocks and qualifies as a Hi-Res scheme.
- For either space scheme above, time stepping can be either Runge–Kutta or implicit Euler. The latter uses Newton solution at each step, with the Jacobian matrix produced by numerical differencing. It attempts to choose as large as possible a time step as the solution proceeds, but is bounded by the necessity to make the initial guess for the Newton iteration sufficiently close. In a sense, this scheme does not try to use information about the solution as it unfolds, except its apparent smoothness and conditioning of the Jacobian matrix.

It is obvious from the solution plots that the explicit schemes follow the waves as they reflect on boundaries and shock – the overall rate of decay constant. The implicit scheme follows a similar pattern initially until it gets the shock in the right place, and then converges in a few iterations.

For explicit schemes, the total work is $O(M^2)$, with M the number of grid points. It follows from the physics that convergence will take place at some "real" time when reflections on inlet and outlet boundary conditions have let out enough errors. The CFL limit on the "real" time step is $O(\Delta x) = O(M^{-1})$, so $O(M^{-1})$ time steps must be taken. Each step requires $O(M)$ work, so the estimate follows.

No such easy estimate is available for the implicit solvers. For them, the issue is robustness: the initial guess is quite incorrect, so they must start by following the physics until the solution is close enough to the steady solution. That is the whole point of the time-stepping technology. Even CFD codes addressing the steady equations by Newton iterations have found it necessary to add an approximation to the time derivative to make the solver more robust, in addition to developing a sophisticated adaptive choice of the (quasi)-CFL number. Thus, in the numerical analyst's vocabulary, a suitably damped Newton scheme is provided.

So which of the four is the winner? As you might expect, the answer is not simple, and it depends on the case at hand. How strong is the shock? How accurate must the solution be? How monotone must it be? How far away can it be from the naive initial guess? The CFD practitioners will gain experience on their chosen class of problems by playing with the controls offered by the code. Now it is your turn to see CFD in action by turning to the DEMOFLOW nozzle simulation tutorial.

We now have an understanding of how to compute shocks waves in 1D inviscid flow. Describing the aircraft shape and generating a grid around it are the next steps to advance to the problem in 3D. This is the subject of the next chapter.

4.5 Learn More by Computing

Gain hands-on experience of the computational tools for the topics in this chapter by working with the on-line resources. Exercises, tutorials, and project suggestions are found on the book website www.cambridge.org/rizzi. Software used to compute many of the examples shown is available from http://airinnova.se/education/aero-dynamic-design-of-aircraft, and for this chapter it is the DEMOFLOW nozzle simulator.

References

[1] J. D. Anderson. *Computational Fluid Dynamics: The Basic with Applications*. McGraw-Hill, Inc., 1995.

[2] S. K. Godunov. A difference scheme for numerical solution of discontinuous solution of hydrodynamic equations. *Math. Sbornik*, 47: 271–306, 1960. Translated US Joint Publ. Res. Service, JPRS 7225.

[3] B. Gustafsson. *High Order Difference Methods for Time Dependent PDE*. Springer Verlag, 2008.

[4] A. Harten. High resolution schemes for hyperbolic conservation laws. *Journal of Computational Physics*, 49: 357–393, 1983.

[5] C. Hirsch. *Numerical Computations of Internal and External Flows, Vol. 1 and 2*. Wiley, 1990.

[6] A. Jameson. Analysis and design of numerical schemes for gas dynamics 1; artificial diffusion, upwind biasing, limiters, and their effect on multigrid convergence. *International Journal of Computational Fluid Dynamics*, 4: 171–218, 1995.

[7] A. Jameson. Analysis and design of numerical schemes for gas dynamics 2; artificial diffusion and discrete shock structure. *International Journal of Computational Fluid Dynamics*, 5: 1–38, 1995.

[8] P. D. Lax and B. Wendroff. Systems of conservation laws. *Communications on Pure and Applied Mathematics*, 13(2): 217–237, 1960.

[9] R. J. LeVeque. *Finite Volume Methods for Hyperbolic Problems*. Cambridge University Press, 2002.

[10] R. W. MacCormack. The effect of viscosity in hypervelocity impact cratering. Technical report. AIAA paper 69-354, 1969.

[11] E. M. Murman and J. D. Cole. Calculation of plane steady transonic flows. *AIAA Journal*, 9: 114–121, 1971.

[12] NASA. Shadowgraph images of re-entry vehicles – gpn-2000-001938.jpg, 1960. Available from https://commons.wikimedia.org/w/index.php?curid=16903011.

[13] A. Rizzi and B. Engquist. Selected topics in the theory and practice of computational fluid dynamics. *Journal of Computational Physics*, 72(1): 1–69, 1987.

[14] P. L. Roe. Approximate riemann solvers, parameter vectors, and difference schemes. *Journal of Computational Physics*, 43: 357–372, 1981.

[15] J. C. Tannehill, D. A. Anderson, and R. H. Pletcher. *Computational Fluid Mechanics and Heat Transfer*. Taylor and Francis, 1997.

[16] B. van Leer. Towards the ultimate conservative difference scheme. II. Monotonicity and conservation – combined in a second-order scheme. *Journal of Computational Physics*, 14(4): 361–370, 1974.

[17] J. Von Neumann and R. D. Richtmyer. A method for the calculation of hydrodynamic shocks. *Journal of Applied Physics*, 21: 232–237, 1950.

5 Airframe Computer-Aided Design and Automated Grid Generation

> During the first decade of the 2000s, mesh generation became bigger than the finite element methods that gave birth to it. Computer animation uses triangulated surface models extensively, and the most novel new ideas for using, processing, and generating meshes often debut at computer graphics conferences. ...Meshes today find heavy use in hundreds of other applications, such as aerial land surveying, image processing, geographic information systems, radio propagation analysis, shape matching, population sampling, and multivariate interpolation. Mesh generation has become a truly interdisciplinary topic.
>
> Jonathan Shewchuk[1]

Preamble

There are two issues to discuss in this chapter: the generation of a geometric model for the outer shape of the configuration and the generation of a computational grid over the flow domain. The following topics will be covered:

- Geometry modeling tools
- Representation and manipulation of curves and surfaces
- The SUMO aeroshape-specific geometry modeler
- Mesh generation generalities and the Delaunay method
- Surface triangle meshes
- Volume tetrahedral meshes
- Mesh control and troubleshooting in SUMO.

For a background on computer-aided geometric design, see Gerald Farin's book [13] and the classic by Carl DeBoor on B-splines [12] or Piegl's "NURBS book" [22]. Mesh generation and its automation is a fast-moving subject, as indicated by Shewchuk's obervations above. Automation requires robust methods that produce "provably good" grids on any geometry. Older attempts were fragile and very dependent on the user's experience. The reader with interest in the generation of structured grids, or hex grids, needs to turn to background literature such as the book by Joe Thompson et al. [30] produced through the US National Grid Project. A more recent comprehensive treatment

[1] *Delaunay Mesh Generation* [9], with permission of Taylor and Francis Group, LLC.

is given in Ref. [29]. For recent developments, see Shewchuk's home page at https://people.eecs.berkeley.edu/jrs/, which lists relevant books, conferences, and symposia.

5.1 Introduction and Overview

We have encountered flow simulation for computing forces on thin lifting surfaces through the vortex lattice method in Chapter 3. The geometry description was defined by the wing planform and the camber surface. The former is defined by a few high-level descriptors such as dihedral, taper, aspect ratio (AR), and sweep, and camber surfaces are lamina defined by the camber curves of airfoils, in turn being defined by interpolation to a set of points. The more advanced flow models required for high-speed flight, for example, are formulated as partial differential equations (PDEs) and need a detailed description of the outer surface of the aircraft, possibly extended by details of propulsion: propeller disks, jet engine intakes, and exhausts. Once the airframe geometry is defined, its exterior volume must be subdivided into small cells – the computational grid or mesh – for the numerical solution of the PDEs. In Chapter 4, this was a trivial task, but grid generation for detailed configurations is very demanding of the engineer's time. Currently, automated grid generation is becoming a reality, and this book uses the SUMO and TETGEN tools that take a geometry description format, specialized for aircraft, into a high-quality grid for Euler computational fluid dynamics (CFD).

5.1.1 Paths to Meshable Models

CAD Modeling Issues

Geometric modeling is often performed in an industry-standard, general-purpose computer-aided design (CAD) system such as CATIA (www.3ds.com/products/catia). Taking the CAD model, a general-purpose mesh generator, such as POINTWISE (www.pointwise.com) can create the computational grids exported to a CFD solver. This approach has some disadvantages. General-purpose CAD systems, mesh generators, and high-fidelity flow solvers are complex. They require expertise in CFD as well as CAD, and this together with licensing and training costs is a significant barrier to their widespread use, especially in educational situations. In addition, the level of details in a CAD model is often excessive for CFD simulations, requiring simplification before export to the mesh generator. Furthermore, unless the CAD model is created appropriately, the geometry description is not watertight and needs "repair" before a volume mesh can be made, which takes extensive effort. Automatic CAD repair is a key enabler for speeding up the simulation process. In addition, although the steps above may all be rather straightforward for an experienced CAD–CFD engineer, the sheer number of cases still makes the work error-prone.

When detailed designs are not called for, such as in early design stages when the configuration is very fluid, the usual practice is to use a *purpose-specific CAD modeler* that is *simpler* than general CAD systems in the sense that fewer parameters are needed for the configuration layout. Examples are the RDS [23], AAA [6], and VSP [15] software systems. VSP supports the export of meshable surface models and/or meshes in standard formats such as IGES [32], STEP [5], and CGNS (http://cgns.sourceforge.net).

Purpose-specific CAD develops the geometry representation of the parameterized design concept and manages, creates, and edits it toward computational discretizations for the different analysis tools. The airframe CAD tool SUMO discussed in this text provides geometry design, IGES output, and high-quality unstructured surface meshing, seamlessly integrated with the TETGEN volume mesh generators, for Euler CFD.

Parameterization

The CPACS extended markup language (XML) schema represents a complete parameterized model that supports trade studies, design space exploration, and optimization. The system handles all types of lifting surfaces and all major controls. Issues requiring more effort include shaping of details such as air intake lips and control surface deflection.

XML formats were first widely used in business for administrative data and are now also accepted by engineering communities for the large, complex data sets encountered in scientific computing. An XML file contains annotated, tree-structured data and is easily interpreted by software. Most text editors or browsers will display the file with indentation, reserved words in color, etc. This makes it readable and also editable by humans.

Many design tools for aircraft use XML formats (SUMO, AAA, DFS, VAMPZERO, VSP, etc.), which simplifies the task of writing interfaces between them to exchange geometry. This has encouraged the development of a data set containing *all* design data for aircraft development from early conceptual to production stages.

The CPACS standard [11] is supported and developed by the German Aerospace Center (DLR) and has gained acceptance in a number of European aerospace companies and research institutes. It supports detailed geometry data for outer skin as well as all structural elements, spars, ribs, stringers, etc., propulsion characteristics, mission profiles, weight and balance data, and compiled simulation (or measurement) results such as aero-force, loads, and stability and control databases. In the current CPACS philosophy, most components, such as the wing or the fuselage, are parameterized through cross sections or profiles. Each cross section or profile is represented by a number of points only. This representation does not in itself admit manipulation via the classical wing geometry parameters (twist, taper, sweep, etc.), but such manipulations can be supported by software layers on top of the CPACS model.

Mesh Generators

Major commercial efforts to speed up the CFD analysis cycle have produced significant progress toward automatic, reliable, physics-aware mesh generation. For external aerodynamics, commercial products are offered for several solver technologies (e.g. POINTWISE www.pointwise.com/about/index.html and ANSYS ICEM-CFD

Figure 5.1 Wireframe representation of Concorde (left); surface representation of Concorde (right).

www.ansys.com/products/icemcfd.asp for unstructured tetra-hexa-prism grids and NUMECA HEXPRESS www.numeca.com/index.php?id=16 for multi-block hexa meshes).

5.1.2 Boundary Representation of the Configuration

Figure 5.1 (left) shows a *wireframe* plot of the curves from which the surface model (right) is built. The surfaces are lofted from the curves, then intersected and trimmed for the wing, the fuselage, and the vertical fin. The resulting surface is closed and orientable (i.e. *manifold*), and hence it defines an interior volume. Thus, it is a boundary representation (*BREP*) of that volume. The surface admits creation of a polyhedral approximation – a surface mesh. The volume enclosed between a large sphere centered at the surface and the surface itself can be subdivided into a volume mesh of nonoverlapping computational cells. Algorithms for creating the volume mesh need to ask the geometry model for the inside-ness of suggested points, and that requires orientability of the surface. A geometry model that admits such automatic meshing will be called *meshable*.

An aircraft surface is piecewise smooth and closed, but may have sharp edges and vertices. Its enclosed volume is most often singly connected, but, for instance, a box-wing configuration, or one with jet engines modeled as open pipes, is not.

5.2 Curve and Surface Geometry Representation

This section gives an overview of computer representation of curves and surfaces as used in CAD for aircraft (airframe CAD).

5.2.1 From Points to Curves

Most, if not all (see Section 5.3.1), curves are represented as parametric; say, as a space curve:

$$\mathbf{p}(s) = (x(s), y(s), z(s)), 0 < s < s_{max}$$

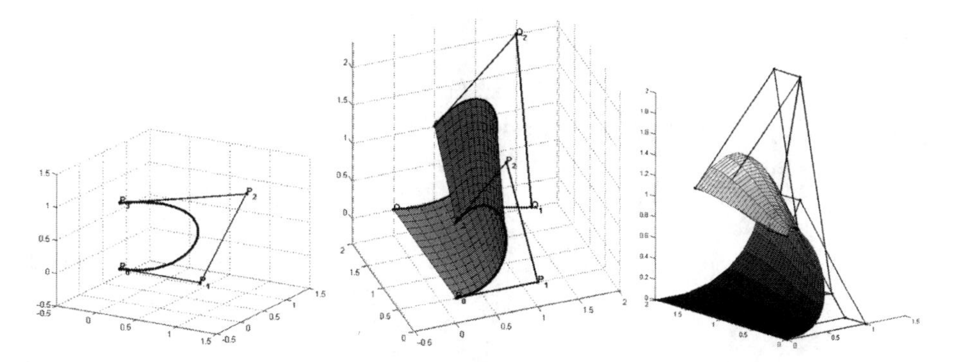

Figure 5.2 Third-degree Bézier space curve (left); ruled surface from two Bézier curves (middle); a bi-cubic Bézier surface (right).

Most readers will be familiar with drawing programs where the user manipulates *control points*, some with attached tangents, to produce a visually pleasing curve. In most CAD programs, such curves are represented by a list of their control points. For evaluation of, for instance, the display of the curve, the functions $(x(s), y(s), z(s))$ are evaluated at many points s_i and the polygon $(x(s_i), y(s_i), z(s_i))$ drawn. As an example, in Figure 5.2 (left), a third-degree Bézier curve $\mathbf{p}(s)$, is defined by the four control points $\mathbf{P}_i, i = 0, \ldots, 3$.

$$\mathbf{p}(s) = \sum_{i=0}^{3} \mathbf{P}_i B_i(s) \quad \text{where the } B_k \text{ are the Bernstein polynomials}$$

$$B_k(s) = \binom{3}{k} s^k (1-s)^{3-k}$$

Note that

- The curve passes through $\mathbf{P}_0$ and $\mathbf{P}_3$
- The line segments $\mathbf{P}_0\mathbf{P}_1$ and $\mathbf{P}_2\mathbf{P}_3$ are tangents to the curve at the ends.

Curve ends can be manipulated by moving $\mathbf{P}_1$ and $\mathbf{P}_2$. Cubic Bézier curves such as this can be strung together to make continuous curves that pass through a given ordered set of points. Tangent continuity obtains when control point triples $\mathbf{P}_{3i-1}, \mathbf{P}_{3i}, \mathbf{P}_{3i+1}, i = 1, 2, \ldots$ are collinear.

Curvature continuity can also be arranged by constraints on the control points. Note that there is no requirement on continuity of the parametric derivatives of $\mathbf{p}(s)$, only on the geometric properties of the curves.

However, representation by control points requires extra information on the degrees of the pieces. Definition of smooth curves by the points on them, and not control points, is a basic and easily interpreted external representation. If the given points are spaced sufficiently close together, many reasonable interpolation formulas give curves that differ only within acceptable tolerances. As we shall see, the SUMO geometry tool uses point set definitions as input and parametric polynomial representation internally.

5.2.2 Lofting Curves to Surfaces

The process of producing a surface from a set of curves is often called *lofting*. In the pre-computer days, shape definitions were drawings managed on large flat surfaces, often the floors of the drawing loft. Given two curves, a *ruled surface* is traced out by moving a straight line over them. Mathematically, a parametric surface is defined by letting x, y, and z be functions of two parameters, say u and s. The first curve $\mathbf{p}_1(s)$ is the one in Figure 5.2 (left), and the second $\mathbf{p}_2(s)$ is likewise a third-degree Bezier curve with control points $\mathbf{Q}_i, i = 0, \ldots, 3$. A point on the ruled surface is as follows.

$$\mathbf{S}(u, s) = (1 - u) \cdot \mathbf{p}_1(s) + u \cdot \mathbf{p}_2(s), 0 \leq u, s \leq 1$$

as shown in Figure 5.2 (middle).

By generalizing the linear interpolation to cubic, carried out by Bernstein polynomials, eight more control points come into play, so we renumber all 16 into a control point grid of four rows and columns $\mathbf{P}_{ij}, 0 \leq i, j \leq 3$. The result is a *parametric bicubic surface patch* as follows.

$$\mathbf{S}(u, s) = \sum_{i,j=0}^{3} \mathbf{P}_{ij} B_i(u) B_j(s)$$

The control point grid and the surface are shown in Figure 5.2 (right).

The surface modeler SUMO offers wing definitions from a set of airfoils (defined by points) by ruled or bi-cubic surface patches. The bi-cubic surfaces are composed from patches such that tangents, but not necessarily curvatures, are continuous across joins.

Historical Remarks

Bézier curves were published in 1962 by Renault engineer Pierre Bézier when drawings became computerized and lofting became a programming sport. Citroën's opposite number was Paul de Casteljau, who developed such curves in 1959. His algorithm for curve evaluation was a repeated linear interpolation and hence was numerically stable even for higher-degree polynomials. The curves have many properties, such as affine invariance, that make them behave in predictable ways when the control points are moved. But they cannot reproduce exact conic sections such as ellipses or quadric surfaces such as paraboloids. The first airplane to be designed with conics was the World War II North American P-51 Mustang, before the advent of Bézier curves. The mathematics of the conic section curves is described in terms of projective geometry of conic sections and (at least one paper) relies on an antique theorem on intersections of multiple tangents to conics, tangency being a property that survives projection. Now, a quadratic Bézier curve is a parabola – a special conic section – that can be projected onto other types of conic sections, keeping the tangency with the projected tangents. This observation opened the way to the generalization to rational quadratic parametric functions, which include conic sections, at the cost of increasing the dimensionality by another set of parameters – the weights.

The industry standard is now the nonuniform rational B-spline (NURBS) curves and surfaces, of arbitrarily high degree. Rational B-splines introduce, in addition to the control points and weights, the concept of *knots*. They offer, by control of the multiplicity of the knots, a systematic way to join the patches of curves or surfaces with control over the continuity so sharp edges can be modeled as easily as curvature continuity. The mathematician Carl de Boor developed theory and algorithms for B-splines and popularized them in a classic textbook [12]. **B** could justly stand for de **B**oor, but the name was coined earlier by Jacob Schoenberg to be short for **B**asis spline.

5.2.3 Free-Form Deformation

The practitioners of geometric modeling for computer animation soon realized that building objects from scratch is time consuming. It is often much faster to modify an already existing model – kneading it, as it were, by movement of control points. The free-form deformation (FFD) technique moves only points, so topologies are not affected. Thus, bodies, surfaces, and curves alike can be deformed. FFD in this generality was described first by A. Barr in 1984 [7] and is now a highly developed interactive technology. FFD is useful in CFD for geometry modification, mesh deformation, and shape optimization, as used in, for example, the SU2 software suite.

Basics of FFD

To modify an object G, we define an enclosing rectangular box B defined by control points $\mathbf{P}_{ijk}$ laid out in an equidistant $(Nx+1)\times(Ny+1)\times(Nz+1)$ grid. Using the Bézier polynomials, a point $\mathbf{p} = (x, y, z)$ inside B is moved to

$$\mathbf{p} + \sum_{i,j,k=0}^{Nx,Ny,Nz} \delta\mathbf{P}_{ijk} B_i^{Nx}(x) B_j^{Ny}(y) B_k^{Nz}(z)$$

where $\delta\mathbf{P}_{ijk}$ is the movement of control point i, j, k from its initial position. G is modified by moving all of its defining points inside the control box, be they vertices in polyhedra or control points that, in turn, define geometric entities.

As an example, Figure 5.3 shows a NACA0012 airfoil with nose at (0.25,0) deformed by movement of the 9×2 control points to bend the foil up 30° from about $x = 0.6$. The rightmost seven pairs of FFD box control points are rotated around $(0.7, 0)$, leaving the left two pairs unmoved. This leaves the nose unchanged with a smooth (C^1) transition to the bent part.

Another application is shown in Figure 5.4, a 1D FFD example with six control points to modify a curve. The task is to modify the thickness-to-chord ratio of wing sections along the span from the baseline to a better aerodynamic shape. CFD on the original wing shows that the root region has unfavorable isobar patterns. As explained in Chapter 9, this can be improved by increasing its thickness and/or moving its location of maximal thickness forward. Figure 5.4 shows the original (thick line),

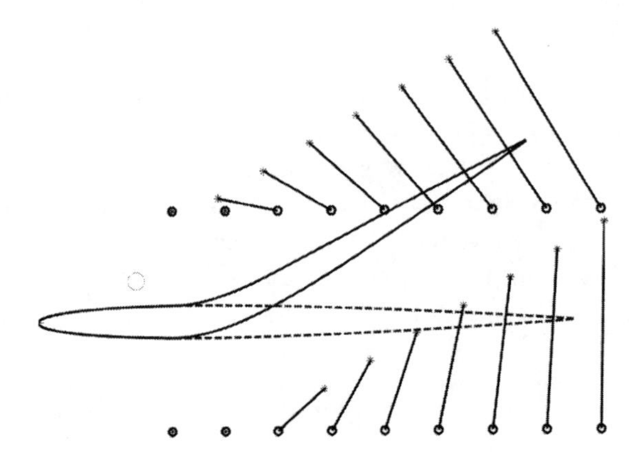

Figure 5.3 Modification of NACA0012 profile by movement of FFD control points.

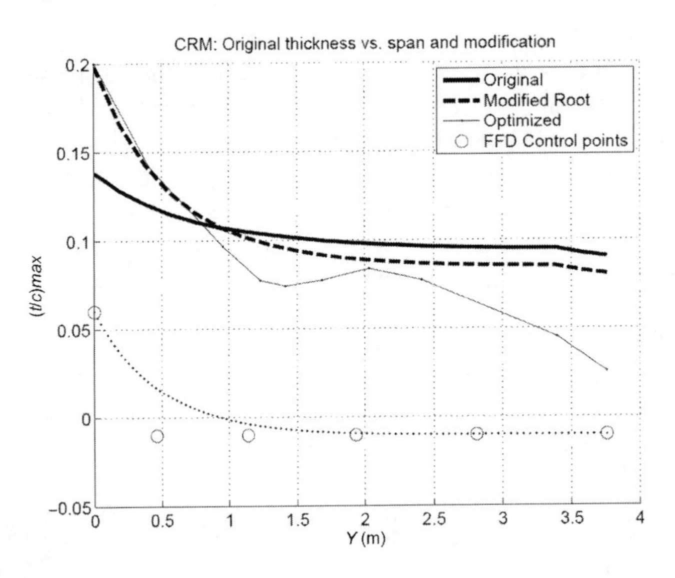

Figure 5.4 Modification of thickness variation along the span of the NASA Common Research Model (CRM) wing.

the modified curve (thick dashed line), the control points (rings), and the increment to the original (dotted line). The thin black curve is a thickness distribution [20] found by nonlinear optimization and with significantly more parameters to vary.

The first point is moved "up" to thicken the root. As a quick calculation shows, that extends the thickening too far out, so we also cluster the remaining points toward the root. Finally, we move them down a little.

5.2.4 Curve Modifiers

When a shape is to be modified, (e.g. to optimize it for some objective), the modification can be done by additive perturbations. The Hicks–Henne bump functions [16] (Eq. (5.1)), are often used to represent airfoil shape variations.

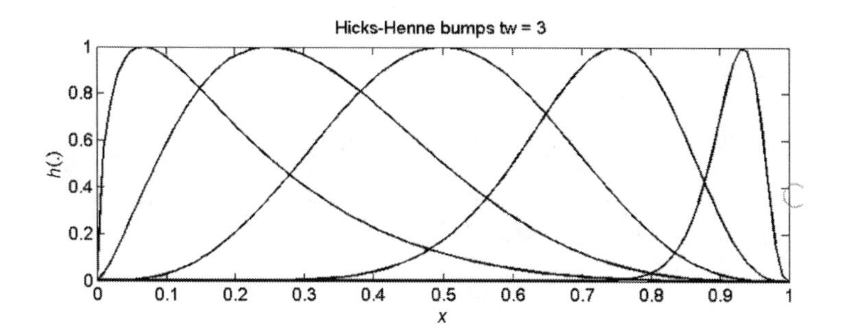

Figure 5.5 Five Hicks–Henne bumps with maxima at the Chebyshev abscissae.

$$H(x,m,tw) = \left(\sin\left(\pi x^{\frac{\log .5}{\log m}}\right)\right)^{tw}, \ 0 \leq x \leq 1. \tag{5.1}$$

Then $H(0) = H(1) = 0$, $0 \leq H \leq 1$, the maximum = 1 is at $x = m$, and the width of the bump is controlled by the exponent tw. The curve $y = f(x)$ is perturbed to the following.

$$f_p(x) = f(x) + \sum_{i=1}^{N} \alpha_i H(x, m_i, tw_i)$$

Figure 5.5 shows five $H(.)$ with maximum at the $N = 5$ Chebyshev abscissae, $m_i = 1/2\left(1 - \cos\frac{i\pi}{N+1}\right)$, $i = 1, \ldots, N$, and $tw = 3$. Examples of shape optimizations with Hicks–Henne bumps are shown in Chapters 7 and 8.

5.3 Airfoils and Surfaces

The lifting surfaces are the object of intense design efforts. Both design and analysis are most often applied to the streamwise cross sections of the wing, the airfoils, and to the wing as built from a set of airfoils. Wind tunnel airfoil testing since the beginning of manned flight has produced a large database of forces and moments, as they depend on angle of attack, airspeed, and Reynolds number. In the days when wings were straight and lofted from two or three airfoils, designers could often find what they needed in the database. But a modern airliner must have the wing shape optimized in great detail for efficient transonic flight, so it cannot be described by the data from just a handful of foils. It follows that new airfoils must be designed with every project, with shapes parameterized for accurate CFD analysis and model manufacture. We discuss below geometric modeling of airfoils and the airframe CAD surface modeler SUMO, which will be responsible for creating the wing shapes in our examples.

5.3.1 Airfoils

Most airfoils are defined by sets of points, available from web repositories such as the University of Illinois at Urbana–Champaign (UIUC) Airfoil Coordinates Database [4].

The points are $(\xi_i, \eta_i), 0 \leq \xi_i \leq 1$. The final shape of a wing section with chord c becomes the following.

$$x_i = c\xi_i, \quad y_i = c\eta_i$$

For some families of airfoils, members with a specified maximum thickness-to-chord ratio $ToC = t/c$ can be produced by the following.

$$x_i = c\xi_i, \quad y_i = c\frac{(t/c)}{t_0}\eta_i$$

where t_0 is the thickness of the template foil.

A number of geometric features are closely related to airfoil performance, and these will be discussed at length in Chapter 8.

1. Leading-edge radius
2. Maximum thickness
3. Location of maximum thickness and curvature around maximum thickness
4. Maximum camber and position of maximum camber
5. Pressure recovery region with trailing-edge angle

Thus, the parameterization should allow manipulation of these geometric features in a transparent way. Two suggestions will be presented here: a cubic Bézier scheme and the *class function/shape function transformation* (CST) scheme, described in Ref. [18], for example.

The Cubic Bézier Scheme

Figure 5.6 shows the control points of four Bézier curves (see Section 5.2.1), which define the airfoil. The trailing edge is assumed closed to a point. The control points can move only along their edges of the bounding rectangle, except $\mathbf{P}_5, \mathbf{Q}_5$, which are free. The shape has tangent continuity except at the trailing edge. For curvature continuity at the nose and the maximum thickness points, three (nonlinear) relations must be satisfied. Thus, there are in all 11 degrees of freedom.

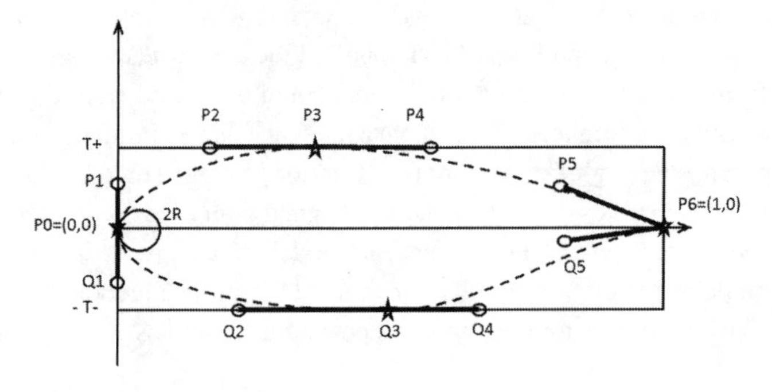

Figure 5.6 Cubic Bezier curve airfoil.

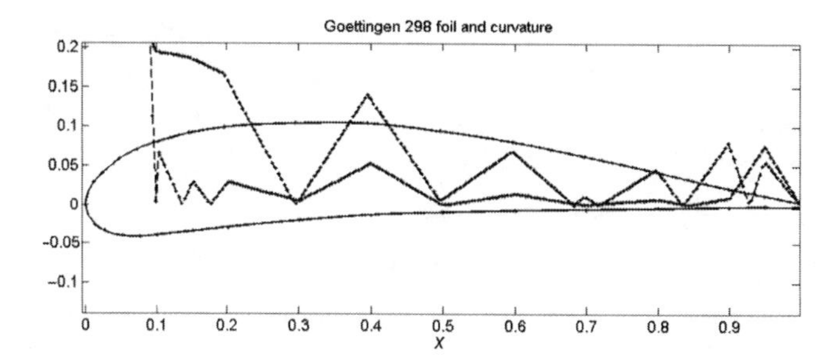

Figure 5.7 Göttingen 298 airfoil and curvature.

The CST Scheme

The control point method represents the shape by parametric curves $x(t), y(t)$. The CST scheme [19] defines the upper and lower sides separately by polynomials in x. The infinite slope at the leading edge is dealt with as follows. Airfoils have smoothly rounded leading edges that can be well approximated by $y = \sqrt{x}$, and they tend to have thin trailing edges, so the CST scheme is as follows.

$$y_{up} = C(x)P_u(x) + \xi_u(1 - x) \tag{5.2}$$

$$y_{lo} = C(x)P_l(x) + \xi_l(1 - x) \tag{5.3}$$

where $C(x) = \sqrt{x}(1 - x)$ is the *class function* and the *shape function* P_u is smooth and can be well approximated by a polynomial of degree N, $P_u(x) = \sum_{i=0}^{N} c_{ui}x^i$, etc.

Quality of Point-Defined Airfoils

It was noted above that the point set defining an airfoil should be dense enough and with sufficient arithmetic precision that curves obtained by any reasonable family of interpolants – parametric or for $y(x)$ – should agree to within strict tolerances for slopes and curvature. This is not always the case.

Example

The Göttingen 298 airfoil was used for the Fokker DR1 triple-decker. It is defined in the UIUC collection by 33 points to five decimals; no mathematical description is available. Figure 5.7 shows the points, the cubic arc-length spline interpolant, and the curvature. At the nose, the curvature goes off the scale. Obviously, the curve has undesirable variations of curvature.

It follows that pressure distributions on this shape will have strong wiggles associated with the curvature, in turn stemming from the sparsity of points. In pre-computer days, the nose radius would be given to help with making a smooth drawing (and hence model). With such complementary data, we could put more points on the nose, but in this case the whole shape must be smoothed. Here, we approximate the points by CST curves of degree 7, which gives a maximum error in position of 0.0004. Increasing the degree of the polynomial does little until degree 17, which can interpolate all of

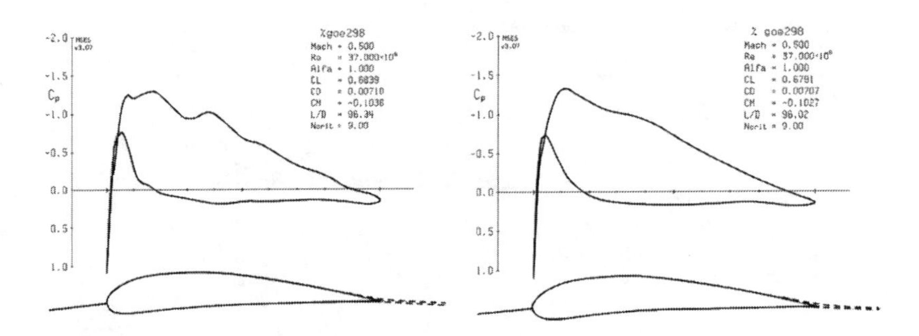

Figure 5.8 Pressure distribution: on original point set (left); on the CST-approximated shape (right).

the points, making the error vanish, with large, undesirable variations in curvature. A balance between goodness of fit to the given points and variation of curvature must be struck. Figure 5.8 shows computed pressure profiles for the original set of points and the smoothed set.

5.3.2 Surfaces by SUMO

Originally, SUMO was developed by David Eller, KTH, as a standalone modeling and mesh generation tool for a Swedish National Aeronautical Research Program. As such, it can create fairly general aircraft geometries and generate quality unstructured surface meshes for use with panel method potential flow solvers. Later, an interface to the tetrahedral volume mesh generator TETGEN [28] was implemented so that unstructured volume meshes suitable for CFD based on the Euler equations could be created from within SUMO. A SUMO mini-manual is found on the book's website, and only a short overview is given here. SUMO works with shapes defined by cross sections. Complex details such as deployed landing gears are beyond its scope. The global coordinate system is a body axis system, with x running from nose to tail, y out the starboard wing, and z vertical.

Lifting Surfaces

SUMO offers lifting surface definitions from sets of cross sections by ruled or bi-cubic surface patches. The bi-cubic surfaces are composed from patches such that tangents are continuous across joins.

Fuselages, tail booms, engine nacelles, and other similar configuration components are also modeled by cross sections along a central curve. The cross-section curves defined by the points are (in general) curvature continuous. The lofting along the central curve is cubic. SUMO does not trim the surfaces by computing intersections of the parametric surface definitions. Instead, it puts surface grids (see below) on the bodies and wings and intersects the grids.

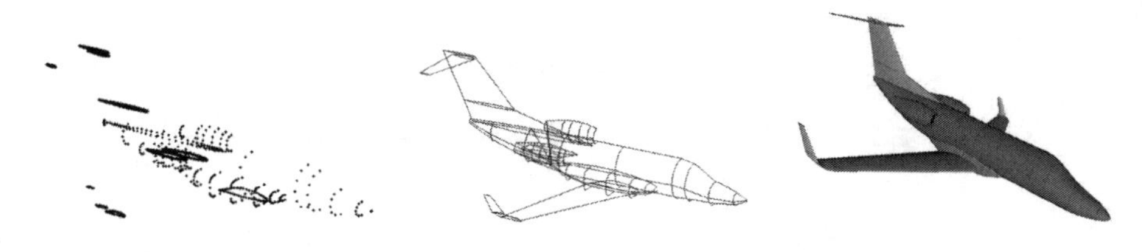

Figure 5.9 SUMO creates curves and surfaces for business jet from set of points.

SUMO Example

Figure 5.9 (left) shows the points in the definition of a small business jet with twin aft-mounted engines, T-tail, body fairing to house the main landing gear, and straight, slightly swept tapered wings with winglets. Symmetry is used where convenient. Actually, a nacelle is right–left symmetric, and it has a symmetric partner on the other side, but SUMO can recognize only one level of symmetry, so both a left and a right nacelle appear in the geometry file. There are hundreds of points in airfoil definitions, but much fewer in fuselage cross sections, which are circular except at the cockpit windshield. Drawing the curves produces the wire-frame picture in Figure 5.9 (middle).

Finally, the curves are lofted into the surfaces (Figure 5.9, right). Note that all of surfaces are displayed in their entirety. Hidden lines and surfaces are managed by the z-buffer algorithm in the graphics engine. Wing tips are not closed, nor are the nose and aft end of the bodies, fuselage, fairing, and nacelles, so one can zoom in to see the wing portion inside the fuselage.

The image is good enough for inspection of curvatures, finding glitches between surfaces, etc. But the geometric model is only a number of surfaces, and much work remains to be done before a computational mesh can be built around it. The holes at the wing tips, fuselage ends, etc., must be closed and intersections computed to make a connected – not necessarily singly connected – interior so that the resulting polyhedral surface does separate the interior from the exterior. The geometry model is saved as an XML text file.

Hierarchical Geometry Model

Although details vary between systems, the principles of geometry representation are general enough to motivate a description of the idea here.

An object is composed of sub-objects, these in turn of sub-sub-objects, etc. The object is then the ancestor of its sub-objects and sub-sub-objects, etc., and these are descendants of the object. Thus, all objects are collected in a tree structure. This structure for the business jet is shown in Figure 5.10.

Each object o is described in its own local coordinate system x_o, in turn connected to the coordinate system x_a of its ancestor a by a linear transformation: scaling in x_o, y_o, z_o by the diagonal matrix D, a rotation R, and a translation Δx_a. Thus, the object cannot

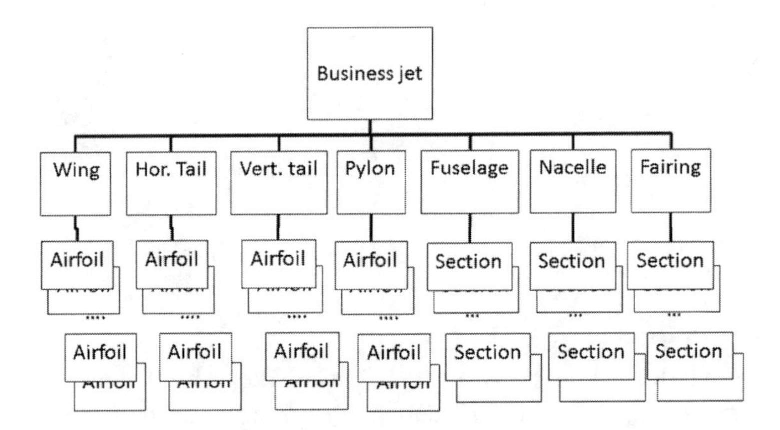

Figure 5.10 Geometry tree for the business jet.

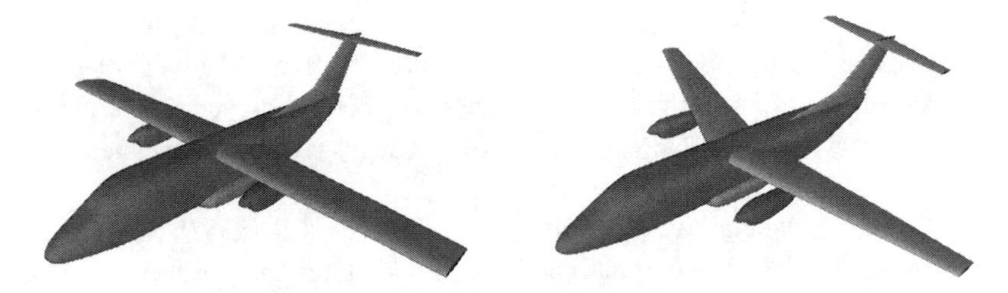

Figure 5.11 SUMO twin prop template (left); adapted with wing swept and tapered (right).

be sheared by the transformation. The scaling is useful because, for instance, airfoil geometries are defined with unit chord length as follows.

$$x_a = R \cdot D \cdot x_o + \Delta x_a$$

where the rotation matrix R is defined from a sequence of rotations α, β, γ around x_a, y_a, z_a axes in that order.

$$R = R(z_a, \gamma) \cdot R(y_a, \beta) \cdot R(x_a, \alpha)$$

Rotations around one axis are common and easy to understand, but the effect of two or three is more taxing. Modification of a component has correct effects for its descendants without further action if they are connected like rigid bodies, and it has no effects on its ancestors.

Templates

Since it is easier to modify a template than to start with a blank piece of paper, SUMO provides a set of template aircraft, from a glider to a four-engine heavy transport. Figure 5.11 shows the twin prop template and its modification by sweep and stronger taper. This is accomplished by editing the wing tip by changing two parameters, since this wing is a ruled surface defined by the root and tip airfoils. Note that the engines do not move; they are not subcomponents of the wing.

Export to CAD

The complete surface geometry representation can be written to Initial Graphics Exchange Specification (IGES) files conforming to the IGES standard, version 5.3 [32]. This format was chosen because it is far less complex than more recent alternatives such as Standard for the Exchange of Product Data (STEP) and therefore offers a reasonable state of implementation with limited resources. SUMO has been adapted to generate a "common denominator" geometry that is understood by all of the CAD systems tried.

Planform Modification of a Wing in SUMO

Consider the standard descriptors for wing planform, sweep, dihedral, and taper. It is desirable to be able to change these high-level parameters by modification of few data items in the SUMO model. As seen below, dihedral is one parameter, but changing sweep and taper may be more involved.

- Dihedral is a rigid body rotation of the wing around the x-axis of the standard body-fixed coordinate system. Thus, it appears as a rotation angle "`dihedral`" in the definition of the wing coordinates in terms of the global coordinates.
- Sweep is similarly a rotation of the whole wing around the body z-axis (i.e. "`yaw`"), and it rotates the wing airfoils too, including the wing tip, which may be undesirable. If one wishes to *shear* the wing planform so airfoils are kept parallel to the x-axis, the yaw of the airfoils must also be changed accordingly:

$$\Delta \mathtt{yaw}(airfoil_k) = -\Delta \mathtt{yaw}(wing), \quad k = 1, 2, \ldots, n$$

- Taper for a wing with straight leading and trailing edges is modified by manipulating the scale factors in the connection of the airfoil coordinate system to the wing coordinate system. Say the wing tip chord is c_{tip} and the wing root chord is c_{root} of a wing with semi-span s. Then, for a wing with planform linearly lofted from root at $y = 0$ to tip, an airfoil located spanwise at y must have a chord as follows.

$$c(y) = c_{root} + y/s \cdot (c_{tip} - c_{root}).$$

SUMO Modeling Concepts

In the following sections, the basic concepts used in the geometry modeling are summarized.

Surface representation and topology

Since SUMO's main capability is to model aircraft configurations and not general mechanical engineering components, a surface modeling approach was taken. A SUMO geometry only provides surface modeling features and does not use the concept of solids. All surfaces created are non-manifold, and only the final surface mesh represents a (faceted) manifold. Figure 5.12 shows a view on the intersecting surfaces representing a T-tail configuration. The vertical fin is visibly open and the stabilizer penetrates the vertical tail plane. In Figure 5.13, the surface mesh for the same

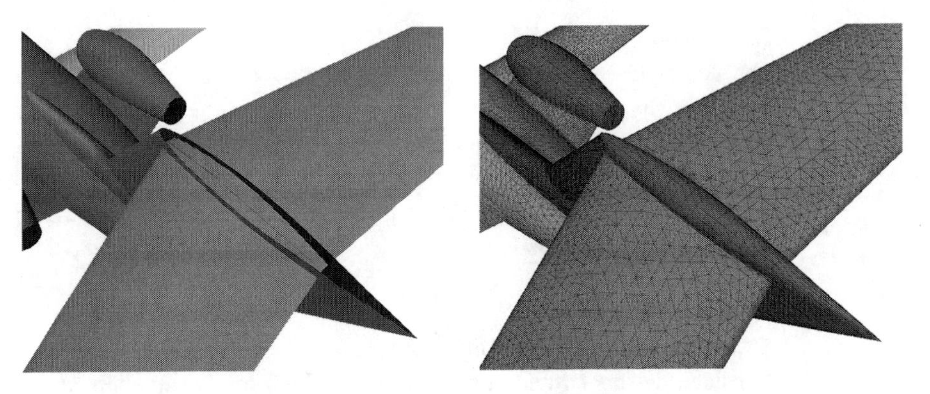

Figure 5.12 Non-manifold surfaces. **Figure 5.13** Manifold mesh.

geometry is shown. Here, the open end curve of the fin has been closed by means of a rounded cap and parts of the mesh in the topological interior of the geometry have been removed.

Numerical Representation

Most surfaces, such as general fuselage bodies, engine nacelles, or curved wings, are represented as nonuniform rational B-spline (NURBS) surfaces with unit weights [22], so they are polynomial nonuniform B-splines (NUBSs) and not rational. This improves performance when evaluating surface derivatives and simplifies interactive modeling considerably by eliminating the need to specify control point weights. The disadvantage is that quadrics cannot be represented exactly and must be approximated. Where relevant advantages can be achieved, different surface representations are used. Piecewise linear (ruled) surfaces, for example, are directly represented as such since that leads to significantly improved performance. As SUMO is an interactive program, performance of surface point and derivative evaluations is crucial for usability.

- Fuselage surfaces are modeled based on top and side views, and cross sections can be modified interactively.
- Lifting surfaces are created from airfoils, which can be read from files or generated. Wings can be linear or cubic B-spline surfaces, and bi-cubic spline transition surfaces to connect main wings and winglets can be generated automatically.
- A control system editor specifies movable wing regions in terms of hinge lines. These control surface definitions are independent of the mesh. Mesh regions that would be affected by flap motion are marked as boundary condition regions in exported mesh files.

Interactive Geometry Modification

Surfaces are constructed from interpolated sections, which, in turn, interpolate sets of points. There is no option to modify spline control points directly since direct manipulation of the control points requires care.

Section curves used to interpolate surfaces are either scaled and rotated airfoil coordinates, which may be generated or loaded from files, or free-form body sections. The latter are usually interactively modified until a certain shape is achieved. As an alternative, free-form sections can also be constrained to basic analytical shapes such as super-ellipses or cubic eggs. At the time of writing, the graphical section editor assumes that sections are symmetrical about the vertical axis. In order to create a surface from a set of interpolation curves, one of two different local interpolation schemes can be selected. Bessel's method achieves curvature continuity, while the default Akima's scheme has the advantage that three identical sections in sequence lead to exactly straight surface segments [22], although continuity drops to slope.

5.4 Grid Generation

Prior to the numerical solution of the governing equations, we have to discretize the surfaces of all boundaries and to generate a volume grid inside the flow domain. In previous chapters, we obtained a hierarchy of flow models from Euler to Reynolds-averaged Navier–Stokes (RANS) and large-eddy simulation (LES). Computational methods to solve them are given in Chapters 4 and 6. First of all, the space where the flow is to be computed – the *physical space* – is divided into a large number of geometric elements called *grid cells*. This process is termed *grid* or *mesh* generation. It can also be viewed as placing first grid points (also called nodes or vertices) in the physical space and then forming cells with them as vertices. Surface grids normally consist of triangles and quadrilaterals, and volume grids are made from tetrahedra, hexahedra, prisms, and pyramids.

The following topics are covered.

- Basics of grid generation, structured and unstructured grids
- Cell-vertex and cell-center grids, grid requirements
- Delaunay triangulation/tetrahedrization
- Grid generation with SUMO and TETGEN

The aim is to make a computational grid covering the volume between the airframe and the outer boundary of the flow domain. For free flight, it is the surface of a sphere, or brick, several wingspans away. For flow in a wind tunnel, it is the walls and inflow and outflow cross sections. The grid must resolve all geometric detail of the boundaries, and needs to be fine on the aircraft, but can be very coarse on the free flight far boundary where velocities are close to free stream. A RANS grid must be extremely fine in the wall normal direction on solid walls, whereas an Euler simulation flow has no boundary layer, so the grid can be more isotropic. Still, accurate representation of shock waves needs a fine grid around the shock. Similarly, modeling of vortices, in RANS as well as Euler flows, needs a fine grid lest the artificial viscosity smear them too much.

We can choose between *structured* grids or *unstructured* grids. After a short discussion of structured grids, we shall focus on unstructured grids.

5.4.1 Structured Grids

The grid can be generated to follow closely the boundaries of the physical space, in which case we speak of a *body-fitted grid* (Figure 5.14). The main advantage of this approach is that the flow can be resolved very accurately in shear layers along solid bodies. The price to pay is the complexity of the grid generation tools.

- A *structured* grid (Figure 5.14) has each grid point (vertex, node) uniquely identified by the indices i, j, k and corresponding Cartesian coordinates $x_{i,j,k}$, $y_{i,j,k}$, and $z_{i,j,k}$. The grid cells are quadrilaterals in 2D and hexahedra in 3D. If the grid is body-fitted, we also speak of a *curvilinear* grid.
- In *unstructured* grids (Figure 5.15), grid cells as well as grid points have no particular ordering. Neighboring cells or grid points cannot be directly identified by their indices (e.g. cell 16 adjacent to cell 23). Unstructured grids usually consist of a mix of quadrilaterals and triangles in 2D and of hexahedra, tetrahedra, prisms, and pyramids in 3D, in order to resolve the boundary layers properly. Therefore, we speak in this case of *hybrid* or *mixed* grids.

5.4.2 Multiblock Structured Grids

The main advantage of structured grids follows from the property that the indices i, j, k represent a linear address space directly corresponding to how the flow variables are stored in the computer memory. This property allows quick access to the neighbors of a grid point, so evaluation of gradients and fluxes and the treatment of boundary conditions are greatly simplified. The same holds for the implementation of an implicit scheme, because of the well-ordered flux Jacobian matrix.

As sketched in Figure 5.16, one possibility is to divide the physical space into a number of topologically simpler blocks (here five), which can be more easily meshed – a *multiblock* approach. Of course, the complexity of the flow solver is increased, since special logic is required to exchange physical quantities or fluxes between the blocks.

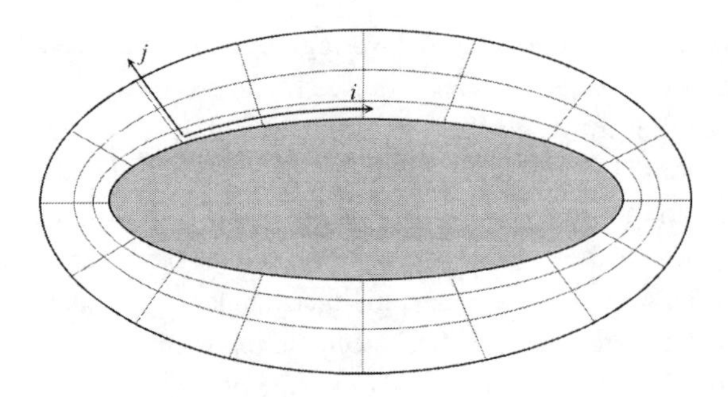

Figure 5.14 Structured body-fitted grid in 2D.

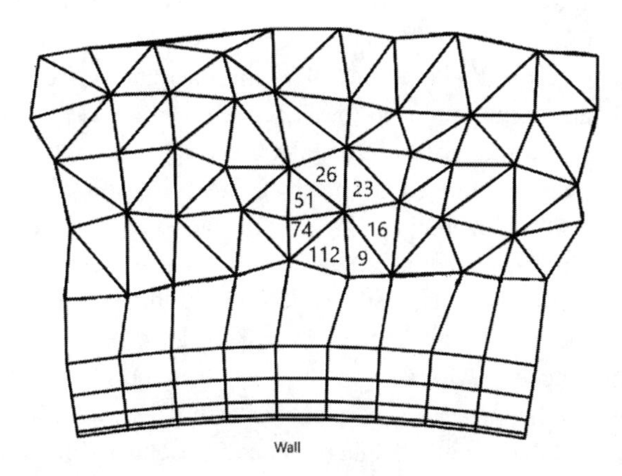

Figure 5.15 Unstructured, mixed-grid approach; numbers mark individual cells.

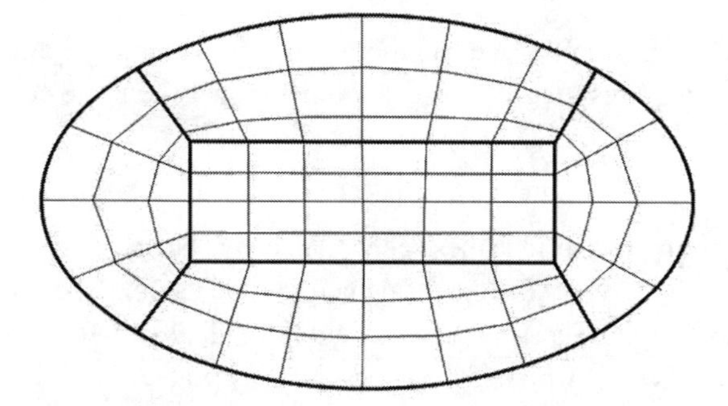

Figure 5.16 Structured, multiblock grid; thick lines represent block boundaries.

While this is a minor nuisance, the generation of multiblock structured grids for complex geometries is very taxing of the engineer's time.

5.4.3 Unstructured Grids

Unstructured grids offer the greatest flexibility in the treatment of complex geometries, and grids can in principle be generated automatically, independent of the complexity of the domain. It is of course still necessary to set some parameters appropriately, in order to obtain a good-quality grid. Furthermore, in order to resolve the boundary layers accurately, it is advisable to employ prismatic or hexahedral elements near solid walls. The generation of mixed grids is nontrivial for geometrically demanding cases, but the time required is still significantly lower than for a multiblock structured grid.

Unstructured grid solvers need to employ sophisticated data structures that require indirect addressing. Depending on the computer hardware, this leads to more or less reduced computational efficiency. In addition, the memory requirements are in general higher as compared to the structured schemes.

Unstructured grids are typically composed of various element types. For example, hexahedra and prisms are often employed to discretize boundary layers. The rest of the flow domain is filled with tetrahedra. Pyramids are used as transitional elements between the hexahedra and the tetrahedra. The advantage of structured hexahedral or prismatic grids on solid walls with boundary layers is the preserved accuracy in the wall normal direction for highly stretched grids. On the other hand, a desirable feature of unstructured tetrahedral grids is the capability to discretize complex geometries such as the volume around the aircraft in Figure 5.17. Mixed grids seek to combine the advantages of both approaches.

Automatic volume tetrahedral grid generators for external flows appeared in the 1980s, pioneered by A. Jameson and T. Baker [17] and R.Löhner [8], among others. Automatic unstructured tetrahedral mesh generation turns out to be a simpler task than the generation of structured hexahedral meshes, because of the following.

- The tetrahedron is a 3D simplex and can more easily fill volumes of any shape.
- A structured mesh requires long-distance coordination between grid cells, whereas the cells of an unstructured grid can be connected much more flexibly to their neighbors.

Figure 5.17 shows a triangular surface mesh generated on the business jet configuration by the default settings in SUMO. Note the refinements at the leading and trailing edges and at surface intersections. The triangles belonging to flaps, ailerons, rudder, and elevators are marked by color, which is not visible in gray scale.

Any 3D point grid admits a unique *Delaunay* tetrahedrization of its convex hull, where the closed circumsphere of a tet contains *only* the four vertices of that tet. There are also efficient algorithms for computing the Delaunay tetrahedrization. This opens the way to create a volume grid by first making unstructured triangular surface grids on its bounding surfaces, then creating a point set in-between, and finally applying the

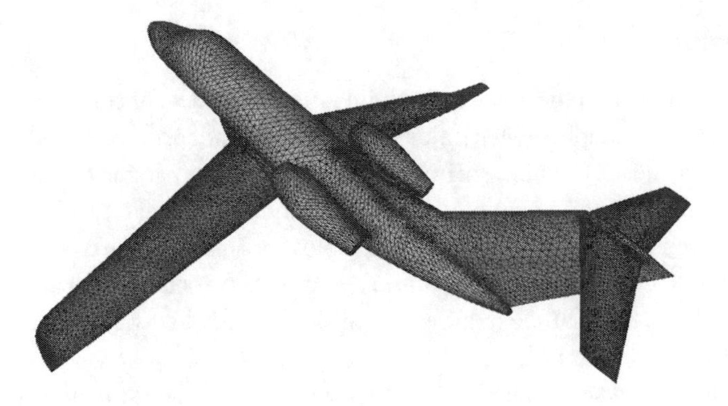

Figure 5.17 Surface unstructured triangular mesh on SUMO business jet template.

automatic tetrahedral volume mesh generator. Tetrahedral mesh generation by Delaunay will be covered in Section 5.4.5, and in Section 5.5.1 we discuss the creation of the triangular surface mesh on the aircraft.

Control Volume

There are several possibilities for defining the shape and position of the control volume with respect to the grid. Two basic approaches can be distinguished as follows.

- *Cell-center* scheme: here, the flow quantities are associated with the centroids of the grid cells. Thus, the control volumes are identical to the grid cells.
- *Cell-vertex* scheme: here, the flow variables are associated with the grid points. The control volume can then either be the union of all cells sharing the grid point or some volume centered around the grid point. In the former case we speak of *overlapping* control volumes, and in the second case of *dual* control volumes.

Chapter 6 discusses the cell-vertex formulations in the EDGE code with regards to discretization and computer implementation.

5.4.4 Grid-Quality Criteria

The quality of the grid strongly influences the accuracy of the simulation. This is particularly true for unstructured grids. The old saying – still very true – is that if the quality of a CFD flow solution seems poor, make a better grid. The most important requirement is that the volume be completely filled and that the cells do not overlap. Of course, the grid must be fine enough to resolve the relevant flow features. Two grid properties can be easily quantified: the size and shape (quantified somehow, see below) of a grid cell (a tetrahedron) and the variation of shape indicators between neighbors. First, the discretization error grows with distortion of the cell from completely regular. This also means that the cells should be identical modulo rotations, but this clashes with the ability for cells to grow toward the outer boundary. Discretizations of spatially first-order PDEs, such as the Euler equations, are not as sensitive to cell distortions as second-order PDEs. It follows that special attention must be paid to the meshing of the boundary layer; fortunately, the viscous effects for attached flows are small outside it.

There are badly shaped tetrahedra to avoid: obtuse elements, "slivers" (i.e. elements with four nearly coplanar points), needles, and wedges. Wedge-like stretched tetrahedra in viscous regions near a wall can only be avoided by using prisms. However, since the gradient of the flow in the wall normal direction (short side of the wedge) is the most important, the numerical error is not disturbing.

Solution-Adaptive Meshing

In many cases, the controls offered by the surface mesh and volume mesh generator suffice to create proper mesh resolution. But economical resolution of high-gradient

(a) Top: SUMO 2M point mesh; Bottom: 6M point mesh from CAD + script tool.

(b) Bottom: 6M point basic mesh; Top: u, ρ, p gradient adapted mesh.

Figure 5.18 SUMO grids, CAD + ANSYS ICEM-CFD mesh generator, and transonic flow-adapted meshes for SACCON. (Courtesy of M. Tomac, PhD thesis [31], reprinted with permission.)

flow features away from the surfaces, such as wakes, shocks, and vortices, cannot be controlled in this way. Solution-adaptive meshing can tailor the grid to the flow pattern at hand, and Figure 5.18 shows mesh adaption for a single transonic flight condition for the SACCON unmanned combat aerial vehicle (UCAV).

The swept leading edges and varying twist produce a complex pattern of shocks on the wing suction side, and mesh refinement of the surface mesh is necessary. The initial mesh (Figure 5.18a) was generated by SUMO and the flow solutions were obtained by EDGE in Euler mode, with grid refinement controlled by gradients of $u, \rho,$ and p. Certain parametric sweeps (e.g. of Mach number or angle of attack) will need different adaptation for each case. Running each case through a sequence of adaptations from an initial mesh may be quite costly. If the flow solver supports not only mesh refinement but also coarsening, each case can start instead from the previous adapted mesh, which is closer to the desired one than the initial.

5.4.5 Delaunay Triangulation

The Delaunay triangulation is an essentially uniquely defined triangulation (tetrahedrization in 3D) of a set of points such that the total edge length is small and the smallest angles are not too small. The circumcircle (circumsphere) of a triangle (tetrahedron) contains only its vertices and no other points, except in degenerate cases where the point set has extraordinary regularity. The triangulation defines a set of triangles (tetrahedra in 3D) that cover the convex hull of the points.

The Delaunay triangulation is related to a methodology proposed by Dirichlet in 1850 for the unique subdivision of space into a set of packed convex regions. Given a set of points, each region represents the space around the particular point, which is closer to that point than to any other. The regions form polygons (polyhedra in 3D) that are known as the Dirichlet tessellation or the Voronoi diagram. If we connect point pairs that share some segment (face) of the Voronoi diagram by straight lines, we obtain the Delaunay triangulation.

Many algorithms have been proposed for the extension of the basic Delaunay triangulation into a mesh generator (e.g. [14]). Figure 5.19 (left) shows the Delaunay triangulation of a set of points on a wing planform clustered toward the leading edge. Figure 5.19 (right) shows the corresponding dual cell-vertex discretization where the cell "centers" are the triangle vertices and the cells are formed from medians of the triangles. The cells are not necessarily convex (near to the leading edge), but a cell belonging to a given triangle vertex contains only parts of the triangles that share the

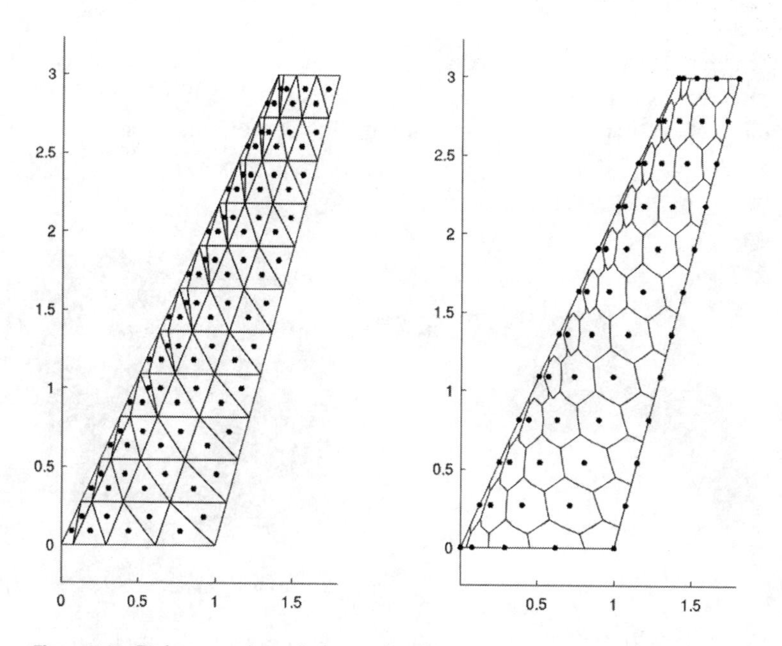

Figure 5.19 Delaunay triangulation and cell barycenters (left); triangle vertices and dual cells (right).

vertex, whereas Voronoi cells may extend into more triangles. The 3D cell-vertex grid is the analog with tetrahedra and median faces, as we show in Chapter 6.

The Delaunay method represents a particular way of connecting grid points. The positions of the points must be determined by some other technique. Therefore, one popular approach for the construction of a Delaunay grid is to insert nodes into an initial triangulation sequentially. The grid is then locally re-triangulated in order to fulfill the empty circumcircle (circumsphere) criterion. Essential mesh generator ingredients include incremental Delaunay algorithms for inserting vertices, algorithms for inserting constraints (edges and triangles), an edge recovery algorithm to ensure that the boundaries are respected, and a refinement algorithm for the improvement of tetrahedron (triangle) quality. The calculations need high arithmetic precision and make use of filtered exact geometric predicates.

The core task of incremental Delaunay grid generation is the insertion of a new point into a valid triangulation. In order to select (or position) the new point, attention must be paid to the quality of the element (minimum angle, aspect ratio, etc.) and its size (volume, edge length) compared to its neighbors. Limits on size can be given as a function of the spatial position; such specification is helped by creating geometric objects, such as a set of boxes or a coarse grid, to carry the information.

5.5 Sumo Mesh Generation

A volume mesh for flow exterior to a configuration is generated in two steps. First, Sumo creates a triangular surface mesh; then, the external program TetGen is called from within Sumo in order to build a quality-conforming tetrahedral volume mesh

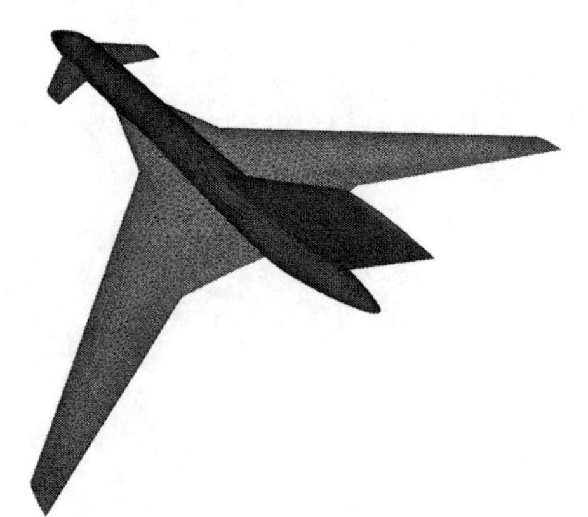

Figure 5.20 Sumo surface mesh, 117,000 triangles.

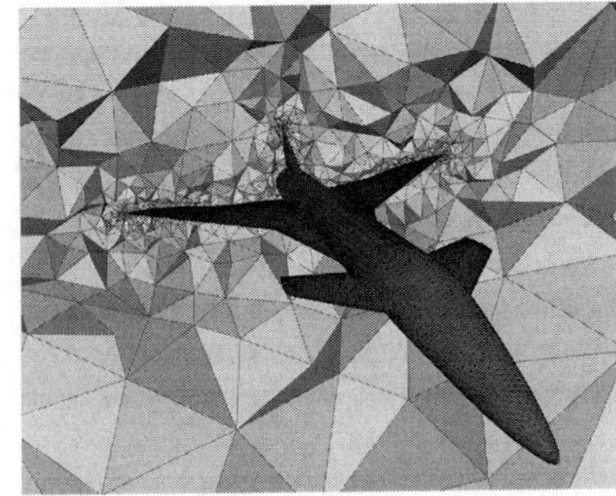

Figure 5.21 Sumo volume-mesh cut, 803,000 tetrahedra.

between the configuration and a large spherical surface. Figures 5.20 and 5.21 illustrate these steps.

5.5.1 Surface Mesh Generation

The accuracy of the numerical solution of the flow problem is affected by the quality of the surface discretization. As most current CFD methods use local low-order polynomial solution representation based on nodal or cell-average values, the surface mesh must do the following.

- Approximate the geometric surface sufficiently accurately.
- Resolve flow features such as pressure peaks and shocks.

While the first requirement is, in most cases, a necessary condition for satisfaction of the second, it is not always sufficient. As an example, consider the pressure recovery near the trailing edge of a lifting surface. Although the surface geometry usually is flat in this region, small elements are needed to capture the steep pressure gradient. The surface mesher automatically ensures that the first requirement above is fulfilled to given tolerances. A set of geometric heuristics described below is used to refine particular regions where large variations in pressure are typically observed.

All surfaces are represented by bi-parametric patches of the form $(x, y, z) = S(u, v)$, where at least the first derivatives with respect to the parameters u and v are continuous. The mesh generation is performed in the parameter space (u, v) by means of a modified Delaunay algorithm, like that presented by Chew [10], where the mesh is defined to be Delaunay if the *circumsphere* of any (x, y, z)-triangle does not enclose vertices of any other triangle. The circumsphere is defined as the smallest sphere that contains the circumcircle in the plane of the triangle.

Unstructured surface meshes can be generated completely without user intervention. Heuristics determine default parameters for the mesh generation code, which usually yield a satisfactory mesh. Manual tuning of these parameters is possible as well.

- Triangulations are based on a 3D in-sphere criterion, which yields better mesh quality than surface Delaunay methods for strongly skewed surfaces, such as thin, swept delta wings.
- Geometric refinement criteria produce a finer mesh in regions of strong curvature, while a limit on the minimum element size can be imposed to avoid resolution of irrelevant geometric detail.
- Unstructured volume meshes can be generated from the surface mesh by, for example, Hang Si's tetrahedral mesh generator TETGEN [1]. The volume mesh can be saved in the CFD General Notation System (CGNS) format [2], for the CFD solvers EDGE [3] and SU2, or in TETGEN's plain American Standard Code for Information Interchange (ASCII) format.

In regions where the flow exhibits strong directionality, anisotropic grids can substantially reduce the number of triangular faces and hence volume elements without impairing the solution's accuracy (see Figure 5.22). For example, a simulation with RANS does not require high grid resolution in the spanwise direction of a lifting surface. As indicated in Figure 5.22, it is important to orient the surface elements properly in areas of high curvature, and SUMO knows how to generate surface grids stretched in the correct direction.

Ruppert [24] and Shewchuk [26] developed algorithms that improve the quality of triangular meshes through a sequence of vertex insertions. Chew shows that every triangular mesh, which initially conforms to the circumsphere criterion, can be refined by means of vertex insertion, such that this property is maintained. This criterion is far better than the Delaunay criterion for the generation of surface meshes in the parameter space of the surface.

Grid Control Parameters

As far as possible, SUMO attempts to determine suitable mesh-quality criteria from the geometric properties of the surface (e.g. the leading-edge radius). As these default parameters should in general ensure a useful mesh, they need to be somewhat conservative. In order to generate a coarse mesh, which may be desired for preliminary computations, some user interaction is required. The user can set the following parameters.

- Dihedral angle: the angle between the normal vectors of two triangles sharing an edge. This parameter mainly affects the resolution of strongly curved regions such as wing leading edges.

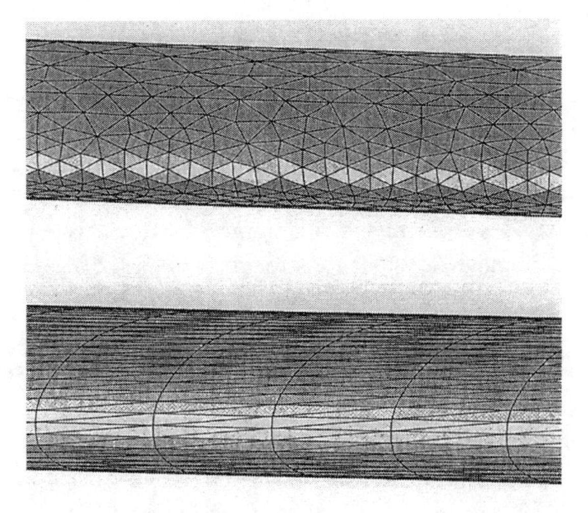

Figure 5.22 Unstructured surface grid at the leading edge: isotropic with no specific orientation (top), stretched and oriented in spanwise direction (bottom).

(a) Initial triangular mesh

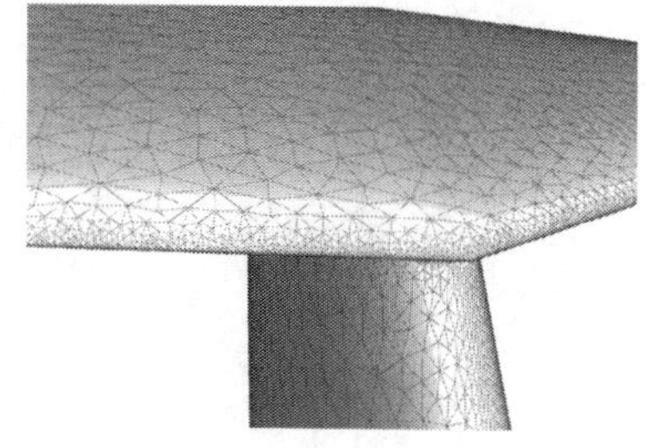

(b) Delaunay mesh

Figure 5.23 Details of surface mesh on Ranger 2000 T-tail.

- Edge length: this can be controlled by enforcing both a minimum and maximum length. The minimum edge length serves to avoid resolution of irrelevant geometric details.
- Triangle stretch ratio: the ratio of the longest to the shortest edge in a triangle. Here, the magnitude of the maximum acceptable ratio depends entirely on the properties of the flow solver used. A high stretch ratio allows fewer elements to discretize large, flat surfaces with small regions of strong curvature, such as thin wings.
- Refinement factors of the leading and trailing edges: setting these factors larger than 1.0 progressively reduces the maximum allowed edge length with the given factor toward the leading/trailing edge. This can help to improve meshes on extremely thin wing surfaces, which otherwise transition very rapidly from coarse to fine. Furthermore, a very fine trailing-edge mesh has been found to be crucial for volume meshes that are intended to serve as background meshes for prismatic mesh generation algorithms.

Mesh improvement for size and shape is performed in an iterative process by refinement. To initialize the process with an approximately Delaunay surface mesh, a structured quadrilateral mesh is first created on each surface and converted into a geometrically adapted triangular mesh. Figure 5.23 shows the mesh just after conversion to triangular elements and the final Delaunay mesh.

In the presence of small surface curvature radii (e.g. near the leading edges of thin wings), triangles are constructed using a cascading method, as illustrated in Figure 5.24. Stretched quadrilateral segments are progressively divided in the span direction as soon as a stretch criterion is violated, leading to a characteristic binary division pattern.

High-quality meshes require quality improvement, involving several Laplacian smoothing iterations (using surface projection) and repeated vertex insertion until all imposed quality measures are fulfilled as far as possible, as illustrated in Figure 5.25. This process makes extensive use of Shewchuk's adaptive precision geometric predicates [25], without which the whole refinement procedure is infeasible.

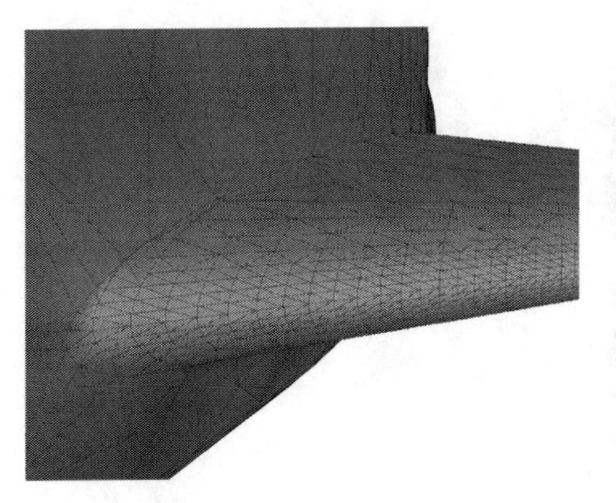

Figure 5.24 Initial cascaded mesh.

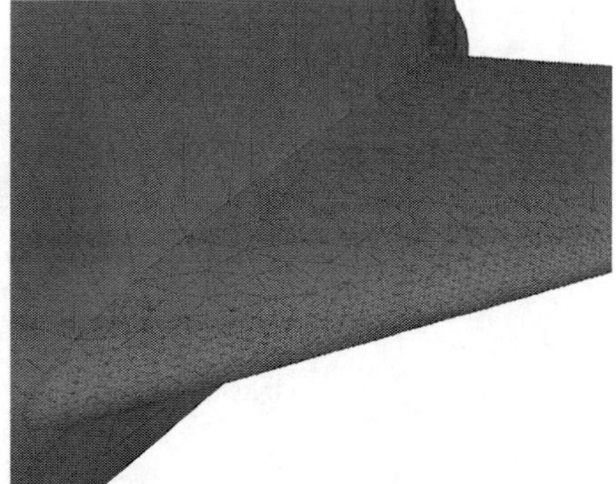

Figure 5.25 Smoothed and refined mesh.

Surface Intersections

SUMO does not compute "trimmed surface patches" from intersections based on the continuous geometry, as this would require additional user interaction in order to determine which parts to trim. Hence, surface intersections are computed based on the triangular meshes on each surface. The current implementation requires that *no more than three surfaces intersect in a single point*. Once every isolated surface is meshed, all discrete triangle intersections are computed by means of a tree bisection procedure followed by application of Möller's triangle intersection algorithm [21] on candidate elements. Following this, the connectivity of the intersection segments is robustly (re)established by marching along intersection lines and identifying branching points. A post-processing stage removes excessively short segments where possible. The subsequent mesh refinement pass must avoid disturbing the intersection lines, and for this reason, mesh-quality criteria may not be fulfilled near surface intersection lines.

Creation of a Wetted Surface Mesh

Once all surfaces are meshed and all intersection constraints are satisfied, the discrete meshes are merged and duplicate vertices along intersections are eliminated. It remains to determine the "wetted" set of triangles through a topological walk over element edges. This procedure is reliable as long as the connectivity of the mesh near the intersection lines is identical to the (conceptual) connectivity of the surfaces themselves.

5.5.2 Troubleshooting

Much effort has been put into transparently handling common geometric difficulties in surface mesh generation, but there are still some difficult cases. As the primary source of meshing problems tends to be degenerate geometric features, the user can very often alleviate such difficulties with little effort.

Slope Discontinuities

Surfaces that display slope discontinuities such as linearly ruled wing surfaces with strong leading-edge sweep and hard kinks are recognized automatically and treated specially. Whenever the slope discontinuity is part of a surface intersection, SUMO can have difficulties in determining the intersection line accurately. Intersection lines with very small loops or tiny isolated circles are the result. The solution for this problem is quite often rather simple, since the geometry causing the problem is not actually intended. Inserting a new section near the kink, for example, reduces the kink angle by half, so that the sharp ridge no longer intersects. The external geometry remains unchanged if the inserted section is appropriately placed inside another body.

Glancing Intersections

Another problem may occur when two coarsely meshed surfaces touch or intersect with an extremely small angle; that is, whenever intersecting triangles are almost coplanar. As the mesh triangles only approximate the true surface, the computed discrete intersection line can be quite far from the actual intersection of the smooth surfaces. To solve this type of problem, it is sufficient to reduce the gap between triangles and surfaces by reducing the maximum allowed edge length or the allowed normal vector difference of the affected surfaces.

5.6 Euler Volume Grids by TetGen

When a high-quality surface mesh has been created, an unstructured tetrahedral volume mesh for the solution of the Euler equations is generated by TETGEN, developed by Hang Si at the Weierstrass Institute in Berlin. TETGEN is a very efficient quality-constrained tetrahedral Delaunay mesh generator. Starting from an initial constrained Delaunay tetrahedrization of the domain, nodes are dynamically inserted until a given set of quality criteria is met. The domain is bounded by surface meshes of the aircraft configuration and the far-field boundary, which are created by SUMO. Tetrahedral-quality criteria available in TETGEN, version 1.4.3, include the following.

- Maximum element volume
- Maximum ratio of circumsphere radius to tetrahedron edge length
- Minimum dihedral angle between faces

Furthermore, the maximum permitted element volume can be defined to vary with (x, y, z) once a volume mesh exists. This feature is not yet exploited by the SUMO calls to TETGEN. In order to comply with imposed element-quality requirements, TETGEN will subdivide boundary triangles if explicitly allowed to do so. Since it currently has no access to the original spline surface description, the subdivision is performed on the polyhedral surface.

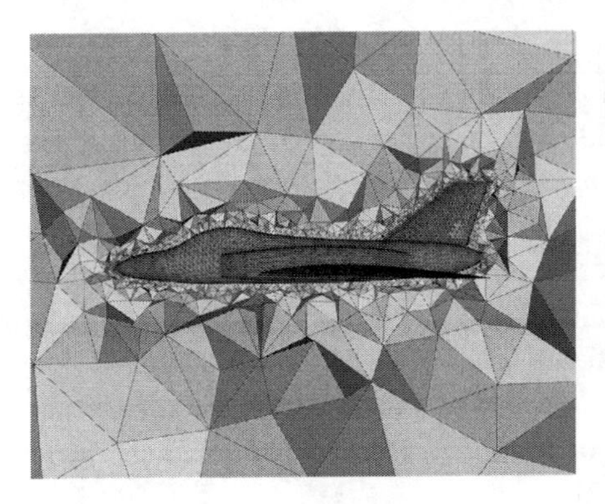

Figure 5.26 Radius-to-edge ratio 1.6.

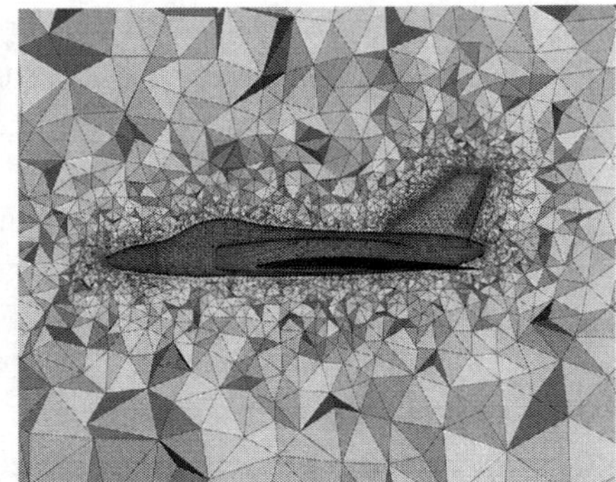

Figure 5.27 Radius-to-edge ratio 1.1.

Once a closed triangular surface mesh is generated, the tetrahedral volume mesh can be created. The aircraft surface is embedded in a spherical domain. The outer boundary of this domain is discretized as a recursively refined icosahedron. Then, TetGen is called to fill the volume with tetrahedra fulfilling certain quality measures [27]. Volume mesh options currently used by Sumo are as follows.

- Tetrahedron radius-to-edge length ratio
- Maximum allowed tetrahedron volume
- Boundary triangle splitting switch

The first of these settings has a strong impact on mesh refinement level, as it prescribes the allowed stretching of tetrahedra in terms of the ratio of circumsphere radius to edge length. The larger this number, the more stretched tetrahedra will be allowed in the mesh. In Figures 5.26 and 5.27, two volume meshes based on the same surface mesh are shown. The coarser mesh in Figure 5.26 has 72,000 vertices, while the finer mesh in Figure 5.27 contains 200,000 nodes. As Figures 5.26 and 5.27 show, the radius-to-edge ratio strongly affects the rate with which small, refined elements near the surface are permitted to grow to larger tetrahedra in the far field.

Maximum tetrahedron volume is rather obvious. It is enabled by default, and it is set such that the maximum element volume is the one determined by the refinement level of the far field boundary. When disabled, the largest tetrahedra will appear in the free space between the far field boundary and the surface mesh, which then may, in some cases, be filled with just a handful of extremely large elements.

Finally, the boundary triangle splitting switch determines whether TetGen is permitted to subdivide triangles on the surface in order to achieve the prescribed radius-to-edge ratio. Since the volume mesher has no information about the underlying surface shape, it is forced to insert points by linear interpolation. Allowing surface element splitting can lead to a substantial improvement in the convergence behavior

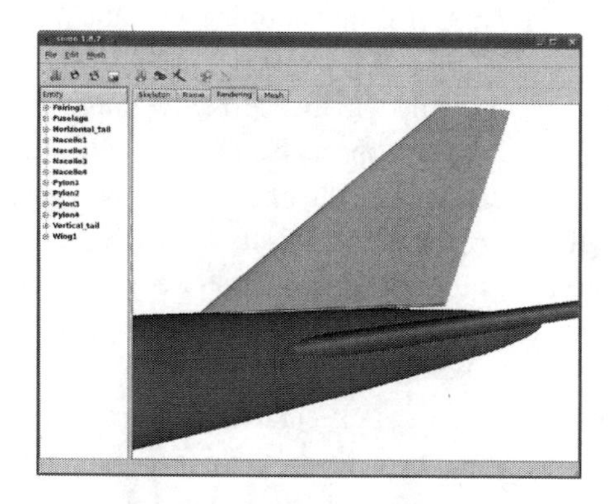

Figure 5.28 Gap between vertical tail and fuselage.

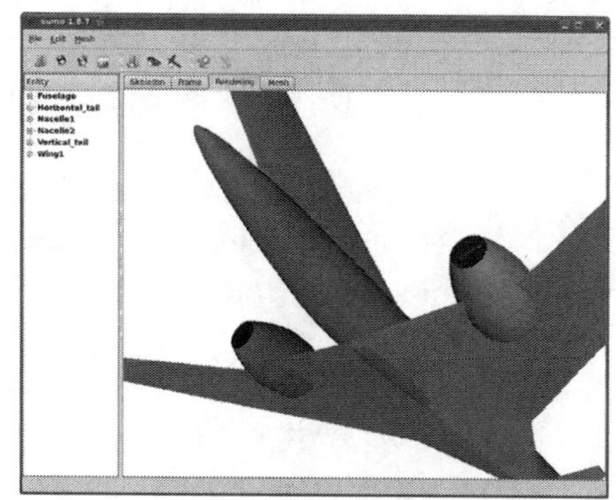

Figure 5.29 Trailing edge/fuselage gap.

of the Edge solver and thereby reduce the runtime for a given accuracy, even though the resulting mesh may contain more elements.

TetGen Limitations

TetGen is based on an iterative refinement procedure that cannot be guaranteed to converge for a finite number of elements [27]. Although TetGen is extremely robust for good-quality surface meshes, it can take a very long time when a small radius-to-edge ratio is required in combination with a coarse, low-quality surface mesh with stretched triangles. In this case, TetGen may be forced to generate extremely numerous tetrahedra in order to achieve the quality required. If TetGen runs too slowly, it is reasonable to interrupt it and improve the surface mesh or reduce quality requirements before trying again.

Finally, communication with TetGen currently involves the writing and reading of very large temporary text files. These are deleted after use, and they may very well grow to hundreds of megabytes for large-volume meshes – approximately 52 MB per million tetrahedra.

Typical Use Case

We assume that a SMX (the Sumo native external model format) file exists with the geometry of the desired configuration. The SMX file is imported into Sumo for visual inspection in the tab labeled "`Rendering`." Figures 5.28 and 5.29 show examples of small modeling errors that are easily detected in this way. In most cases, model features of this type are not intended. A quick visual inspection of the geometry can therefore prevent costly simulations on a bad mesh. The fastest remedy is to modify the geometry inside Sumo, such as by changing the position of the tail in Figure 5.28: right click on the "`Vertical Tail`" entry in the component tree to the left, choose "`Edit properties`" and change the z-coordinate of the component location. The modified model can be saved as SMX. If the source of the model is a

CPACS file or other format translated into SMX, the modification should be performed on the original file.

Once the surface geometry is determined to be correctly imported, a surface mesh can be generated. Usually, the default mesh parameters will allow a useful surface mesh to be created. This will, however, neither be the coarsest possible nor the best-quality mesh, so that, if either of these is the objective, manual modification of the surface mesh generation parameters described in Section 5.5.1 is necessary.

After inspection of the surface mesh, TETGEN can be called using the "`Generate volume mesh`" option from the "`Mesh`" menu. When started for the very first time, SUMO will ask for the location of the TETGEN executable, unless it is automatically found. Again, choosing the default values will likely yield a usable, but not necessarily optimal mesh. If the volume mesh is intended for numerous (or high-quality) simulations, it may be worth modifying the default settings.

To generate the mesh file for the EDGE solver, select "`bmsh`" as the output format. Along with the mesh file, a boundary condition file will be created with the same name prefix. Currently, the boundary condition definition file marks wall and far field conditions appropriate for Euler simulations and controls surface elements for the use of transpiration boundary conditions (see Chapter 6). Engine inlet and exhaust conditions must be added outside SUMO.

5.6.1 Tutorial: Mesh Generation for a Sears–Haack Wing–Body Configuration

The tutorial on the book's website introduces the Sears–Haack body shaped for minimal wave drag. The task is to verify that, when completed by a swept wing, the body needs reshaping to minimize the wave drag of the wing–body configuration. This is done by simulations with the EDGE or SU2 CFD packages. Now is the time to start getting hands-on experience of SUMO – TETGEN mesh generation for Euler simulation.

We now can model 3D geometry and construct a grid around it. The next chapter outlines the various difference schemes that solvers normally apply on these grids for approximating the inviscid and viscous terms of the governing equations. A demonstration computation of typical transonic flow past an airfoil exemplifies the working of a modern solver for both steady and unsteady flow.

5.7 Learn More by Computing

Gain hands-on experience of the computational tools for the topics in this chapter by working with the on-line resources. Exercises, tutorials, and project suggestions are found on the book website www.cambridge.org/rizzi. Software used to compute many of the examples shown is available from http://airinnova.se/education/aero-dynamic-design-of-aircraft. Work with SUMO and its template models to create surface and volume grids.

References

[1] http://wias-berlin.de/software/tetgen/.

[2] www.grc.nasa.gov/WWW/cgns/CGNS_docs_beta/user/index.html.

[3] www.foi.se/en/Customer--Partners/Projects/Edge1/Edge/.

[4] UIUC airfoil coordinates database. Available from http://m-selig.ae.illinois.edu/ads/coord_database.html.

[5] Industrial automation systems and integration product data representation and exchange part 21: Implementation methods: Clear text encoding of the exchange structure, 2016. Available from https://en.wikipedia.org/wiki/ISO-10303-21.

[6] W. A. Anemaat and B. Kaushik. Geometry design assistant for airplane preliminary design. Presented at 49th AIAA Aerospace Sciences Meeting including the New Horizons Forum and Aerospace Exposition, Orlando, FL, January 2011.

[7] A. Barr. Global and local deformations of solid primitives. *Computer Graphics*, 18(3): 21–30, 1984.

[8] J. Cabello, R. Löhner, and O. P. Jacquotte. A variational method for the optimization of two- and three-dimensional unstructured meshes. Technical report. AIAA paper 92-0450, 1992.

[9] S. W. Cheng, T. K. Dey, and J. R. Shewchuk. *Delaunay Mesh Generation*. CRC Press, 2012.

[10] L. P. Chew. Guaranteed-quality mesh generation for curved surfaces. In *Proceedings of the Ninth Annual Symposium on Computational Geometry*, San Diego, May 1993. Available from: portal.acm.org/citation.cfm?id=161150.

[11] P. D. Ciampa, B. Nagel, P. Meng, M. Zhang, and A. Rizzi. Modeling for physics based aircraft pre-design in a collaborative environment. Presented at 4th CEAS Air & Space Conference, Linköping, Sweden, September 2013.

[12] C. de Boor. *A Practical Guide to Splines*. Springer, 1978.

[13] G. Farin. *Curves and Surfaces for CAGD: A Practical Guide*, 5th edition. Elsevier, 2001.

[14] L. Guibas, D. Knuth, and M. Sharir. Randomized incremental construction of delaunay and voronoi diagrams. *Algorithmica*, 7: 381–413, 1992.

[15] A. S. Hahn. Vehicle sketch pad: a parametric geometry modeler for conceptual aircraft design. Presented at 48th AIAA Aerospace Sciences Meeting, Orlando, FL, January 2010.

[16] R. M. Hicks and P. A. Henne. Wing design by numerical optimization. Technical report. AIAA paper 79-0080, 1979.

[17] A. Jameson and T. Baker. Solution of the euler equations for complex configurations. Technical report. AIAA paper 83-1929, 1983.

[18] B. M. Kulfan. A universal parametric geometry representation method – "CST". Technical report. AIAA paper 2007-0062, 2007.

[19] B. M. Kulfan. Universal parametric geometry representation method. *Journal of Aircraft*: 142–158, February 2008.

[20] Z. Lyu, G. K. Kenway, and J. R. R. A. Martins. Aerodynamic shape optimization studies on the common research model wing benchmark. *AIAA Journal*, 53(4): 968–985, 2015.

[21] T. Möller. A fast triangle-triangle intersection test. *Journal of Graphics Tools*, 2(2): 25–30, 1997.

[22] L. Piegl and W. Tiller. *The NURBS Book*. Springer, 1997.

[23] D. P. Raymer. *Aircraft Design: A Conceptual Approach*. AIAA Education Series, 5th edition. AIAA, 2012.

[24] J. Ruppert. A delaunay refinement algorithm for quality 2-dimensional mesh generation. *Journal of Algorithms*, 18(3): 548–585, 1995.

[25] J. R. Shewchuk. Adaptive Precision Floating-Point Arithmetic and Fast Robust Geometric Predicates. *Discrete & Computational Geometry*, 18(3): 305–363, 1997.

[26] J. R. Shewchuk. Delaunay refinement algorithms for triangular mesh generation. *Computational Geometry*, 22: 21–74, 2002.

[27] H. Si. On refinement of constrained delaunay tetrahedralizations. Presented at Proceedings of the 15th International Meshing Roundtable, September 2006. Software available from http://tetgen.berlios.de, July 2009.

[28] H. Si and K. Gaertner. Meshing piecewise linear complexes by constrained delaunay tetrahedralizations. Presented at Proceedings of the 14th International Meshing Roundtable, September 2005. Software available from http://tetgen.berlios.de, July 2009.

[29] J. F. Thompson, B. K. Soni, and N. P. Weatherill, editors. *Handbook of Grid Generation*. CRC Press, 1998.

[30] J. F. Thompson, Z. U. A. Warsi, and C. W. Mastin. *Numerical Grid Generation: Foundations and Applications*. Elsevier North-Holland, Inc., 1985.

[31] M. Tomac. *Towards Automated CFD for Engineering Methods in Airfcraft Design*. KTH School of Engineering Sciences, 2014.

[32] US Product Data Association, Charleston, SC. Initial Graphics Interchange Standard 5.3, September 1996. Available from www.uspro.org.

6 Computational Fluid Dynamics for Steady and Unsteady Flows

When there are disputes among persons, we can simply say: Let us calculate, without further ado, and see who is right.

Gottfried Wilhelm Leibniz, *The Art of Discovery* (1685)

All the mathematical sciences are founded on relations between physical laws and laws of numbers, so that the aim of exact science is to reduce the problems of nature to the determination of quantities by operations with numbers.

James Clerk Maxwell

Preamble

Chapter 4 taught us how numerical schemes accurately compute shock waves in the 1D Euler equations. With geometry description and grid generation from Chapter 5, we now want to leverage that knowledge to procedures suitable for computing solutions to the Navier–Stokes equations in 2D and 3D and to illustrate the workflow involved in carrying out a simulation aerodynamic problem. The finite-volume formulation is the numerical model. Standard physical modeling for turbulence yields the Reynolds-averaged Navier–Stokes equations used in most computational fluid dynamics (CFD) codes directed toward compressible flow aeronautical applications. This chapter takes us through the process that an informed CFD user needs to know for applying a typical code of this genus to a very central problem in high-speed aerodynamics: that of transonic flow past an airfoil. Speed of convergence is vital to all methods solving steady-flow problems, and methods to accelerate this are discussed. Some airfoil flows, however, are unsteady, so a time-accurate methodology is also presented, illustrated by an example.

6.1 Introduction: Scope and Objectives

Chapter 4 went into the details of modern computational techniques to solve the Euler equations for flows containing shock waves, offering hands-on experience in the tutorial for solving the quasi-1D nozzle problem. That example was comparatively simple, as a solution can be found by analytical methods. The purpose was to demonstrate how the various methods work in the nonlinear model problem. This chapter takes the same approach but now applies modern computational fluid dynamics (CFD) tools, a Reynolds-averaged Navier–Stokes (RANS) solver, to a more realistic problem, the classical canonical transonic airfoil, which drove CFD development starting in the 1970s.

Even with the background of Chapter 4 for computing flow with shock waves, much more material is needed to cover all of the functions of a modern CFD code. For example, we must progress from the 1D situation in Chapter 4 to 3D, from Euler's equations to the RANS equations with modeling for the physics of turbulence, etc. The inviscid convective terms are the same and the spatial discretization schemes we have studied are applicable, but new are the viscous terms and their discretization.

The question is how deep that discussion should be. Our answer, and the scope of this chapter, is as follows to the depth needed for the reader to become an *informed user* of a modern CFD code. Thus, we do not go into details necessary to *write* such a code from scratch.

The discussion is intended to make the reader achieve the following.

(1) Understand the user manual of an aerospace CFD code.
(2) Understand the background to the various choices involved.
(3) Select wisely from the options available.
(4) Solve the particular problem at hand effectively.

This said, one should be aware of the wealth of video tutorials available, in particular for the commercial codes. Definition of the computational model in many cases means clicking one's way through a graphical user interface with a plethora of pull-down menus and buttons. Software developers are hard at work to create "wizards" that employ rules of thumb to define many parameters in order to reduce the steep learning curve. Commercial virtual wind tunnels are now available that require only a geometry model as an STL file and a handful of data, such as angle of attack, sideslip, and velocity.

We illustrate the power of a modern CFD code on the transonic airfoil problem by exploring the nonlinear phenomena of flight through the transonic range, as was done with wind tunnel testing during the early 1950s.

Chapter 5 presented a general discussion of airframe geometry modeling and computational grid generation. The nozzle problem in Chapter 4 introduced through time stepping to reach a converged steady-state solution using convergence-acceleration techniques when time accuracy is of no concern. There are, however, examples of unsteady transonic flow when time accuracy is indeed of concern. The last part of this chapter explains how this is carried out. But before getting into these topics, let us take a look at likely ways to access a CFD RANS code.

6.2 RANS Software

RANS solvers have become a standard tool in industry to study steady transonic aerodynamics because of the following.

1. The numerical models and software have reached sufficient maturity and robustness for use on a daily basis.
2. The costs of computer simulations have been reduced by advances in computer hardware, so RANS modeling becomes feasible as a design tool.
3. Wind tunnel testing is costly and does not match the flight Reynolds number.

Two decades ago, Vos et al. [28] surveyed the capabilities of RANS model CFD methods and software for application to aircraft design. Prediction of separated flows was then, and is still, a challenge, as is the physical modeling of turbulence. Rizzi and Luckring [19] present a recent review and outlook for CFD simulations used in aircraft design in attached as well as separated flow conditions.

Major categories of various features that distinguish the character of a particular code are: structured or unstructured grids; the available physical models of turbulence and transition; the schemes offered for space discretization; and the choice of time-integration method and its high-performance computing programming structure and language. Another distinguishing feature is the platforms the code runs on (e.g. clusters or laptops, running either Linux, iOS, or Windows).

The intended use of this text assumes access to a validated and reliable RANS solver. Quite a number of aerospace RANS codes exist today, and there are three broad avenues to access and use.

- *Commercial software*, the NUMECA codes, CFD++, ANSYS FLUENT, STAR-CD, etc.
- *Partnership software* based on agreement in a group of joint contributors/developers; for example, in Europe:

  TAU code between the German Aerospace Center (DLR) and certain German universities

  ELSA code between the French national aerospace research center (ONERA) and certain French universities and institutes

  EDGE code between the Swedish Defence Research Agency (FOI) and Swedish universities

  NSMB code between Computational Fluid and Structures Engineering (CFSE) and agreed partners, etc.

  There are similar partnerships in the USA; for example:

  FUN3D code between NASA and certain universities

  KESTREL code between the US Air Force and certain universities
- *Open-source software*, SU2, OPENFOAM, etc.

In this book, we work on the examples, exercises, and tutorials with either the EDGE or SU2 codes, and we give a brief overview of the EDGE code.

6.2.1 Typical Aerospace RANS Solver: The EDGE Code

The Swedish flow solver EDGE [3, 9] has been developed by the FOI in collaboration with external national and international partners. EDGE solves the RANS (or unsteady

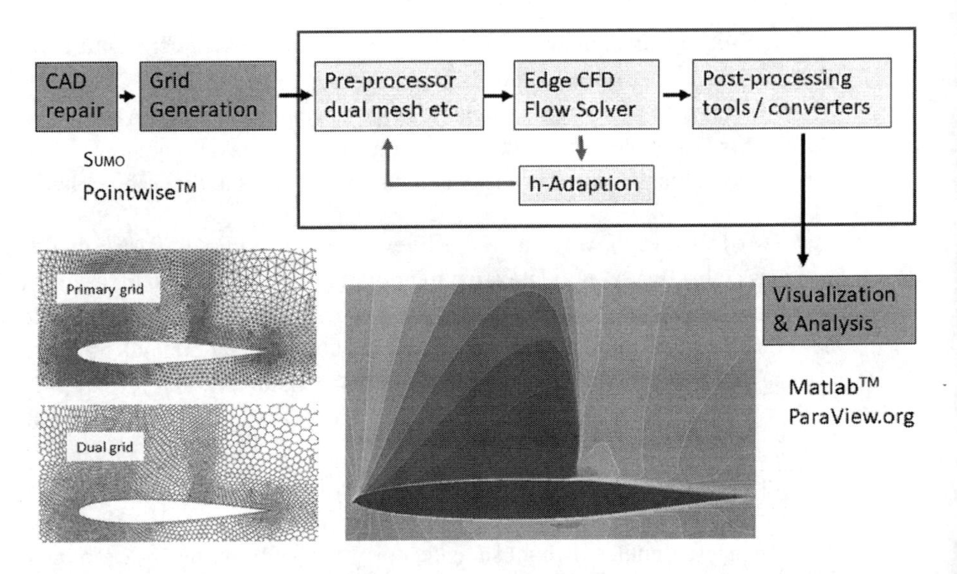

Figure 6.1 Workflow for EDGE simulations.

RANS (URANS)) equations of compressible flow on unstructured grids for different gas models. Turbulence can be modeled with one- or two-equation eddy viscosity models and explicit algebraic or differential Reynolds stress models. Hybrid RANS – large-eddy simulation (LES) models are also provided.

The solver adopts an edge-based node-centered finite-volume technique to solve the governing equation on control volumes formed by a dual grid obtained from the primal grid coming from a mesh generator (see Figure 6.1, left). All elements must be connected by matching faces: no "hanging nodes" are allowed.

Runge–Kutta time stepping for steady problems is accelerated with full approximation scheme (FAS) agglomeration multigrid [2], implicit residual smoothing, and low-speed preconditioning. A partial line implicit scheme also admits large time steps for thin boundary layers. A few examples of runtimes for RANS with multigrid acceleration, the one-equation Spalart–Allmaras (SA) model, and 1000 steps are as follows.

- Airfoil, 50 K nodes grid, one core: 5 minutes
- Wing alone, 1 M nodes grid, two cores: 2 hours
- Wing–body, 5 M nodes grid, two cores: overnight

For time-accurate analysis in URANS or hybrid URANS–LES mode, the convergence procedures are used inside an implicit "dual" time-stepping scheme.

Having highlighted the main features of a representative RANS flow solver, such as EDGE, let us look a little further at putting it to use. Section 6.5.1 demonstrates transonic flow around an airfoil for $0.7 \leq M_\infty \leq 1.2$ computed by EDGE. Figure 6.1 broadly outlines the steps involved in running the computation. Airframe computer-aided design (CAD) and mesh generation as per Chapter 5 makes up the primal grid, the EDGE preprocessor makes up the dual grid, the solver carries out the time-stepping solution procedure, and finally the computed data are post-processed to visualize the flow, compute forces and moments, etc.

An adaptive mesh refinement (AMR) module allows for local h-refinement of the hybrid grid by cell subdivision based on a sensor derived from the flow solution. There are sensors for identifying vortices based on the total pressure ratio, the production of entropy, and an eigenvalue analysis of the velocity gradient tensor [11].

A few more words about physical modeling follow.

6.2.2 Standard Physical Modeling in CFD Software

CFD codes offer a variety of turbulence models for RANS. Their rough characterization will be recognizable from the discussion in Chapter 2, but for details, the software documentation must be consulted. In particular, a turbulence model may depend critically on the computational mesh close to walls, and the documentation will hopefully give hints on good meshes. It is always a good idea, and is prescribed by due diligence, to compare results from different models in order to assess uncertainty from physical modeling. A few more details follow on the three most used models in EDGE.

6.2.3 SA One-Equation Model

The SA model from 1994 [24] solves a modeled transport equation (Eq. (6.1)), for a quantity $\tilde{\nu}$ that is equivalent to the eddy kinematic viscosity. Turbulence is characterized by two scales (e.g. for velocity and length). Since the SA model solves for only one property, additional information is needed, and to this end it uses the wall distance. The model equation is typical of modeled transport equations for turbulent quantities. For instance, it contains a wall destruction term, which reduces the turbulent viscosity close to the wall where the flow is laminar. That term is active through the complete boundary layer, proportional to $(\tilde{\nu}/d)^2$, with d being the distance to the nearest wall. Of course, $\tilde{\nu}$ vanishes at the wall.

$$\underbrace{\frac{D\tilde{\nu}}{Dt}}_{\text{advection}} = \underbrace{c_{b1}\tilde{S}\tilde{\nu}}_{\text{production}} + \underbrace{\frac{1}{\sigma}[\nabla\cdot((\nu+\tilde{\nu})\nabla\tilde{\nu}) + c_{b2}(\nabla\tilde{\nu})^2]}_{\text{diffusion}} - \underbrace{c_{w1}f_w(r)\left(\frac{\tilde{\nu}}{d}\right)^2}_{\text{dissipation}} \tag{6.1}$$

When this term is balanced with the production term, the eddy viscosity becomes proportional to $\hat{S}d^2$, where $\hat{S}$ is the local strain rate $\sqrt{\sum_{i,j} S_{ij}^2}$.

The detailed definition of the model is given in Ref. [24]. Similarly to other models presented here, it is integrated all the way to the wall, which requires the first grid points off the wall to have wall distances that satisfy $y^+ \sim 1$. See the Nomenclature section for a definition of the nondimensional wall distance y^+. The model was designed specifically for aerospace applications involving wall-bounded flows and has been shown to give good results for boundary layers subjected to adverse pressure gradients.

6.2.4 Eddy Viscosity Two-Equation Models

These turbulence models solve two additional transport equations for the turbulent kinetic energy k and some auxiliary quantity, such as turbulent kinetic dissipation rate ε or turbulent fluctuation rate frequency ω. There is a family of $k-\omega$-models, and the standard one is described below. The Hellsten variant was specifically tailored to work with the Wallin–Johansson Reynolds stress model.

The Wilcox Standard $k-\omega$ Turbulence Model

The standard $k-\omega$ turbulence model by Wilcox from 1988 [33] can be put in the form of Eq. (6.5), so it remains to discuss the source terms for the transport equations of k and ω.

$$\begin{pmatrix} q_k \\ q_\omega \end{pmatrix} = \begin{pmatrix} P_k - \beta^* \rho k \omega \\ \gamma \dfrac{\omega}{k} P_k - \beta \rho \omega^2 \end{pmatrix} \tag{6.2}$$

where P_k is the production of turbulent kinetic energy given by the following.

$$P_k = \frac{\partial \tilde{v}_i}{\partial x_j} \left(\mu \tilde{S}_{ij} - \frac{2}{3} \bar{\rho} k \delta_{ij} \right) \tag{6.3}$$

The averaged strain rate is $\tilde{S}_{ij}$ (Eq. (2.10)) and the eddy viscosity is given by the following.

$$\mu_T = \rho \frac{k}{\omega} \tag{6.4}$$

The Wilcox $k-\omega$ model has an unphysical free stream dependency, which, for example, the Menter baseline and shear stress transport $k-\omega$ models overcome. These $k-\omega$ models are basically derived for a standard eddy viscosity relation, but the requirements from an explicit algebraic Reynolds stress model (EARSM) are different. A new $k-\omega$ model derived with the EARSM [29] and, thus, completely consistent with it, is the Hellsten model [8].

6.2.5 The Wallin–Johansson EARSM model

EARSMs can be viewed as a combination of lower-level models and the differential Reynolds stress model (DRSM) approach. EARSMs employ only two transport equations, mostly for the turbulent kinetic energy and the dissipation rate. The components of the Reynolds stress tensor are related to the transport quantities by nonlinear algebraic relations. This model is capable of predicting rotational turbulent flows with accuracy similar to the DRSM approach. The EARSM in Ref. [29] is a rational approximation of a full Reynolds stress transport model in the weak equilibrium limit where the Reynolds stress anisotropy may be considered constant in time and space. The Reynolds stress tensor is explicitly expressed in terms of the velocity gradient and the turbulence scales. The EARSM in EDGE has been extensively tested in different flows and is the *recommended default model.*

6.3 RANS Finite-Volume Numerical Modeling

The discussion focuses on the EDGE software and should also be useful for other external aerodynamics flow solvers such as SU2 using cell-vertex formulation.

The RANS equations for compressible flow are solved by EDGE [3] in conservation form with an integral formulation for correct capturing of discontinuities such as shock waves. Moreover, the system offers a choice of turbulence models. The conservation-law form of the RANS equations for mean mass, momentum, and energy over a volume Ω contained in the surface S reads as Eq. (6.5). The flux functions are given in Eq. (6.6) and Eq. (6.7).

$$\int_{\Omega} \frac{\partial \mathbf{u}}{\partial t} d\Omega + \int_{S} (\mathbf{F}_I - \mathbf{F}_V)\mathbf{n} dS = \int_{\Omega} \mathbf{q} d\Omega \tag{6.5}$$

where $\mathbf{n}$ is the outward unit normal vector to S.

$\mathbf{u}$ is the vector of the conserved variables $(\rho, \rho u, \rho v, \rho w, \rho E, \rho k, \rho\omega)^T$ with E the total energy, ρ the density, and $\mathbf{V} = (u, v, w)^T = (v_1, v_2, v_3)^T$ the velocities. The pressure p is needed and computed from the conserved variables. Equation (6.5) is written here for a two-equation $k - \omega$-type turbulence model with turbulent viscosity μ_T. $\mathbf{q} = (0, 0, 0, 0, 0, q_k, q_\omega)^T$ is the vector of source terms, which comes from the turbulence model. The current zeros in $\mathbf{q}$ can be exchanged for volume forces as appear, for example, in rotating coordinate systems and heat sources as appropriate.

The flux function has been additively split into an inviscid flux $\mathbf{F}_I$ and a viscous contribution $\mathbf{F}_V$. $\mathbf{F}_V$ requires computation of space derivatives, whereas $\mathbf{F}_I$ requires only evaluation. If the cell is polyhedral with faces S_k, the surface integral turns into a sum of integrals over the plane faces, so that each has a constant unit normal $\mathbf{n}_k$.

$$\mathbf{F}_I = (\mathbf{F}_{Ix}\ \mathbf{F}_{Iy}\ \mathbf{F}_{Iz}) = \begin{pmatrix} \rho u & \rho v & \rho w \\ \rho u^2 + p & \rho v u & \rho w u \\ \rho u v & \rho v^2 + p & \rho w v \\ \rho u w & \rho v w & \rho w^2 + p \\ u(\rho E + p) & v(\rho E + p) & w(\rho E + p) \\ \rho u k & \rho v k & \rho w k \\ \rho u \omega & \rho v \omega & \rho w \omega \end{pmatrix} \tag{6.6}$$

$$\mathbf{F}_V = (\mathbf{F}_{Vx}\ \mathbf{F}_{Vy}\ \mathbf{F}_{Vz}) = \begin{pmatrix} 0 & 0 & 0 \\ \tau_{11} & \tau_{21} & \tau_{31} \\ \tau_{12} & \tau_{22} & \tau_{32} \\ \tau_{13} & \tau_{23} & \tau_{33} \\ (\mathbf{TV})_1 + h_1 & (\mathbf{TV})_2 + h_2 & (\mathbf{TV})_3 + h_3 \\ \mu_k \frac{\partial k}{\partial x_1} & \mu_k \frac{\partial k}{\partial x_2} & \mu_k \frac{\partial k}{\partial x_3} \\ \mu_\omega \frac{\partial \omega}{\partial x_1} & \mu_\omega \frac{\partial \omega}{\partial x_2} & \mu_\omega \frac{\partial \omega}{\partial x_3} \end{pmatrix} \tag{6.7}$$

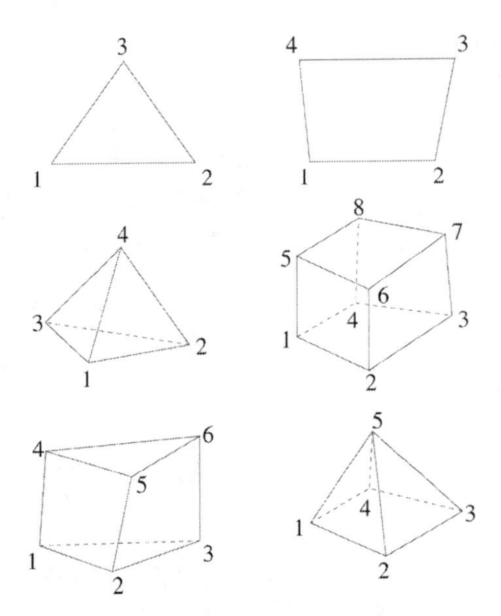

Figure 6.2 Typical polyhedral computational cells: triangle, quadrilateral, tetrahedron, hexahedron, prismatic element, and pyramid.

with $\tau_{ij} = (\mu + \mu_T)\left(\frac{\partial V_i}{\partial x_j} + \frac{\partial V_j}{\partial x_i} - \frac{2}{3}(\nabla \cdot \mathbf{V})\,\delta_{ij}\right)$ the elements of the turbulent stress tensor $\mathbf{T}$ and $\boldsymbol{h} = (h_1, h_2, h_3)$ the conductive heat flux according to Fourier's law, $-(\kappa + \kappa_T)\nabla T$, where κ_T is the turbulent contribution to the heat conductivity. Usually, it is modeled from μ_T by a turbulent Prandtl number Pr_T.

$$\frac{\kappa_T}{C_p} = Pr_T\,\mu_T$$

This depends on the gas, and values between 0.7 and 0.9 have been used for air.

6.3.1 Finite-Volume Cells

The finite-volume discretization balances differences in flux values across the faces of the computational cell. Chapter 4 applied the finite-volume formulation to the Euler equations in 1D. Creating the computational cells there was trivial. Now considering 2D or 3D, we must pack a large number of control volumes – the *grid cells* – around the object(s) to completely fill the region of flow. Chapter 5 describes how to generate such a grid, and here we address characteristics of the particular finite-volume discretization that is used. In 2D, the grid network typically contains triangles and/or quadrilaterals, and in 3D, it contains tetrahedra, prisms, pyramids, and hexahedra. The surface grid on the boundary of the body (solid surface) is composed in 2D by line segments, as in Chapter 7 focusing on airfoils, and in 3D by triangular and quadrilateral facets. Figure 6.2 depicts these various polyhedra with numbering for their vertices that are the *cell vertices* (i.e. the *grid points*).

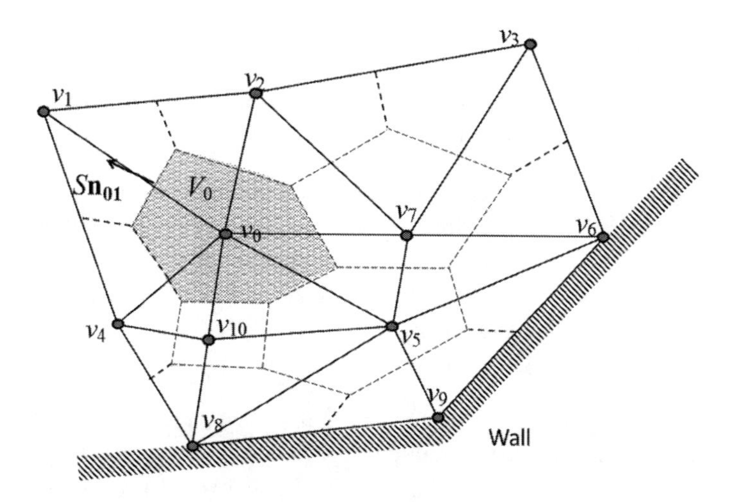

Figure 6.3 Triangular (solid line) input grid and its dual grid (dashed line). Control volumes for an interior node v_0 and a boundary node v_9 in gray.

The Cell-Vertex Median Dual Grid

A finite-volume method works with cell averages. Often, the cell average is associated with a point, the natural one being the centroid of the cell volume. In the *cell-center* method, the control volumes are the input grid cells themselves, while in the *cell-vertex* method, the solution is associated with the input cell vertices (i.e. the grid points). The control volumes now form a *dual grid* such that the nodes are (close to) the cell centroids. To make the description of the multidimensional discretization process as clear as possible, we present most graphics for 2D and avoid the clutter of the third dimension. One idea is to let the control volumes be the Dirichlet tesselation of the set of nodes. A cell around a node consists of all points closer to it than to any other node; see the discussion of the Delaunay triangulation in Chapter 5. The EDGE code uses dual cells whose edges are the medians of the input grid triangles: a *median dual* grid. The dual cell of a node is then a polygon included in the N_0 cells joined at the node. However, since medians on either side of a triangular edge are not necessarily colinear, the polygon has $2N_0$ vertices. A little extra arithmetic is necessary, exemplified by Eq. (6.8), to produce the sum of the oriented intercell faces. A preprocessor produces the dual grid as an input to the flow solver. Figure 6.3 illustrates an example of the input grid and its median dual in 2D. The typical control volume around the node is gray and its surface-normal $\mathbf{n}S_{10}$ with node 1 is shown.

Control surfaces positioned at the boundary close the control volumes there. In Figure 6.3, the control surfaces to node v_0 are given at all edges connected to v_0. The control volumes of boundary nodes, such as v_9, are closed by separate boundary faces of the primal grid cells.

6.3.2 Finite-Volume Cell Face Fluxes

The volume in Eq. (6.5) is now a computational cell. The recipe for evolution of the set of discretized equations becomes the following.

- Reconstruct a field inside the cells from the cell averages so that flux functions can be evaluated and integrated over the cells and cell faces, as required by the scheme.
- Evolve the dynamic system over the time Δt by either explicit Runge–Kutta or implicit Euler time integration.
- Average the evolved fields into cell averages.

This recipe, which stems from Godunov, is not so apparent in cell-vertex schemes because of their association of cell-average values with point values at the nodes of the primal grid. Thus, averaging is trivial and reconstruction, as described in Section 4.2.4, is carried out in the evaluation of flux functions.

Spatial Discretization

The integral form (Eq. (6.5)), is applied in a finite-volume discretization, (Eq. (6.9)), to the control volume V_i surrounding the node v_i. The solver needs the scalar product of a mean flux vector with the sum of the oriented surfaces for an edge as follows.

$$\mathbf{n}_{01} S_{01} = \mathbf{n}_{01,1} S_{01,1} + \mathbf{n}_{01,2} S_{01,2} \tag{6.8}$$

$$\frac{d}{dt}(\boldsymbol{u}_i V_i) + \sum_{j \in N(i)} \hat{\mathbf{F}}_{Iij} \boldsymbol{n}_{ij} S_{ij} - \sum_{j \in N(i)} \hat{\mathbf{F}}_{Vij} \boldsymbol{n}_{ij} S_{ij} = \boldsymbol{q}_i V_i \tag{6.9}$$

with $N(i)$ the set of neighboring nodes to node i, $\hat{\mathbf{F}}_{Iij}$ and $\hat{\mathbf{F}}_{Vij}$ the discretized inviscid and viscous flux matrices between nodes i and j, $\boldsymbol{n}_{ij}$ the unit normal vector to the surface between node i and node j, and S_{ij} the surface area between nodes i and node j. Equation (6.9) is the semi-discrete system of governing equations.

Inviscid Flux

The schemes for the inviscid flux $\mathbf{F}_{I0k}$ considered here are based on a central discretization with dissipation terms of either artificial dissipation type or upwind flux difference splitting type [3], as discussed in Chapter 4.

Viscous Fluxes Need Gradients

The reconstruction is required for determination of the viscous flux from gradients of the velocity components and the temperature on the control volume faces and also for the artificial dissipation terms in the Jameson scheme and the limited quantities in high-resolution (Hi-Res) upstream schemes. In a structured grid, the Jameson scheme requires third-order differences at cell boundaries, and upstream schemes require comparison of differences between neighbor cells. In the following, a few hints are provided on the generalization to unstructured cell-vertex grids.

Gradients

The Green–Gauss theorem approximates the gradient of a scalar function ϕ as the surface integral of the product of ϕ with an outward-pointing unit normal vector over a control volume $\mathcal{V}$ with surface S as follows.

$$|\mathcal{V}| \cdot \nabla\phi \approx \int_{\mathcal{V}} \nabla\phi dV = \int_S \phi \boldsymbol{n} dS \tag{6.10}$$

The resulting formula for the gradient at a node reads as follows (see Figure 6.3).

$$\nabla\phi_i \approx \frac{1}{|\mathcal{V}_i|} \sum_{j \in N(i)} \frac{1}{2}(\phi_i + \phi_j)\boldsymbol{n}_{ij} S_{ij} \tag{6.11}$$

Artificial Second- and Fourth-Order Dissipation for the Jameson Scheme

First, let us note that repeated use of the Green–Gauss formula on a structured 1D grid leads to a noncompact second derivative formula as follows.

$$d^2 f/dx^2 \approx (f_{i-2} - 2f_i + f_{i+2})/(4dx^2)$$

which is insensitive to a sequence $\ldots -1, 1, -1, 1, \ldots$ and hence clearly is not helpful in suppressing short-wavelength oscillations. The standard compact formula is as follows.

$$d^2 f/dx^2 \approx (f_{i-1} - 2f_i + f_{i+1})/dx^2$$

The corresponding difference formula, less the division by dx^2, in the cell-vertex grid is as follows.

$$\nabla_U^2 \phi_i = \sum_{j \in N(i)} (\phi_j - \phi_i)$$

and ∇_U^2 is called the "undivided" Laplace operator.

The inviscid flux across the cell face between nodes ν_0 and ν_1 is computed as follows.

$$\mathbf{F}_{I_{01}} = \mathbf{F}_I\left(\frac{\mathbf{u}_0 + \mathbf{u}_1}{2}\right) - \mathbf{d}_{01} \tag{6.12}$$

where $\mathbf{d}_{01}$ denotes the artificial dissipation. The Jameson blend of second- and fourth-order differences corresponds to a blend of first- and third-order differences for the fluxes (Section 4.2.2). The following form has been shown to be suitable.

$$\mathbf{d}_{01} = \left(\varepsilon_{01}^{(2)}(\mathbf{u}_1 - \mathbf{u}_0) - \varepsilon_{01}^{(4)}(\nabla_U^2\mathbf{u}_1 - \nabla_U^2\mathbf{u}_0)\right)\varphi_{01}\lambda_{01} \tag{6.13}$$

with ∇_U^2 the "undivided" Laplace operator.

The factor φ_{01} is introduced to account for the stretching in the grid, since the cells are anisotropic. It is defined from eigenvalues and cell geometries to converge to a similar factor for structured, orthogonal grids, defined in Ref. [13]. This gives a dissipation proportional to the maximal characteristic velocity scaled by cell face area in the direction of the stretching and smaller in other directions.

The function $\varepsilon_{01}^{(2)}$ is the pressure switch to turn second-order dissipation on at shocks and off in smooth regions of the flow.

$$\varepsilon_{01}^{(2)} = \kappa^{(2)} s_2 \frac{\left|\sum_{k\in N(0)}(p_k - p_0)\right|}{\sum_{k\in N(0)}(p_k + p_0)} \tag{6.14}$$

where $\kappa^{(2)}$ is an input constant (= VIS2 of Section 4.2.2) and s_2 is a scaling factor to reduce the dependency on the number of neighbors.

The fourth-order difference dissipation is switched off in the vicinity of shocks (see Chapter 4).

The Hi-Res Flux Splitting Scheme

In addition to the central scheme, Hi-Res schemes are available. The upwind scheme is of Roe flux difference splitting type, slightly different from the monotonic upstream-centered scheme for conservation laws (MUSCL)-type upwind schemes. As in the central scheme, the inviscid flux is computed as a central part with additional dissipation (see also Section 4.2.5; cf. Eq. (6.12)).

$$\mathbf{F}_{I01} = \frac{1}{2}\left(\mathbf{F}_I(\mathbf{u}_0) + \mathbf{F}_I(\mathbf{u}_1)\right) - \mathbf{d}_{01} \tag{6.15}$$

The upwind dissipation $\mathbf{d}_{01}$ is here computed as follows.

$$\mathbf{d}_{01} = \frac{1}{2} R\tilde{\Lambda}R^{-1} d\mathbf{u}_{01} = \frac{1}{2} R\tilde{\Lambda}L^{-1} d\mathbf{v}_{01} = \frac{1}{2} R\tilde{\Lambda} d\mathbf{W}_{01}, \tag{6.16}$$

$$d\mathbf{W}_{01} = \mathbf{W}_1 - \mathbf{W}_0, \text{ etc.}$$

where $\mathbf{u}, \mathbf{v}$, and $\mathbf{W}$ denote the conservative, primitive, and characteristic variables, respectively, the primitive variables being the ones used in the code. The matrices R and L are right eigenvector matrices to the flux Jacobian.

$$\frac{\partial \mathbf{F}_I}{\partial \mathbf{u}} = R\Lambda R^{-1} \text{ and } \frac{\partial \mathbf{F}_I}{\partial \mathbf{v}} = L\Lambda L^{-1} \tag{6.17}$$

$\tilde{\Lambda}$ is as follows.

$$\tilde{\Lambda} = |\Lambda^*|(I - \Phi) \tag{6.18}$$

where Φ is a diagonal matrix with limiters acting on the primitive variables and $\Phi = 0$ for a first-order upstream scheme. For a Roe flux difference splitting scheme, the components of R, L, and Λ should be computed from the Roe-averaged variables.

The diagonal matrix $|\Lambda^*|$ contains the absolute values of eigenvalues Λ adjusted with an entropy fix to bound them away from zero (at sonic velocity) and avoid unphysical solutions such as expansion shocks.

The limiters in a structured grid use ratios of slopes (or variable differences) of neighbor cells (see Chapter 4). Here, the slopes are computed from the cell gradients (Eq. (6.11)), as shown below. EDGE uses differences of the characteristic variables $\mathbf{W}$ to avoid oscillations in the pressure, although it is computationally more expensive.

Gradients of all primitive variables are needed to compute the characteristic variable differences in the nodes, with $d\mathbf{W}_0$ and $d\mathbf{W}_1$ for neighbor nodes 0 and 1 obtained as follows.

$$\begin{aligned} d\mathbf{W}_0 &= 2L_0(\nabla\mathbf{V}_0 \cdot (\mathbf{x}_1 - \mathbf{x}_0)) - d\mathbf{W}_{01} \\ d\mathbf{W}_1 &= 2L_1(\nabla\mathbf{V}_1 \cdot (\mathbf{x}_1 - \mathbf{x}_0)) - d\mathbf{W}_{01} \end{aligned} \tag{6.19}$$

which correspond to the left and right compact differences for the cell face $i+1/2$ flux in the 1D scheme. The more obvious expression becomes analogous to the noncompact central finite difference, which leads to solution wiggles.

EDGE offers the MINMOD, VAN LEER, and SUPERBEE limiters. See Chapter 4 for a general overview of limiters. At boundaries, no particular modification to the scheme is made. The numerical flux due to the upwind dissipation is zero, and the fluxes through cell face areas on boundary surfaces are computed from the boundary conditions.

6.3.3 Boundary Conditions

Weak Conditions

Most boundary conditions are implemented by a *weak* formulation (i.e. the boundary conditions are imposed through the flux and all unknowns on these boundaries are updated like any interior unknown).

A few exceptions exist, however (e.g. the no-slip condition on viscous walls, which is imposed strongly by prescribing the velocity variables instead of treating them as unknowns).

Some of the boundary conditions are described theoretically below using the notation from Figure 6.3. Note that only the convection terms and the viscous terms contribute to the fluxes on the boundaries – the dissipative fluxes in the energy equation are set to zero.

Inflow and Outflow

The boundary conditions indicated in Section 4.4.4 can be used by projection on the boundary normals. Instead of employing the compatibility relations for the outgoing waves, one can often simply extrapolate the corresponding Riemann invariants (Figure 6.4), where c_b is the local speed of sound at the boundary b of the cell. Consider a subsonic outflow boundary at b, $0 < \mathbf{u}_b \cdot \mathbf{n} < c_b$, adjacent to cell i with outer unit normal vector $\mathbf{n}$. Assume that the time step Δt is chosen such that the characteristics C_+ with $\frac{dn}{dt} = \mathbf{u}_b \cdot \mathbf{n} + c$ and C_0 with $\frac{dn}{dt} = \mathbf{u}_b \cdot \mathbf{n}$ come from cell i. Then the corresponding Riemann invariants can be approximated, neglecting tangential derivatives and source terms, by the following.

$$\frac{p_b}{\rho_b^\gamma} = \frac{p_i}{\rho_i^\gamma} \text{ Entropy}$$

$$\mathbf{u}_b - (\mathbf{u}_b \cdot \mathbf{n})\mathbf{n} = \mathbf{u}_i - (\mathbf{u}_i \cdot \mathbf{n})\mathbf{n} \text{ Tangential velocities} \tag{6.20}$$

$$\mathbf{u}_b \cdot \mathbf{n} + \frac{2}{\gamma - 1}c_b = \mathbf{u}_i \cdot \mathbf{n} + \frac{2}{\gamma - 1}c_i \text{ Riemann variable} \tag{6.21}$$

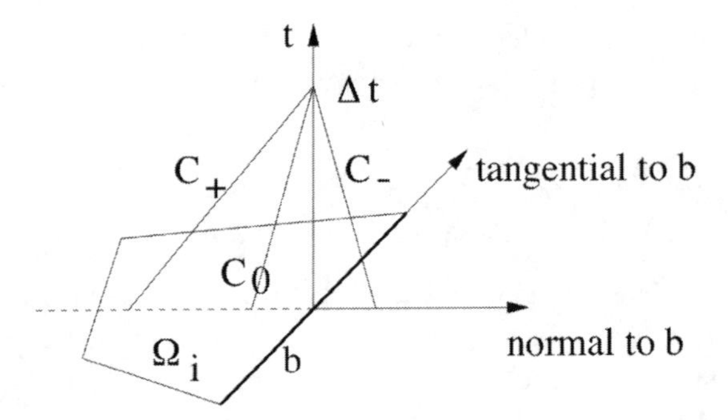

Figure 6.4 Subsonic outflow boundary cell.

Using the boundary condition $p_b = p_a$, where p_a is the ambient pressure, we can determine $\mathbf{V}_b$ and thus compute the inviscid boundary flux $\mathbf{F}_I(\mathbf{U}_b)$. The viscous boundary flux $\mathbf{F}_V(\mathbf{U}_b)$ is usually neglected.

6.3.4 Time Integration to Steady State

The discretization turns the partial differential equation (PDE) into a set of ordinary differential equations (ODEs), and it remains to find a steady solution or follow its evolution over time accurately. This proceeds as for the nozzle quasi-1D model in Chapter 4, with a few new issues. The inclusion of the viscous terms requires a refined analysis of the choice of time step for the Runge–Kutta schemes, and implicit time stepping becomes a challenge because the linear systems appearing are so large that Gaussian elimination is not a viable option.

Convergence Acceleration

Over the years, various methods have been developed in order to accelerate the computation of steady flows by time stepping. The acceleration techniques are also applicable to the inner iteration of the dual time-stepping scheme for computing unsteady flows. Among them are local time stepping, residual smoothing, and the more generic technique multigrid. Virtually all aerospace RANS solvers will offer multigrid. Starting in the 1980s, Antony Jameson has been the most proficient developer of these methods, and he has demonstrated that RANS solutions converged to steady state in fewer than 20 multigrid cycles. Multigrid acceleration implementation requires the generation of a hierarchy of grids, which is a challenge for unstructured grids, and the prolongation and restriction operators are more complicated. The theory behind these techniques will not be further explained here, and P. Wesseling's book [32] is recommended as an overview and introduction.

The catalog of methods for accelerating convergence to steady state is augmented by residual smoothing, low Mach number preconditioning, and partially implicit schemes. Time-accurate simulations are done using implicit time-stepping schemes. They provide temporal damping and act as a low-pass filter for the dynamics, so the choice of time step must be considered carefully. Inside each time step, the algebraic equations can be solved by (quasi or dual) time stepping, just like for a steady problem, so the convergence acceleration devices are brought to bear again.

6.4 Due Diligence CFD

Validation deals with comparison between computational results and experimental data and does not specifically address *how* the computational model can be changed to improve the agreement with the experiment, nor does it specifically address the inference that the model is accurate for cases different from the validation comparison.

The process of determining how accurate a model is as a representation of the real world from the perspective of the intended uses of the model is a very demanding task. It involves the identification and quantification of the error and uncertainty in the conceptual and computational models.

6.4.1 Sensitivity Analysis

Uncertainties are ubiquitous in mathematical modeling, and uncertainty quantification (UQ) is a research field that is in rapid development, relying largely on advanced mathematical statistics. In general, it is beyond the scope for this text.

However, when the number of "uncertain" parameters involved is small, the problem can be approached by sensitivity analysis. This is a quantitative investigation of the influence of parameters on the quantities of interest QoI_j. Although limited information is obtained with sensitivity analyses, sensitivity analyses are much less demanding computationally and in terms of required information as compared to uncertainty analyses. Sensitivity analyses are also referred to as a *what-if* or perturbation analyses.

Multiple simulations are carried out to determine the effect of the variation of some component of the model, such as an input parameter p_k (e.g. the angle of attack) or modeling assumptions (e.g. the turbulence model) on the selected QoI, say $\lambda_k = \frac{\Delta QoI}{\Delta p_k}$. When the sensitivity of a QoI is computed for a set of input quantities $p_k, k = 1, 2, \ldots$, one can rank its sensitivities λ_k.

Methodology

Informed users must reassure themselves that the solution is reasonable and trustworthy. One way to do this is through sensitivity studies of the following.

- Numerical modeling:

  Systematically vary the *grid size* and see its effect on the solution. With decreasing cell size, the numerical errors decay and the QoI should converge, demonstrating *grid convergence*.

- Physical modeling:
  Systematically vary governing equations and turbulence closure modeling.
  For example:
  Turn off all viscous terms to see how much the boundary layer effects overall flow.
  Run series of cases through the hierarchy of turbulence models.

After application of due diligence to judge the quality of CFD solutions follow the last steps of the CFD workflow.

- Analyze and interpret results.
- Visualize flow fields.
- Draw conclusions for solution of the aerodynamics problem.

6.4.2 Numerical Modeling

The starting assumption is that the solver has been previously verified and you are applying it to a problem that is within the span of its validation base, in this case for transonic airfoil flow. Given a numerical procedure that is stable, consistent, and robust, the three major sources of errors due to numerical models in CFD solutions are insufficient spatial discretization, temporal discretization, or convergence of iterative procedures.

These error sources (spatial discretization error, temporal discretization error, and iterative error) are considered to be controllable by objective measures. It is possible, although computationally costly, to automate their reduction to user-defined levels, and significant efforts are spent by software developers in this pursuit.

Convergence to Steady–State and Grid Convergence

Figure 6.5 plots the residuals and the QoI – the forces and moments – to indicate convergence to steady state. It is easier to put a number on tolerable error in forces and moments than on the number of orders of magnitude reduction of the residual required. Figure 6.5 (right) demonstrates mesh convergence of an airfoil pressure distribution through a sequence of three grid resolutions (coarse–medium–fine). C_p^* is shown as a dashed line, and the bottom surface is just subcritical.

6.4.3 Physical Modeling and Turbulence Model

Euler and RANS CFD solutions, and experiment for a Mach sweep are compared in Figure 6.6 (left).

The solutions for $M = 0.85$ and 0.875 are not really steady because the residuals oscillate, due to shock-boundary layer interaction. This indicates that a time-accurate solution may be called for. Section 6.6 explains how this can be achieved.

Figure 6.6 (right) compares computed pressure distributions with different turbulence models. The case studied is known to be challenging for CFD. A number of turbulence models are evaluated: the one-equation SA model [25], the two-equation

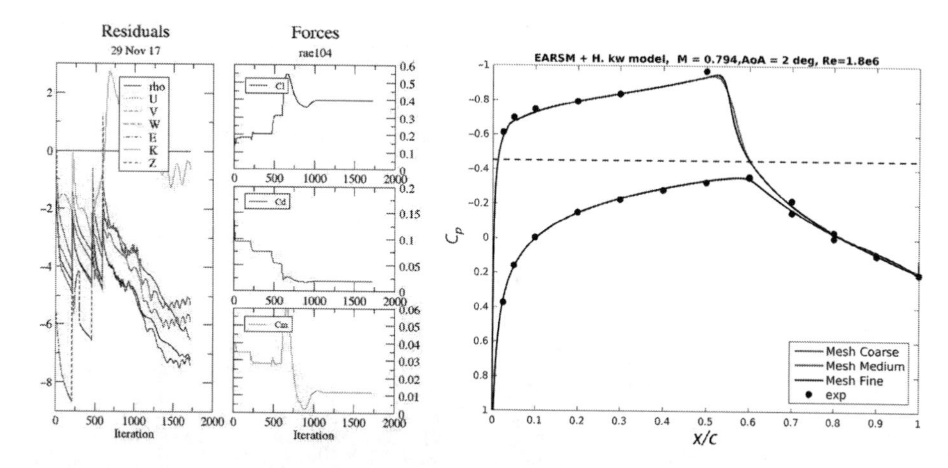

Figure 6.5 Residual and forces/moments (QoI) convergence to steady state (left); grid convergence (right) EDGE code EARSM + H. $k - \omega$ model, RAE104 airfoil, $M_\infty = 0.794, \alpha = 2^\circ, Re = 1.8 \times 10^6$. AoA = angle of attack.

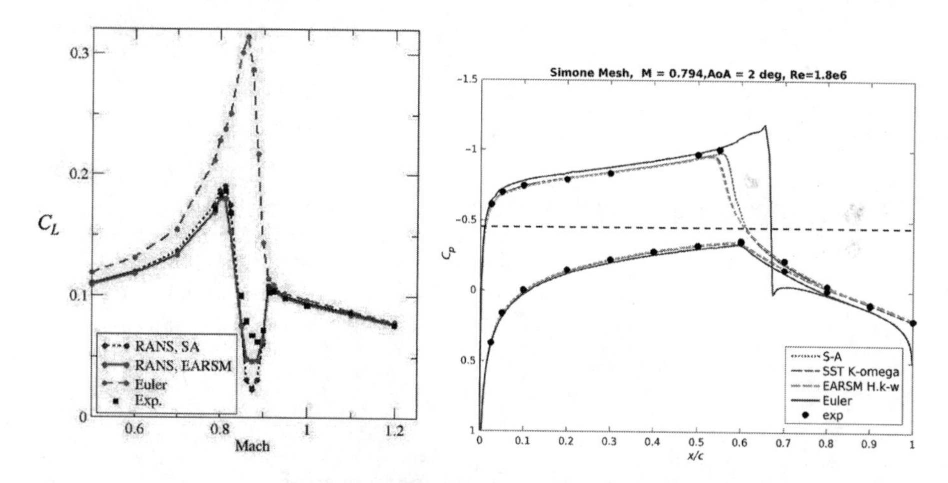

Figure 6.6 Mach sweep through transonic-flow regime, lift vs Mach number at $\alpha = 0.9^\circ$ and $Re = 1.8 \times 10^6$, for Euler, RANS-SA, and RANS-EARSM + H. $k - \omega$ models compared with experiment (left); RAE104 airfoil. Pressure-distribution sensitivity to physical modeling: Euler and RANS with a range of turbulence models (right); RAE104 airfoil, EDGE code $M_\infty = 0.794, \alpha = 2^\circ, Re = 1.8 \times 10^6$.

$k - \omega$ SST variant, and the EARSM [29] with the Hellsten $k - \omega$ model, all of which are common in the CFD community today. Although the models are very different in their approaches, no significant difference is seen in their results. Figure 6.6 also shows Euler results without shock–boundary layer interaction with erroneous shock position. The RANS solutions are close together, with the SA model being slightly different in terms of shock position. The two-equation model and EARSM agree very closely.

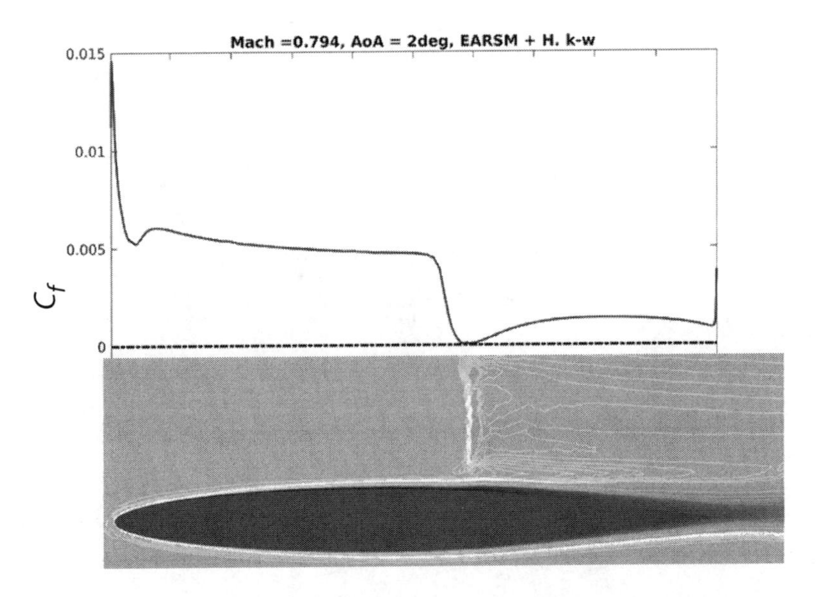

Figure 6.7 Skin-friction coefficient, C_f, profile (top); and contours of total pressure, p_{tot} (bottom). RAE104 airfoil, EDGE code EARSM + H. $k - \omega$ model $M_\infty = 0.794, \alpha = 2°, Re = 1.8 \times 10^6$. AoA = angle of attack.

6.4.4 Visualizing and Understanding the Results

Figure 6.7 shows the friction coefficient profile on the upper airfoil surface and total pressure contours. There is no recirculation, and the thickening of the boundary layer at the foot of the shock is obvious. The total pressure loss through the shock, with its associated wave drag, is obvious.

6.5 Nonlinear Aerodynamics of Increasing M_∞

The lift on an airfoil increases with speed, even at a fixed angle of attack, until at a certain speed it drops markedly with further speed increase, while the drag continues to rise. Parameters such as the thickness ratio, camber, and nose radius also influence the magnitude of the compressibility effects. Transonic flows are very sensitive to the contour of the body surface, since changes in the surface slope affect the location of the shock wave and, therefore, the flow field outside the boundary layer, as well as the downstream boundary layer itself. Furthermore, the shock wave–boundary layer interaction and the possible development of separation downstream of the shock wave are sensitive to the character of the boundary layer, to its thickness, and to its velocity profile at the interaction location. Since a turbulent boundary layer can negotiate higher adverse pressure gradients than can a laminar one, the shock wave–boundary layer interaction is smaller for a turbulent boundary layer.

Compressibility Effects in Transonic Flow

In general, the air density changes very little in low-speed flight, $M_\infty < 0.2$–0.3, and one speaks therefore about incompressible subsonic flow. There are exceptions, however. In the high-lift system deployed at takeoff and landing, locally large flow velocities can arise with associated compressibility effects, although the aircraft flies very slowly.

At flight Mach numbers $M_\infty = 0.3$ to ≈ 0.7, it is necessary to account for compressibility effects. The Prandtl–Glauert correction permits statements concerning aerodynamic wing characteristics at high speeds based on results obtained at low speeds, such as in a low-speed wind tunnel or computed by linearized potential theory. Similar corrections, due to J. Ackeret [1], can be applied in purely supersonic flow past slender aircraft configurations.

When the flight speed approaches $M_\infty = 1$, the abovementioned corrections become meaningless. The Prandtl–Glauert correction for lift and pitching moment, $c_{..} \sim 1/\beta$, holds for $0 < M_\infty \lesssim 0.75$. It does not hold for drag, since it does not model wave drag. The Ackeret rule $c_{..} \sim 1/\beta$ (note the definition in Eq. 2.17) holds in the supersonic speed regime, as well as for wave drag, with $1.2 \lesssim M_\infty \lesssim 5$: the supersonic range here is too close to $M1$ for the Ackeret rule to apply. The Mach number range $0.80 \lesssim M_\infty \lesssim 1.2$ is called the *transonic* regime.

6.5.1 Airfoil Transonic Flow at $0.7 \lesssim M_\infty \lesssim 1.2$ by CFD

Figure 6.8 shows lift-curve slope $c_{L,\alpha}$ and zero-lift drag for a Mach sweep between 0.5 and 1.2. The Prandtl–Glauert model is the dashed line extrapolated from $M_\infty = 0.5$.

Pure subsonic flow obtains for $M_\infty \lesssim 0.75$. The transonic regime begins at the (lower) critical Mach number M_{crit}, where, for the first time, if the flight Mach number is increasing, an embedded supersonic flow region occurs, which is terminated by a shock wave. Once the upper critical Mach number of about 1.2 has been exceeded, the

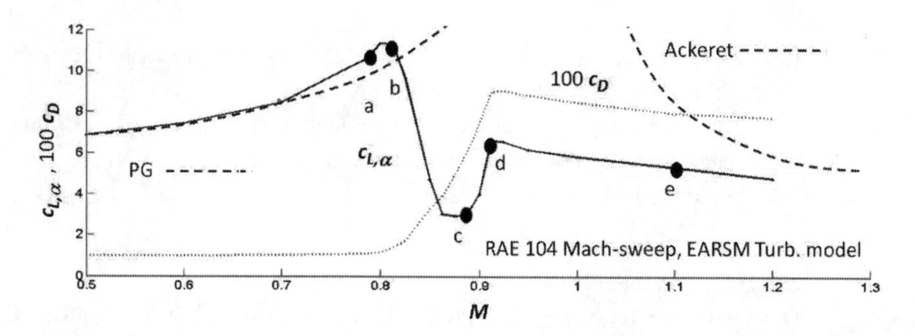

Figure 6.8 RANS computation of RAE104 lift-curve slope $c_{L,\alpha}$ and 100 × drag coefficient c_D vs Mach number show strong nonlinear effect of compressibility. Refer to Figures 6.9–6.13 for the flow fields corresponding to the five lettered points (a)–(e) on this graph. (Courtesy of P. Eliasson, personal communication)

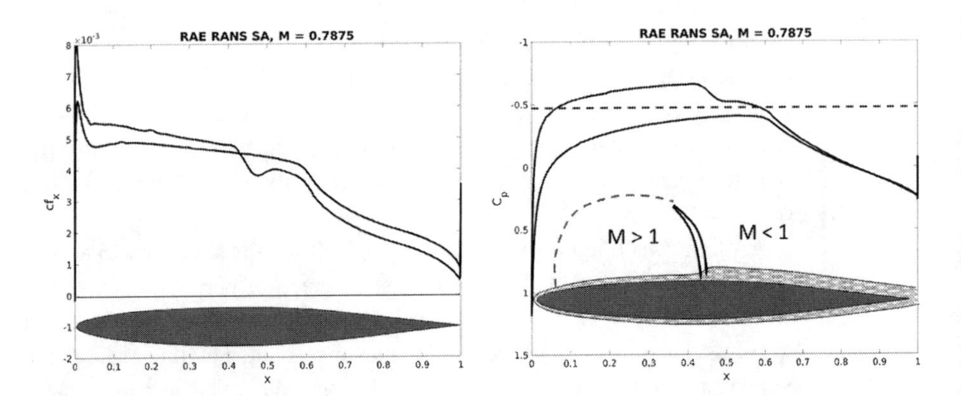

Figure 6.9 Case (a), $M_\infty = 0.7875$. Skin-friction coefficient (left); pressure coefficient and boundary layer (right). EDGE code RANS-SA model, RAE104 airfoil, $\alpha = 0.9°$, $Re = 1.8 \times 10^6$. (Courtesy of P. Eliasson, personal communication)

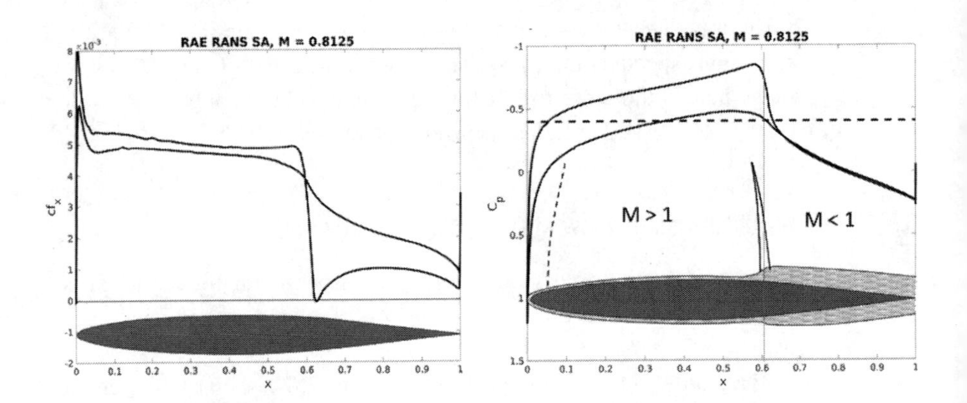

Figure 6.10 Case (b), $M_\infty = 0.8125$. Skin-friction coefficient (left); pressure coefficient and boundary layer (right). EDGE code RANS-SA model, RAE104 airfoil, $\alpha = 0.9°$, $Re = 1.8 \times 10^6$. (Courtesy of P. Eliasson, personal communication)

bow shock appears, there is supersonic flow everywhere except in a small pocket at the nose, and the Ackeret correction is applicable.

Figure 6.8 illustrates *shock stall*, the name given to the behavior exhibited by early high-speed aircraft when they encountered highly nonlinear compressibility effects throughout the transonic flight range.

Shock Stall on the RAE104

During the 1950s, the National Physical Laboratory in the UK studied shock stall with extensive wind tunnel experiments, reported in e.g Ref. [4], for example. Here, we repeat that exercise with CFD on the 10% thick symmetric RAE104 airfoil throughout the entire transonic speed range $0.50 \lesssim M_\infty \lesssim 1.2$ using EDGE. We present the following for five Mach numbers.

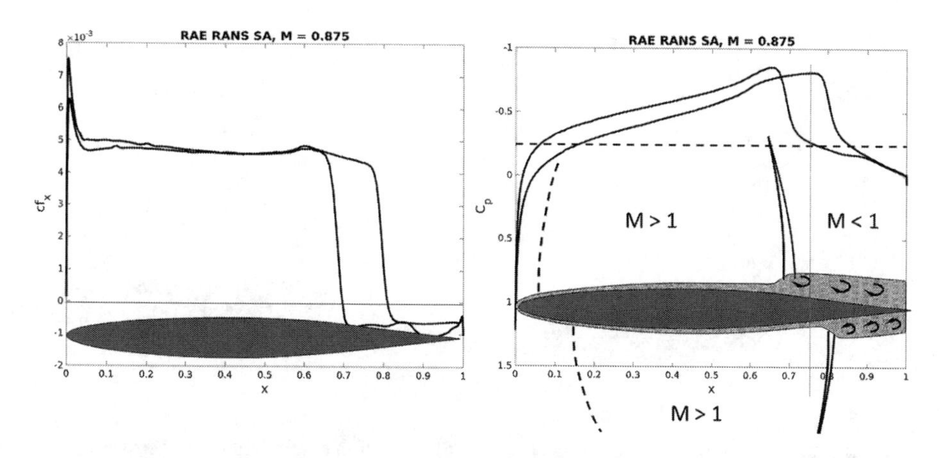

Figure 6.11 Case (c), $M_\infty = 0.8750$. Skin-friction coefficient (left); pressure coefficient and boundary layer (right). EDGE code RANS-SA model, RAE104 airfoil, $\alpha = 0.9°$, $Re = 1.8 \times 10^6$. (Courtesy of P. Eliasson, personal communication)

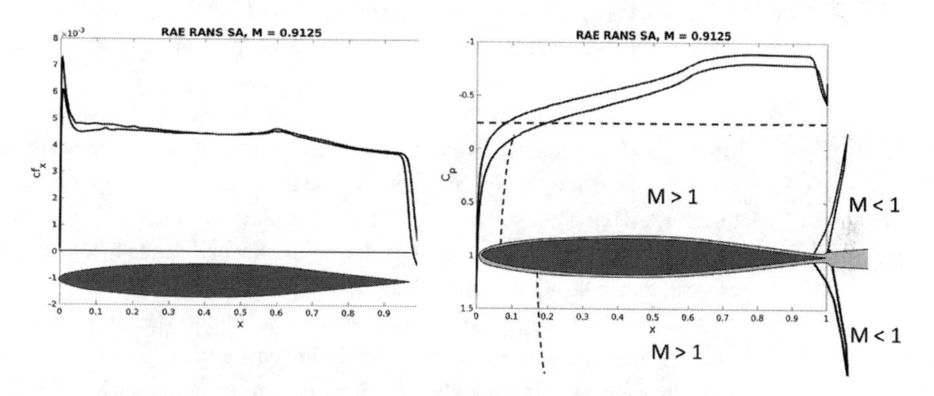

Figure 6.12 Case (d), $M_\infty = 0.9125$. Skin-friction coefficient (left); pressure coefficient and boundary layer (right). EDGE code RANS-SA model, RAE104 airfoil, $\alpha = 0.9°$, $Re = 1.8 \times 10^6$. (Courtesy of P. Eliasson, personal communication)

- Lift-curve slope and drag coefficient in Figure 6.5
- Pressure and friction coefficient distributions $C_p(x)$ and $C_f(x)$

The computed flow fields also allow investigation of details of the boundary layer under and behind the shock wave. In the intermediate Mach region, the shocks move quickly aft with increased M, with the exact position being determined by the interaction with the boundary layer, as described in Section 2.5.2.

The computations indicate that the flow is essentially unchanged up to approximately one-third the speed of sound – the incompressible flow regime. The variation in the lift coefficient with Mach number indicates complex changes in the flow field through the transonic speed range, identified by the letters (a)–(e). Significant differences exist between the flow fields at these five Mach numbers.

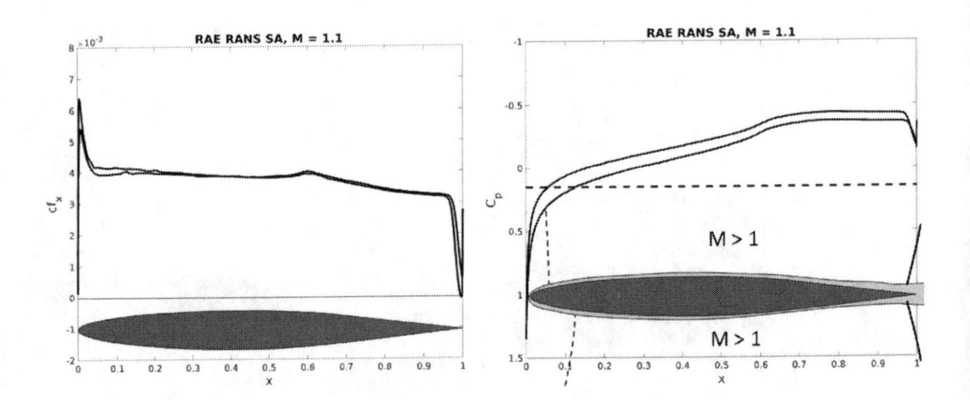

Figure 6.13 Case (e), $M_\infty = 1.100$. Skin-friction coefficient (left); pressure coefficient and boundary layer (right); EDGE code RANS-SA model, RAE104 airfoil, $\alpha = 0.9°$, $Re = 1.8 \times 10^6$. (Courtesy of P. Eliasson, personal communication)

(a) At $M_\infty = 0.7875$, Figure 6.9, just supercritical flow with a single shock on the upper surface appears, with the boundary layer thickening under its foot. This is a typical condition for a transonic airliner in cruise.

(b) At $M_\infty = 0.8125$, Figure 6.10, the lift coefficient reaches its maximum value, which is approximately twice the low-speed value. A shock wave sits on the upper surface, and the abrupt pressure rise through it causes the boundary layer to thicken, giving a deeper wake aft of the airfoil, as well as an increase in drag due to wave drag. Meanwhile, the velocity on the lower surface has just become critical, causing the abrupt drop in lift with further increase in M_∞.

(c) At $M_\infty = 0.8750$, Figure 6.11, lift reaches a minimum (i.e. shock stall) and the flow has become supersonic over nearly the entire lower surface terminated by a shock wave at the trailing edge. Meanwhile, on the upper surface, the shock has not yet reached the trailing edge, resulting in a higher pressure aft of the shock on the upper surface compared to the lower surface. As a result, the lift is drastically reduced. Separation at the foot of the upper surface shock wave is more conspicuous and the turbulent wake is wide.

(d) When $M_\infty = 0.9125$, Figure 6.12, the second maximum lift, the shock wave on the upper surface has now also reached the trailing edge, adding more suction, and thus increasing lift again. The local Mach number is supersonic for most of the upper and lower surfaces. The friction coefficient shows the flow to be attached all the way along the trailing edge, where the shock and the pressure rise widen the boundary layers at their confluence into the wake.

(e) When the free stream flow is supersonic, Figure 6.13, a bow shock wave (i.e. the detached shock wave in front of the leading edge) appears to the left of the visible area in the plot. The flow around the airfoil is supersonic everywhere except near the rounded nose. The "fishtail" shock waves at the trailing edge remain, but they have become weaker.

Drag Rise and Divergence
The transonic flight regime is where today's subsonic commercial aircraft of Airbus or Boeing type fly. Up to M_{crit}, the total drag is almost independent of the Mach number, as seen in Figure 6.5. The total drag rises strongly shortly after M_{crit} has been exceeded. This drag divergence at $M_{dd} > M_{crit}$ is due to the fact that now wave drag is added to the body drag. In addition, the viscosity-induced pressure or form drag is increased since the terminating shock wave thickens or even separates the boundary layer due to the pressure rise through the shock wave. A quantitative definition of M_{dd} may be taken as the lowest Mach number where $\frac{dC_D}{dM} \geq 0.1$.

Shock–boundary layer interaction can be highly unsteady and produce shock oscillations and buffet. An example is given in Section 6.8.2. Moreover, in the transonic regime, the pitching moment changes rapidly as the center of pressure moves a little downstream and the aerodynamic center shifts aft from a 1/4-chord to a mid-chord location. The pilots describe this as pitch-down increment and tuck-under. The aft movement of the aerodynamic center increases the pitch stability, and the aircraft responds more slowly to elevator commands. This will be discussed in Chapter 10.

An aircraft accelerating to exceed $M_\infty > 1$ must first overcome the drag rise associated with the transonic range. Efficient cruise at high-subsonic speed, as is the case for commercial aircraft, requires M_{crit} and M_{dd}, to be as high as possible.

The shock stall characteristic out-of-step movements of the shock waves and their influence on the wing's pitching moment caused many of the stability and control difficulties afflicting the early transonic fighters. Thicker airfoils were found to aggravate the above behavior and cause it to occur at lower speeds. It was conclusively established in the late 1940s that thinner sections could greatly alleviate these effects and postpone the drag rise to higher Mach numbers. Since the wave drag due to thickness is proportional to the thickness ratio squared, the demand for aircraft capable of supersonic flight caused the thickness ratio to tumble even further, being stopped short of structural limits only by increases in sweepback.

6.6 Time-Accurate Simulations

The mathematical models introduced in Chapter 2 focus on finding steady flows for various simplifications of the full compressible Navier–Stokes PDE. The RANS models with associated turbulence models were discussed at some length, and Chapter 4 shows how time stepping of the unsteady equations has become a standard solution method. However, as we have seen above, weak convergence to steady state with oscillatory residuals is an indication that the case gives unsteady flow and must be computed time accurately. Time-accurate simulation models, in particular the LES and its hybridization with RANS, is described in Section 6.7, and Section 6.7.1 shows how the time stepping has been adapted.

6.7 Hybrid RANS–LES for Unsteady Flow

RANS methods have demonstrated an ability to predict attached flows very well at a relatively low computational cost. LES methods, on the other hand, have been shown to compute separated flow fields accurately. Although the computing cost of LES of turbulent flows is significantly less than that of direct numerical simulation (DNS), it is still too expensive for engineering applications involving thin boundary layers near surfaces, since the resolution needed to capture these layers results in exorbitant demands on central processing unit power and memory.

Spalart et al. [26] proposed a hybrid RANS–LES turbulence model based on the one-equation SA eddy viscosity model as an alternative. This hybrid approach, also called detached-eddy simulation (DES), employs traditional RANS turbulence models to overcome the near-wall resolution problem. This robust approach is aimed at high-Reynolds number separated flows. It switches smoothly from RANS turbulence modeling in the wall boundary layer to LES farther away, improving results noticeably over pure RANS models. This is critical to obtaining accurate estimates of aerodynamic loads for massively separated flows, such as the flow over delta wings and delta-wing configurations at high angle of attack.

In the DES approach, the length scale d in the destruction term of the SA model is modified so that the eddy viscosity crosses over from the usual SA RANS eddy viscosity near the wall to a proposed LES eddy viscosity, similar to that defined by Smagorinsky for LES, away from the wall. The SA wall destruction term, which reduces the turbulent viscosity in the laminar sublayer, is proportional to $(\tilde{\nu}/d)^2$ (see Eq. (6.1)), where $\tilde{\nu}$ is the eddy viscosity and d is the distance to the nearest wall. When this term is balanced with the production term, the eddy viscosity becomes proportional to $\hat{S}d^2$, where $\hat{S}$ is the local strain rate. The Smagorinski LES model, on the other hand, varies its subgrid-scale (SGS) turbulent viscosity with the local strain rate and the grid spacing as follows.

$$\nu \propto \hat{S}\Delta^2 \tag{6.22}$$

where Δ is the diameter of the cell. Thus, if d is replaced with Δ in the wall destruction term, the SA model will act as a Smagorinski LES model. Consequently, the DES formulation is obtained by replacing in the SA model the distance to the nearest wall, d, by $\tilde{d}$, where $\tilde{d}$ is defined as follows.

$$\tilde{d} \equiv \min(d, C_{\text{DES}}\Delta) \tag{6.23}$$

Thus, the switch from RANS to LES depends on the spatial discretization. When the length scale d (distance to the wall) is smaller than the wall-parallel grid spacing Δ, which is typically the case for the highly stretched cells in the boundary layer, the model acts in RANS mode. When d is larger than Δ, the model acts in Smagorinsky LES mode. This is illustrated in Figure 6.14. This approach introduces only one additional model constant ($C_{\text{DES}} = 0.65$) in the one-equation SA model.

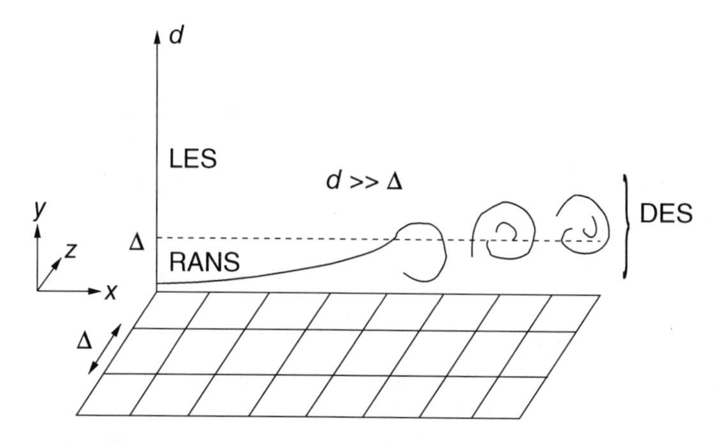

Figure 6.14 Schematic of detached-eddy simulation. (Courtesy of S. Görtz, PhD dissertation [7], reprinted with permission)

A powerful feature of DES is that it directly resolves turbulent eddies with increasing fidelity as the grid is refined [15]. Note that in RANS it is the mean flow that is computed. The role of grid refinement is to ensure convergence of the numerical solution and to minimize (or eliminate) the influence of the grid. In the fine-grid limit, the accuracy of RANS predictions is controlled by the turbulence model. In LES and DES, on the other hand, grid refinement resolves additional physical features: a wider range of turbulent eddies are represented as grid spacings are decreased. Correspondingly, the contribution of the turbulence model to the solution decreases as the grid is refined. The fine-grid limit of LES is a solution free of turbulence modeling errors to become DNS.

Although DES is not a zonal method, flow regions with very different gridding requirements emerge. Spalart [23] has given some guidelines for creating grids suitable for DES. He points out that it is desirable to have isotropic grid cells (cubic for structured grids) in the "LES region," in which unsteady, time-dependent features are resolved. Isotropic cells are desired because they ensure the lowest value of Δ for a given cell volume, lowering the eddy viscosity and allowing more fluctuations to be resolved on the grid. In addition, the orientation of turbulent structures is not known a priori, so isentropic cells represent a logical approach to resolving turbulent length scales.

6.7.1 Numerical Methods for Time-Accurate Simulations

The LES model and its hybridizations with RANS are dynamic models very similar to the full Navier–Stokes model, with terms added to model the effects of Reynolds stresses or SGS eddies. It follows that much the same numerical methods can be applied as for time-stepping RANS to steady state (Chapters 4 and 6). All acceleration techniques of Section 6.3.4 can be used also to accelerate time-accurate computations through the technique of dual time stepping.

To avoid restrictive time-step conditions, implicit methods are preferred for *unsteady* problems with very different timescales (e.g. high-Reynolds number flows). The class of backward differentiation formulas was popularized by W. Gear ("Gear's method") [6] for general systems of ODEs, and the second-order member of the family is widely applied for unsteady-flow problems because it also provides strong damping for short-timescale modes.

$$\frac{3}{2}\mathbf{W}^{n+1} - 2\mathbf{W}^{n} + \frac{1}{2}\mathbf{W}^{n-1} = \Delta t \mathbf{R}(t_{n+1}, \mathbf{W}^{n+1}) \tag{6.24}$$

The temporal damping is a low-pass filter, so the time step must be set to pass all of the modes that are important for the phenomena under study. It remains to solve the nonlinear system of equations for $\mathbf{W}^{n+1}$. For ease of notation, we rewrite Eq. (6.24) as follows.

$$\frac{3}{2}\mathbf{W} - \Delta t \mathbf{R}(t_{n+1}, \mathbf{W}) = \mathbf{r} \tag{6.25}$$

where $\mathbf{r} = 2\mathbf{W}^{n} - \frac{1}{2}\mathbf{W}^{n-1}$, to be solved for $\mathbf{W}$.

Newton's Method

Newton's method requires a linearization. If the linearization is accurate and the initial guess is good, such as $\mathbf{W}^{\nu=0} = \mathbf{W}^{n}$, the iteration converges quickly.

$$\left(\frac{3}{2}\mathbf{I} - \Delta t \frac{\partial \mathbf{R}(\mathbf{W}^{\nu})}{\partial \mathbf{W}}\right) \Delta \mathbf{W}^{\nu} = \mathbf{r} - \frac{3}{2}\mathbf{W}^{\nu} + \Delta t \mathbf{R}(t_{n+1}, \mathbf{W}^{\nu}), \; \nu = 0, 1, 2, \ldots \tag{6.26}$$

Here, $\Delta \mathbf{W}^{\nu} = \mathbf{W}^{\nu+1} - \mathbf{W}^{\nu}$. The iteration continues until a convergence criterion $\|\Delta \mathbf{W}^{\nu}\| \leq \epsilon$ is reached. Then, $\mathbf{W}^{\nu+1}$ is accepted as $\mathbf{W}^{n+1}$.

Often, only one iteration of Newton's method is computed (i.e. Eq. (6.26) for $\nu = 0$ only).

Solution of Linear System

For each ν, Eq. (6.26) represents a sparse linear system. For 1D and small 2D flow problems, the linear system has a small bandwidth and can be solved directly by Gaussian elimination. For 3D flow problems, suitable iterative methods must be devised for Eq. (6.26). Popular methods have been the following.

- The lower–upper symmetric Gauss–Seidel (LU-SGS) method by Yoon and Jameson [34], later revisited by, for example, Langer [10] and Otero, [16, 17]. The Gauss–Seidel relaxation method solves sequentially for the unknowns at one gridpoint, holding all others fixed. For 3D flow with a two-equation turbulence model, there are seven unknowns per point, so 7×7 systems have to be solved. The "symmetric" procedure sweeps over all points of the grid and back again. The convergence rate does depend on the ordering of points. The process should move from a point to a neighbor instead of jumping to points far away. This is easy to arrange for structured grids, but unstructured grids need ordering algorithms. Criteria for optimal ordering are not known.

- The generalized minimum residual (GMRES) method by Saad and Schultz [21]. GMRES can be seen as a generalization to unsymmetric, indefinite systems of the conjugate gradient method for positive definite matrices. The solution of the linear system $\mathbf{A}\,\mathbf{x} = \mathbf{b}$ is sought in the Krylov subspace spanned by the residual $\mathbf{r}_0 = \mathbf{b} - \mathbf{A}\mathbf{x}_0$ of the initial guess $\mathbf{x}_0$ and powers of $\mathbf{A}$ applied to $\mathbf{r}_0$. Only matrix–vector and vector–vector operations are involved, and the matrix–vector product can be approximated by differencing of $\mathbf{R}(\mathbf{W})$, since the matrix is the Jacobian $\partial\mathbf{R}(\mathbf{W})/\partial\mathbf{W}$.

Dual Time Stepping

To solve Eq. (6.25) for unsteady problems, Jameson (1991) proposed time stepping of an artificial unsteady problem in dual time τ until steady state.

$$\frac{d\mathbf{W}}{d\tau} = -\frac{3}{2}\mathbf{W} + \mathbf{r} + \Delta t\mathbf{R}(t_{n+1}, \mathbf{W}) \tag{6.27}$$

Note the appearance of the term $-\frac{3}{2}\mathbf{W}$, which moves the poles to the left compared to those of the original $\frac{\partial\mathbf{R}(\mathbf{W}^\nu)}{\partial\mathbf{W}}$. The $\mathbf{W}(\tau)$ is expected to steady in a dual time τ covered in about 100 steps.

For all of the convergence accelerations described above for the steady-state solutions (i.e. local time stepping), the Runge–Kutta methods of Gary type ([5]) with residual smoothing, multigrid, etc., are now available. Typically, they enable $\mathbf{W}^{n+1}$ for the unsteady Euler equations to be computed in $O(10)$ multigrid cycles.

6.8 Steady and Unsteady Separated Flows

Flow separation must be under control for the flow to be "healthy." Vortices separating from wing leading edges can be steady and contribute to wing performance, but most shock–boundary layer and aft-wing separations lead to undesirable unsteady flows. If too large, they produce buffet. Two examples are given here: (1) transonic steady flow over a cranked delta wing with vortices of different origins; and (2) definitely unsteady, limit cycle periodic shock–boundary layer interaction flow over a zero-lift airfoil.

6.8.1 Example 1: Sensitivity of Steady Vortex Shedding to the Turbulence Model

As part of due diligence, one must check the sensitivity to physical modeling, especially when the flow separates. Let us look at an example when the turbulent boundary layer lifts up and away from the wing surface and sheds coherent vortices.

Vortex Liftoff on the F-16XL Configuration

Extensive flight data, including pressure maps, are available from a comprehensive program for the experimental F-16XL modification presented in Chapter 1 in Figure 1.9.

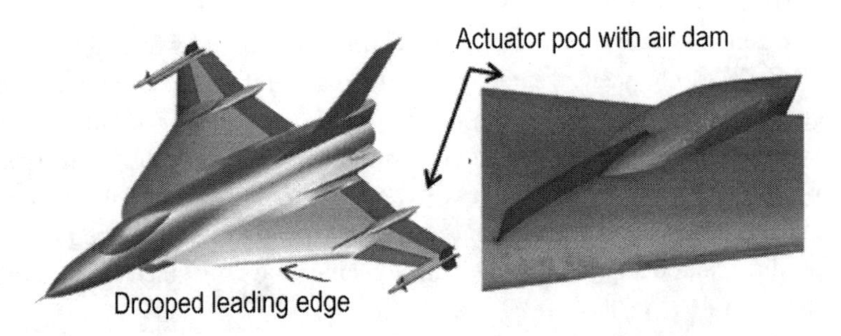

Figure 6.15 F-16XL wing geometry, with enhanced image of the air dam and actuator pod. (Courtesy of M. Tomac, PhD thesis [27], reprinted with permission)

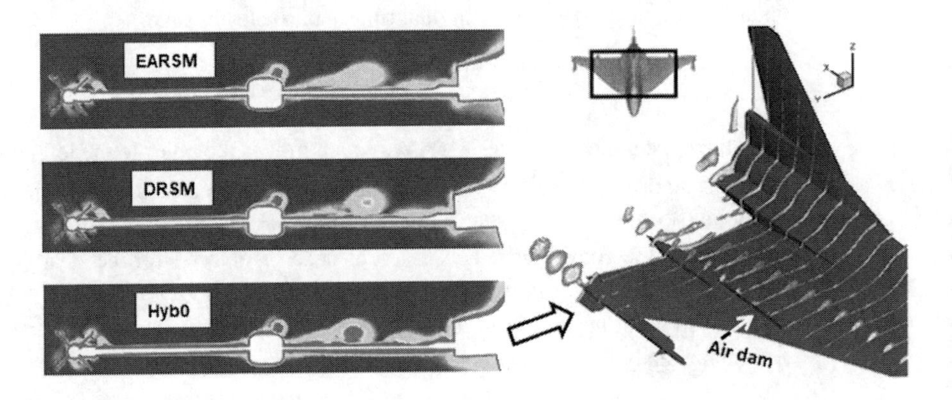

Figure 6.16 Vortex-strength sensitivity to physical modeling, EDGE code. Vorticity magnitude in FS492 spanwise cut plane (left): EARSM + H. $k - \omega$ model (top); DRSM (center); Hyb0 (bottom). Iso-curves of total pressure in several body axis normal cuts (right), FS492 station indicated by large arrow. $M_\infty = 0.97, \alpha = 4.3°$, and $Re = 88.8 \times 10^6$. (Courtesy of M. Tomac, PhD thesis [27], reprinted with permission)

Figure 6.15 shows the geometry of the F-16XL cranked–delta hybrid wing. The inner delta is highly swept with a drooped leading edge to improve lift at low speed. The S-curved shape blending the front of the inner delta into the fuselage is less swept with a sharp leading edge, and the outer wing panel is again less swept and thin. The pod houses the control surface actuators. The thin plate extending upstream of it, similarly to a boundary-layer fence, is called an air dam here. Its function is to stabilize the actuator pod induced vortices, which, if unsteady or bursting, would make the control surfaces inefficient. Figure 6.16 shows several vortical structures over the wing that will be studied further in Section 9.4.3. The flight condition we consider is the high subsonic free stream Mach number $M_\infty = 0.97$ at the small angle of attack $\alpha = 4.3°$ and $Re_{MAC} = 88.8M$. At this low angle of attack, the vortices do not dominate the flow field as they would at high α. The iso-total pressure contours in Figure 6.16 (right) show vortical layers over the wing, indicating that separation does occur, but the lifted-off shear layer appears to remain close to the upper surface, just above the boundary layer.

The case is known to be challenging for CFD, and numerical (grid) effects were investigated to isolate the physical modeling effects. Several well-researched RANS turbulence models in common use were compared with unsteady hybrid RANS–LES modeling. Overall, the differences were not very significant, except in certain flight conditions, such as the one reported here.

Figure 6.16 (left) compares the vorticity magnitudes of the solutions from the EARSM [29], the $k - \omega$ DRSM [30], and the hybrid RANS–LES model Hyb0 [18] on the spanwise cut plane "FS492." The cut is indicated by the arrow in Figure 6.16 (right), which shows Hyb0 total pressure iso-curves in several body axis normal planes. Three vortices appear. The leading-edge vortex lifts off from the boundary layer between the fuselage and actuator pod on the inner wing panel, the actuator pod vortex is shed from the air dam, and the wing tip missile vortex is associated with the multiple lifting surfaces there.

The models show little difference on missile and actuator pod vortices, but they do show difference on the inboard wing panel vortex. It separates from a smooth surface, so its "birth" is sensitive to turbulence modeling. The actuator pod vortex is created by the air dam, whose sharp edges define the separation. Similar effects of the fins create the missile vortex.

It is clear that the Hyb0 model preserves the vortex much further downstream compared to the other two models. The time-accurate Hyb0 simulation uses a total of 10,000 physical time steps of 0.1 ms with up to 100 dual time steps per physical time step. This gives a better preserved surface pressure footprint and a higher suction peak over the inner wing compared to the RANS models tested. Section 9.4.3 provides further remarks on CFD due diligence for this case.

6.8.2 Example 2: Transonic Buffet on Circular-Arc Airfoil

Next, we consider the periodic self-excited unsteady buffet turbulent flow around an 18% thick circular-arc airfoil of chord c in free flow at $M_\infty = 0.76$, $\alpha = 0°$, and $Re_c = 11M$ with computed results from the PhD thesis of S. Goertz [7]. At these conditions, the flow is a periodic, with 180° out-of-phase motion of the shocks over the upper and lower surfaces of the airfoil. The unsteadiness is driven by the interaction between the shocks, the boundary layer, and the vortex shedding in the wake. Experimental data [12, 14, 22] as well as previous computations [20, 31] are available for comparison.

Grid and Case Setup

A 2D hybrid grid with 12,000 nodes was generated using the commercial grid generation software ICEM CFD" (see Figure 6.17).

There are 216 grid points on the airfoil and 50 layers of stretched quadrilateral cells to resolve the boundary layer (Figure 6.17). The wall distance of the first off-wall grid point is 2 μm, giving $y^+ < 1$. The cell size growth ratio out from the wall is 1.2

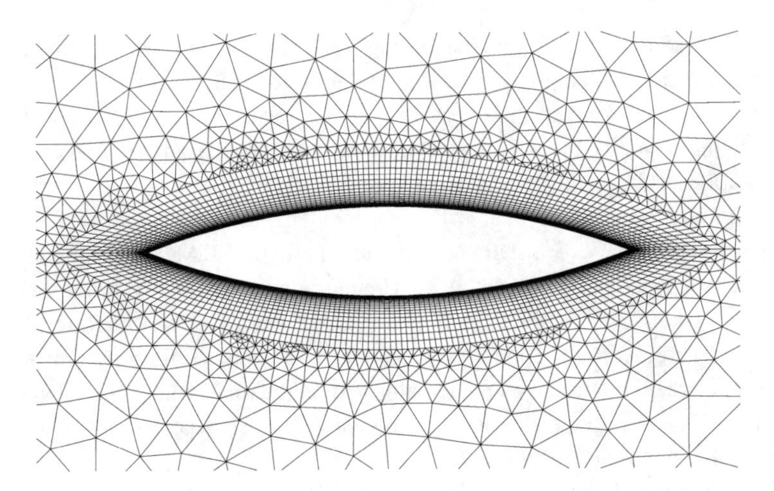

Figure 6.17 Close-up of the computational grid. (Courtesy of S. Görtz, PhD thesis [7], reprinted with permission)

(see Chapter 5). The circular outer boundary is 25 chord lengths away from the airfoil. The boundary condition on the airfoil is a solid wall. Characteristic variable free stream conditions are used on the outer boundary.

The unsteady flow was calculated with EDGE in URANS mode using dual time stepping and the EARSM turbulence model. Previous computational results [31] for the same geometry and flow conditions demonstrated that the EARSM predicts the unsteadiness due to strong shock–boundary layer interaction better than both algebraic and two-equation eddy viscosity turbulence models. Fully turbulent flow was assumed. The turbulence equations were discretized by a first-order upwind scheme.

The outer (physical) time step was set to about 250 time steps per period of oscillation. The dual time stepping inner-loop Courant–Friedrichs–Lewy (CFL) number was set to 0.7. Three multigrid levels and residual smoothing were used in the inner loop.

The unsteady numerical simulation was initialized by a steady-state solution. A total of 2000 outer time steps, corresponding to 0.2 s of real time, were computed. For every outer time step, the max-norm density residual of the inner loop was decreased by 2.5 orders of magnitude, sufficient to converge the aerodynamic coefficients within the inner loop. The number of inner iterations was not limited.

Computed Buffet

The simulation reproduced the time-dependent aspects of the onset of buffet. Figure 6.18 shows instantaneous Mach number contours at six different times through a period once the limit cycle has established itself. Coalescence of the near-vertical contours over the aft half of the airfoil indicates the formation of a shock wave.

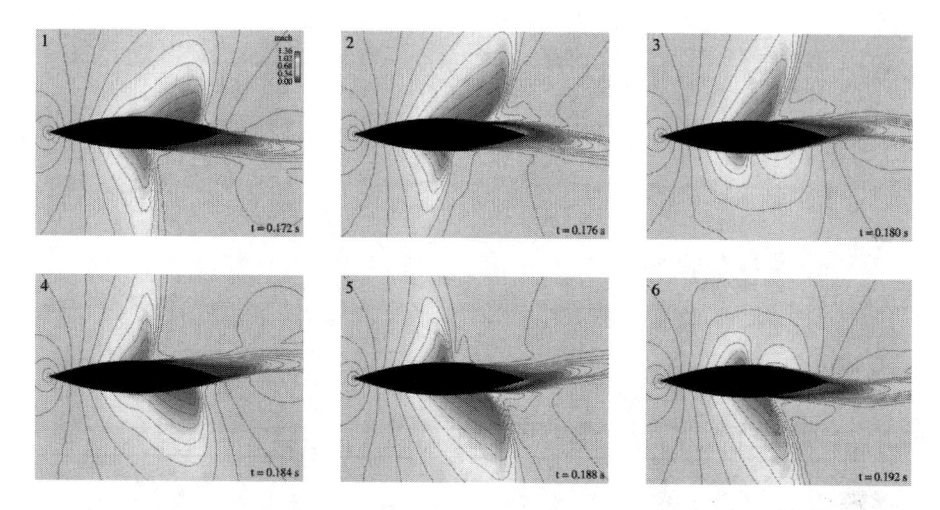

Figure 6.18 Instantaneous Mach number contours in the flow field about the circular-arc airfoil, showing oscillatory separation; $M_\infty = 0.76$, $Re = 11M$. (Courtesy of S. Görtz, PhD thesis [7], reprinted with permission)

It forms near the trailing edge just above a region of trailing-edge separation. Its strength increases as the local velocity ahead of the shock increases. The increased strength gives rise to shock-induced separation, and the shock wave and separated region begin to move forward. The local surface velocities upstream of the shock continue to increase and stabilize in a maximum velocity distribution. As the shock continues forward into a region of lower velocities, it diminishes in strength and vanishes as the separation point reverts to the trailing edge to complete the cycle. Meanwhile, the analogous process is occurring on the 180° out-of-phase lower surface.

This periodic phenomenon of frequency f causes oscillations in the aerodynamic forces. The time histories of the lift coefficient c_L and the drag coefficient c_D are shown in Figure 6.19.

After the transients have decayed, the lift coefficient oscillates almost sinusoidally around a zero mean value with an amplitude of about 0.35. The drag coefficient oscillates with twice the frequency around a mean value of 0.067 and an amplitude of 0.008. The reduced frequency of the computed lift coefficient is $k = \frac{\pi f c}{V_\infty} = 0.485$. The experimental value is $k_{\text{exp}} = 0.49$ [14]. Figure 6.18 shows one period toward the end of the cycle in Figure 6.19.

We have just witnessed the application of a RANS solver for transonic flow. Designing an airfoil requires solving the flow problem many times, as its shape is changed by small amounts as the optimum shape is approached. The next chapter describes how a special-purpose solver can be constructed to solve the airfoil problem much more rapidly than a general-purpose RANS solver, thus being an ideal tool for airfoil analysis and design.

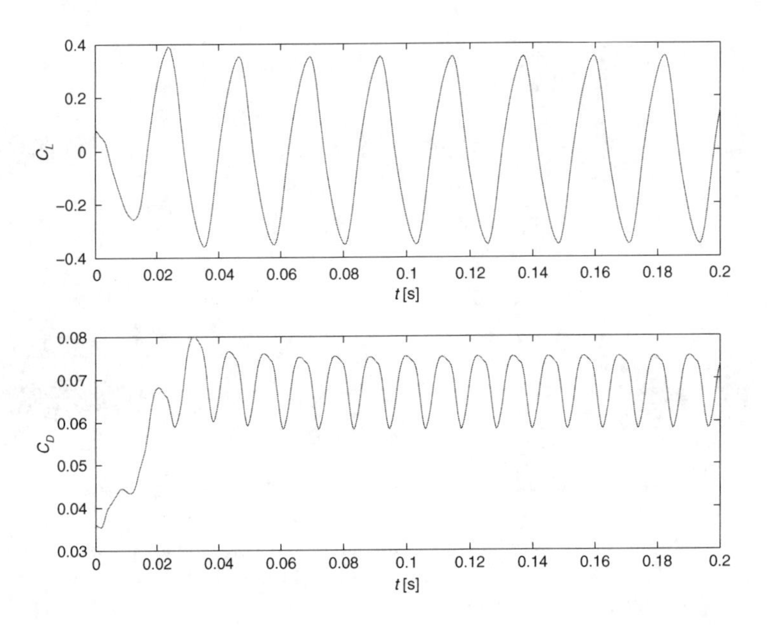

Figure 6.19 Lift and drag coefficient vs time; $M_\infty = 0.76$, $Re_c = 11M$, $\Delta t = 0.1$ ms. (Courtesy of S. Görtz, PhD thesis [7], reprinted with permission)

6.9 Learn More by Computing

Gain hands-on experience of the computational tools for the topics in this chapter by working with the on-line resources. Exercises, tutorials, and project suggestions are found on the book website www.cambridge.org/rizzi. For example, run the RAE100 airfoil through the "capricious" transonic speed regime, and compare to the results in Figure 6.6 for the RAE104. Software used to compute many of the examples shown is available from http://airinnova.se/education/aero-dynamic-design-of-aircraft

References

[1] J. Ackeret. Luftkräfte auf flügel, die mit größerer als schallgeschwindigkeit bewegt werden. *Zeitschrift für Flugtechnik und Motorluftschiffahrt*, 16: 72–74, 1925.

[2] A. Brandt. Multi-level adaptive solutions to boundary-value problems. *Mathematics of Computation*, 31: 333–90, 1977.

[3] P. Eliasson. EDGE, a Navier–Stokes solver for unstructured grids. In *Proceedings of Finite Volumes for Complex Applications III*. Hermes Penton Ltd, 2002, pp. 527–534.

[4] W. Farren. The aerodynamic art. *Aeronautical Journal*, 60: 431–449, 1956.

[5] J. Gary. On certain finite difference schemes for hyperbolic systems. *Mathematics of Computation*, 18: 1–18, 1964.

[6] C. W. Gear. *Numerical Initial Value Problems in Ordinary Differential Equations*. Prentice-Hall, Inc., 1971.

[7] S. Görtz. *Realistic Simulations of Delta Wing Aerodynamics Using Novel CFD Methods*. KTH School of Engineering Sciences, 2005.

[8] A. Hellsten. New advanced $k - \omega$ turbulence model for high-lift aerodynamics. *AIAA Journal*, 43(9): 1857–1869, 2005.

[9] P. Eliasson and P. Weinerfelt. High-Order Implicit Time Integration for Unsteady Turbulent Flow Simulations. *Computers and Fluids*, 112: 35–49, 2015.

[10] S. Langer. Application of a line-implicit method to fully coupled system of equations for turbulent flow problems. *International Journal of Computational Fluid Dynamics*, 27(3): 131–150, 2013.

[11] Y. LeMoigne. Adaptive mesh refinement sensors for vortex flow simulations. Presented at ECCOMAS 2004, Jyväskylä, Finland, July 2004.

[12] L. L. Levy Jr. Experimental and computational steady and unsteady transonic flows about a thick airfoil. *AIAA Journal*, 16(6): 564–572, 1978.

[13] L. Martinelli and A. Jameson. Validation of a multigrid method for the Reynolds averaged equations. Technical report. AIAA paper 88-0414, 1988.

[14] J. B. McDevitt and R. A. Taylor. An investigation of wing–body interference effects at transonic speeds for several swept-wing and body combinations. Technical report. NACA RM A57A02, NACA, 1957.

[15] S. A. Morton, J. R. Forsythe, J. M. Mitchell, and D. Hajek. Des and rans simulations of delta wing vortical flows. Technical report. AIAA Paper 2002-0287, 2002.

[16] E. Otero. *Acceleration of Compressible Flow Simulations with Edge Using Implicit Time Stepping*. PhD thesis, KTH Kungl Tekniska Högskolan, December 2014.

[17] E. Otero and P. Eliasson. Parameter investigation with line-implicit lower-upper symmetric Gauss–Seidel on 3D stretched grids. Technical report. AIAA Paper (2014–2094), 2014.

[18] S.-H. Peng. Hybrid RANS–LES modelling based on zero- and one-equation models for turbulent flow simulation. Presented at 4th International Symposium on Turbulence and Shear Flow Phenomena, Williamsburg, NY, 2005.

[19] A. Rizzi and J. M. Luckring. Evolution and use of cfd for separated flow simulations relevant to military aircraft. In *Separated Flow: Prediction, Measurement and Assessment for Air and Sea Vehicles*. Neuilly-sur-Seine, 2019, pp. 11-1–11-58.

[20] C. L Rumsey, M. D. Sanetrik, R. T. Biedron, N. D. Melsom, and E. B. Parlette. Efficiency and accuracy of time-accurate turbulent Navier–Stokes computations. Presented at 13th AIAA Applied Aerodynamics Conference, San Diego, CA, June 1995.

[21] Y. Saad and M.B. Schultz. GMRES: A generalized minimal residual algorithm for solving nonsymmetric linear systems. *SIAM Journal on Scientific and Statistical Computing*, 7(3): 856–869, 1986.

[22] H. L. Seegmiller, J. G. Marvin, and L. L. Levy Jr. Steady and unsteady transonic flow. *AIAA Journal*, 16(12): 1262–1270, 1978.

[23] P. R. Spalart. Young-person' guide to detached eddy simulation grids. Technical report. CR-2001-211032, NASA, 2001.

[24] P. R. Spalart and R. S. Allmaras. A one-equation model for aerodynamic flows. Technical report. AIAA paper (92-439), 1992.

[25] P. R. Spalart and R. S. Allmaras. A one-equation turbulence model for aerodynamic flows. *La Recherche Aerospatiale*, 1: 5–21, 1994.

[26] P. R. Spalart, W.-H. Jou, M. Strelets, and S. R. Allmaras. Comments on the feasibility of LES for wings and on a hybrid RANS/LESapproach. In C. Liu and Z. Liu, editors, *Advances in DNS: Proceedings of the First AFOSR International Conference on DNS/LES*. Greyden Press, 1997, pp. 137–147.

[27] M. Tomac. *Towards Automated CFD for Enginerring Methods in Airfcraft Design*. KTH School of Engineering Sciences, 2014.

[28] J. Vos, A. Rizzi, D. Darracq, and E. Hirschel. Navier–Stokes solvers in European aircraft design. *Progress in Aerospace Sciences*, 38(8): 601–697, 2002.

[29] S. Wallin and A. Johansson. An explicit algebraic Reynolds stress model for incompressible and compressible turbulent flows. *Journal of Fluid Mechanics*, 403: 89–132, 2000.

[30] S. Wallin and A.V. Johansson. Modelling streamline curvature effects in explicit algebraic Reynolds stress turbulence models. *International Journal of Heat and Fluid Flow*, 23: 721–730. 2002.

[31] D. Wang, S. Wallin, M. Berggren, and P. Eliasson. A computational study of unsteady turbulent buffet aerodynamics. Technical report. AIAA-2000-2657, December 2000.

[32] P. Wesseling. *An Introduction to Multigrid Methods*. R.T. Edwards, Inc., 2004.

[33] D. C. Wilcox. Reassessment of the scale determining equation for advanced turbulence models. *AIAA Journal*, 26(11): 1299–1310, 1988.

[34] S. Yoon and A. Jameson. Lower–upper symmetric-Gauss–Seidel method for the euler and Navier–Stokes equations. *AIAA Journal*, 26(9): 1025–1026, 1988.

7 Fast Computation of Airfoil Flow

> Before Prandtl, there existed no closed-form solution of the Navier–Stokes Equations and no theoretical solution for the drag on an airfoil. Then, in 1904, a breakthrough was obtained when Prandtl postulated the idea of simplifying these equations for the thin region of flow adjacent to a surface, leading to the concept of the boundary layer and tractable mathematics. These boundary layer equations led to solutions for the skin-friction on an airfoil.
>
> John D. Anderson Jr.[1]

> To iterate between the inviscid outer flow and the boundary-layer equations ...tends to be unstable ...and will also fail outright if separation is encountered. ...The most reliable approach has been to solve the inviscid and viscous equations simultaneously by the Newton method.
>
> Mark Drela[2]

Preamble

This chapter presents a fast method for computing flows over wing sections (i.e. plane flows over airfoils). Examples include airfoils for low Reynolds numbers and modern airfoils for high speed, as are found on current transonic airliners. As we know, the vortex lattice model is incapable of describing flows at Mach numbers approaching 1, as well as the boundary-layer behavior responsible for stall and shock–boundary layer interaction, for example. Such computations can be done by CFD codes solving the Reynolds-averaged Navier–Stokes equations, but this can be achieved much more rapidly by schemes that model an inviscid outer flow coupled to an inner flow modeled by the boundary-layer equations, thus exploiting Prandtl's idea of a zonal approach. The MSES [3] and XFOIL (https://web.mit.edu/drela/Public/web/xfoil/) software systems include provisions for finding shapes to produce given pressure distributions and are airfoil design tools. The price to pay is the inability to cope with massive separation and flow with supersonic free stream. XFOIL is a potential flow solver in the public domain limited to flows without supersonic pockets, whereas MSES deals with flows nearly up to Mach 1 and is a commercial product.

[1] With his permission, personal communication.

[2] With his permission, personal communication.

7.1 Introduction

The various airfoil analysis and/or design methodologies that have been developed in the past decade have employed one of two distinct formulations of the mathematical problem. The first approach is to employ a Reynolds-averaged Navier–Stokes (RANS) computational fluid dynamics (CFD) code such as SU2 or EDGE. The second approach adopts the Prandtl zonal approach to solve the steady inviscid-flow equations coupled with a steady model for the boundary-layer flow. As a rule, RANS is too slow for routine design work and has not yet shown any accuracy advantages over the much faster zonal approaches. But it is more robust with respect to Mach number and flow separation and can compute the entire shock stall phenomenon as we saw in the EDGE demonstration in Chapter 6. The MSES and XFOIL codes use *integral* formulations for the boundary-layer and wake regions, which also enables computation of separation bubbles with reattachment.

7.2 Zonal Approach: Physical Observations

Consider the flow of air around a slender object such as an airfoil set at a small angle of attack, as shown in Figure 7.1. The flow sticks to the solid airfoil wall – the no-slip condition – and it is retarded in comparison to the more distant flow. Away from the airfoil, the velocity field approaches that in an inviscid medium. The region containing flow retarded by viscosity is called the *boundary layer*. At the bottom of the layer, the velocity component parallel to the surface increases rapidly in the normal direction. As the friction force is proportional to this gradient of velocity, this force is largest close to the airfoil. Outside of the boundary layer, the friction force disappears. Prandtl [9] reasoned it was rational to divide the flow field into the following.

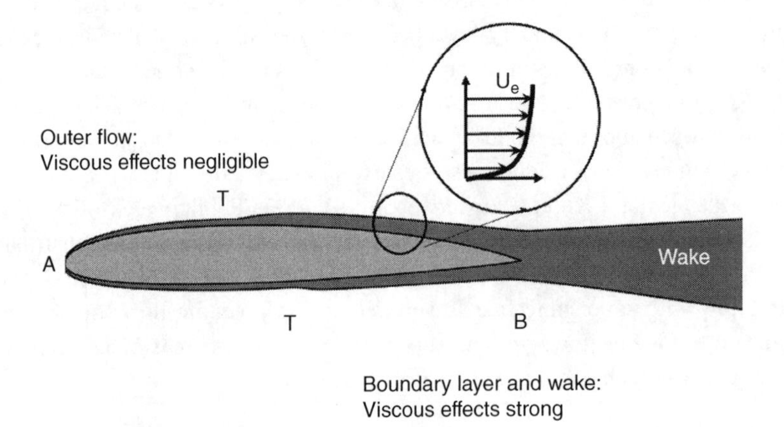

Figure 7.1 Flow regions – outer effectively inviscid and inner viscous – in a subsonic high-Reynolds number flow.

- The region of the boundary layer where the viscosity of the flow is important
- The region outside of the boundary layer with practically inviscid flow

7.2.1 Thin Boundary Layers

Before proceeding any further, it is of interest to have an idea of the magnitude of the thickness, δ, of the boundary layer. Assume that in the boundary layer the viscous terms and the inertia terms are of equal order of magnitude. The inertial force per unit volume is $\rho u \partial u/\partial x$, where ρ is the density. For a plate of length l, the gradient $\partial u/\partial x$ is proportional to U_e/l, where U_e is the velocity outside of the boundary layer. Then, the inertia force is seen to be of the order of $\rho U_e^2/l$. The friction force per unit volume is $\partial \tau/\partial z$; assuming laminar flow, this equals $\mu \partial^2 u/\partial^2 z$. The velocity gradient perpendicular to the wall $\partial u/\partial z$ is of the order of U_e/δ, and so the friction force per unit volume equals $\mu U_e^2/\delta^2$. The assumption of equality of the inertial and viscous forces gives the following.

$$\mu \frac{U_e}{\delta^2} \sim \frac{\rho U_e^2}{l} \text{ or } \delta \sim \sqrt{\frac{\mu l}{\rho U_e}} = l/\sqrt{Re_l}$$

It remains to establish the numerical factor. Blasius [2] solved the case of incompressible laminar flow over a flat plate at zero angle of attack. This case has vanishing longitudinal pressure gradient and gives the following.

$$\delta = 5.836 \frac{x}{\sqrt{Re_x}}$$

The boundary-layer thickness is of the order of 0.6% of the plate length at $Re_l = 10^6$. As the Blasius expression shows, the thickness increases with reduced Reynolds number. It also does so with an adverse pressure gradient, as shown by, for example, the Falkner-Skan [6] boundary-layer model. A realistic model must also take into account the laminar–turbulent flow transition, shown at points T in Figure 7.1. Turbulent boundary layers tolerate stronger adverse pressure gradients before separating due to the mixing with the higher-velocity outer flow.

As we saw in Section 2.3.2, turbulent boundary-layer flows are time dependent, even if the geometry boundary conditions are not. The unsteadiness is the unstable response to small disturbances in the flow. The steady boundary-layer model, like the RANS model, considers the time-averaged velocities.

Flat Plate

Measured velocity profiles in laminar and turbulent boundary layers on a flat plate appear in Figure 2.10 (right), showing the difference between laminar and turbulent velocity distributions at a given Reynolds number. Figure 7.2 shows a laminar boundary-layer profile with thickness δ and displacement thickness δ^*.

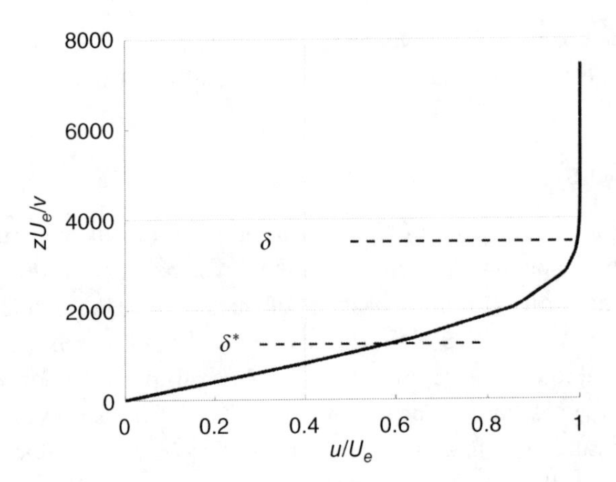

Figure 7.2 Laminar velocity profile on a flat plate with thickness δ and displacement thíckness δ^*.

7.2.2 Inner/Outer Zone Coupling

In the following, we consider steady, 2D, incompressible flow. In the boundary layer, the momentum equation in the direction normal to the surface degenerates because the velocity components normal to the surface are smaller than the ones tangential to the surface. Since there is no variation of pressure across the boundary layer, the pressure determined in the exterior flow region also prevails at the airfoil contour.

Consider Figure 7.2, for the case in Figure 2.10. The exterior contour-parallel flow velocity is U_e. Given this, the boundary-layer calculation produces the thickness of the boundary layer δ and the displacement thickness $\delta^*(x)$ to represent the mass-flow deficit in the boundary layer, precisely defined in Eq. (7.2).

The displacing effect of the boundary layer can be included in the region outside by modification of the airfoil shape or as a mass source on its surface. The outcome of the outer flow calculation is the velocity $U_e(x)$. When δ^* is very small, the calculations can be repeated alternately and the process may converge.

The interaction must consider the presence of the trailing wake. However, upstream of the trailing edge there is a friction force acting on the flow and downstream there is not. This singularity at the trailing edge causes difficulties in the integration of the flow equations into the wake. There will be no convergence for thicker boundary layers, even if the flow is attached. For these cases, it is necessary to solve the equations for the two regions simultaneously, as is done in MSES.

The reliability of any method based on boundary-layer equations decreases with increasing thickness. Mild separations, also known as "separation bubbles," may still give reasonable results, but severe separations require RANS solutions. Cases such as the one in Figure 2.22 are too violent for MSES, as may also be the case for RANS.

7.3 MSES: Fast Airfoil Analysis and Design System

MSES is a numerical airfoil development system. It includes capabilities to analyze, modify, and optimize single- and multi-element airfoils for a wide range of Mach and Reynolds numbers. The range of validity includes low Reynolds numbers and transonic Mach numbers. Flows with transitional separation bubbles, shock waves, trailing edge, and shock-induced separation can be predicted. Separation bubbles are discussed in Section 7.5.3, and a shock–boundary layer interaction case is shown in Figure 7.10 with experimental results. Surface pressure and aerodynamic force predictions are accurate up to stall ($C_{L,max}$). Transition can be forced or predicted as part of the flow calculation. Automated calculation of angle-of-attack and Mach-number sweeps is provided. More of the fundamentals behind the program can be found in Ref. [5], and Ref. [3] is the user guide.

Airfoil design is accomplished by interactive specification of surface pressures, with the resultant airfoil geometry being computed. Analysis calculations may be performed at any time during the design process. The boundary layer influences the outside flow through its displacement thickness δ^*, and the pressure distribution of the exterior flow determines the character of the boundary layer. The *steady* Euler equations described in Section 7.4 model the outer flow, and integrated boundary-layer equations (Section 7.5.2) model the boundary-layer displacement thickness.

Mark Drela, now Professor of Aeronautics at MIT, wrote the MSES code together with Mike Giles as part of his MIT doctoral dissertation in the mid-1980s. He is an original thinker, and he has incorporated a number of novel features in this airfoil analysis and design package.

7.4 Outer Euler Flow Solver

The flow is to be computed in the domain shown in Figure 7.3, in the inviscid region and the boundary-layer and wake region. The MSES features are summarized as follows.

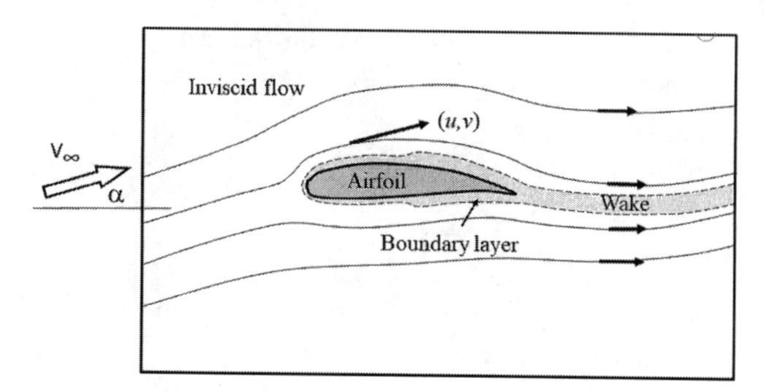

Figure 7.3 Flow domains.

1. The Euler equations are discretized on a conservative streamline grid. They are strongly coupled to a two-equation integral compressible boundary-layer formulation, using the displacement thickness concept and the velocity at the edge of the boundary layer.
2. Speed upwinding, equivalent to an artificial bulk viscosity, is used to stabilize the solution in supersonic regions.
3. The subsonic far field is represented by vortex, source, and doublet singularities placed at the airfoil.
4. A transition prediction formulation of the $e^{N_{crit}}$ type is incorporated into the boundary-layer model.
5. The entire discrete equation set, including the viscous and transition formulations, is solved as a fully coupled nonlinear system by a global Newton method. This is a rapid and reliable method for dealing with strong viscous–inviscid interactions, which invariably occur in transonic and low-Reynolds number airfoil flows. It is particularly efficient when a parameter sweep, such as a drag polar, is being calculated.

The results presented below demonstrate the following.

- Prediction of low-speed transitioning separation bubbles and their associated losses. The rapid airfoil performance degradation with decreasing Reynolds number is thus accurately predicted.
- Transonic airfoil calculation involving shock-induced separation, showing the robustness of the global Newton solution procedure. Good agreement with experimental results is obtained, further demonstrating the performance of the present integral boundary-layer formulation.

The intrinsic streamline grid eliminates diffusion between adjacent streamtubes and gives a completely nondissipative scheme in subsonic regions. This permits the drag to be calculated entirely from the far entropy and viscous wakes, a procedure less prone to numerical errors than airfoil surface force integration.

7.4.1 Discretization of Homenthalpic Flow on a Streamline Grid

The inviscid region of the flow field is described by the steady-state Euler equations in integral form where the integration is around a closed curve $\partial\Omega$ with normal $\mathbf{n}$.

$$\oint_{\partial\Omega} \begin{pmatrix} \rho\mathbf{u} \\ \rho\mathbf{u}\,\mathbf{u} + p\mathbf{I} \\ \rho H\,\mathbf{u} \end{pmatrix} \cdot \mathbf{n}\, ds = 0 \tag{7.1}$$

For an ideal gas $p = R_{gas}\rho T$ with constant specific heats, specific total energy $E = \frac{R_{gas}}{\gamma-1} T + \frac{1}{2}u^2$ and total enthalpy $H = E + \frac{p}{\rho}$.

In the free stream ahead, the total enthalpy is constant $= H_\infty$. The energy equation says that it remains constant along every streamline – a *homenthalpic* flow. Thus,

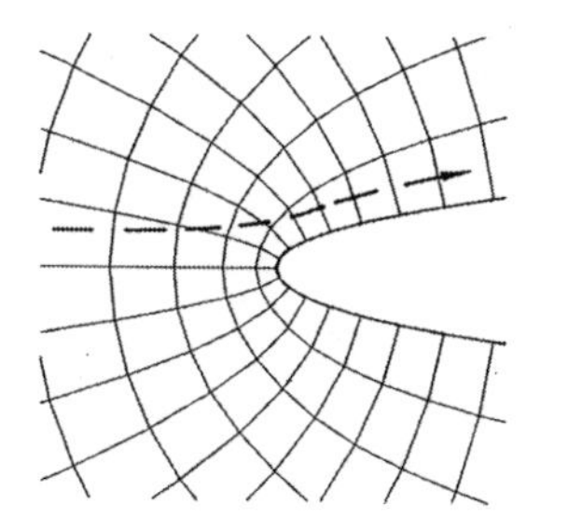
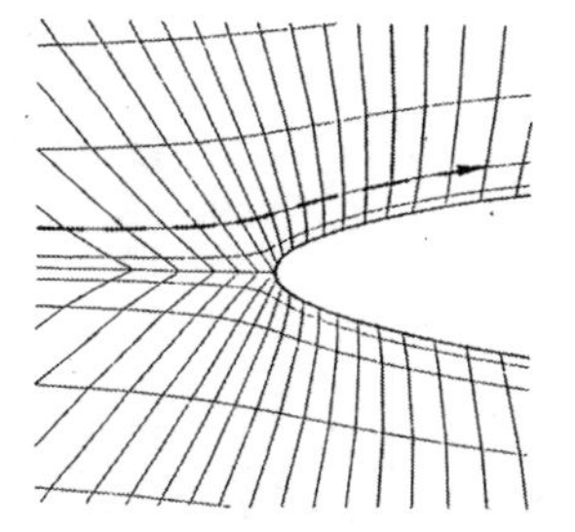

Figure 7.4 MSES uses a streamline-based grid in which one family of grid lines corresponds to streamlines.

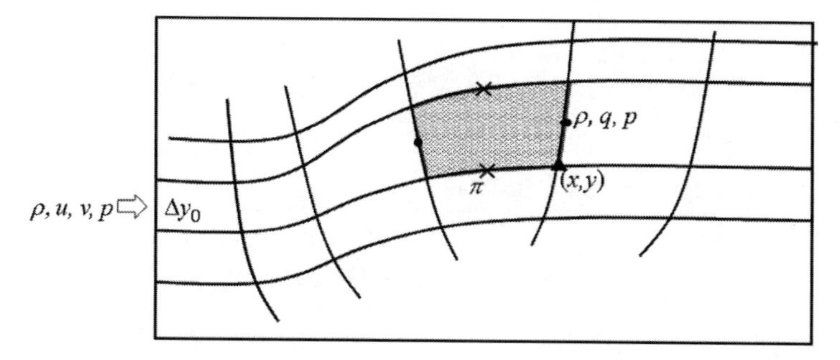

Figure 7.5 A streamtube in the grid, with the unknowns indicated.

$$p = \rho \frac{\gamma - 1}{\gamma}(H_\infty - 1/2u^2)$$

Eq. (7.1) reduces to one for mass conservation and two for the momentum conservation – three in all. This holds for an arbitrary control volume, hence for every cell in any type of grid.

The conventional type of grid (Figure 7.4, left), is stationary through the solution process. Instead, Drela chooses to use a *streamline-based* grid that adapts to the flow (Figure 7.4, right), with one family of grid lines corresponding to streamlines and the other family nearly normal to the streamlines. This reduces the number of equations to be solved, a novel feature for a finite-volume scheme, although it has long been used in streamline curvature schemes.

Figure 7.5 shows a streamtube between two neighbor streamlines, a control volume, and the variables used. The quadrilateral grid is defined by the coordinates (x, y) of one corner per cell and the Euler variables ρ, u, v, p are associated with the midpoints of the grid line family transversal to the streamlines and the pressure π at mid-streamline points. There are thus seven unknowns per cell, but only four Euler equations. Now, one variable is immediately eliminated by the constant enthalpy, as described above, and by the streamline grid, the velocity direction is defined by the cell corner coordinates. Moreover, the mass flow in a streamtube is constant along the tube, and is known from boundary conditions at inflow once the grid points there are known. So, the

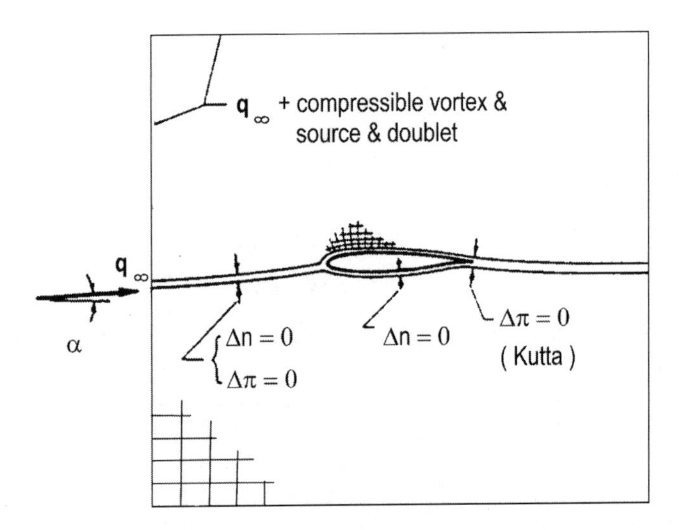

Figure 7.6 Boundary conditions used for computations.

product of velocity $\sqrt{u^2 + v^2}$ and density ρ becomes known from the mass balance, and we are left with the (x, y) per corner and density ρ and pressure at streamtube edges $\pi()$ as flow unknowns. All of the other variables can be computed from them by straightforward assignment statements. Thus, the two remaining momentum balances are complemented by a relation between π and p expressing that they should have the same average over the cell, and by fixing the transversal grid lines, so a single y determines the corner.

The final tally is two unknowns per cell and two momentum balance equations. Taking edge and corner cells and the boundary conditions on the airfoil and far field into account, the number of equations and unknowns will be found to match exactly.

Boundary Conditions

In order to keep the number of unknowns as small as possible, one must make the computational domain as small as possible. For an airfoil in free flight, this means moving the far field in as close as possible to the airfoil. This will require some special techniques, explained below.

Wall

The surface boundary conditions required to close the discrete Euler equations are very simple (see Figure 7.6). At a solid surface, only the position of the adjacent streamline needs to be constrained. The surface pressure is a result of the calculation. For a viscous case, the surface streamline is simply displaced normal to the wall by $\delta^*(x)$ (see Section 7.5.2). In an inverse design calculation, the pressure over all or part of the airfoil surface is imposed, and the wall streamline position there is a result of the calculation. Special treatment is necessary to ensure that the resulting geometry is continuous and/or smooth at the blend points between the direct and inverse boundary condition regions. The simple direct/inverse boundary condition switching results in full compatibility between the analysis and design calculations and enhances the overall effectiveness of the MSES code as a design tool.

Inflow/outflow and far field

The subsonic far-field flow is subsonic, so the Euler equation system requires two boundary conditions per node at inflow and one boundary condition per node everywhere else. One of the inlet boundary conditions is the prescription of the stagnation density at each streamtube.

The velocity of a uniform free stream plus a vortex, source, and doublet inside the foil is imposed on the outermost streamlines of the subsonic domain to construct small disturbance potential flows at nonvanishing Mach numbers. The vortex strength is defined by the lift, the source accounts for the boundary-layer displacement of the outer flow, and the doublet is related to the moment on the foil.

The boundary conditions specify the inflow and outflow angles and the outer streamline pressures based on this far-field expansion. The doublet strengths are determined by minimizing the deviation of the two discrete outermost streamlines from the direction of $\mathbf{V}_\infty$.

This boundary treatment needs the outer boundary to be only about $2/\sqrt{1-M_\infty^2}$ above and below an airfoil of unit chord, just large enough to comfortably contain any embedded supersonic regions.

Global Variables

The boundary conditions introduce quantities, such as a vortex strength, with global influence on the solution. One final boundary condition required for the inviscid equations is the Kutta condition, which is enforced by letting the outlet flow angle be free and requiring vanishing pressure difference across the trailing edge of the airfoil. MSES recognizes a set of such global variables and constraints in addition to the conservation laws and incorporates them as selected into the system to be solved.

Wind tunnel equipment and standard CFD packages such as EDGE condition one to think of the angle of attack as something given and the lift coefficient to be computed or measured. MSES does not make this distinction. You can specify a desired C_L as a global constraint and α as a global variable, or vice versa. Such freedom increases the versatility of MSES greatly, and it is well worth the effort to master the choice of variables and constraints, although the user interface is devoid of "syntactic sugar" to guide the neophyte.

Wake

The wake is treated as two boundary layers with zero wall shear. Hence, the gap between the two streamlines adjacent to the wake is set equal to the sum of the upper and lower wake displacement thicknesses. A zero pressure jump across the wake is also imposed. With these wake jump conditions, the wake trajectory defined by the two streamlines bounding it evolves as part of the overall solution scheme.

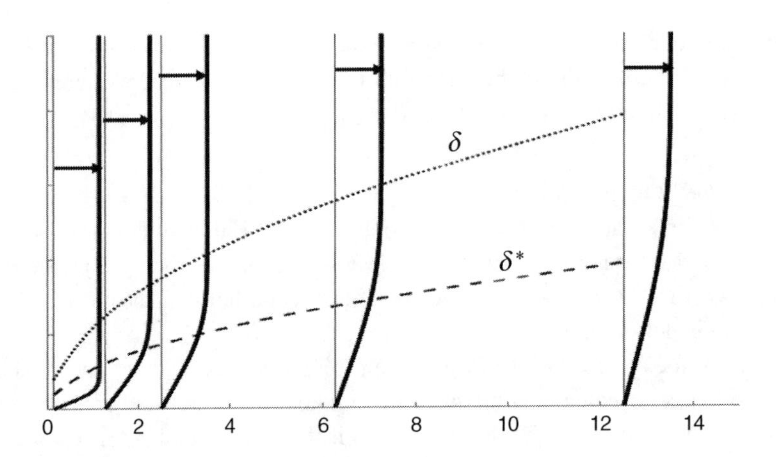

Figure 7.7 Boundary-layer velocity profile u/U_e development.

7.5 Boundary-Layer and Integral Boundary-Layer Models

The following provides a background for the boundary-layer model in MSES.

- The boundary-layer characteristics: thickness, displacement thickness, momentum thickness, and friction coefficient will be defined and discussed.
- Laminar and turbulent flow and transition criteria will be discussed.
- Interaction between the boundary layer and exterior flow will be discussed.
- Viscous and wave drag will be discussed.

7.5.1 Boundary-Layer Partial Differential Equation Model

Consider the viscous flow over a flat or slightly curved plate as in Figure 7.7, which shows the velocity profiles u/U_e vs. wall distance for growing x. The wall distance length scale is $L = \nu/U_e$ with ν kinematic viscosity.

The medium is air, and the speed of the air stream is considerable, so the Reynolds number is high, which makes the boundary layer thin. At the surface of the object, the flow velocity is zero – the no-slip condition. Away from the surface, the flow velocity increases rapidly and approaches the free stream velocity. At the distance δ, the velocity has reached 99% of the free stream velocity. Because the boundary layer slows down, it will displace the exterior flow away from the surface. The amount of displacement is called the boundary-layer displacement thickness δ^*, which, with velocity U_e, would give the correct mass flow deficit as follows.

$$\delta^* = \int_0^\delta \left(1 - \frac{u}{U_e}\right) dz \tag{7.2}$$

The definition is insensitive to the precise value of δ, as long as it is large enough, since $u \to U_e$ as z increases.

The reduced velocity also carries less momentum. The momentum thickness, θ, multiplied with U_e gives the correct momentum deficit over δ as follows.

$$\theta = \int_0^\delta \frac{u}{U_e}\left(1 - \frac{u}{U_e}\right) dz \tag{7.3}$$

The local shear stress force τ_w is normally given in wall friction coefficient form c_f as follows.

$$\tau_w = \mu \left(\frac{\partial u}{\partial z}\right)_{z=0}, c_f = \frac{\tau_w}{\frac{1}{2}\rho U_e^2} \tag{7.4}$$

The governing equations for incompressible laminar flow are given in Eq. (2.23), repeated here for convenience.

$$\frac{\partial u}{\partial x} + \frac{\partial w}{\partial z} = 0 \tag{7.5}$$

$$u\frac{\partial u}{\partial x} + w\frac{\partial u}{\partial z} = U_e\frac{dU_e}{dx} + \nu\frac{\partial^2 u}{\partial z^2} \tag{7.6}$$

$$\frac{\partial p}{\partial z} = 0 \tag{7.7}$$

where ν is the kinematic viscosity, $\nu \equiv \mu/\rho$.

It is possible to march the solution from the leading-edge stagnation point ($x = 0$) downstream. But the solution blows up when $u(x,0)$ approaches zero, so separated flow cannot be described by this model alone. Separation in general requires the full RANS model, but mild cases can be treated by a model integrated across the boundary layer, as is described next.

7.5.2 Integral Boundary-Layer Model

Theodore von Kármán derived an ordinary differential equation (ODE) model to replace the partial differential equation model above. The derivation is given in Ref. [4], and it produces the von Kármán integral boundary-layer equation as follows.

$$\frac{d\theta}{dx} + (2 + H - M_e^2)\frac{\theta}{U_e}\frac{dU_e}{dx} = \frac{c_f}{2} \tag{7.8}$$

where the shape factor $H = \delta^*/\theta$ and the edge Mach number $M_e = U_e/a_e$ appear. The equation and its ingredients can be used for both laminar and turbulent flows. The model can be closed by relations for H and c_f, as was done in the classical Thwaites method (see Ref. [4], chapter 4.5). Note that u appears nowhere, so only vanishing U_e threatens to blow the solution up. MSES uses another ODE, a kinetic energy shape parameter equation (Eq. (7.9)).

$$\frac{1}{H^*}\frac{dH^*}{dx} = \frac{2C_\Delta}{H^*\theta} - \frac{c_f}{2\theta} + (H - 1)\frac{1}{U_e}\frac{dU_e}{dx} \tag{7.9}$$

A number of new quantities appear here.

$$\text{Kinetic energy shape parameter: } H^* = \frac{\theta^*}{\theta} \tag{7.10}$$

$$\text{Kinetic energy thickness: } \theta^* = \int_0^\delta \left(1 - \frac{u^2}{U_e^2}\right) \frac{u}{U_e} dz \tag{7.11}$$

$$\text{Dissipation integral: } \Delta = \int_0^\delta \tau du \tag{7.12}$$

$$\text{Dissipation coefficient: } C_\Delta = \frac{\Delta}{\frac{1}{2}\rho U_e{}^3} \tag{7.13}$$

The two ODEs are valid for both laminar and (time-averaged) turbulent flows. They must be closed by empirical correlations relating, for example, H^* and H and c_f and C_Δ to H and the boundary-layer thickness Reynolds number Re_θ. The correlations may be different for the laminar and turbulent regions, so a smooth transition in the numerics where the laminar–turbulent transition takes place must be ensured. The ODEs are discretized by implicit central differencing, a second-order scheme, which reads for the system of ODEs $\frac{d\mathbf{w}}{dx} - \mathbf{f}(\mathbf{w}) = \mathbf{0}$.

$$\mathbf{w}(x + dx) - \mathbf{w}(x) - dx/2\,(\mathbf{f}(\mathbf{w}(x + dx)) + \mathbf{f}(\mathbf{w}(x))) = \mathbf{0} \tag{7.14}$$

The whole set of equations is adjoined to the set from the outer flow.

7.5.3 Transition to Turbulence

Boundary layers usually start out as steady and laminar and transition to turbulent flow at some downstream station. In reality the transition occurs within a region and is unsteady, but in our simulation it is assumed that the change from laminar to turbulent flow happens instantly at a point on the airfoil contour. Since the momentum thicknesses on both sides of the transition point must be the same and as the velocity profiles in laminar and turbulent flows are different, there must be a jump in displacement thickness as well as in shape factor.

MSES offers a choice of forced transition, when the user specifies the transition point. This is used for comparison with experiments using devices such as sandpaper strips to trigger the transition. Natural transition uses the growth of unstable wave motion – the Tollmien–Schlichting waves. The transition takes place where the most unstable wave amplitude has risen to $e^{N_{crit}}$ with a user-specified value for N_{crit}, often set to 9, hence being known as the e^9 criterion. Laminar and turbulent flows have quite different (mean) $u(z)$ profiles, as seen in Figure 2.10, which will be reflected in the H, c_f models used.

The e^9 method was used in Ref. [1] to predict transition for a large number of cases, and it was found that the transition Reynolds number $Re_{x,tr}$ is related to the shape factor H by the following semi-empirical formula.

Table 7.1 Boundary-layer characteristics for flow over a flat plate. Blasius laminar data, 1/7th-power law turbulent data.

	Laminar	Turbulent
δ_{99}/x	$5.00/\sqrt{Re_x}$	$0.37/\sqrt[5]{Re_x}$
δ^*/x	$1.721/\sqrt{Re_x}$	$0.046/\sqrt[5]{Re_x}$
θ/x	$0.664/\sqrt{Re_x}$	$0.036/\sqrt[5]{Re_x}$
c_f	$0.664/\sqrt{Re_x}$	$0.0576/\sqrt[5]{Re_x}$
H	2.59	1.28

$$\log_{10}(Re_{x,tr}) = -40.4557 + 64.8066H - 26.7538H^2 + 3.3819H^3; \;\; 2.1 < H < 2.8 \tag{7.15}$$

This can be used in laminar boundary-layer calculation by marching from the leading-edge stagnation point to predict the transition point (if any). The transition point is located where the two expressions become equal. The boundary-layer characteristics for a flat plate with laminar data according to Blasius and turbulent data from the 1/7th-power law in Ref. [13] are reproduced here in Table 7.1. We note in passing that the 1/7th-power law is wrong close to the wall, since its gradient becomes infinite; $c_{f,turb}$ is based on measurement. There is a "laminar sublayer" where the fluctuations are small, so wall stress follows the laminar formula as in the log-law-of-the-wall profile in Figure 2.10.

Separation

Consider the flow over an airfoil with a region of very low pressure. In such a region, the velocity exterior to the boundary layer is very high. So is also the velocity gradient at the airfoil surface, $(\partial u/\partial z)_{z=0}$, and consequently the friction force, τ. There is an adverse pressure gradient in the pressure recovery region on the aft part of the airfoil. As the fluid particles have already spent some of their kinetic energy in overcoming the friction, they may not be able to proceed all the way to the trailing edge, but instead stop somewhere. The oncoming flow then has to deviate there and flow away from the airfoil contour (see Figure 7.8). A turbulent case with more mixing of high-speed flow from the exterior can advance further than a laminar flow can. The initial value problem for the boundary-layer equations with specified U_e will have a singularity when u vanishes, so some remedy must be found to treat cases with separation bubbles (i.e. cases where the flow is steady and the boundary layer is not too thick).

Laminar Separation Bubble

Figure 7.9 illustrates the phenomenon of a laminar bubble in the boundary layer over an airfoil. The boundary layer at the airfoil's leading edge is laminar, but, typically for $Re > 0.2 \cdot 10^6$, it transitions to turbulent somewhere along the upper surface

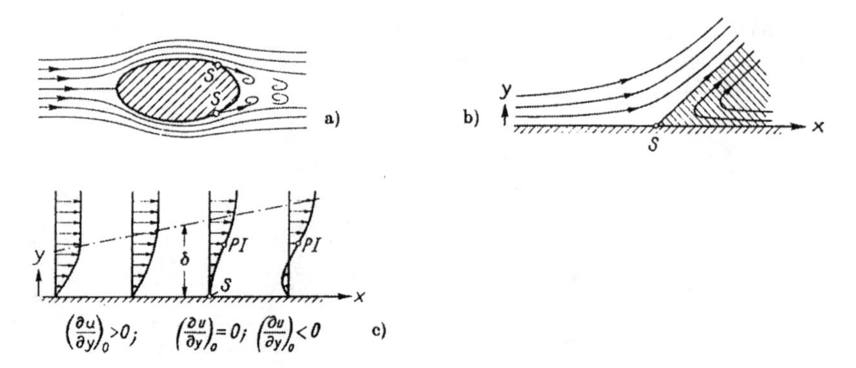

Figure 7.8 Separation of a boundary layer. (a) Flow past a body with separation (S = point of separation). (b) Shape of streamlines near the point of separation. (c) Velocity distribution near the point of separation (PI = point of inflection). (From H. Schlichting [10], reproduced with permission)

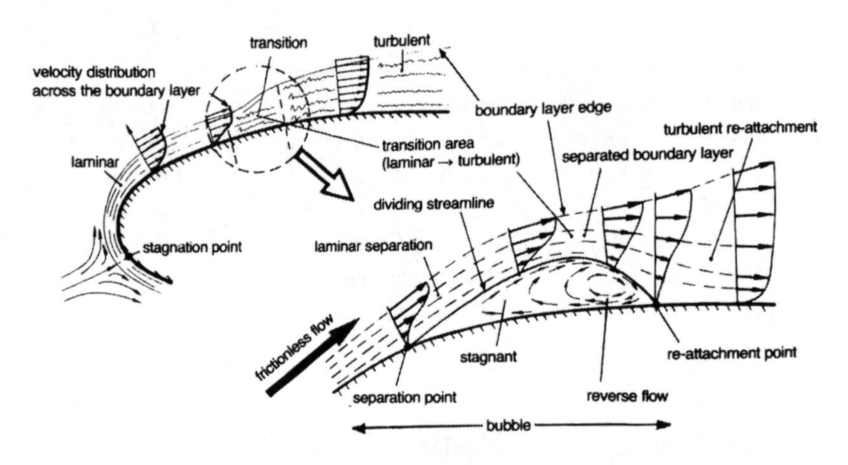

Figure 7.9 Schematic of a laminar separation bubble. (Courtesy K. Huenecke [7])

(suction side). For airfoils with highly cambered upper surfaces (or less cambered upper surfaces but at high angles of attack), the laminar boundary layer starts to separate. However, the increased thickness of the boundary layer results in a transition to a turbulent boundary layer, which is less sensitive to stall. Consequently, the flow reattaches, creating a bubble with an enclosed area of recirculating flow.

A low-speed example is shown in Figure 7.10, comparing wind tunnel data from Ref. [8] and an MSES calculation for the Eppler 387 airfoil at $Re_c = 200,000$, $\alpha = 2.0°, M_\infty = 0.04$ with free transition. The circle symbols as well as the MSES dots depict the sharp local drop in the pressure distribution due to this bubble. The agreement is excellent.

The effect of this laminar bubble on airfoil performance is significant for two reasons in addition to the obvious influence on the pressure distribution. First, the laminar bubble area is sensitive; the flow may separate entirely without a reattachment, resulting in a considerable drag increase. Second, the laminar bubble appears in the

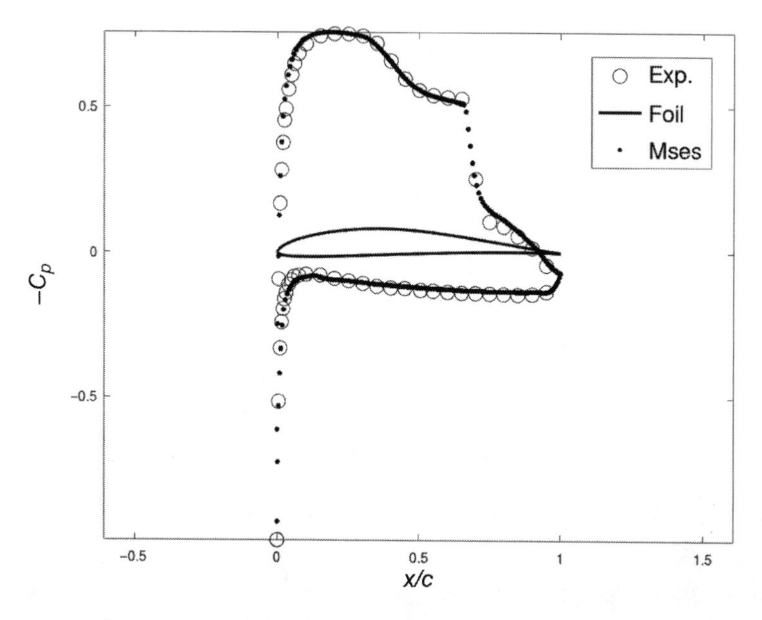

Figure 7.10 Measured and MSES pressure distribution for the Eppler 387 airfoil at $Re_c = 200{,}000,\ \alpha = 2.0^\circ,\ M_\infty = 0.04$

low-Reynolds number range (10^4–$0.2 \cdot 10^6$) and may disappear as the vehicle speed is increased. This causes severe discrepancies in aerodynamic data when comparisons are made over a wide speed range, such as using small-scale wind tunnel models to develop a full-scale vehicle.

Transition to turbulence – "tripping the boundary layer" – can be forced by introducing disturbances. It can be done by small vortex generators (little wedges of the height of the boundary layer), or even by a strip of coarse sand paper on the desirable transition line. Since a turbulent boundary layer has a tendency to stay attached longer, some drag benefits due to a reduction in separated flow can be gained by using this technique.

7.6 Drag Calculation

In many cases the determination of total drag is difficult. Integrating a computed or measured pressure distribution over the surface. means adding contributions of opposite signs, yielding poor accuracy. In wind tunnel testing, the suspension wires or other model support arrangements may introduce large errors. Airfoils are usually tested as wings spanning the tunnel. Then, the tunnel–wall boundary layers cause deviations from two-dimensionality. Flight testing suffers from the difficulty of knowing the flight status and the separation of drag and thrust. A useful alternative is the pitot traverse method of determining the profile drag from the pressure distribution in the wake. Here, the 2D case will be described. The method has also been extended to bodies of revolution.

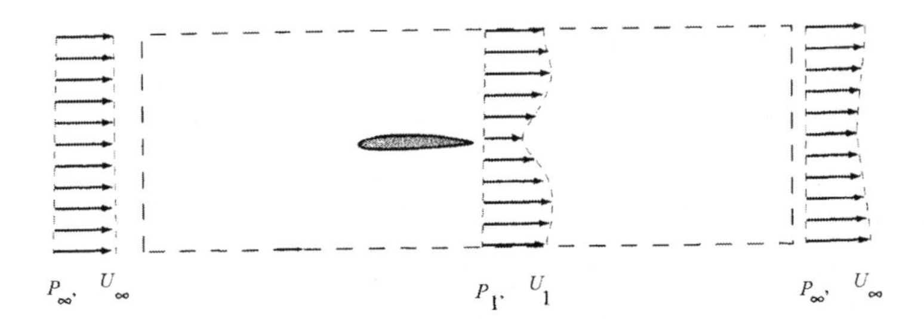

Figure 7.11 Control box in flow past a wing section.

The Squire–Young formula [12] is a common tool for this kind of airfoil drag estimation used in MSES and XFOIL. For transonic flow, the airfoil is subject to viscous drag and wave drag. Their sum can, of course, be obtained from the momentum defect far downstream. In MSES, the two components can actually be distinguished: viscous drag from the wake and wave drag from the outer inviscid flow region.

7.6.1 Momentum Defect Considerations

In Figure 7.11, a control box is outlined around an airfoil. Far upstream, the pressure is p_∞ and the velocity is U_∞; far downstream, the conditions are p_∞ and $u(z)$. At the boundary-layer edges, the velocity is written U.

The net momentum that enters and exits the control box in the free stream direction equals the drag force D.

$$D = \rho \int_{z=-\infty}^{\infty} u\,(U_\infty - u)\,dz$$

where z is the vertical coordinate along the box side normal to the flow and ρ is the density of the fluid. The integration must be done in a cross section so far downstream that the streamlines are parallel and thus the static pressure has returned to p_∞. The drag so computed is the total drag, and in general CFD computations, it is difficult to separate the viscous drag from the wave drag. Note that there is no lift-induced drag for an airfoil.

7.6.2 Wave Drag

Wave drag is easy to extract in MSES, since it appears only in the region outside of the boundary layer with the inviscid flow model. The streamline grid streamtubes cover the inviscid flow region. The mass flow through a tube is known exactly from the inflow conditions. It remains only to extrapolate the state at the outflow isentropically to the free stream state and to sum the results from all streamtubes.

7.6.3 Drag Due to Viscosity

Consider a 2D flow past an airfoil, as shown in Figure 7.1. With a given pressure distribution, the development of the boundary layers can be analyzed from the stagnation point A as laminar layers up to T, and from there on as a turbulent boundary layers as far as B.

For airfoils, methods can be based on momentum defect (Section 7.6.1) or boundary-layer properties, as is shown below.

The boundary layer displaces the outer flow from the airfoil surface. The displacement thickness of the boundary layer and the wake cause a change in the pressure distribution due to the changed shape, as seen in the outer inviscid flow. The viscosity thus produces not only a friction drag, but also a pressure drag. However, the computation of the pressure drag from wall pressure is difficult, as pointed out above.

The total pressure is constant in the fluid except in the boundary layer and the wake, shown shaded in Figures 7.1 and 7.3. At the trailing edge, the boundary layers on the upper and lower surfaces coalesce into the wake, extending downstream. The wake has a minimum thickness just downstream of the trailing edge, broadening gradually downstream. The static pressure in the wake is greatest at the trailing edge and decreases downstream, eventually becoming equal to the free stream static pressure.

The momentum thickness θ_∞ and the drag coefficient c_D in the wake are defined as follows.

$$\theta_\infty = \int_{-\infty}^{\infty} \frac{u}{U_\infty}\left(1 - \frac{u}{U_\infty}\right) dz$$

$$c_D = \frac{D}{\frac{1}{2}\rho {U_\infty}^2 c} = \frac{2\theta_\infty}{c} \tag{7.16}$$

where c is the chord length. Thus, the airfoil drag can be calculated from the θ value obtained from an integral boundary-layer calculation. However, for flight testing or in wind tunnels, the momentum thickness at downstream infinity is unavailable. A relation for the variation of θ with downstream position along the wake is looked for, so the thickness far downstream can be predicted from its value just behind the trailing edge or at the outflow end of the computational grid.

The wake is treated as one viscous layer. At the trailing edge, the momentum thickness of the wake equals the sum of two momentum thicknesses: one from the upper surface and one from the lower. Thus, the momentum thickness is continuous at the trailing edge, while other boundary-layer characteristics may not be.

The broadening of the wake, progressing downstream, is caused by turbulent mixing with the external flow, so that the shape factor $H = \delta/\theta$ falls from its value at the trailing edge to the value unity far downstream, as U approaches U_∞. The integral momentum equation for the wake is Eq. (7.8) with the surface friction c_f put to zero:

$$\frac{d\theta}{dx} + \frac{U'}{U}(H + 2)\theta = 0$$

where x is the coordinate downstream along the centerline of the wake and θ is the momentum thickness of the wake. This needs to be integrated over x from station 1 at the trailing edge to a station very far downstream. Replacing $H(x)+2$ by its algebraic mean $\frac{H_1+1}{2}+2$ produces the following.

$$\theta_\infty = \theta_1 \left(\frac{U_1}{U_\infty}\right)^{(H_1+5)/2}$$

where H_1, the shape factor at the trailing edge, is known from the boundary-layer calculations. Combination with Eq. (7.16) gives the Squire – Young formula for airfoil drag.

$$c_D = \frac{2\theta_\infty}{c} = \frac{2\theta_1}{c}\left(\frac{U_1}{U_\infty}\right)^{\frac{H_1+5}{2}} \tag{7.17}$$

where subscript ∞ refers to downstream infinity and subscript 1 refers to the station slightly downstream of the trailing edge. U_1 is the velocity at the edge of the wake layer at that station and H_1 is the boundary-layer shape factor.

Experimental drag values and calculations with Eq. (7.17) for a large number of airfoils have been compared in Ref. [11]. For incidences below 6°, the agreement is good, with errors of only a few percent. With increasing angle of attack, the error increases.

7.7 Newton Solution Method

The collection of equations $\mathbf{R}=\mathbf{0}$, also referred to as the set of residuals to be driven to zero, is compiled from the discretized Euler equations, the boundary-layer model (including relations for transition from laminar to turbulent flow), the extra Kutta conditions, and other global constraints. The set of unknowns $\mathbf{Q}$ come from flow variables in the outer flow and the boundary layer and wake and the global variables. The complete nonlinear system of equations $\mathbf{R}(\mathbf{Q}) = \mathbf{0}$ is solved by a damped Newton scheme as follows.

For $n = 0, 1, \ldots$

Solve the linear system $J(\mathbf{Q}_n)\mathbf{h}_n = \mathbf{R}(\mathbf{Q}_n)$ for $\mathbf{h}_n$

Set $\mathbf{Q}_{n+1} = \mathbf{Q}_n - \lambda \mathbf{h}_n$

The derivatives making up the Jacobian matrix J, $J_{ij} = \frac{\partial R_i}{\partial Q_j}$ are computed analytically. The part of J related to the outer flow variables and residuals is sparse, but the global variables create columns with few zeros. The time for solution can be brought down substantially for the sparse system when the eliminations and unknowns are ordered properly. The Newton scheme is known to converge quadratically when $\lambda = 1$; in other words:

$$\frac{||\mathbf{h}_{n+1}||}{||\mathbf{h}_n||^2} \to K$$

as $n \to \infty$, if the initial guess $\mathbf{Q}_0$ is good enough. Unless the residuals are nondimensional(ized), K is a dimensional quantity. A potential solution with vanishing boundary layers offers a good initial guess for low Mach numbers and high Re, but it may be quite far from the solution (if there is one without massive separation!) in strongly interacted cases with shocks or low Re. The standard approach is then to adapt the step size λ by reducing it when residuals do *not* decrease quickly and increasing it when they do, so that the quadratic finale is achieved. $O(10)$ iterations is a guideline for nondifficult cases.

Difficult cases are treated by having MSES perform a *sweep* of, say, Mach number, or some other global quantity such as angle of attack.

For $M_\infty = M_0, M_1, \ldots, M_{desired}$, run MSES. M_0 is chosen to make an easy case, and the sequence of M values leads up to (and past?) the desired value. MSES will then use the result of M_n as an initial guess for M_{n+1}, hopefully close enough to make the $n+1$ solution easy as well. If not, MSES will take a smaller increment and try again and again until either success is achieved, enabling a move to the next M, or failure must be admitted, when the user-specified increment has been reduced by a factor 32 without success.

Such continuation methods will be successful when the solution path as parametrized (in the example by M_∞) is continuous and the increments are small enough. The natural tendency to try with large increments must be curbed. Small increments will require only a few Newton iterations and increase robustness at little risk of extra cost.

7.8 Airfoil Computations

Chapter 8 presents many MSES results for subsonic and transonic airfoils. We limit the discussion here to results for the RAE104 (symmetric) airfoil at $M_\infty = 0.794$.

7.8.1 Transonic Flow over the RAE104 Airfoil

Experiment, EDGE, and MSES

Figure 7.12 shows pressure distributions for the RAE104 airfoil at $M0.794, \alpha = 2°$, and $Re = 1.8 \cdot 10^6$. The flow is attached all the way to the trailing edge, and the MSES and EDGE results with the explicit algebraic Reynolds stress model (EARSM) Hellsten $k - \omega$ (see Chapters 2 and 6) turbulence model agree very well with the experiment.

Selection of MSES Results

Typically, the C_p distribution and drag polars from M or α sweeps are desired. Profiles of the boundary-layer properties such as c_f, H, and δ^*, as well as transition stations, are useful to help us understand the coupling effects in shock–boundary layer interaction (Figure 7.13).

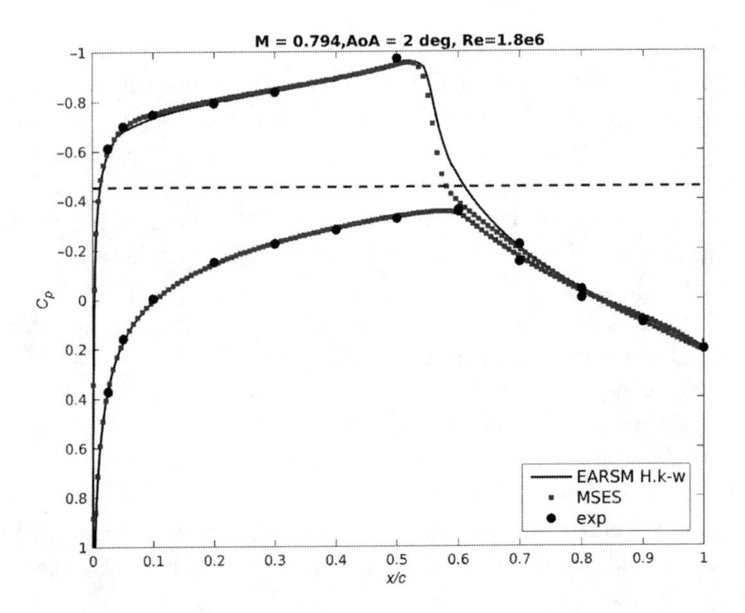

Figure 7.12 MSES, EDGE, and experimental pressure distributions on the RAE104 airfoil. AoA = angle of attack $\alpha = 2$ deg.

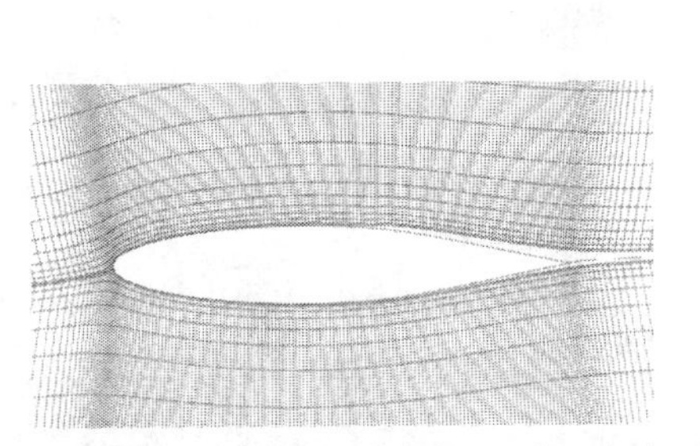

(a) RAE104 airfoil, boundary-layer edges, and MSES grid

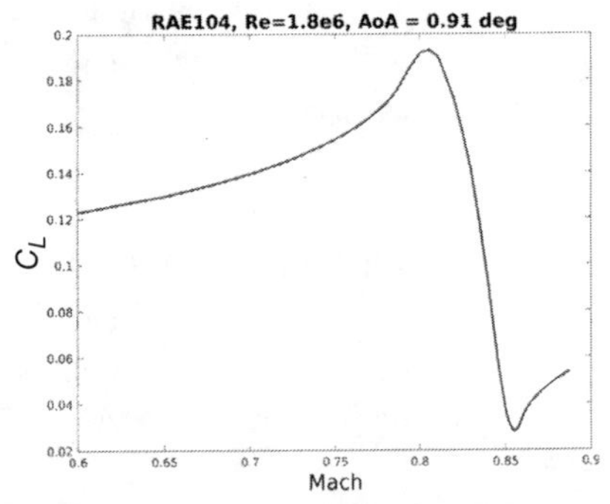

(b) M sweep of C_L at fixed Re and α

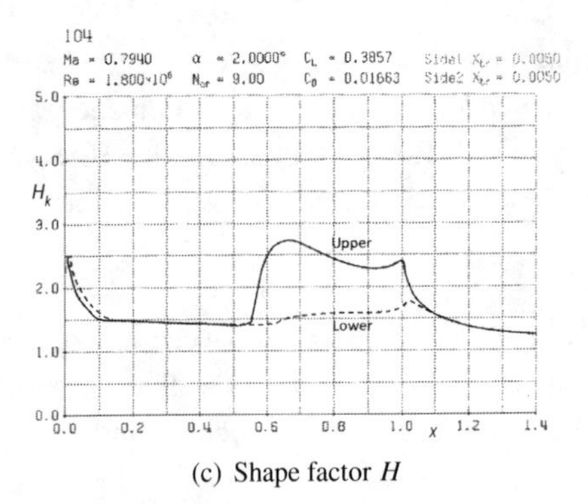

(c) Shape factor H

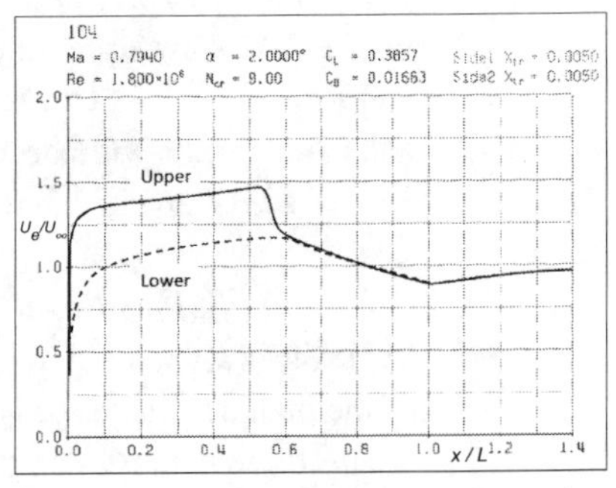

(d) Velocity ratio U_e/U_∞ at edge of boundary layer

Figure 7.13 Selected MSES results.

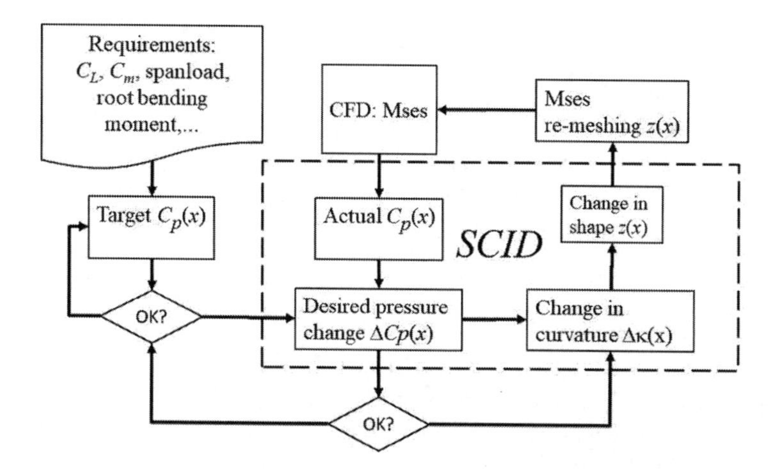

Figure 7.14 The SCID algorithm with MSES as the CFD tool.

7.9 MSES Design Application

7.9.1 Inverse Design

Inverse design works by first finding a pressure distribution that fulfills the design requirements and then determining a geometry that yields this target pressure. Note the difference with the vortex lattice method design task of Chapter 3, where the wing *camber surface* is determined from the *lift distribution*. Formulation of a good target pressure distribution requires a knowledgeable engineer to be in the loop. Inverse design is treated in Chapter 8.

Figure 7.14 illustrates a generic inverse design approach for the streamline curvature iterative displacement (SCID) scheme described below.

Choice of Target Pressure Distribution

Aerodynamicists have substantial know-how on the undesirable features of wing pressure distributions. For instance, too strong suction peaks on the leading edge should be avoided, since the adverse pressure gradient is inducive to boundary-layer separation. Similarly, shocks should be weak to reduce wave drag. Isobars should be swept with the wing to avoid normal shocks. Pressure gradients in the pressure recovery region downstream of the wing thickness maximum should be mild, again to avoid separation of the boundary layer. Strong loading of the wing tip increases the risk for tip stall.

However, the class of feasible pressure profiles (i.e. those that can be generated by a wing shape that fulfills the geometric and structural constraints) is not known. Thus, the procedure must be exercised iteratively by modification of the original target pressure profile, as the inverse designs indicate what is possible. The pressure distribution alone determines the pressure forces but not the friction. In addition, pressure needs to be supplemented by another thermodynamic quantity such as

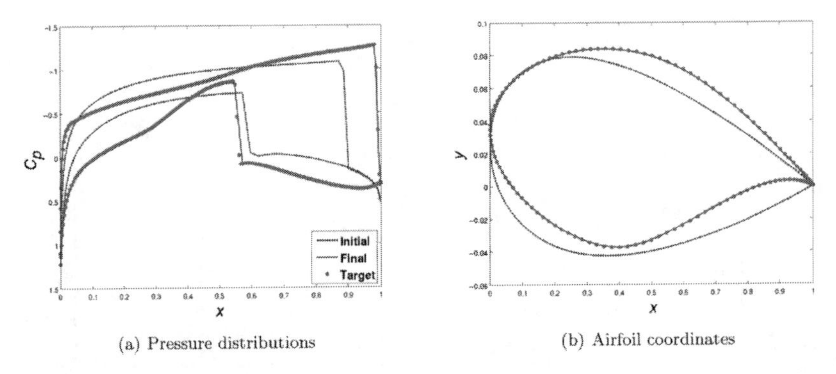

Figure 7.15 Morphing a NACA0012 airfoil using the SCID algorithm into one with the pressure distribution of RAE2822, inviscid flow at $M_\infty = 0.8$. (Courtesy of M. Zhang, personal communication)

entropy in order to fully characterize a compressible flow with shocks. This, again, puts a constraint on the class of feasible pressure profiles. Friction drag can be estimated by computing the boundary layer, since its forcing is the streamline pressure gradient.

The Algorithm

Figure 7.15 shows the result of morphing a NACA0012 profile at a given α into a shape that gives the (precomputed) pressure profile of RAE2822 in inviscid flow at the same M_∞. The target pressure profile is thus obviously feasible. The shape was iteratively modified by the SCID algoritm using the relation between pressure normal gradient and the streamline (i.e. airfoil surface) curvature κ.

$$dp/dn = (p - p_\infty)/l = \rho \kappa u^2$$

The length scale l is a parameter to adjust. The difference between actual and desired surface pressure modifies the curvature, and the new streamline shape is then integrated from the arclength description of a curve $\mathbf{x}(s)$ in the plane with $\hat{\mathbf{e}}_s$ as the unit tangent and $\hat{\mathbf{e}}_n$ as the unit normal.

$$d\mathbf{x}/ds = \hat{\mathbf{e}}_s$$

$$d\hat{\mathbf{e}}_s/ds = \kappa(s)\hat{\mathbf{e}}_n \tag{7.18}$$

The 3D relations are the Frenet–Serret formulas that also need a unit binormal $\hat{\mathbf{e}}_b$ to complete a local orthogonal coordinate system and the torsion $\tau(s)$ of the curve.

7.9.2 Design Example: Increasing M_{dd} for the RAE104

Figure 7.16 shows an exercise to design an extension of the RAE10x family to RAE104mod to increase the drag divergence speed M_{dd} by moving the crest downstream. The crests are given in Table 7.2.

Table 7.2 Crest locations for the RAE10x airfoil family.

Foil	100	101	102	103	104	104mod
x/c crest	0.26	0.30	0.36	0.40	0.42	0.4402

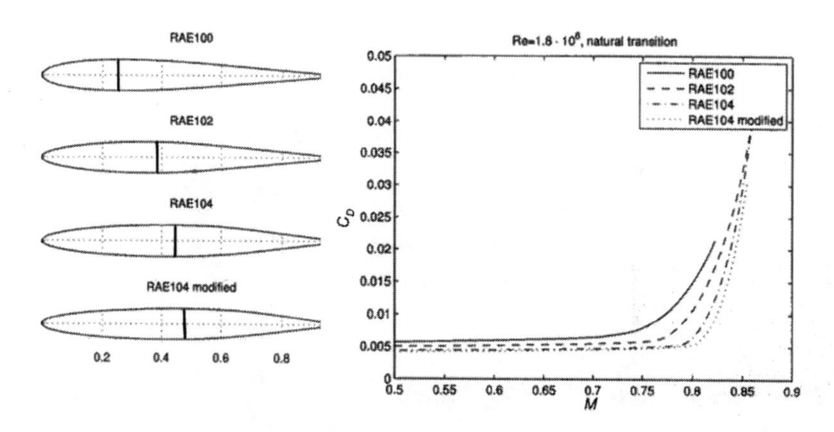

Figure 7.16 Increase of M_{dd} for the RAE100, 102, 104, and 104mod with increasing crest location.

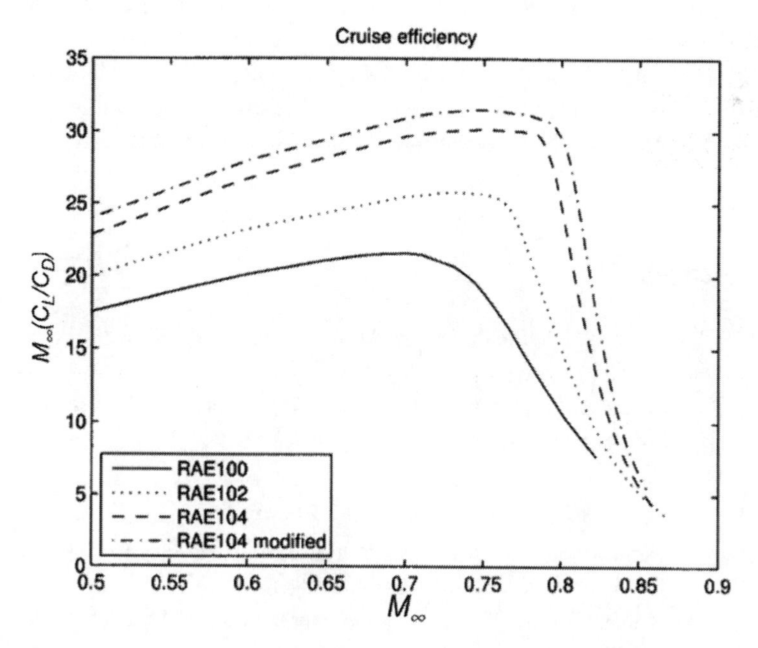

Figure 7.17 Cruise efficiency gain of RAE10x extended to RAE140mod.

The zero-lift drag for $M_\infty \geq 0.5$ (as high as MSES would converge) shows how M_{dd} grows with crest position, with just a minor change from 104 to 104mod. Moving the crest downstream of 0.42 is very sensítive. The 104 foil is essentially wedge-shaped downstream of 80%, so one expects the curvature to decrease after the crest.

Continuity of curvature is very desirable to discourage premature separation. But even the 104 has curvature increasing a little past the crest. We shall also see that airfoils on the early jets were symmetric and had crests in the 0.40–0.45 range.

The shape change was carefully manually implemented by the shape modifiers available (e.g. Hicks–Henne bumps) in the MSES shape optimizer to make sure that the maximum thickness does not change more than tolerably, that curvature varies smoothly, and that separation is avoided, shock-induced or otherwise. Only the upper side was modified, and 104mod actually has reflex camber at the trailing edge. As can be seen, the improvement in M_{dd} is about $M0.01$. Figure 7.17 shows the cruise efficiency gain.

Our toolset is now complete in terms of the general RANS code and the airfoil code MSES. The next chapter applies both of these in the consideration of airfoil design.

7.10 Learn More by Computing

Gain hands-on experience of the computational tools for the topics in this chapter by working with the on-line resources. Exercises, tutorials, and project suggestions are found on the book website www.cambridge.org/rizzi. For example, run the Eppler airfoil at different angles of attack to see the development of the laminar separation bubble. Software used to compute many of the examples shown is available from http://airinnova.se/education/aero-dynamic-design-of-aircraft

References

[1] J. J. Bertin and M. L. Smith. *Aerodynamics for Engineers*, 3rd edition. Prentice-Hall, Inc., edition, 1998.

[2] H. Blasius. Grenzschichten in Flüssigkeiten mit kleiner Reibung. *Zeitschrift für Angewandte Mathematik und Physik*, 56:1–37, 1908.

[3] M. Drela. *A User's Guide to MSES 3.05*. MIT Press, 2007.

[4] M. Drela. *Flight Vehicle Aerodynamics*. MIT Press, 2014.

[5] M. Drela and M. B. Giles. Viscous–inviscid analysis of transonic and low Reynolds number airfoils. *AIAA Journal*, 25(10): 1347–1355, 1987.

[6] V. M. Falkner and S. W. Skan. Some approximate solutions of the boundary-layer equations. *Philosophical Magazine*, 12: 865–896, 1931.

[7] K. Huenecke. *Modern Combat Aircraft Design*. Naval Institute Press, 1987.

[8] R. J. McGhee, B. S. Walker, and B. F. Millard. Experimental results for the eppler 387 airfoil at low reynolds numbers in the langley low-turbulence pressure tunnel. NASA Technical Memorandum 4062. NASA, 1989.

[9] L. Prandtl. Über flüssigkeitsbewegung bei sehr kleiner reibung. In *Proceedings Third International Congress of Mathematicians, Heidelberg*, 1904, pp. 484–491; English translation: "On the motion of fluids of very small viscosity." NACA-TM-452, March 1928.

[10] H Schlichting. *Boundary-Layer Theory*. McGraw-Hill, 1979.

[11] A. M. O. Smith and T. Cebeci. Remarks on predicting viscous drag. In *AGARD Conference Proceedings No. 124*. AGARD, pp. 3.1–3.26.

[12] H. B. Squire and B. A. Young. The calculation of the profile drag of aerofoils. ARC Technical Report R. & M. No. 1838. ARC, 1937.

[13] D. C. Wilcox. *Turbulence Modeling for CFD*. DCW Industries, Inc., 1998.

8 Airfoil Design Considerations

> A theoretical airfoil design method enables the design of airfoils that fall outside the range of applicability of available catalogs, the design of airfoils that exactly match the requirements of the application, and the economic exploration of various airfoil concepts.
>
> Dan M. Somers, President, Airfoils, Inc.[1]

Preamble

As assumed by Prandtl's lifting-line theory, a section of the wing parallel to the free stream can be considered to act independently of the other sections, except for an induced angle of attack varying along the span. The flow over any section is therefore plane with no component along the span. The aerodynamic characteristics of wings of infinite span are accordingly called "wing section characteristics," "airfoil section characteristics," or simply "airfoil characteristics," and they determine the boundary-layer flow and compressibility effects. The study of wing characteristics simplifies by decomposition into two weakly coupled systems – section and planform – which may be considered separately; planform variations were treated in Chapter 3.

Even for wings of arbitrary planform and twist, characteristics such as angle of zero lift, lift-curve slope, span loading, and drag for high-speed and cruise conditions can be predicted from airfoil data. This actually works very well at low- and moderate-lift coefficients if no spanwise discontinuities or rapid changes of section, chord, or twist are present, if the wing has no pronounced sweep, and if the wing is of sufficient aspect ratio. It follows that building a knowledge base for airfoil characteristics is an effort well spent.

This chapter explains and illustrates the connection between the airfoil geometry and its resulting pressure distribution in low- and high-speed flight. Deployment of slats and flaps increase lift for takeoff and landing, and multielement airfoils demonstrate the effects of these high-lift devices. An example of drag minimization of the RAE2822 airfoil outlines the mathematical procedure that changes the geometry in order to optimize its performance.

[1] With his permission, personal communication.

8.1 Introduction to Airfoil Design

For wings of large aspect ratios (*AR*s) and small sweep angles, the flow at a given chordwise wing station resembles the 2D flow about that airfoil. It is therefore useful to study such 2D flows.

Airfoil design is based on knowledge of the boundary-layer behavior and the relation between geometry and pressure distribution. There are many objectives for the design of an airfoil. At a given lift, most sections produce a low drag. Some of them may not have to generate any lift at all. For others, the maximum lift is all-important. The section may have to achieve this performance with constraints on thickness (for strength, storage, and fuel), pitching moment, off-design performance, or sufficient thickness at the trailing edge to allow for a high-lift mechanisms, etc. The selection should also consider the ease of manufacture.

In the past, a designer would find an airfoil from an airfoil catalog (e.g. [3–5]). The advantage to this approach is the availability of test data. With today's capability to compute airfoil characteristics, it is common to custom design airfoils. There are design considerations that lead to airfoils producing a substantial load over a great portion of the chord, which, for a given lift force, permits a reduced wing area and yields higher wing loading. Another consideration is the desire to have laminar flow over the greatest possible part along the airfoil.

It is necessary to obtain a general idea of the flow around an airfoil and how it is influenced by the shape, airspeed, and incidence. As a simple example of the complexity of the mapping from shape to flow, consider the airspeed. Its influence comes from two mechanisms. One is characterized by the Reynolds number, as shown in Chapter 7, through the behavior of the boundary layer, its thickness, and its tendency to separate. The other, characterized by the Mach number, is the effect of compressibility. Of course, as expected, even these two different issues will interfere with each other and must be considered together. Chapter 2, for example, introduced the shock-stall phenomenon brought about by shock wave interaction with the boundary layer. Figure 2.7 suggests how widely varying the airfoil's chord size and speed (Reynolds number) leads to orders of magnitude differences in the lift-to-drag ratio L/D. This chapter concerns airfoils mainly in the Re_c range from 10^6 to 10^8.

8.1.1 Airfoil Parameters Map Shape to Flow

The intended use of an airfoil determines its shape. Size, speed range, and altitude determine its Reynolds number and ultimately its performance.

Chapter 2 indicated that thickness and camber are two fundamental geometry parameters of an airfoil, and Chapter 5 showed how the wing's geometry is built up as a stack of airfoils. This chapter will further develop the details of the shaping parameters. Recall Figure 2.5, which presented the geometry for airfoils. The parametrization of the airfoil shape offers a handle to describe the mapping from shape to flow. Simple airfoil families are completely characterized by a few parameters. The National Advisory Committee on Aeronautics (NACA) four-digit family is an example, but its nose radius and thickness cannot be varied individually. Such a given thickness distribution can be scaled to different maximal thicknesses t/c. Then, the leading-edge radius $r_{nose}/c = const. \cdot (t/c)^2$.

We shall therefore work with different families to illustrate a few key features of the shape-to-flow map. The following are geometric features with identifiable influence on the pressure distributions.

- Leading-edge radius of curvature
- Maximum thickness, its location, and upper surface curvature
- Camber curve, with features such as maximum camber, position of maximum camber, drooping of nose and tail regions, etc.

We attempt below to quantify these influences. It is also necessary to consider structural issues such as the volume and height of the wing box and the "thinness" of the trailing edge. Usually, these are taken into account by adhering to a lower limit on allowable thickness, etc.

Geometric Influence on Separation Pattern

There are several types of separation in plane flow. The geometric properties of wing sections correlate with the stalling properties and the shape of lift and pitching-moment curves. Two representative types of airfoil stall are considered: trailing-edge stall and leading-edge stall.

Trailing-edge stall is characteristic for most airfoil sections with thickness-to-chord ratios of approximately 15% and higher, characterized by progressive thickening of the turbulent boundary layer on the upper surface, as indicated in Figure 2.4. As the angle of attack is increased, flow separation moves gradually forward and the lift-curve slope decreases. The peak of the lift curve is rounded. The variation of the pitching moment with lift is smooth; there is no sudden break at the stall.

Thin airfoils may suffer *leading-edge stall*, an abrupt separation of the flow near the leading edge. Sections with thickness-to-chord ratios of less than about 6% have a small separation bubble on the nose, even at very small angles of attack, as shown in Figure 8.1 (top). The anatomy of the bubble is shown in Figure 7.9. If the laminar

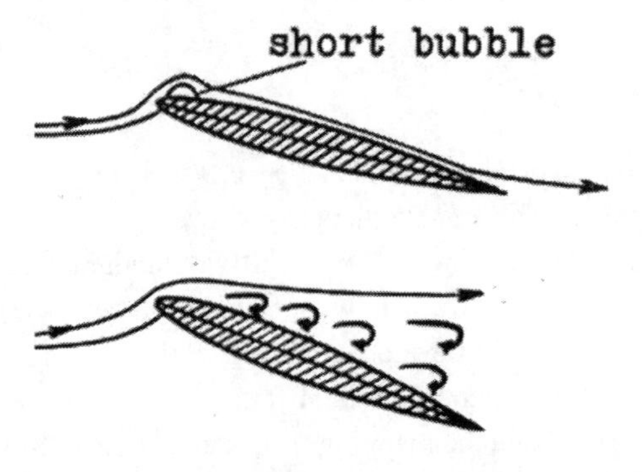

Figure 8.1 Character of leading-edge airfoil stall: thin airfoil leading-edge separation bubble before stall (top) and after stall (bottom).

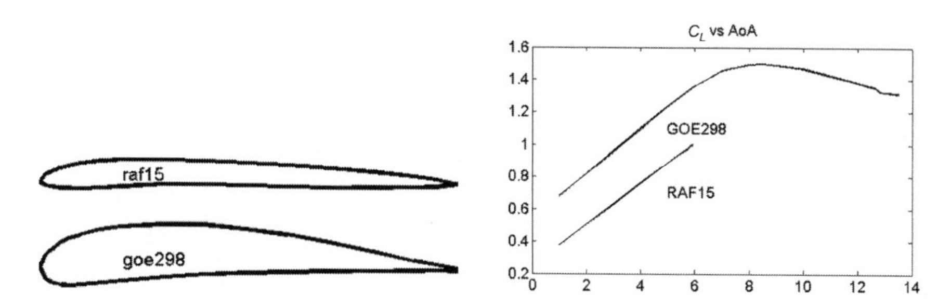

Figure 8.2 Göttingen 298 and RAF15 airfoil shapes (left); c_L vs. α (right).

boundary layer separates prior to transition to turbulence in the shear layer, reattachment of the flow quickly occurs, enclosing a "short bubble" and subsequently forming a turbulent boundary layer. The pressure distribution is not much affected by the short bubble. Increased angle of attack moves the separation point into a region of sharp airfoil curvature under the suction peak's adverse gradient, so that the turbulent shear layer fails to reattach: the bubble bursts and the flow over the entire upper surface separates. The lift and moment curves exhibit abrupt changes when the angle of attack for maximum lift is exceeded. There is little or no rounding of the lift curve, and a sudden negative shift of the pitching moment resulting from the rearward shift in the center of pressure is observed. The thin-airfoil stall is typical for sharp-nosed thin sections, such as the NACA64A206. There is a relationship between stalling characteristics, airfoil nose geometry, and Reynolds number. Various combinations of trailing- and leading-edge stall can be observed. The different types of stall should not be too closely associated with the thickness ratio. Stalling characteristics are difficult to predict and are very sensitive to minor variations in airfoil geometry.

Example: Fokker DR1

The German World War I Fokker DR1 triplane had superior climb performance to most other fighters of the time, such as the Sopwith Camel biplane flown by the Royal Air Force (RAF). The DR1 Göttingen GÖ298 airfoil was thicker and had a larger nose radius than the RAF15, as seen in Figure 8.2 (left). When the r_{nose}/c is still small, $c_{L,max}$ grows substantially with thickness. The airfoil points obtained from the University of Illinois at Urbana–Champaign (UIUC) airfoil database are too sparse for accurate shape definition, and they were preprocessed as described in Chapter 5 for flow calculations by the MSES airfoil simulation program described in Chapter 7. The lift curves are shown in Figure 8.2 (right). The GÖ298 foil has much higher $c_{L,max} = 1.5$ and a docile stall: the lift decays only slowly when the angle of attack is increased past stall at 8°. The RAF15 foil stalls abruptly at 6° with $c_{L,max} \approx 1$.

Table 8.1 displays a historical view of the development of airfoil shapes. Consider now the evolution from the last big prop airliner Douglas DC7 through a more modern commuter, the SAAB SF-340, to a modern transonic transport design example, such as the Common Research Model (CRM). The SAAB JAS-39 Gripen is the example of a fourth-generation Mach 2 multipurpose fighter.

Table 8.1 Historical use of airfoil shapes on aircraft, 1930 to present.

Aircraft	Speed	Airfoil	First flight
Douglas DC-3	333 km/h	NACA2215	1935
Douglas DC-7	653 km/h	NACA23012(tip)	1953
SAAB 340	470 km/h	NASAMS(1)-0313	1983
CRM	Transonic M 0.85	CRM 65%	Designed 2012
Messerschmitt 262	900 km/h	NACA00011-0.25-35	1941
NAA F-86 Sabre	1068 km/h	NACA0008.5-64 (tip)	1947
SAAB J-29 Tunnan	1060 km/h	FFA5106mod	1948
Concorde	M2.04	3% thin airfoil	1969
GD F-16 Falcon	M2	NACA64A204	1974
SAAB JAS-39 Gripen	M2	NACA64A206mod	1989

Table 8.2 Performance of four airfoils at low speed.

	NA23012	MS10313	CRM65	NA64A206
c_L	0.50	0.75	0.79	0.53
$c_D \times 10^4$	63	67	66	56
L/D	78	110	118	94

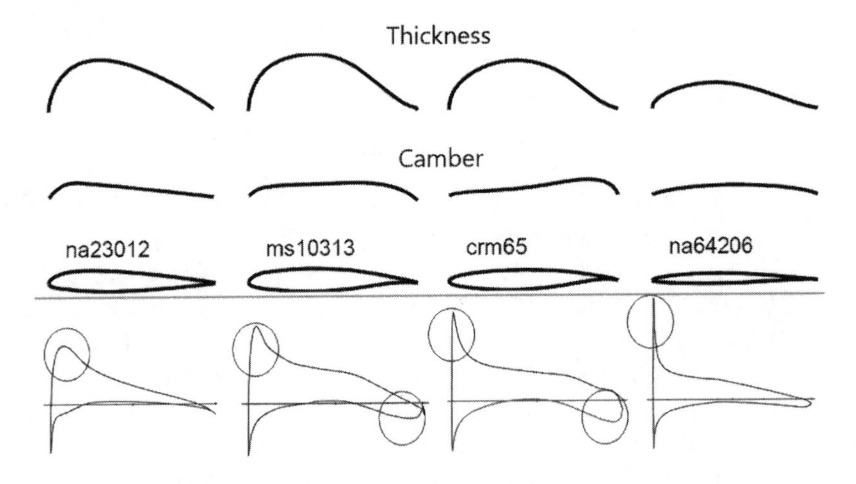

Figure 8.3 Evolution of airfoil shapes and associated pressure distributions. Note changes to suction peak and effect of trailing edge droop.

Figure 8.3 shows, from the top, the thicknesses, camber curves, profiles, and pressure distributions for the four airfoils for $M_\infty = 0.3, Re_c = 37 \cdot 10^6$, and $\alpha = 3°$. The transport wing foils are 12% and 13% thick and the supersonic fighter wing is 6% thick. Their performance is given in Table 8.2. The chosen flow case may not be the design point for the airfoil. For instance, the MS10313 foil has a design c_L of 0.6. The trends for the airliners show how the maximum camber position moves aft, as does the position of maximum thickness. The transonic foil CRM65 has strong reflex camber and the thickness is nonconvex for the SAAB SF-340 and the CRM. The fighter airfoil is designed for low wave drag, hence the small thickness.

8.1.2 Select Airfoil after Intended Use

The criteria for airfoil selection depend on the specific case, but there are some common desirable features.

- Docile stall (may be partially compensated with wing planform and twist)
- High $c_{L,max}$ with flaps out for low takeoff and landing speed (related to the basic airfoil's $c_{L,max}$)
- Low drag coefficient at:
  - Cruise c_L and Re_c for low fuel consumption and high cruise speed
  - Climb c_L and Re_c for good second- and third-segment climb
- Critical Mach number M_{crit} sufficiently high
- Low (absolute value) moment coefficient c_m for low trim drag

8.1.3 Designing for Optimum L/D

The aircraft should be designed to maximize the aerodynamic efficiency (i.e. to fly its design mission with L/D as large as possible). As a first approximation, the wing lift coefficient, C_L, may be set to the same value as the airfoil lift coefficient, c_L. The first consideration is then the design lift coefficient, the lift coefficient for which the best L/D occurs. In practice, the choice is also based upon past experience. For most types of airplanes, it will be around 0.5.

Contour Shaping for Optimum Pressure Distribution

Figure 2.8 pointed out some elementary properties of desirable pressure profiles, such as peak suction and pressure recovery.

The design task becomes: first, determine a pressure distribution for maximum lift; then, find the corresponding shape of the airfoil yielding this pressure – so-called *inverse design*. The airfoil is shaped to obtain desirable pressure gradients that drive the boundary layer over the surface. For example, premature separation of the boundary layer on the upper contour of the airfoil limits the amount of lift that can be achieved and must be avoided.

Boundary-layer separation is governed by the following.

- The adverse pressure gradient profile
- How much kinetic energy the boundary layer has lost at the start of the adverse pressure gradient

Skin friction is much smaller in laminar than in turbulent flow, as discussed in Section 2.2. Maintaining laminar flow then becomes a design goal. Factors influencing shape contouring are as follows.

- Nose radius influences the slope of the favorable pressure gradient and the level of minimum pressure reached.
- The pressure gradient, which maintains the flow along the wall, always exerts an important influence on stability and transition to turbulence. A favorable pressure gradient stabilizes the flow and an adverse pressure gradient destabilizes it. The pressure gradient strongly influences the curvature of the laminar velocity profile and can delay the occurrence of an inflection point. This stabilizing influence was the underlying physical principle exploited for the development of the laminar-flow airfoils shortly before World War II.
- Displacing the position of maximum thickness downstream moves the point of minimum pressure, as well as the inflection point in the velocity profile, downstream.
- For transonic speed, the strategy is to reduce the strength of the shock wave over the airfoil to minimize the wave drag. The nose radius as well as the profile shape after the point of maximum thickness strongly influence the results.

Example: Achieving High Lift Coefficient

An "ideal" optimal pressure distribution is schematically shown in Figure 8.4 (left).

- At the lower airfoil contour, pressure should be as large as possible, with the upper and lower streams meeting smoothly at the trailing edge to satisfy the Kutta condition.
- At the upper contour, pressure should be as low as possible, with constraints on the minimal pressure, $C_p > C_p^*$, to avoid supersonic flow and on the pressure gradient to avoid flow separation.

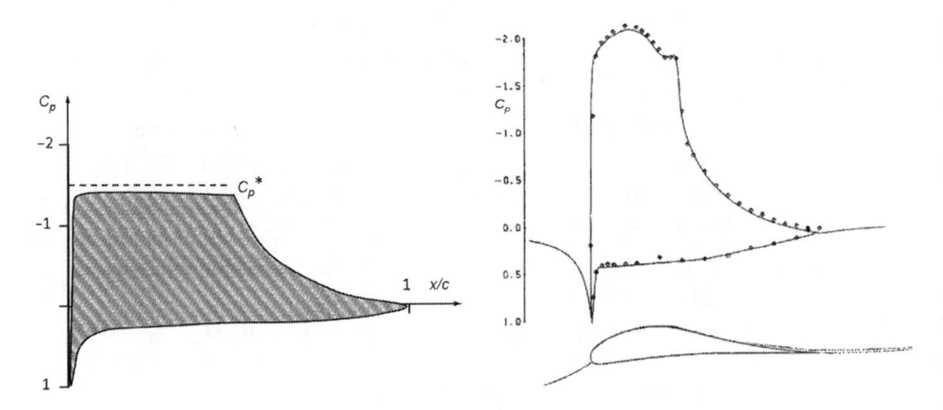

Figure 8.4 "Ideal" pressure distribution for maximal lift (left). Liebeck LNV109A airfoil (right) computed with MSES at $Re = 0.5M$, $\alpha = 7.4°$, $c_L = 1.234$.

A separation criterion was formulated by Stratford [16], used here to plot the pressure of a suction-side boundary layer on the verge of separating all of the way. This limits the length of low-pressure plateau to the start of pressure recovery. The plot motivates the "rooftop" denomination.

This strategy of airfoil design, together with the Stratford pressure recovery criterion, was successfully used by Liebeck [11] to develop a high-lift airfoil for low *Re* (Figure 8.4, right). Note the telltale kink, a sign of a laminar separation bubble – not included in the shape optimization. The recovery profile agrees well with Stratford theory.

Similar reasoning applies for the other design cases.

1. Optimum pressure for maximal extent of laminar flow
2. Optimum pressure for transonic speed (e.g. make shape symmetrical, nose radius not too small, position of maximum thickness further back, etc., to weaken shock wave)
3. Optimum pressure for supersonic speed (a thin foil with a small nose radius, little camber to minimize drag, yet at a much higher level than transonic)

The following sections of this chapter treat airfoils for each of these speed regimes as well as multielement airfoils for high lift, and the chapter concludes with a shape optimization example.

8.2 Subcritical-Speed Airfoils $M_\infty \leq 0.7$

The aircraft in Table 8.3 were designed in the 1950s and 1960s. Powered mostly by propeller, with a few powered by turbo-prop, they indicate the standard usage of the conventional classical NACA airfoils. For example, the NACA2412 flies on several of today's light aircraft. NACA230xx airfoils, or small modifications thereof, lift general aviation's higher-performance twins, such as the Cessna Citation and the Beech King Air. The notation NACA23018-23009 indicates the airfoil at the root and tip, respectively.

8.2.1 Classical NACA Airfoil Taxonomy/Catalog

The NACA Langley Memorial Aeronautical Laboratory in the early 1920s undertook the systemization of airfoil design and the prediction of their aerodynamic characteristics. Various avenues for airfoil improvement were vigorously explored. Correlating the aerodynamic characteristics of a consistent family of airfoils on the basis of the amount and position of maximum camber and the maximum thickness of the airfoil shape was one of its first achievements. It gave the aircraft designers ready access to a large bank of aerodynamic performance information for selecting an airfoil best suited to their particular airplane and its mission. The NACA catalog of four-digit airfoils won prominence in low-speed design.

Table 8.3 Airfoil selection for a few aircraft from the 1950s and 1960s.

Aircraft	**Airfoil**
Cessna Citation II	NACA23014-23012
Beech King Air	NACA23018-23012
Aero Commander 500	NACA23012
Piaggio P166	NACA230XX
Dornier Do 28 D Skyservant	NACA23018

Table 8.4 NACA four-digit airfoil interpretation.

Example: NACA 2412, Profile for the Cessna 150

Digit number	Meaning	Example%
1	Maximum camber	2
2	Position of maximum camber	40
3 and 4	Thickness ratio	12

Extensions of this research in the early 1930s resulted in its modified four-digit airfoils, while at about the same time the investigators were developing a new camber-line family to increase the maximum lift. The 230 camber line is the best known of this family. Combining these new camber lines with the standard NACA thickness distribution created the NACA five-digit airfoil family.

An important theoretical breakthrough in about 1930 abetted this work through calculation of the inviscid flow about an arbitrary airfoil for the first time [18]. The new tool, coupled with new wind-tunnel results on turbulence, transition, and boundary-layer growth, now permitted researchers to mathematically design airfoils with prescribed pressure gradients in order to obtain long runs of laminar flow at design conditions. During the late 1930s and the 1940s, NACA designed, tested, and reported on laminar-flow airfoils [3]. The six-series foils are probably the most successful and most used. A short summary of the most common NACA airfoil series is presented.

NACA Four-Digit Airfoils (XXXX)

These wing sections basically constitute a synthesis of early Göttingen and Clark Y sections, empirically developed on a basis of prewar experimental data. Data on Göttingen and other older airfoils are found in Ref. [14]. Both the thickness distributions and the mean lines are defined in the form of polynomials; the sections have a near-elliptical nose shape. The maximum camber is at approximately mid-chord position. Table 8.4 explains the notation for the NACA four-digit series airfoils with an example.

Although the sections in the four-digit series are by no means low-drag airfoils, the drag increase with lift is fairly gradual. The cambered sections have relatively high

Table 8.5 NACA five-digit airfoil interpretation.

Example: NACA 23015

Digit number	Meaning	Example
1	Design lift coefficient	$c_{L_D} = 3/20 \cdot (2) = 0.3$
2 and 3	Position of maximum camber	$1/2 \cdot 30\% = 15\%$
4 and 5	Thickness ratio	15%

section lift and the stalling is fairly docile. These properties have marked the 2412 and 4412 sections, for example, as being suitable tip sections for wings of light aircraft and for tailplanes. In view of the gradual changes in drag and pitching moment with lift, the four-digit sections are frequently used for light trainers, which often fly in different conditions.

However, as part of the greatly increased research activity prior to and during World War II, German aerodynamicists investigated these NACA four-digit series airfoils more systematically for high speed than was done at NACA Langley, which had moved on to the five-digit series.

NACA Five-Digit Airfoils (XXXXX)

These wing sections have the same thickness distributions as the four-digit series, but the camber curves are different, having their maximum ordinate further forward. Table 8.5 explains the notation for the NACA five-digit series airfoils with an example. The well-known 230XX airfoils have their maximum camber at the 15% chord point. These sections have the highest maximum lift of the standard NACA sections, but stalling behavior is not particularly favorable and rather sensitive to scale effects. For wings where high lift performance is a prerequisite, the 230-series sections have been frequently used, sometimes combined with a four-digit section at the tip.

NACA Six-Series Airfoils (XX_X-XXX)

This series is the outcome of a succession of attempts to design airfoils by (approximate) theoretical methods (Theodorsen's among others), aimed at achieving low-profile (form) drag in a limited range of lift coefficients: the "sag" or the "bucket" in the low-drag range. The form of the laminar bucket is a result of drag reductions due to laminar flow simultaneously on both the upper and the lower surface. Table 8.6 explains the notation for the NACA six-series airfoils with an example.

In a modified series (e.g. NACA64_2A415), the A before the fourth digit means that the airfoil contour is essentially straight aft of 80% chord. A great advantage of the six-series is that the section properties have been tested extensively and reported systematically. The designer is thus provided with a tool for establishing the best sectional shape by systematically varying the shape parameters. The NACA 63- and 64-series airfoils had tendencies to stall abruptly at the leading edge.

Table 8.6 NACA six-digit series designation interpretation.

Example: NACA 64_2-415

Digit number	Meaning	Example
1	Series designation	
2	Rooftop pressure-plateau length	40% from leading edge
Subscript	Half of laminar bucket width	$\Delta c_L = 0.2$
3	Design lift coefficient	$c_{L_D} = 1/10 \cdot 4 = 0.4$
4 and 5	Thickness ratio	15%

Laminar-Flow Airfoils

The laminar boundary layer over the forward part of the section is stabilized by avoiding pressure peaks, keeping the local velocities low and applying a favorable pressure gradient over the forward part of the upper surface. The extent of the laminar layer is limited by the separation of the turbulent boundary layer over the rear part. The low supervelocities on these airfoils also favor the attainment of a high critical Mach number.

Due to the relatively sharp nose of the thinner so-called laminar-flow sections, their maximum lift is notably below that of the four- and five-digit series, although the difference for the thicker cambered sections is negligible; these sections also exhibit a docile stall. The profile (form) drag, although very low under ideal conditions, is sensitive to surface roughness, excrescences, and contaminations. Special structures are therefore needed to maintain laminar flow, and on practical wing constructions of transport aircraft, the potentially large extent of the laminar boundary layer will not normally be realized.

On NACA six-series airfoils, decreasing the thickness ratio and moving the maximum thickness point further aft produce a velocity distribution with a narrow and deep laminar bucket.

8.2.2 Modern NASA Airfoils: New Technology

When aircraft began to enter the high-subsonic-speed region where local sonic velocities become a problem, the six-series came into its own. It was thought at the time that it was necessary to avoid sonic velocity anywhere on the airfoil. The six series tried to maintain a relatively flat top pressure distribution, and this was exactly the type of pressure distribution needed to get the greatest possible lift with a given maximum negative pressure coefficient anywhere on the airfoil. Therefore, many early jet aircraft used six-series airfoils.

In the early 1950s, it became apparent that the six series was not the best for high-speed airfoils and that it is perfectly permissible and even desirable to have considerable supersonic velocity on the forward part of an airfoil. Therefore, the design criterion changed from avoiding supersonic velocities everywhere to avoiding supersonic velocities at or behind the crest. It was also necessary to have the peak negative pressure coefficient very far forward so that the slope of the pressure coefficient versus chordwise position was gentle for some distance ahead of the crest. This led to airfoils that were really closer to the NACA four-digit airfoils with nose peaks, but they differed from them in having rather sharp nose peaks. These *peaky* airfoils are found on a number of transport aircraft of that period.

Many general aviation aircraft were designed with six-series airfoils starting in the late 1940s, either to take advantage of the alleged laminar flow or just to look modern. The laminar flow was never attained, and the six-series airfoils had poor maximum lift. Even if laminar boundary layers had been attainable in general, the propeller slipstream would probably have caused transition on the wing immersed in the slipstream anyway.

For almost 15 years, no systematic airfoil research went on in the USA. The resurgence of interest came largely through the persistent, dedicated efforts of Richard T. Whitcomb and his research in the Langley 8-foot transonic pressure tunnel, which resulted in the now-famous NASA supercritical airfoil breakthrough in 1965.

Excellent research elsewhere by people such as Pearcy working on the "peaky" airfoils in the UK and F. X. Wortmann and his high-performance sailplane airfoils in Germany also served to call attention to the need for a new thrust in airfoil design.

In 1973, NASA responded with its Airfoil Research Program, with the technical objectives of developing computational methods for predicting the complete flow about an arbitrary airfoil and ultimately optimizing its shape for operation over a complete flight regime in either low-speed or supercritical viscous flows. Basic experiments and selected verification tests played a vital part in this effort. Today, practically all airfoils are tailor-made for their particular design requirement. The old NACA four-digit, five-digit, and six-series airfoils are seldom used except by general aviation aircraft manufacturers. The NASA-MS-0313 airfoil was used on the Saab SF340 commuter.

NASA Laminar-Flow Airfoils

In its Natural Laminar Flow (NLF) Program, NASA designed the NLF airfoils for general aviation aircraft. The results showed that the new NLF airfoil, even with transition fixed near the leading edge, achieved the same maximum lift as the NASA low-speed airfoils. At the same time, the NLF airfoil, with transition fixed, exhibited no higher cruise drag than comparable turbulent-flow airfoils. Thus, if the new NLF airfoil was used on an aircraft where laminar flow was not achieved, nothing was lost relative to the performance of the NASA low-speed airfoils. If laminar flow was achieved, however, a very substantial drag reduction would result (e.g. see Figure 8.5). NACA six-series airfoils are members of the laminar-flow airfoil family. Deviations from the design point will cause the boundary layer on either the upper or lower contour to go turbulent. These airfoils thus show a "laminar bucket" in their drag polars, as indicated in Figure 8.6.

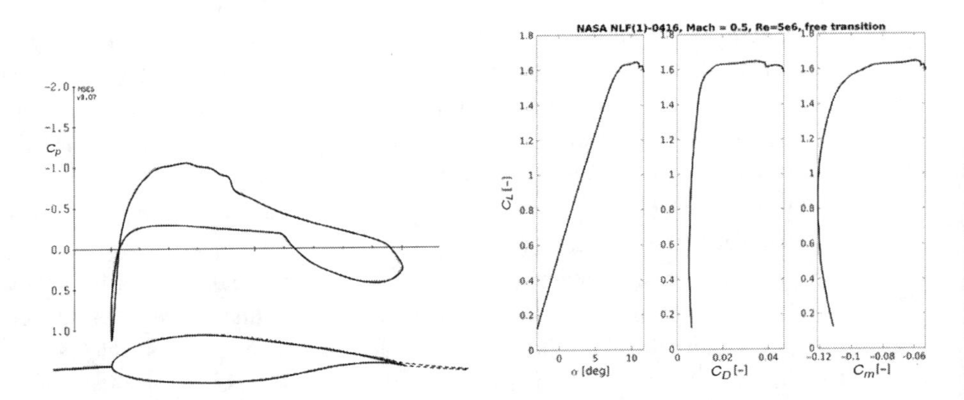

Figure 8.5 Laminar-flow airfoil at $M0.5$, $Re = 5M, \alpha = -0.8°, c_L = 0.5$, MSES code.

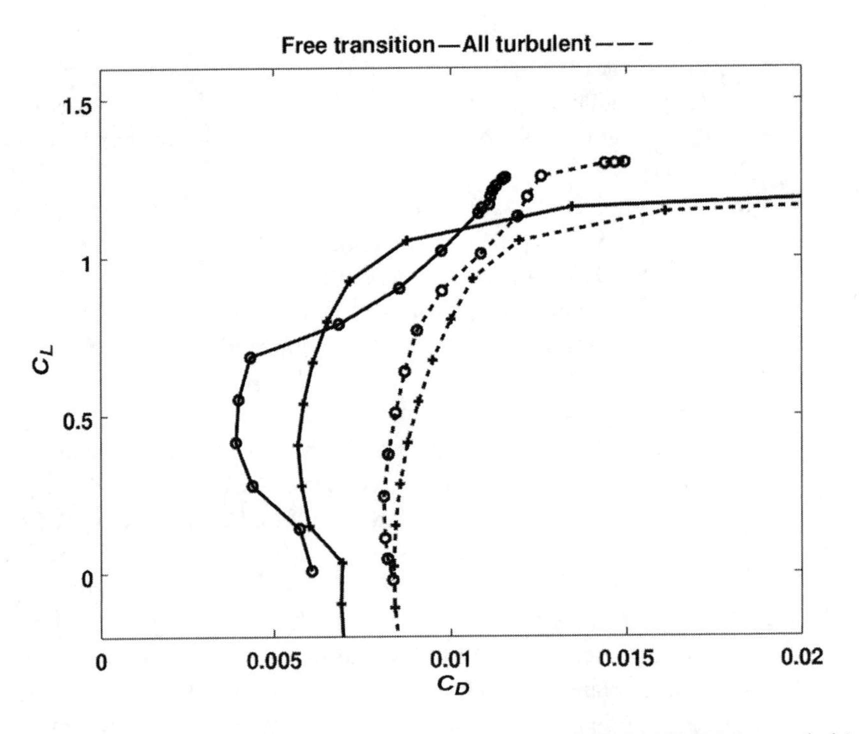

Figure 8.6 Drag polars. NACA65$_1$-412 (circle symbols) and NACA23012 (plus symbols); MSES computation with fully turbulent (dashed line, and free transition (solid line).

NASA Modern Low- and Medium-Speed Airfoils

Earlier airfoils were often created using the old NACA system of parametric geometry definitions, which, of course, have their aerodynamic consequences, rather than using the aerodynamics as the primary variable. A modern example is the NASA General Aviation (Whitcomb) number 1 airfoil, GA(W)-1, alternatively designated NASA-LS(1)-0417 and modified for higher *Re* as the NASA-MS(1)-0313

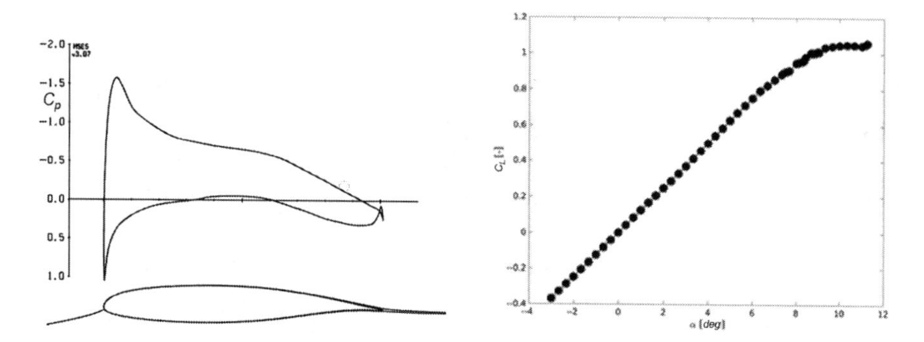

Figure 8.7 NASA-MS(1)-0313. (Left) Geometry and design pressure distribution; (right) drag polar.

(see Figure 8.7). The airfoil was not designed to obtain a high value of maximum lift, but to be a high-quality general airfoil. The design lift coefficient is 0.4 and the thickness ratio is 17%. The airfoil has a large nose radius to make it tolerant to changes in incidence, a thin trailing edge to avoid a high trailing edge pressure, an upper contour pressure distribution (see the discussion above), and some camber at the aft end of the airfoil to generate more lift. The payoff of this design strategy over a traditional airfoil is found in the results.

Significant improvements have been obtained both in $c_{L,max}$ and in L/D for higher lift coefficients. The latter is important, for example, in order for underpowered airplanes to be able to climb.

8.3 Transonic Airfoils $0.7 < M_\infty \leq 0.9$

There are regions in the airfoil vicinity where the local airspeed is higher than the flight speed, and at some high flight speed, though lower than the speed of sound (say $M_\infty = 0.8$), there may be pockets of supersonic flow. Such flight cases are said to be transonic. As the retardation from supersonic to subsonic local flow in general is connected with a shock wave, the airfoil experiences an increase in drag. Often, the flight Mach number for beginning shock wave drag is defined as the critical Mach number, M_{crit}, the speed at which supersonic flow appears, or the drag divergence Mach number, M_{dd}, the (smallest) Mach number for which $dc_D/dM_\infty = 0.1$. An alternative defines M_{dd} by a wave drag increase to 20 drag counts. Both experimental and theoretical determination of the aerodynamic characters of airfoils at supercritical speeds are difficult. However, as it is economical for the big airliners to fly just above drag divergence, the design of airfoils for this speed regime is an enormously important subject.

Transonic airfoil design seeks to limit shock drag losses at a given transonic speed. This effectively limits the minimum pressure coefficient that can be tolerated.

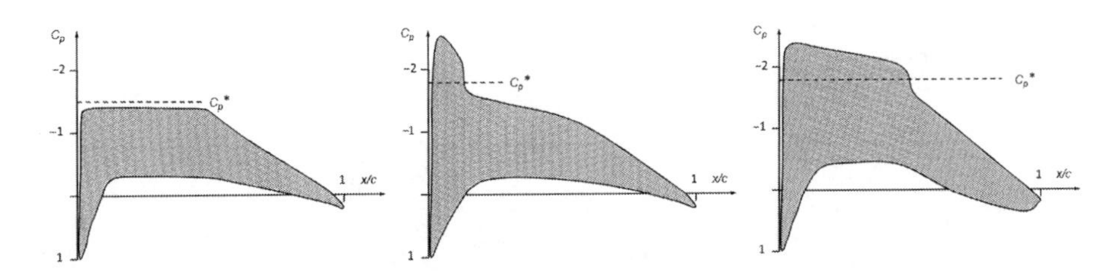

Figure 8.8 Pressure-distribution design goals: "sonic rooftop," "peaky," and "supercritical" airfoils.

Since both lift and thickness reduce (increase in magnitude) the minimum C_p, the transonic design problem is to create an airfoil section with high lift and/or thickness without strong shock waves. Some supersonic flow is tolerated with only a small drag increase, so that most sections of the wing can operate efficiently in supercritical flow. A rule of thumb is that the maximum local Mach numbers should not exceed about 1.2–1.3 on a well-designed supercritical airfoil. This produces a considerable increase in available c_L compared with entirely subcritical designs. The attempts at creating efficient airfoils for transonic flow can be classified as "rooftop," "peaky," and "supercritical" types (Figure 8.8).

It should be clear that airfoil maximum t/c is a major influencer on transonic airfoils and *the* most important shape parameter for supersonic wave drag. Wing thickness, however, can be reduced only so much before structural weight and wing volume issues become critical. The foils discussed here are around 10% thick.

To avoid very low pressures and thus high velocities above the airfoil, designers instead spread out the upper-contour low-pressure region, generating *rooftop airfoils*. However, when the design Mach number is exceeded, the supersonic low-pressure region will act on the rearward-sloping part of the airfoil and the drag rises rapidly.

Increasing the nose radius increases the velocity over the forward profile region. The airfoil nose was designed on a peaky airfoil so that near-isentropic compression and a weak shock is obtained: the suction forces have a large forward component and the drag rise is postponed to high speeds. If the design Mach number is exceeded for this airfoil, the supersonic speed region develops *forward* of the maximum thickness and the drag does not rise rapidly (Figure 8.8, middle). Compared to conventional sections of the same thickness ratio, the value M_{crit} for a peaky airfoil is approximately 0.03–0.05 higher and the off-design is improved. This type of airfoil has been used on the BAC 1-11, VC-10, and DC-9 aircraft. The development of these airfoils was highly empirical.

The *supercritical airfoils* developed by NASA's R. Whitcomb were also designed largely by experimental means. The emphasis was on obtaining a long region of not too high, slowly retarding, supersonic flow, so the final shock would be weak (Figure 8.8, right). These wing sections delayed drag divergence more than peaky or rooftop shapes. The cambered aft portion of the pressure side creates lift without excessive suction on the upper surface.

8.3.1 Early Attempts to Design Transonic Airfoils

A retrospective look at the historical development of transonic wings is enlightening.

Early German Work: The Me 262

The Messerschmitt 262 Schwalbe (Swallow) was the first operational jet fighter in history. There is no doubt this was an aircraft before its time, as the aircraft itself was a vision of the future for fighters. However, the engines were not reliable. The requirements for high-temperature alloys and other materials in short supply in the war kept the aircraft on the ground instead of in the air for most of World War II.

The Germans foresaw the great potential for the jet engine constructed by Hans Joachim Pabst von Ohain in 1936. After the successful test flights in August 1939 of the world's first jet aircraft, the Heinkel 178, they adopted the jet engine for an advanced fighter aircraft. As a result, the Me 262 was already under development as Projekt 1065 (P.1065) before the start of World War II.

The project aerodynamicist was Ludwig Bölkow. He initially designed the wing using NACA four-digit airfoils modified with an elliptical nose section. Later in the design process, these were changed to the Deutsche Versuchsanstalt der Luftfahrt (DVL) derivatives of NACA airfoils, with the NACA00011-0.825-35 being used at the root and the NACA00009-1.1-40 being used at the tip.

Adolf Busemann, who together with Albert Betz pioneered and obtained a German patent for the swept wing for high-speed flight, suggested a wing swept at 30°, but severe pitch-up stability problems were encountered at high speed [6, 12] and the actual sweep built was 14° – more to ensure stability by accommodating the position of the center of gravity than to delay transonic drag rise.

Dive tests determined that the Me 262 went out of control in a dive at Mach 0.86, and that higher Mach numbers would cause a nose-down trim that the pilot could not counter. The resulting steepening of the dive would lead to even higher speeds, threatening structural disintegration of the airframe.

Messerschmitt made no attempt to exceed the Mach 0.86 limit. After the war, the Royal Aircraft Establishment (RAE), at that time one of the leading institutions in high-speed research, retested the Me 262 to help with British attempts at exceeding Mach 1. The RAE achieved speeds of up to Mach 0.84 and confirmed the results from the Messerschmitt dive tests. During 1943, RAE made extensive flight tests that exceeded Mach 0.86 with the Spitfire XI, and in the 1950s, RAE developed the RAE100–104 series of symmetric high-speed airfoils, which we studied in Chapter 6.

Early Transonic Airfoils

Table 8.7 lists the key characteristics of the foil shapes for early swept-wing jet fighters, as well as those of their contemporary Lockheed P-80, with its unswept wing, and the modern (2012) CRM wing kink section. As the F-80, the P-80 became the first operational US jet fighter. The first four used symmetric airfoils, similar to the DVL-modified NACA four-digit foil as used on the Me 262 (see below). The table shows quarter-chord sweep, thickness of section measured normal to leading edge

Table 8.7 Characteristics of early jet airfoils and a modern supercritical airfoil.

Airplane	Sweep (°)	t/c (%)	Position of max t/c (%)	Nose radius (%)
Me 262	15.3	9.0	40.0	0.89
	Streamwise	8.7	41.6	0.83
NA F-86	37.6	8.5	40.0	0.79
	Streamwise	7.1	43.1	0.55
SAAB J-29	24.9	10.0	45.0	0.55
	Streamwise	9.2	47.8	0.46
LM P-80	0.0	13.0	40.9	1.12
CRM	Streamwise	10.5	40.2	0.69

Table 8.8 Interpretation of NACA DVL notation.

2	100 × maximum camber
2	10 × x-station of maximum camber
09	100 × maximum thickness
-1.1-	Nose radius/$(t/c)^2$
40	100 ×x-station of maximum thickness

and streamwise. Table 8.8 gives the interpretation of the five-digit groups in the DVL NACA designation 2 2 09 -1.1- 40. All lengths are in units of airfoil chord length c. The nose radius parameters are multiples of 0.275, seemingly with only 0.55, 0.825, and 1.1 being used: a NACA four-digit foil has parameter 1.1. The NACA DVL camber curve is the same as that of the NACA four-digit series, but the thickness distribution is different.

The P-80 and CRM foils are cambered, the P-80 being a NACA six-series airfoil and the CRM being a custom supercritical design arrived at by shape optimization. Corresponding to Table 8.7, the Mach-sweep plot in Figure 8.9 shows the drag rise at a lift coefficient 0.3, $Re_c = 37M$, and fixed transition at 2% chord, as computed by MSES. The plane flow idealization was made on a wing section parallel to the free stream, taking both sweep and taper into account (see Table 8.7). The CRM foil has been defined in the streamwise direction, so the sweep is immaterial here.

The effective thickness-to-chord ratio is *the* decisive parameter for wave drag, so the F-86 has the largest drag divergence Mach number (see Figure 8.9), followed by the Me 262. The P-80 NACA65-213 is by far the thickest, and its early drag rise stands out. But its large nose radius makes for a docile stall, and the design served for decades on the T33 jet trainer in a number of versions. Note also the large reduction in nose radius that accompanies the slimming by the wing sweep; it scales as $(t/c)^2$. The 10.6%-thick modern CRM foil does as well as the 9.2%-thick J-29 foil designed 65 years earlier, at a time when there were no computational tools for transonic aerodynamics.

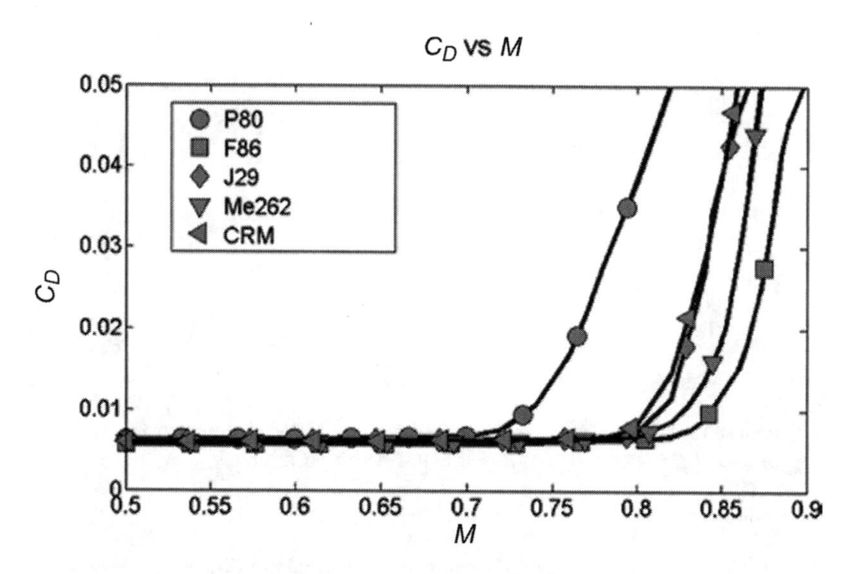

Figure 8.9 Drag rise for five airfoils computed by MSES at $c_L = 0.3$, $Re_c = 37M$, and transition fixed at 2% chord.

8.3.2 Rooftop Airfoils

These airfoils have a gradually changing or approximately constant upper surface pressure over the forward part of the section, thus delaying the critical Mach number, M_{crit}, by virtue of a uniform velocity at the design condition. Since the pressure distribution is selected primarily with the aim of obtaining low supervelocities rather than special high-speed characteristics, a strong shock and a rapid drag rise occur soon after M_{crit} is exceeded. The associated suction forces then occur near the crest (the highest point relative to the free-flow direction) or behind it.

The NACA six-series airfoils have this type of pressure distribution at subcritical speed for a limited range of c_L values.

The drag-divergence Mach number of an airfoil with a rooftop pressure distribution can be increased by extending the rooftop further back. This is the development path for the RAE100, 101, 102, and 104 foils studied in Chapter 7. For example, a crest of about 60% chord results in a ΔM_{crit} value of about 0.04 relative to a section with a crest at 30%. This effect cannot be exploited too far, since the boundary layer may not be able to cope with the adverse pressure gradient aft of the rooftop without separating. The wing of the Aerospatiale Caravelle, originating from about 1950, is based on the 65-series airfoils; the operating speeds of this aircraft are relatively low and allow a reasonable thickness ratio of 12%.

The "sonic rooftop" refers to the sonic flow over the first part of the airfoil. A sonic rooftop airfoil is designed for transonic speeds, but such that the flow at the design condition just reaches sonic speeds over the flat region of the upper surface.

The design of the thickest possible profile for one given design point is governed essentially by the following conditions.

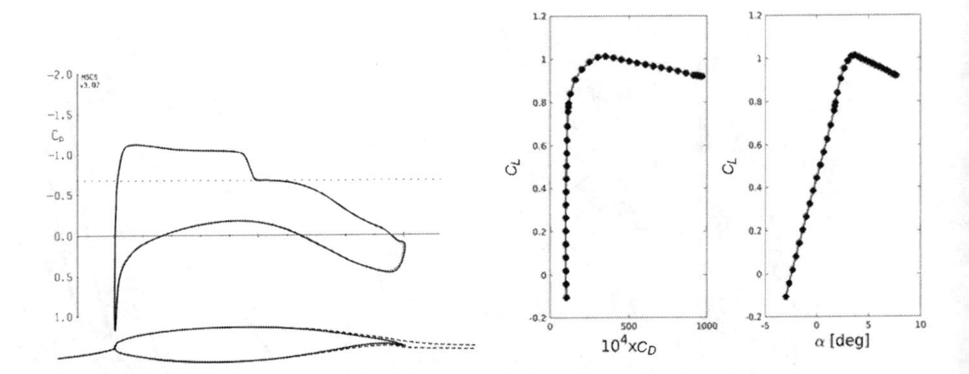

Figure 8.10 C_p distribution and drag polar for the CRM airfoil at 65% span, MSES computation; $M_\infty = 0.725$, $C_L = 0.784$, $Re = 3.19M$, free transition.

(1) The suction pressure must not reach critical condition.
(2) The pressure distribution must just deliver the given c_L value.
(3) The crest location may be pushed downstream only so far that the pressure rise in the rear profile area does not cause any separation.

The consideration of constraints outside the design point (e.g. $c_{L,max}$ may create additional constraints). The first Airbus model, A300 (first flight October 1972), used a sonic rooftop airfoil with a plateau-like C_p distribution limited to just under critical C_p* values.

Rooftop Example: The RAE104 Airfoil

Designed by H. B. Squire in 1945 for high speed, the RAE104 airfoil has a maximum thickness-to-chord ratio of 10%, occurring at 42% chord. The rear section of the airfoil beyond about 80% of the chord is wedge shaped. At zero lift, the subcritical pressure distribution is almost rooftop in shape and constant back to 60% chord and then falls uniformly to the trailing edge.

Figure 6.10 showed lift on the RAE104 10%-thick airfoil varying in a dramatic way with change in Mach number over the transonic range, and Figures 6.11–6.15 showed associated pressure distributions, shock positions, friction coefficients, and (sketched) boundary-layer development.

8.3.3 Modern Supercritical Airfoils

The most successful and lasting high-speed airfoil development is called the *supercritical airfoil*. For any given thickness ratio, this airfoil provides a higher M_{dd} than conventional airfoils, about 0.06 above the best peaky airfoils and 0.09 above the NACA six-series. With these sections, a much greater extent of shock-free supersonic flow can be created than in the peaky design. Figure 8.10 (left) demonstrates the C_p plateau at $M0.725, Re = 3.19 \cdot 10^6, \alpha = 1.77°$, which gives $c_L = 0.784$,

$c_D = 115 \times 10^{-4}$. The drag polar is shown to the right with $c_{L,max} \approx 1.1$. The drag rise for the CRM airfoil is shown in Figure 8.9.

The supercritical airfoil was pioneered by Richard Whitcomb at NASA's Langley Research Center. The "supercritical" title is a misnomer since present peaky airfoils also cruise with large regions of supercritical ($M > 1.0$) flow with little drag penalty. The supercritical airfoils exploit this effect to a greater extent by having very small curvature over much of the upper surface so that the aft-facing surface has very little vertical projected area for a considerable distance behind the crest.

Furthermore, cambering the aft portion of the airfoil to carry more load aft can lower the velocity at the crest for any given C_L, thereby raising M_{dd}. Another feature is the tangency of the upper and lower surfaces at the trailing edge. This reduces the high adverse pressure gradient at the rear and permits the aft cambering without excessive pressure drag.

As in so many technical advances, there are some negative aspects to the supercritical wing. The very thin trailing edge is a structural problem, although honeycomb construction can provide the required strength and stiffness without excess weight penalty. The aft camber leads to a large negative pitching moment, which must be balanced by the tail and causes trim drag.

At the wing root, the large adverse pressure gradient affects not only the airfoil boundary layer, for which the airfoil has been designed, but also the fuselage boundary layer. Thus, wing root flow separation is a problem at higher angles of attack. On swept wings, this is less serious because the fuselage–wing interference due to sweep requires less camber on the root airfoils anyway.

In spite of these problems, the supercritical wing can either permit greater speed with a given wing sweep and thickness or allow less sweep and/or greater thickness for a given M_{dd}. Less sweep and greater thickness reduce wing weight, with benefits of smaller wing area, reduced fuel consumption, and lower operating costs.

The inherently large nose radius gives docile stall characteristics. The penalty due to a large negative pitching moment has not been significant at high Mach numbers. At lower Mach numbers, the advantages of supercritical flow cannot be realized and trim drag is a drawback.

Supercritical Airfoil Technology: Trade-Off of M_{dd} vs Thickness

Mason [13] discusses a rather simple empirical relation for the performance of supercritical airfoils.

$$M_{dd} + 0.1c_L + t/c = \kappa_A$$

This formula, known as the Korn equation, contains M_{dd}, the drag divergence Mach number, c_L, the lift coefficient, t/c, the relative thickness, and κ_A, an airfoil technology number. The number equals 0.87 for a NACA six-series airfoil and 0.95 for a supercritical airfoil. Naively, a verbal interpretation could be that there is, for a given airfoil, a balance between flight Mach number, lift coefficient, and relative thickness. The relation makes it simple to make trades between the quantities. Figure 8.11 is a graphical presentation of this.

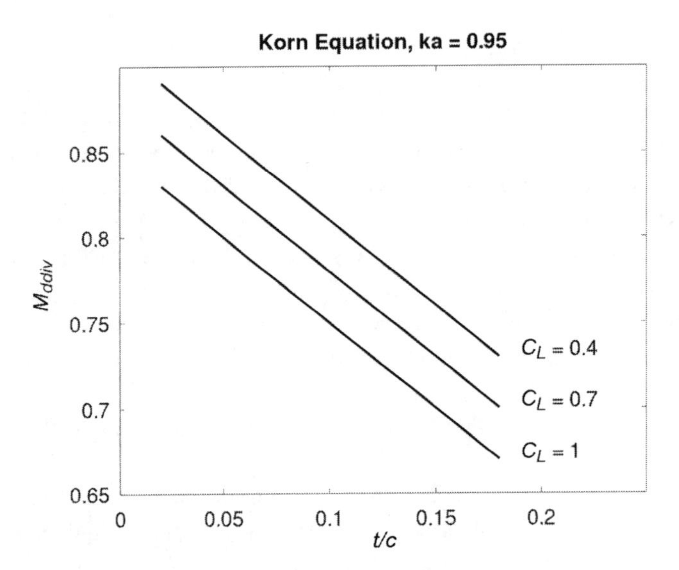

Figure 8.11 Korn equation for transonic airfoil performance, $\kappa = 0.95$ for supercritical airfoils.

8.4 Supersonic-Speed Airfoils $0.9 < M_\infty \leq 2$

Airfoils for supersonic flight are very thin, since C_D grows substantially with t/c. A highly swept supersonic wing with subsonic leading edges is a very complex cambered and twisted surface responding to sophisticated mathematical analyses. An example – the $M2$ Concorde wing – is shown in Figure 9.19. It is about 3% thick, with drooped nose, and spanwise as well as chordwise camber. Its S-curved leading edge shape is also known as an *ogee* or *sigmoid* curve.

In the higher transonic and supersonic range design is all about drag, and the way to deal with this is to make the airfoil thin and the configuration slender. Figure 1.14 showed the planform evolution with Mach number. Figure 8.12 indicates how wave drag increases dramatically with airfoil thickness for the NACA64A2xx airfoils. The calculation is for inviscid flow and the shock–boundary layer interaction may change the values somewhat, but the trend is correctly captured.

The F-104 Starfighter M 2 interceptor

The development of supersonic aircraft was pushed forward and enabled by continuous rapid improvements of jet engine thrust. Two strains emerged. One uses highly swept or delta wings, such as the Concorde and the Saab J-35 Draken double-delta configuration. The other branch relies on a very thin and low-*AR* wing with the Lockheed early 1950s Starfighter design, to become the F-104, as the main exponent. Its overriding design goal was speed and climb: it cruised easily at $M2$ and climbed faster than contemporaries to higher altitude. It held numerous speed and altitude records.

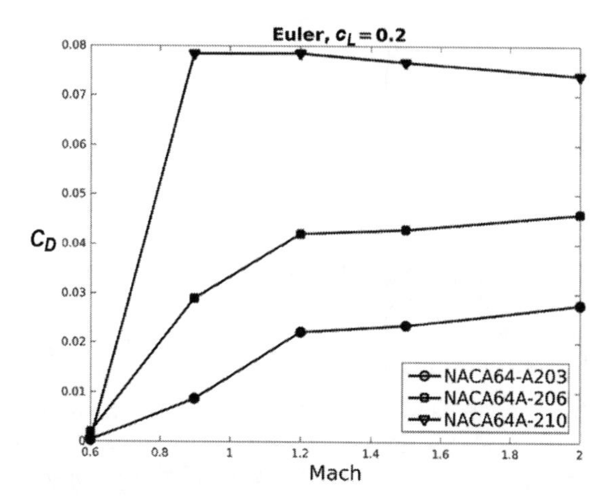

Figure 8.12 Mach sweep at $c_L = 0.2$ showing drag increase with thickness for NACA64A2xx high-speed airfoils, EDGE Euler computations.

Slated initially as air superiority interceptor of strategic high-speed bombers, it was later recast as an attack bomber and became an export success.

The wing is an *AR* 2.45, 26° sweep leading edge planform with a 3.3%-thick biconvex airfoil. The leading edge is 0.4 mm (!) thick to minimize the bow shock drag. The supersonic performance exceeded expectations, but low-speed handling and turn capabilities predictably suffered. Slats and blown flaps were retrofitted to decrease landing speed, but the aircraft was not easy to fly. Some 24% of all Starfighters built were lost, mostly due to engine problems and "pilot errors" in low-altitude flight.

8.5 Multielement Airfoils for High Lift

To fly an airplane in different conditions, the pilot must vary the wing geometry. A practical option is the flap. Figure 8.13 is a sketch of an arrangement with a main element as well as a leading-edge slat and trailing-edge flap. The slots between the elements are designed to assist the upper-contour boundary layer, as indicated in Figure 8.13. The trailing-edge flap system directs the air downward and increases the lift, while the leading-edge flap modifies the shape so that the incoming flow can manage the large turning over the leading-edge region. Some typical results appear in the sketched curves in Figure 8.14.

The solid curves in Figure 8.14 apply to a case without leading-edge devices, while the dashed curves show the same configuration with added leading-edge slats. For small α, the slats reduce the lift a little bit, but their main effect is to assist the flow over the leading edge and so increase the range of α, so higher lift coefficients become available.

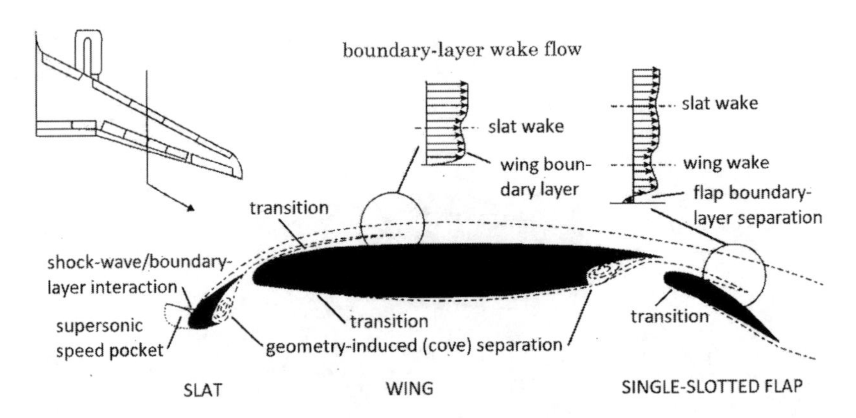

Figure 8.13 Examples of flow phenomena occurring on a three-element airfoil. (Courtesy of Axel Flaig [7] and Ernst Hirschel [10], reprinted with permission)

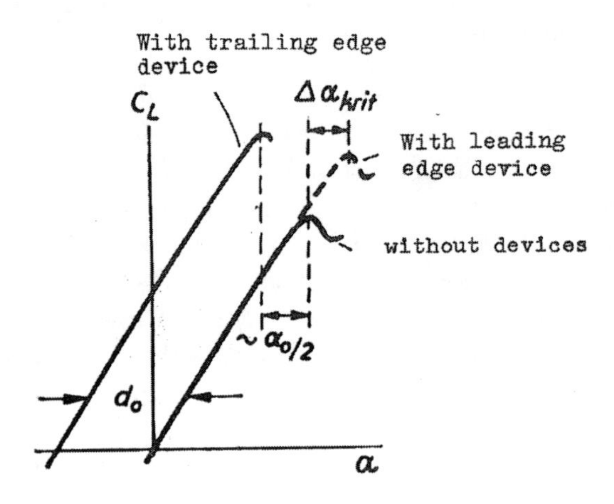

Figure 8.14 Typical effects of leading- and trailing-edge flaps.

The accurate calculation of flows about multielement high-lift airfoil configurations remains a challenge. Maximum lift, drag, and pitching moment are difficult to calculate consistently with accuracy, especially in the presence of local flow separations on one or more of the airfoil elements. The mixing of wakes and boundary layers from the airfoil elements further complicates the problem.

8.5.1 L1T2 Three-Element Airfoil Test Case

The L1T2 test case has a number of important features.

1. It is a practical configuration of interest to the aircraft industry.
2. It possesses complex flow features, including flow separations and wake–boundary layer interaction.
3. Reliable experimental data are available.

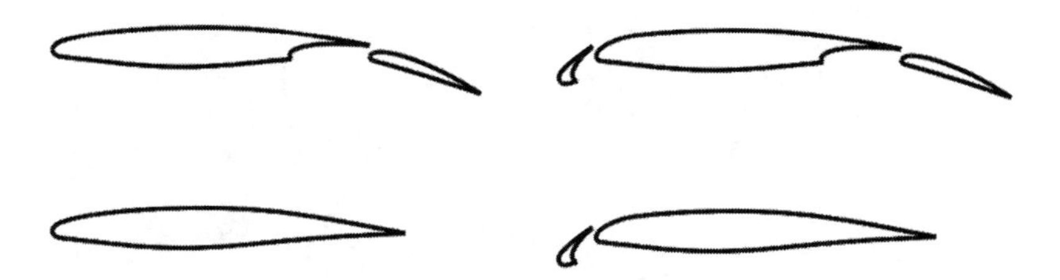

Figure 8.15 L1T2 geometry of all four deployments.

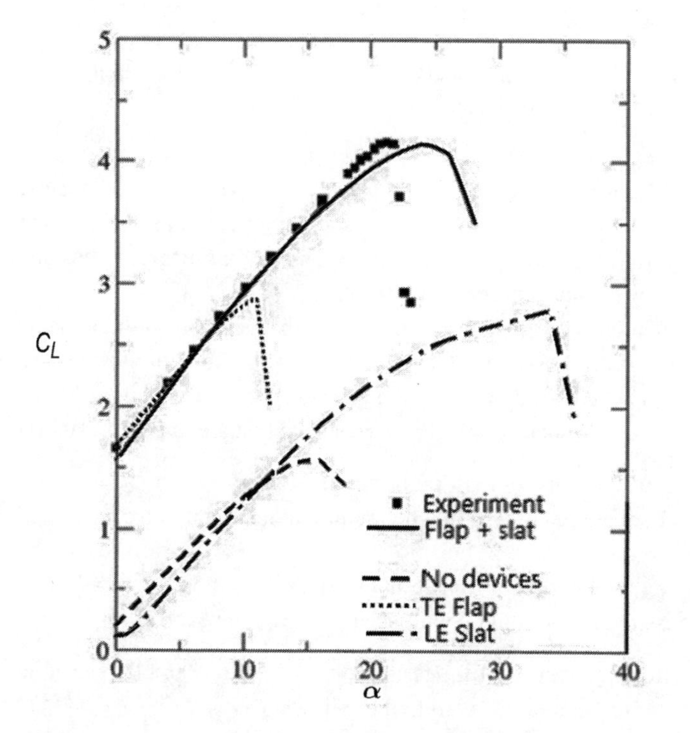

Figure 8.16 Effect of extending the slat and/or flap, c_L vs α. LE = leading edge; TE = trailing edge. (Courtesy of P. Eliasson, personal communication)

The experimental data are particularly valuable for computational fluid dynamics (CFD) code validation because of the accuracy of the measurements, which is due mainly to the high degree of two-dimensionality of the wind tunnel flow field.

The configuration is comprised of a slat, main foil, and single-slotted flap. The L1 slat is positioned at an angle of 25°, and the T2 flap is a single-slotted flap with a deflection angle of 20°, typical of a takeoff configuration.

Figure 8.15 shows the four deployment cases. The test conditions are $M0.197$, $Re_{chord} = 3.52 \times 10^6$. Transition to turbulence is specified as follows: slat and flap; free transition on upper and lower surface, main element; and fixed transition at $x/c = 0.125$ on upper and lower surfaces. Figure 8.16 shows the computed and measured effects of extending the slat and/or flap as c_L vs α and drag polars for the

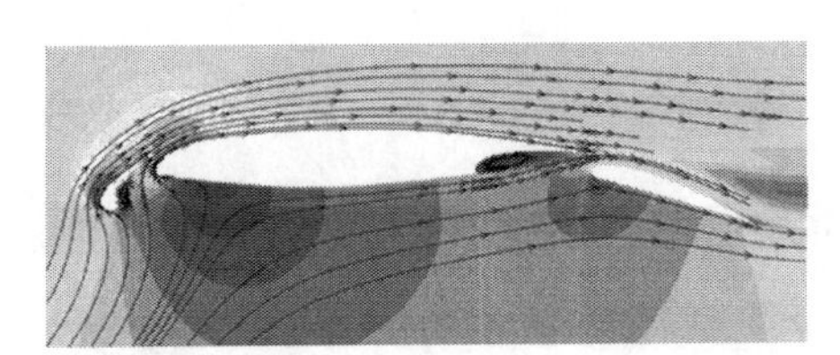

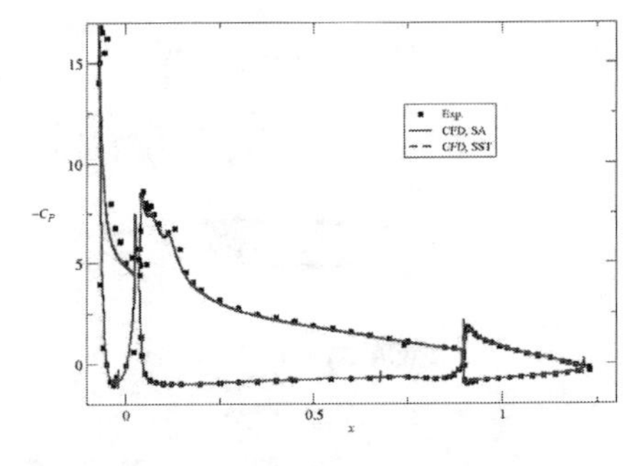

Figure 8.17 Computed streamlines and C_p over L1T2 three-element airfoil. (Courtesy of J. Vos and P. Eliasson, personal communication)

four deployment cases. These results can be compared to the "generic" sketch in Figure 8.14 – experiment with flap and slat deployed – and the CFD results for all four configurations. The effects of trailing- and leading-edge devices are substantial, moreover, they are approximately additive.

This validation exercise appears to indicate that accurate multielement airfoil high-lift calculations are possible, at least for takeoff configurations where flow over the flap is fully attached. Prediction of drag is less accurate. The ability to consistently and accurately calculate such flow fields is predicated on a number of important issues.

1. Suitable grid density and gridpoint distribution
2. Far-field boundary correction when required due to the proximity of the grid outer boundaries
3. Accurate specification of transition locations/conditions
4. Compressible RANS model with accurate turbulence model

It appears from the results of the test case that, in order to obtain accurate drag solutions, more effort should be given to the following.

(1) Ensuring that the results are grid independent
(2) Attempting to accurately predict transition locations

The flow incorporates separation and wake–boundary layer interaction and is presumably sensitive to the position of laminar–turbulent transition.

Figure 8.17 (right) shows a comparison of measured C_p with a RANS calculation using the Spalart–Allmaras turbulence model. The left-hand image shows the streamlines for this configuration, and the recirculation zones on the slat and below the trailing edge of the wing can be clearly seen.

8.6 Optimization Example: ADODG Test Case RAE2822

The American Institute of Aeronautics and Astronautics (AIAA) Aerodynamic Design Optimization Discussion Group (ADODG) [1] works to establish well-researched cases for airfoil and wing-shape optimization. The idea is to encourage researchers to test their algorithms on cases with readily available geometry definitions, computed results, and wind tunnel data. The latter serve to highlight the quality of the CFD models used in the optimization. We report here on an exercise run with EDGE and a number of optimization algorithms provided by the MODEFRONTIER [2] package for the analysis and optimization of parametric designs [8]. Not surprisingly, the work shows that different algorithms may produce different results, some obviously pointing to a common optimum and others probably on their way to another local optimum. This is to be expected for any comparison of complex algorithms, and further discussion of this point is not within the scope of this text. We focus instead on the resulting shape modification.

8.6.1 RAE2822 Single-Point Drag Minimization with Edge

The task is to modify the baseline RAE2822 airfoil to minimize a figure of merit (FoM) chosen to be drag at $M_\infty = 0.734$, chord $Re = 6.8M$ with very little allowed change in lift or moment.

$$\begin{aligned} &\text{Minimum } c_D \text{ subject to} \\ &\left|\frac{c_L}{c_{L,base}} - 1\right| \leq 0.02, \\ &\left|\frac{c_m}{c_{m,base}} - 1\right| \leq 0.02, \\ &\alpha_l \leq \alpha \leq \alpha_u,\ \beta_l \leq \beta \leq \beta_u \end{aligned} \tag{8.1}$$

The parameters X are coefficients of Hicks–Henne bump functions (Section 5.2.4) $H_i(\ldots, x)$, $0 \leq x \leq 1$ for the upper (suction) side shape $\alpha_i, i = i, \ldots, a$ and the lower side $\beta_i, i = i, \ldots, b$. The geometry constraints are expressed as limits on the parameters, so it is expected that the optimization will make the foil thinner. The forces and moments are obtained at $\alpha = 2°$. An alternative formulation is to let the optimizer change the angle of attack to keep the lift at exactly the base value. The bump function centers are chosen more closely spaced toward the leading and trailing edges, as in Figure 5.5. The results given here are for $a = 5$, $b = 6$, so $dim(X) = 11$.

Optimized Shape

The final changes are mostly on the suction side. The pressure side shows only slight deviations from the initial shape. On the suction side, the thickness of the profile is decreased in the first two-thirds of the chord length. In the last third, the profile is slightly thickened. This annihilates the shock wave on the suction side (Figure 8.18).

The friction coefficient plot (Figure 8.19, left) shows no separation on either baseline or optimized shape, so there is no total pressure loss due to separation. The y^+ plot (Figure 8.19, right) shows that the grid cells in the boundary layer are small enough to resolve it. Thus, the turbulence model should be accurate and the friction forces well predicted. The friction drag increases by 1.2%, but the gain due to the annihilated

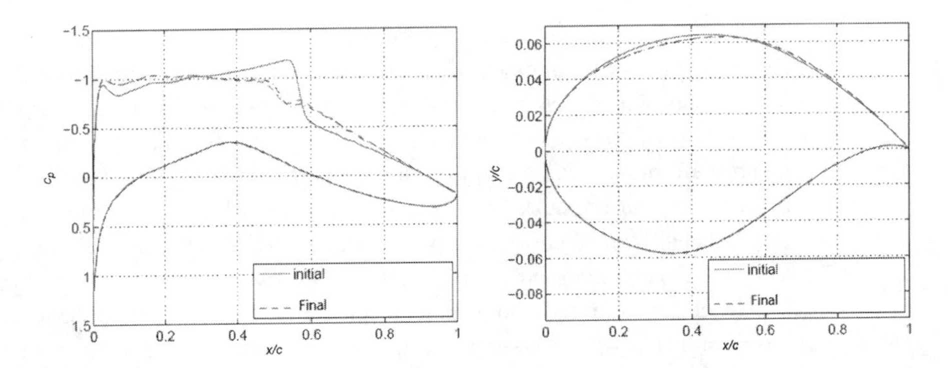

Figure 8.18 Optimization of RAE2822 airfoil, EDGE code RANS. Pressure coefficient distribution (left), surface shape (right). (Courtesy of D. Franke, MSc thesis [8], reprinted with permission)

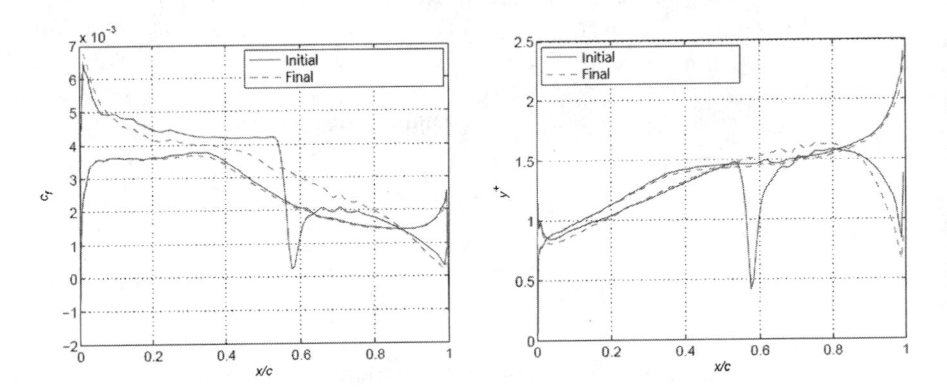

Figure 8.19 Optimization of RAE2S22 airfoil, EDGE code RANS: Friction coefficient (left); y^+ distribution (right). (Courtesy of D. Franke, MSc thesis [8], reprinted with permission)

shock outweighs it by far. In all, the exercise cuts the drag from 165 counts to 136. This is significant, but one must keep in mind that the improvement is for a single operating point at the suggested cruise Mach number. A more realistic exercise must consider several operating points, and then the choice of FoM is an issue.

The shape modification by such a small number of Hicks–Henne bumps can only create smooth profiles. This hinders the optimizer from exploiting very local shape changes, such as raising a bump at the shock foot: if the optimizer is given too much rope, it may hang itself.

The airfoil is a fundamental building block for the wing. With this knowledge of airfoil design in hand, we turn in the next chapter to the general consideration of wing design with a RANS solver.

8.7 Learn More by Computing

Gain hands-on experience of the computational tools for the topics in this chapter by working with the on-line resources. Exercises, tutorials, and project suggestions are found on the book website www.cambridge.org/rizzi. For example, scale the P-80 NACA65-213 airfoil to 10% thickness, compute its transonic drag rise and compare to that of the other airfoils in Figure 8.9. Software used to compute many of the examples shown is available from http://airinnova.se/education/aero-dynamic-design-of-aircraft

References

[1] AIAA ADODG. Available from http://mdolab.engin.umich.edu/content/aerodynamic-design-optimization-workshop.

[2] ModeFrontier ESTECO SpA home page. Available from www.esteco.com/modefrontier.

[3] I. H. Abbot and A. E. von Doenhoff. *Theory of Wing Sections, Including a Summary of Airfoil Data*. Dover Publications, Inc., 1958.

[4] D. Althaus. Stuttgarter profilkatalog i. Technical report. Institut für Aerodynamik und Gasdynamik der Universitet Stuttgart, 1972.

[5] D. Althaus. *Profilpolaren für den Modellflug, Windkanalmessungen an Profilen im Kritischen Reynoldszahlsbereich*. Neckar-Verlag VS-Villingen, 1980.

[6] M. Blair. Evolution of the F-86. Technical report. AIAA-Paper-80-3039, 1980.

[7] A. Flaig and R. Hilbig. High-lift design for large civil aircraft. In *High-Lift System Aerodynamics, AGARD Conference Proceedings*. AGARD, 1993, pp. 31.1–31.12.

[8] D. M. Franke. Interfacing and testing the coupling of the CFD solver EDGE with the optimization package `modeFRONTIER`. Diploma thesis, KTH School of Engineering Sciences, 2007.

[9] C. D. Harris. NASA supercritical airfoils. Technical report. TP 2969, NASA, 1990.

[10] E. H. Hirschel, A. Rizzi, and C. Breitsamter. *Separated and Vortical Flow in Aircraft Aerodynamcs*. Springer, 2021.

[11] R. H. Liebeck. A class of airfoils designed for high lift in incompressible flow. *Journal of Aircraft*, (10): 610–617, 1973.

[12] H. Ludwieg. *Pfeilflügel bei hohen geschwindigkeiten*. Lilienthal-Ges. Ber., 1940.

[13] W. H. Mason. AOE 4124 configuration aerodynamics. www.dept.aoe.vt.edu/mason/Mason_f/ConfigAero.html.

[14] F. W. Riegels. *Aerofoil Sections – Results from Wind-Tunnel Investigations and Theoretical Foundationss*. Butterworths, 1961. Transl. D. G. Randall.

[15] R. S. Shevell. *Fundamentals of Flight*, 2nd edn. Prentice-Hall, 1989.

[16] B. S. Stratford. The prediction of separation of the turbulent boundary layer. *Journal of Fluid Mechanics*, 5: 1–16, 1959.

[17] T. Theodorsen. Theory of wing sections of arbitrary shape. Technical report, NACA-TR-411. National Advisory Committee for Aviation, 1931. Available from https://ntrs.nasa.gov/citations/19930091485.

9 Wing Design Considerations

Definition of an airplane – a device that almost won't work.
. . . backbone of an airplane is its wing.

A. M. O. Smith[1]

Preamble

Section 1.1, particularly Subsections 1.1.4 and 1.1.5, described Cycle 1 as a broad-ranging exploration over the parameter space that results in a complete layout of the baseline design sized to the mission. Chapter 3 exemplified how a key tool, the vortex lattice method, is used to study quantitatively the aerodynamics of the planform shape appropriate for the varying flight regimes discussed in Chapter 2. If the design concept is to be developed further, the next step is to check to what extent its characteristics and performance meet the design requirements.

As detailed in Section 1.2, this is advanced wing design – Cycle 2. Carried out by aerodynamic specialists using the tools we have developed in Chapters 4–8, it is the main focus of this chapter. The presentation spells out in practical detail the aerodynamic tasks to improve the baseline performance, as Figure 1.10 suggested. Since it is the backbone of the aircraft, we begin by refining the clean-wing shaping, namely camber, thickness, and twist, using the tools from the previous chapters to spell out the protocol for the design parameters, shape variables, and performance measures. Cycle 3 develops the clean wing further by adding fuselage, wing tips, high-lift elements, tail plane, etc., to advance the layout.

Acknowledging the seemingly endless considerations required to finalize the ultimate final layout, we decided the best way to describe this procedure is by studying the work of professional aerodynamicists in six cases of industrial configuration layout that investigate the real design choices made in bringing an aircraft to life. As a means to elucidate that knowledge base, we specifically chose these studies in order to follow the design decisions that industrial aerodynamicists made in producing their wings. The intention is to study what the masters have done, retrace their steps, compute our own data, and come up with a "second opinion" on their designs in order to provide a coherent explanation of why these aircraft are the shapes they are for the tasks they have to perform.

[1] *Wing Design and Analysis – Your Job* [29], with permission of Springer Nature.

9.1 Introduction to Aerodynamic Wing Design

In very general terms, Chapter 1 explained the process of aerodynamic design of wings, breaking it down into three distinct Cycles: 1, 2, and 3. The discussion in Chapter 2 of wing-flow physics indicated how matching the planform to the speed regime led to beneficial flow patterns, and Chapter 3 showed how L1 methods such as the vortex lattice method (VLM) make useful predictions of aerodynamic performance in Cycle 1. Tools for Cycle 2 and 3 analysis are explained and demonstrated in Chapters 4–8, so that now we have the background to go into more advanced aerodynamic design. The work in those first eight chapters now comes to use in this and the following chapters.

9.1.1 Three Design Cycles

Cycle 1: Sizing the Wing for the Mission

The baseline sizing procedure, Cycle 1, is a broad-ranging exploration over the parameter space embracing much cross-disciplinary-type analysis ranging from aerodynamics to manufacturing and costs. As described in Section 1.1 and illustrated in Figure 1.4, Cycle 1 of conceptual design sizes the wing to the mission of the aircraft with the required performance. It provides the geometric layout of the configuration along with basic performance measures, thus giving us the generalized wing planform parameters and other design characteristics. These primary parameters are: aspect ratio AR, span b, wing area S, average section thickness $(t/c)_{ave}$, taper ratio λ, sweep angle Λ, cruise speed M_{cruise}, and design lift coefficient C_L.

Wing sizing uses L1 methods such as the vortex lattice method (VLM; Chapter 3) to determine the spanwise lift distribution. The result is a first guess for the wing twist distribution obtained from an aero-structural trade-off with constraints on wing bending moments. The discussion here addresses an isolated *clean* wing devoid of engine nacelles, high-lift devices, and control surfaces not mounted on the fuselage.

Cycle 2: Refining the Clean-Wing Design

If the concept is to be developed further, the next step is to check the extent to which the characteristics and performance of the speculated design will meet the design requirements. Thus, following Section 1.2, Cycle 2 of wing design concerns the shaping of the wing and airfoil sections to define the thickness characteristics, camber, leading- and trailing-edge radii, twist, and spanwise variation of the thickness-to-chord ratio. Cycle 2 demonstrates the feasibility of the baseline design by improving its performance, exemplified in a quantitative way by Figure 1.10. Here, the protocols for the design parameters, shape variables and performance measures are spelled out.

Table 1.3 lists the *secondary parameters* for detailed aerodynamic wing-shape design, namely *section shape*: thickness chordwise, camber line chordwise, leading-edge radius R_n, and trailing-edge angle τ_{te}; and *spanwise shape*: twist, camber, and $(t/c)_{max}$ variation.

Cycle 3: Aerodynamic Configuration Design

From the family of configurations emerging from Cycle 2, Cycle 3 develops the wing further by adding, for example the fuselage, out-of-plane wing tips, high-lift devices such as slats and flaps, engine integration, etc.

The work continues until the design is sufficiently mature. There are always uncertainties involved in such a decision, so it is common practice in Cycle 3 to carry out further detailed investigations of *a few* – usually two – alternative configurations to differentiate between the evolving designs before an irrevocable choice is made. Final judgment then results in *freezing the configuration* as the best design among all the candidates investigated. Despite all efforts, flight testing often uncovers aerodynamic problems that must be remedied. Chapter 10 illustrates this with a few examples from actual aircraft projects.

Section 1.1.1 indicated the enormity of the work that Cycle 3 implies on an industrial level; this chapter can only scratch the surface of it. Even on a research-project level such as the Transonic Cruiser (TCR) design, Cycle 3 involves the work of many hundreds of person-hours and would be way beyond what could be undertaken in a semester-long course and/or project.

9.1.2 Cycle 2 Procedure for Clean-Wing Design

From among the procedures recommended by a long list of industrial aerodynamic designers (Ryle [26], Lynch [19], Obert [23], Mason [20], etc.), we can synthesize a prescription for the Cycle 2 clean-wing design process (see also Section 1.2.1). Underlying it all is the core of aerodynamic design: a clear idea of achievable $C_{L,max}$ – the assertion that the design delivers the lift required over the flight envelope at acceptable drag levels.

The procedure becomes the following.

1. To start the chordwise shaping of the wing, select/shape the airfoils in 2D to meet the design requirements, as described in Chapters 7 and 8.
2. Given the planform from Cycle 1, place airfoil sections in the wing to define its surface (Chapter 5). Foils at root and tip give a *ruled surface* that usually suffices for straight wings. High-speed wings typically require more sections. The lofted surface then sets the chordwise thickness shape, camber-line definition, and leading- and trailing-edge radii.
3. Apply L2 and L3 methods for full 3D computational fluid dynamics (CFD) analysis to optimize the spanwise shape for the cruise condition(s). This includes the spanwise variation of $(t/c)_{max}$, spanwise twist, and camber schedule.
4. Pay special attention to root and tip modifications of transonic swept wings in order to maintain isobar sweep. Check spanwise loading; if needed, apply washout to relieve outboard load and avoid premature tip stall.
5. Check off-design conditions: buffet onset and overspeed M_{dive} to sense the boundaries in the V–n diagram (see Figure 1.9).

As pointed out in Chapter 1, wing design uses optimization tasks in a number of ways, but it is not carried out by solving a few very large complex optimization problems by numerical algorithms. It is customary to group the design parameters and use each group in turn in a coordinated iterative process (e.g. twist and thickness distributions), as we shall see below.

9.1.3 Wing-Flow Physics Dictates Planform

Understanding of the flow achieved in the speed regime of interest and its control drives the wing-design process. One case in point is to find the shape for a large-*AR* wing that restricts flow separation to the trailing-edge area only. The approach looks at the wing as a series of 2D elements with the corrections and procedures necessary to convert the flow phenomena over them into 3D space.

Figure 2.1 mapped three different planforms – straight, swept, and slender delta wings – to three distinct flight regimes – subsonic, transonic, and supersonic, with their respective slenderness ratios $s/\ell \approx 0.5, 0.35$, and 0.2. This mapping achieves the high aerodynamic efficiency offered by the respective planform's *flow pattern* that is stable and predictable, resulting in low drag. The prescription of a particular type of flow pattern is in itself a powerful factor in defining the shape of an airplane.

Chapter 2 showed us how low-speed aerodynamics phenomena differ markedly from those encountered at high speed. Thus, our discussion begins with wings for low speed, $M_{cruise} \lessapprox 0.7$, and continues for high-speed wings, $M_{cruise} \gtrapprox 0.7$. The planform is one output of Cycle 1 design.

Subsonic Wing Design

Wings are straight for this flight regime. The overall design strategy is much like that for subsonic airfoil design, namely to keep the flow attached all the way to the trailing edge. Lift is maximized by applying a Stratford-like profile for the upper surface pressure recovery from maximum suction to the trailing edge.

Thus, identify useful airfoils for the intended mission, and follow procedure steps 1–5 described in Section 9.1.2.

Transonic Wing Design

The transonic flight regime is of great interest for today's commercial aircraft of Airbus or Boeing types. In becoming the mainstay of the aerospace industry, the swept-wing aircraft has achieved a high degree of sophistication. But much of the design knowledge – accumulated over 50 years of practical experience – required to continue advancing these designs is held by only a few experienced practitioners in industry.

Up to M_{crit}, the total drag coefficient is almost independent of the Mach number. Shortly after M_{crit} has been exceeded, the total drag of the aircraft rises strongly (drag divergence) at $M_{dd} > M_{crit}$ due to wave drag. In addition, the viscosity-induced pressure or form drag is increased because the pressure rise at the terminating shock wave of the embedded supersonic region thickens or even separates the boundary layer.

This shock–boundary layer interaction can be very unsteady and produces shock oscillations and consequently buffeting. Moreover, in the transonic regime, the pitching moment is changed (pitch-down increment, tuck-under), since the center of pressure moves further downstream. For high-subsonic cruise, typical of commercial aircraft, M_{crit} and M_{dd} should be as high as possible, since this is a matter of flight effectiveness.

The overriding design goal is to sweep the isobars as much as possible so that the upper surface streamlines do not meet a normal shock wave that imparts higher drag than an oblique shock. Truncating the swept wing at the root and the tip naturally removes the advantages of sweep from these regions, and special remedies need to be taken in step 4 of the design procedure laid out in Section 9.1.2.

Supersonic Wing Design

Because wave drag becomes so large at $M > 1$, supersonic wings must have low AR and be slender $s/\ell \approx 0.2$ and thin $t/c \lessapprox 5\%$. The flow is fully 3D, and the quasi 2D approximation, useful for straight subsonic wings, no longer holds; nor does the strategy of maintaining attached flow.

Instead, as explained in Sections 2.1.3 and 2.5, the wing is designed for a separated flow by making the leading edges aerodynamically sharp. The vortex shed from the highly swept edges gives coherent vortical flow over the upper surface that enhances lift and makes the lift curve nonlinear. This strategy applies to both subsonic and supersonic flight. A slender aircraft can be designed for the same type of flow pattern throughout its flight range up to when the vortex breaks down and chaotic flow takes over. The natural outcome is a planform with highly swept and aerodynamically sharp leading and possibly side edges and a nearly unswept and sharp trailing edge (i.e. some variant of the delta wing). The wing sweep must be large enough to ensure that the component of M_∞ normal to the leading edge is subsonic, otherwise the wave drag is higher.

9.1.4 Industrial Design Studies

Six case studies illustrate how the above design strategies are applied in practical clean-wing shaping. These cases are chosen to follow the steps taken by professional aerodynamic designers and to demonstrate the reasoning behind the design decisions.

The first study is for the subsonic straight wing of the Saab-Fairchild SF340 regional commuter, referred to henceforth as Saab 340 or 340. The second is the development in the late 1940s of one of the first transonic swept-wing fighters, the Saab J 29. The third is the minimization of wave drag on the modern Boeing-NASA Common Research Model (CRM) transonic transport wing.

The last three concern sonic and supersonic aircraft. The fourth, the TCR industry–academic concept design study, features throughout this book. It illustrates the flight regime just under $M_\infty = 1$, where challenges abound and no operational aircraft yet cruises. Here, we can see how the interactions of wing slenderness and thinness on wave drag play out.

For the fifth, we recount a few aerodynamic highlights of a current Swedish fighter, the Saab JAS 39 Gripen. The Aerospatiale–BAE Concorde is the only operational supersonic transport. The sixth study investigates the aerodynamic *finesse* of its wing.

9.2 Subsonic Straight-Wing Design

Classical aircraft, such as the prop airliners at the top of Table 8.1 with speed under 700 km/h, operate without significant clean-wing shocks. However, propeller slipstream may give local supercritical flow as on the Saab 2000, a stretched version of the Saab 340 studied next. The design proceeds on the various components as if they are isolated from each other, the Cayley paradigm truly holds for this category of airplanes, and quasi-2D analysis methods prevail. The components are sized and positioned to provide a balanced configuration with the required stability and control characteristics as well as the desired cruise, takeoff, and landing performance. Typically, the span of the wing is about equal to the length of the fuselage, and Figure 1.13 suggested the approximate cruise efficiency.

The absence of shock waves sets the basic design strategy and allows wing shaping to follow directly from the constituent airfoils. There are often just two such airfoils: one at the root and one at the tip. The overriding requirement is to maintain attached flow to the wing trailing edge. Designed in isolation, each lifting surface is positioned by considerations of balance, takeoff conditions, and stability and control issues, as addressed in Chapter 10. The basic airfoil shapes selected set the limit for the airplane speed. The new-technology airfoils with superior performance (Chapter 8) give corresponding wing performance. Optimizing the wingspan is one way to minimize the induced drag. It remains to integrate the engines into the wing without significant performance loss from propeller slipstream and nacelle aerodynamic interference.

Presented in the next section, the study of the Saab 340 exemplifies this design strategy. It briefly discusses how the central geometric parameters that determine the gross shape of its clean wing influence its aerodynamic properties by tracing some of the steps taken and choices made by the Saab design team during Cycles 1 and 2.

9.2.1 Case Study I: Saab 340 Commuter

Saab designed the 340 series in the 1970s as a cost-effective aircraft to enter the commuter market. This venture produced a successful regional airliner for 30–40 passengers, with more units being sold than any other during its 40-year service history. The structural, aerodynamic, and propulsion features, although commonplace in aviation, demonstrate the design of the 340 fitting its mission efficiently. Turboprop engines were chosen due to their inherent fuel economy compared to turbofan or turbojet engines. The wings are tapered, mounted low with dihedral, and cantilevered from the fuselage with a single spar. The turboprops are elegantly and aerodynamically efficiently integrated into the configuration by local wing-shape adaption.

Turboprop Efficiency

Advanced turboprops are significantly more fuel-efficient (10–30%) at (slightly) lower cruise speeds than turbojets and high-bypass turbofans. The speed disadvantage is of no importance on regional flights. In fact, at slightly lower speeds than the maximum cruise speed, turboprops have longer range and endurance. Moreover, there are environmental and economic benefits. The favorable propulsive efficiency of turboprops also has advantages regarding takeoff and landing being explicitly useful for steep descents to reduce the noise footprint and to allow operation from city airports with short runways.

Requirement: Fast Climb and High Wing Loading

In the design process of a transport aircraft, customer inputs define a mission in terms of payload, range, cruise, airport compatibility, and so forth. These goals in combination with design objectives and regulatory requirements, as well as producibility requirements, lead to the basic wing design and in turn to the development of target design parameters such as clean-wing $C_{L,max}$ and L/D.

Saab and Fairchild engineers carried out design trade studies that included low and high wing configurations, high-lift devices, passenger capacity, seating arrangements, nacelle locations, and other pertinent parameters. Let us look more closely at their wing design.

Wing Aerodynamics: Airfoil Selection

Advanced NASA airfoils were candidates studied for their reduced cruise drag, increased low-speed lift, and superior lift-to-drag ratio during takeoff and climb. In the airfoil selection process, the key design considerations are the profile shape and thickness to obtain superior aerodynamic characteristics and to provide sufficient internal volume for fuel.

The NASA medium-speed (MS) advanced airfoils feature a blunter leading edge than does the conventional NACA airfoil and a cusped lower surface at the trailing edge. They show approximately 31–55% improvement in climb L/D over the conventional NACA airfoils while maintaining a slight advantage in cruise L/D and providing a significant increase in maximum lift. Chapter 8 presented some comparative results with conventional NACA airfoils. The MS section, with superior $C_{L,max}$ and climb L/D, yet with tolerable initial pitching moment, was deemed optimum for the 340.

Wing Thickness and Twist

For transport aircraft, wing thickness has a strong influence on the available fuel volume because all of the fuel is stored in the wing. Trade studies determined an average thickness ratio of 14% as optimum, which would provide the fuel volume required for long-range missions with full passenger load while maintaining good cruise characteristics, low wing weight, and minimizing direct operating cost (DOC).

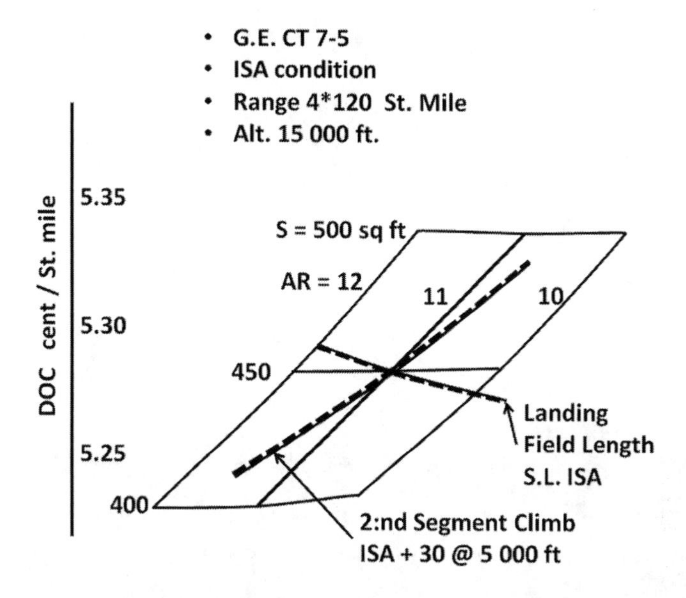

Figure 9.1 Saab MDO Phase 1 study to determine wing area and *AR*. (Courtesy of T. Jonsson, Saab AB, reproduced with permission)

An important factor affecting spanwise lift distribution and stall is wing twist. The 340 wing is defined by the following straight-wing design procedures (see Section 9.1.3).

- Straight generators (lines connecting points at equal chord percentage on root and tip sections; i.e. a ruled surface)
- The root airfoil section MS(1)-0316 twisted up 2°
- The tip airfoil section MS(1)-0312 twisted down 1°

Optimize Planform (i.e. the MDO Trade to Set Spanload)
Optimized for aircraft operation over short to medium sectors and multi-stop routes, the chosen 340 wing planform yields the lowest operating cost for the required cruise speed, payload range, and field. The selection procedure of the wing planform parameters is documented in carpet plots such as Figure 9.1. Positive wing dihedral produces favorable yaw–roll coupling $C_{l\beta}$ for low-wing aircraft with straight wings (generally 5–7°) and 7° dihedral was chosen.

The foregoing procedure resulted in the following 340 wing geometric characteristics.

- Wing area 450 ft^2,
- *AR* 11
- Taper ratio 0.375
- Thickness ratio 16% root, 12% tip
- Incidence angle 2° root, −1° tip
- Dihedral angle 7°

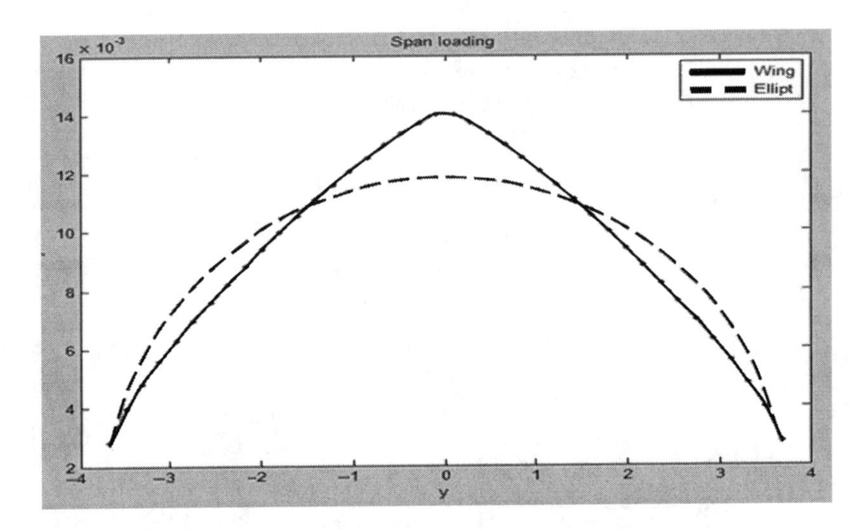

Figure 9.2 Spanloading of the Saab 340 wing and elliptic distribution.

With these parameters, one can run the L1 methods of Chapter 3 to determine, for example, the spanload at the design cruise conditions. Figure 9.2 presents the spanload computed with TORNADO. It is triangular and departs from elliptic loading more than any spanloadings shown in Chapter 3. The difference is due to the twist, and this can imply higher induced drag. In any case, the designers concluded that other benefits in the trade-offs favored this choice.

$C_{L,max}$: Add Washout for Docile Stall

The Saab 340 cruises at $M_\infty = 0.5$, $C_L = 0.6$, and $Re_{MAC} = 10M$. A Reynolds-averaged Navier–Stokes (RANS) computation produces the drag polar in Figure 9.3 and confirms that the wing twist makes the stall behavior substantially more docile. The clean untwisted wing has $C_{L,max}$ at 10° and the lift-curve slope starts dropping off at 8°. The numbers for the twisted wing are $C_{L,max}$ at 12° and slope drop-off at 10°. The two wings have the same $C_{L,\alpha}$, but the twisted wing curve is shifted 1–2°.

Compare these curves to the drag polar for the NASA-MS(1)-0313 airfoil shown in Figure 8.7. As we have seen before, larger thickness gives larger $C_{L,max}$; the 340 root foil is 16% thick, so the wing achieves higher $C_{L,max}$ than the MS-0313 foil. The foil stalls at low Mach at around 10°. Cruise at 15,000 ft. at maximum take-off weight requires $C_L \approx 0.31$, but $(L/D)_{max}$ obtains for $C_L \approx 0.6$, which would seem nonoptimal. But, as a regional airliner, the 340 spends a large fraction of its time and fuel climbing at higher C_L to service altitude, so it is important to climb at high L/D.

Buffet Margin: Upper Surface Oil Flow

The Federal Aviation Regulations (FAR) 23 condition requires that wing buffet due to separation does not occur in a 1.3g pull-up maneuver in cruise. RANS computations verify that there is no hint of unsteadiness at the required α. However, climb is another

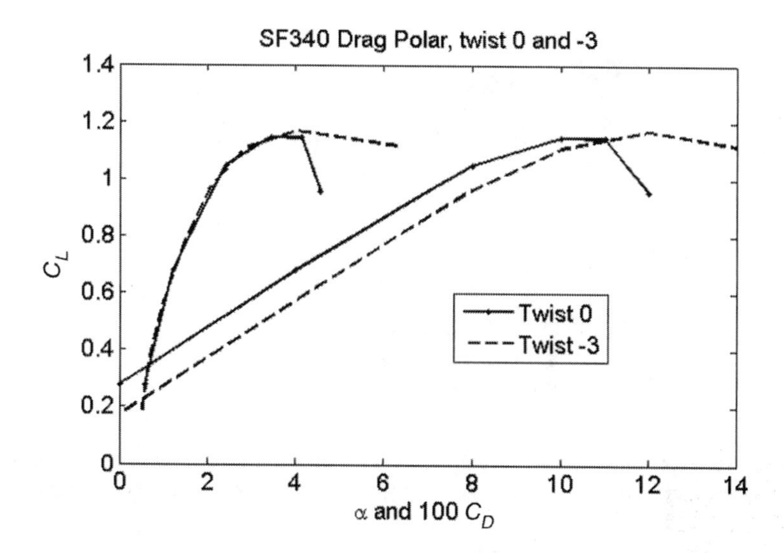

Figure 9.3 Saab 340 clean-wing drag polar computed for cruise condition $M_\infty = 0.5$. $C_L \approx 0.6$ where L/D is maximum, $\alpha \approx 4°$. Two wings are computed, with and without twist. EDGE RANS computation. (Courtesy of P. Eliasson, personal communication)

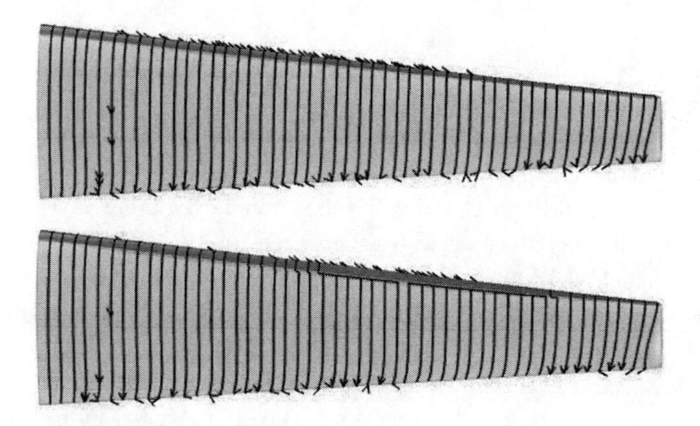

Figure 9.4 Digital oil-flow patterns show attached flow on the upper surface for both wings, with and without twist, $M_\infty = 0.5$, $\alpha = 8°$, and $Re = 10M$. (Courtesy of P. Eliasson, personal communication)

matter at higher α. We know the airfoil and wing(s) stall at 8–10° and show digital oil flow on the wing upper surface of the computed skin friction vectors for flow conditions $M_\infty = 0.5$, $\alpha = 8°$, and $Re_{MAC} = 10M$ (Figure 9.4). The flow remains attached for both twisted and untwisted wings.

Zero-Twist Wing Stalled at $\alpha = 12°$

Similar to Figure 9.4, Figure 9.5 showed the digital oil flow on the upper surface for $\alpha = 12°$, where the zero-twist wing is stalled, as indicated by the lift curve

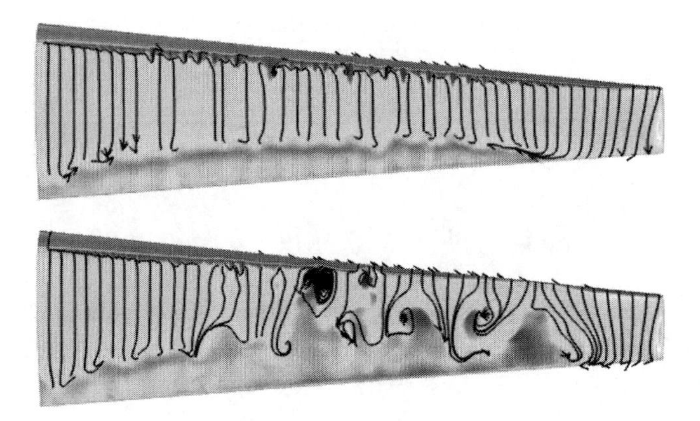

Figure 9.5 Post-stall skin friction computed at maximum lift, $\alpha = 12^\circ$ with twist (top) compared to without twist (bottom). (Courtesy of P. Eliasson, personal communication)

and the digital oil-flow pattern. The twisted wing is at $C_{L,max}$ with flow separation only around mid-span, but the tip flow remains attached, and this testifies to the well-chosen twist distribution.

9.3 Transonic Swept-Wing Design

The Me 262 was the first operational jet fighter. Its wing leading edge was swept 18.5°, and at the end of World War II, the Allies discovered even more advanced German prototypes awaiting flight testing with wing sweeps of 40° or more, as described in Polhamus [17] and Wänström [32]. The reaction to this discovery was a frenzied rush into swept-wing designs by the Allies as well as by the Swedes, as the historical record indicates; for example, Blair [1] terms it "Germanizing our design."

George S. Schairer [27] claimed that, in 1943, no manufacturer in the world knew how to build a successful transonic swept wing. By early 1945, the Germans were ready to test fly prototypes, followed by the other major manufacturers quickly responding to the challenge with first flights of the North American Aviation F-86 in October 1947, the Boeing B-47 and the Mikoyan-Gurevich MiG-15 in December 1947, and the Saab J 29 "Tunnan" in September 1948. Both the F-86 and J 29 used airfoil shapes closely related to the Me 262 (see Figure 8.9 showing their airfoil drag-rise plots).

The euphoria around sweep technology outpaced its design know-how, which was still lodged with the straight-wing development experience of fighters such as the Spitfire and Mustang. The late 1940s saw wings with swept but straight edges (only the J 29 had a crank in the leading-edge sweep), little or no washout, and sparing the application of "flow control devices" such as leading-edge devices, vortex generators, or boundary-layer fences for the "tailoring of details," to use Schairer's words.

Flight testing slowly brought reality to the euphoria when drawbacks such as tip stall, pitch-up instability, and loss of aileron control became evident and required remedying. Applying sweepback and jet engines to transport aircraft proved to require very extensive development of new concepts for many design details [27]. Jet fighters and transports have required special and different tailoring of details.

Sweep theory for a constant-chord wing is simply, but not necessarily properly, explained by the suggestion that it is the component of Mach number normal to the leading edge that governs critical Mach (see Section 3.4.1). How valid is this? What merit does it have when applied to an airplane? These questions beg for answers in order to understand the need for tailoring transonic swept-wing design.

9.3.1 The Design Challenge

As introduced in Section 3.4, sweep theory is in fact a simplification, since the flow over a typical swept wing of relatively low AR is a very complex 3D one. The simple $\cos\Lambda$ relationship applies only to a yawed wing of infinite span, and the delay in reaching M_{crit} is much less than predicted because of the effects at the wing root and tip.

Isobars Are Highly Informative

The criterion for critical conditions can be generalized beyond simple sweep theory and applied to local conditions by studying the *isobars* (i.e. surfaces or curves joining points with equal pressure (or velocities)). On a wing of arbitrary shape, critical conditions obtain when the velocity component normal to the isobars (i.e. in the direction of the local pressure gradient) reaches the local velocity of sound. Once the critical velocity is exceeded, a continuous steady irrotational flow is impossible and discontinuities in the form of shock waves can be expected to occur.

On a swept wing the isobars are curved, and nonuniformities in the isobar pattern are mainly due to root and tip effects. The undesirable features of a finite swept wing become as follows.

1. Unnecessarily low critical Mach number owing to loss of isobar sweep
2. Premature flow separation near the wing tips owing to high-suction peaks and steep adverse pressure gradients with a consequent tendency of the wing to pitch up and make matters worse

Buschner [2] in 1944 was the first to realize these shortcomings. He suggested that wing shapes should be modified to pull the isobars straight at the full angle of sweep, and he demonstrated that this is possible. This is one step toward loading up the whole wing as uniformly as possible. It cannot be said, however, that straight isobars are a necessary condition for minimum wave drag of the finite wing. Governed by the physics of transonic flow, the design space is highly nonlinear and can only truly be explored by mathematical optimization techniques.

Sudden Tip Stall

More troublesome than nonuniform isobars was the tendency for sudden tip stall causing nonlinear pitch-up instability. Realized first in 1940 by Ludwieg [18], it grew in severity with increasing sweep angle Λ and AR, and this problem delayed swept-wing development.

Nevertheless, the benefits of sweepback remain well worth having, so much so that the designers of the early jet fighters were overly eager, even euphoric, in combining them with the higher thrust of the evolving turbojets. Among other things, sweep enabled contemporary wing thickness ratios to be retained while allowing an increase in M_{crit}. For example, the F-86 Sabre wing of 10% thickness provided space for fuel and relatively uncomplicated flaps.

Two Design Goals

Thus, the design challenges became the following.

1. Sweep the wing to reduce drag by delaying drag rise to a higher Mach number.
2. Mitigate the loss of sweep effect at the root and tip by judiciously tailoring the wing shaping.
3. Mitigate tip stall and pitch-up instability by controlling flow separation at the tip, often caused by shock–boundary layer interaction and usually occurring most severely at the tip. Pressure increases through the shock wave: the adverse pressure gradient across it causes the boundary layer to separate.

The challenge lies in judicious shape tailoring to mitigate these unwanted effects.

Maintain Isobar Sweep

Normally in Cycle 2 design, the aerodynamicist is given the planform and maximum thickness of the wing and told to design its twist and camber, as well as shifting the thickness envelope appropriately, to obtain *swept parallel isobars* over the wing. The natural tendency is for the isobars to unsweep at the root and tip. So the designer tries to reduce this tendency to obtain an effective aerodynamic sweep (i.e. in the pressure distribution) as large as the geometric sweep. Generally, the wing has a weak shock wave. Giving more sweep to the shock wave reduces its strength and also the wave drag.

Design modifications at the designer's disposal for counteracting the unsweeping effect at root and tip are as follows.

1. Root region
   a. Leading-edge effect: extend wing root by leading-edge extension (LEX; often called strake) to draw the isobars afore.
   b. Thickness effect: to move the isobars at the root forward, the thickness should be increased over the front part and reduced over the rear part.
   c. Lifting effect: increase the lift over the forward portion of the root section and reduce it on the rearward portion by including positive twist and negative camber.

2. Tip region: the planform may be altered from the streamwise cutoff to a curved form (i.e. locally rake the tip leading edge to push the isobars further astern).

All of these shape modifications apply to a specific design point with given values of lift coefficient and Mach number. Notice also that the prescription is not quantitative – it does not tell one *how much* to change the shape. One must do this either by trial and error manually or by mathematical algorithm in an optimization procedure.

Mitigating Tip Stall and Pitch-Up Instability

In addition to higher loading, sweep sets up a velocity component along the wing that thickens the boundary layer toward the tip, and both effects make the tip prone to premature flow separation (see Section 2.7). This can occur asymmetrically to generate a roll moment too, and if the ailerons lie in separated flow there may not be sufficient authority to control the roll. Figure 9.6 (right) schematically indicates how this can happen and how a leading-edge fence mitigates the effects.

Such tip stall makes for unacceptable flying quality. After much flight testing, various alleviating measures were devised to delay separation or at least make it predictable. The most common among these are as follows.

1. Leading-edge devices (leading-edge slats that automatically open and close under aerodynamic forces alone (pioneered in the Me 262 development and refined on the F-86), sawteeth (pioneered in the J 29 to become a hallmark of later Saab aircraft), notches, etc.).
2. Distributed devices (fences chordwise over the wing (MiG-15), vortex generators, etc.). Vortex generators effectively cured the pitch-up instability on the B-47, as explained by Cook [4].

Most of these work by producing vortices, as depicted for the fence in Figure 9.6. But their precise functioning is not easily predicted and must be studied by computation in each specific case.

Wing fences were used on the earliest swept-wing fighters. The MiG-15 had two large fences on each wing. The fence does present a physical barrier to the spanwise boundary-layer flow and, at higher incidence, to the inward spread of separation

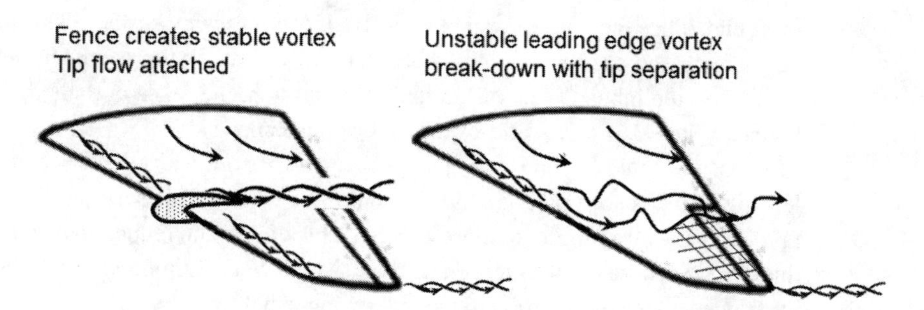

Figure 9.6 The leading-edge fence creates a stable vortex flowing downstream inboard of ailerons (left) to avoid the unsteady, separated flow (right).

Figure 9.7 Two versions of the Saab J 29 Tunnan. J 29A (left) and J 29F with sawtooth and fence on the wing (right). (Photos by Swedish Air Force, reprinted with permission by Archive of Saab Veteran's Club)

(Figure 9.6). More importantly, however, the fence's projection forward of the leading edge distorts the flow locally and causes the formation of a separated flow region just inboard of the fence and a vortex just outboard of it. This separated flow region generally incurs neither significant lift loss nor severe drag penalty.

The effect of the flow outboard of the fence on the wing pressure distribution is profound, since the vortex formed flows back over the wing surface. The vortex produces increased suction on the surface beneath its path, which compensates for the lift lost on the inboard side of the fence. In addition, the fence curtails the spanwise boundary-layer drift and relieves the tip loading somewhat.

Usually located beyond mid-span, a sawtooth is formed by locally extending the leading edge, as seen on the J 29F in Figure 9.7 (right). This serves a dual function, producing a vortex similar to those generated by a fence and reducing the wing section thickness-to-chord ratio to minimize drag at high Mach numbers. In addition, it can introduce leading-edge droop to the outer wing panel.

Two Transonic Case Studies Exemplify Procedure

The following sections present two case studies to demonstrate the design procedures. The first treats the swept wing of a very early jet fighter, the Saab J 29, and the second looks at a contemporary transonic transport wing, the CRM. We shall also see unwanted side effects of sweepback on the original J 29 wing and their mitigation, comparing the original untreated A-wing with the F-version with sawtooth and stall fence (Figure 9.7).

9.3.2 Case Study II: Early Jet Fighter Swept-Wing Design – Saab J 29

For the J 29 case, we outline below the swept-wing problems that the designers encountered in a very early application of sweep theory. The very clean planform of the original A-version and the later modified F-version are shown in Figure 9.7.

As we have discussed in Chapter 3, tip stall on a backward-swept wing is associated with the pitch-up phenomenon. In addition, the separated flow usually is unsteady, which creates buffet.

The wing modification with a "sawtooth" leading edge and a plate covering the break in the planform, much resembling a boundary-layer fence, improved the maneuverability at high speed and also the maneuver-critical Mach number by three-hundredths. As is illustrated below, the devices changed the stall pattern completely. A similar effect on the B-47 contemporary jet bomber was created by the engine pylons, which mitigated the pitch-up. It is explained below with CFD how the fence changes the flow field.

J 29A and J 29F Wing Geometry

The wing has been lofted according to the classical tradition as a ruled surface between root (15% half-span) and tip (85%) airfoils. The Saab engineers selected the FFA airfoil 124-5112, presented in Chapter 8, 10% t/c at the root and 11% at the tip. Notice how the leading edge has been extended at the root – a recipe to draw forward the isobars. The LEX radius is small, presumably to generate vortex lift at high angle of attack. The tip leading edge is similarly slightly raked aft to draw back the isobars.

Later Saab fighters from J 32 to JAS 39 have sawteeth and fences, too. The exception is the $M2$ double-delta J 35 "Dragon," which featured "vortilon" fences on the lower side of the leading edge supposed to mitigate lift force buffeting in turns.

In the 1940s, the only tool available to the engineers was the wind tunnel, and that had problems with Reynolds number and compressibility effects of the limited tunnel cross section. Details of shock–boundary layer interaction and separation were hard to observe and even harder to quantify. We shall make a CFD investigation of the A-wing for transonic flight and the effects of the F-modification, which raised the maneuver-critical Mach number from 0.86 to 0.89.

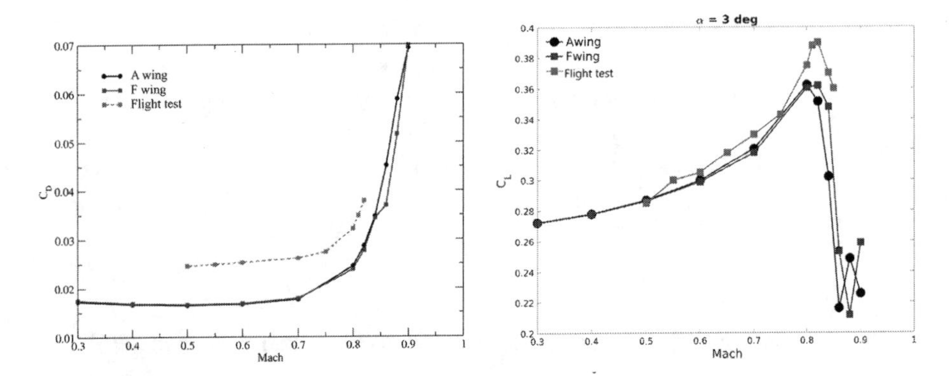

Figure 9.8 Mach sweeps of C_D (left) and C_L (right) for A- and F-wings computed with EDGE RANS at constant $\alpha = 3^\circ$, and compared with flight test data.

Reconstruction of J 29 Geometry

There is no digital geometry for the J 29. It is defined in a combination of analytical shape definition and coordinate measurements developed by Saab's geometry office. The shape is determined by a series of parallel, diagonal, and cylindrical sections. The wing itself can be accurately constructed from profiles and planform. Cross sections of body, hood, and tail boom were digitized from the sections in a large drawing from Saab's archive.

The wing profiles bear marks of their relationship with those developed for Me 262 by Deutsche Versuchsanstalt für Luftfahrt (DVL) as modifications of symmetrical National Advisory Committee on Aeronautics (NACA) profiles. Profile coordinates are presented at the stations at 15% and 85% of the span, respectively. The wing tip between 85% and 100% was simplified. For the LEX, we estimated the nose radius from a wing section. Usually, swept wings are "washed out" (i.e. have tips twisted nose down), but the J 29 had root and tip chords parallel.

Drag Rise

A number of computations have been made to illustrate some basic transonic characteristics of the wing for high-speed cruise conditions $\alpha = 3^\circ$ and $Re = 9M$ over a sweep of Mach numbers; see Figure 9.8 (left) for a comparison with flight tests. Note that the computations are for a wing and body alone, so the whole configuration has more drag. The comparison shows that drag rise begins substantially at and beyond $M_\infty = 0.8$, and the flight tests verify this. The wing stalls because of shock waves forming over the wing at approximately $M_\infty = 0.84$.

J 29 A and F Drag Rise

Figure 9.8 shows drag rise with Mach number. There is a tiny difference between A- and F-wings, which is somewhat consistent with the minor increase in high-speed performance for F wings. A moment plot would show that something happens over a small α-span around $M0.84$, but this we already know. More precise information

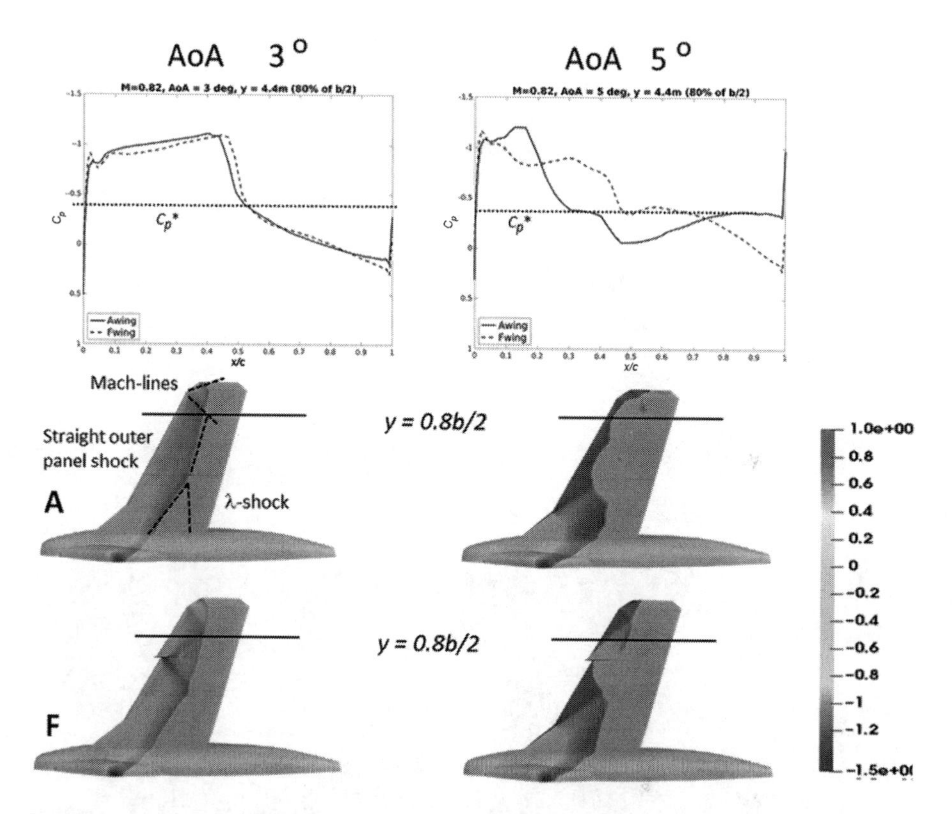

Figure 9.9 Pressure distributions on A- and F-wing suction sides and along 80% span chord at 3° and 5°. AoA = angle of attack. (Courtesy of M. Zhang, personal communication)

can only be extracted if the moment curve were plotted with aircraft center of gravity as a reference point, but that is not necessary for the current discussion. To move on, we consider pressure distributions in Figure 9.9. Consider first the middle-left plot in Figure 9.9 on the A-wing. The features resemble those on a transonic airliner wing (see Figure 2.23, left). The four RANS simulations show that the isobars are not swept across the span, and that with increasing Mach the shock wave on the upper surface moves rearward.

The differences between A- and F-wings are clearer at 5° than at 3°, especially on the outer wing. The dogtooth/leading-edge fence creates Mach waves, and the separation and hence pressure distributions are very different on the outer wing.

This represents early work on swept wings, and the above shows that the countermeasures taken at the root and tip were not very effective. It also indicates that the classical procedure of selecting an airfoil at root and tip with linear ruling in-between was not sufficient. To see more quantitatively what the sawtooth and fence contribute to forces, consider Figure 9.10, which showed lift (actually C_z) along chords at 20, 40, 60, 80 and 95% half-span at 5° angle of attack. The F-wing has lift force everywhere, but the A-wing has significant negative contributions on the outer portions, which are typical of pitch-up.

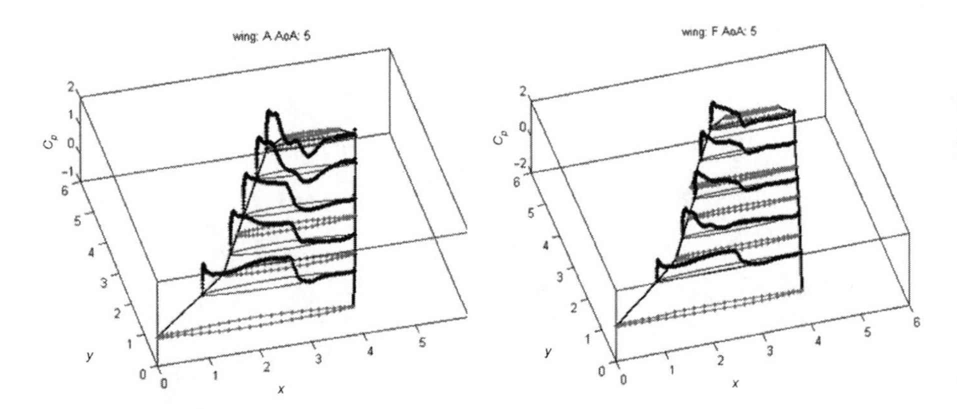

Figure 9.10 Normal force c_Z along selected chords, A- and F-wings at 5°. AoA = angle of attack.

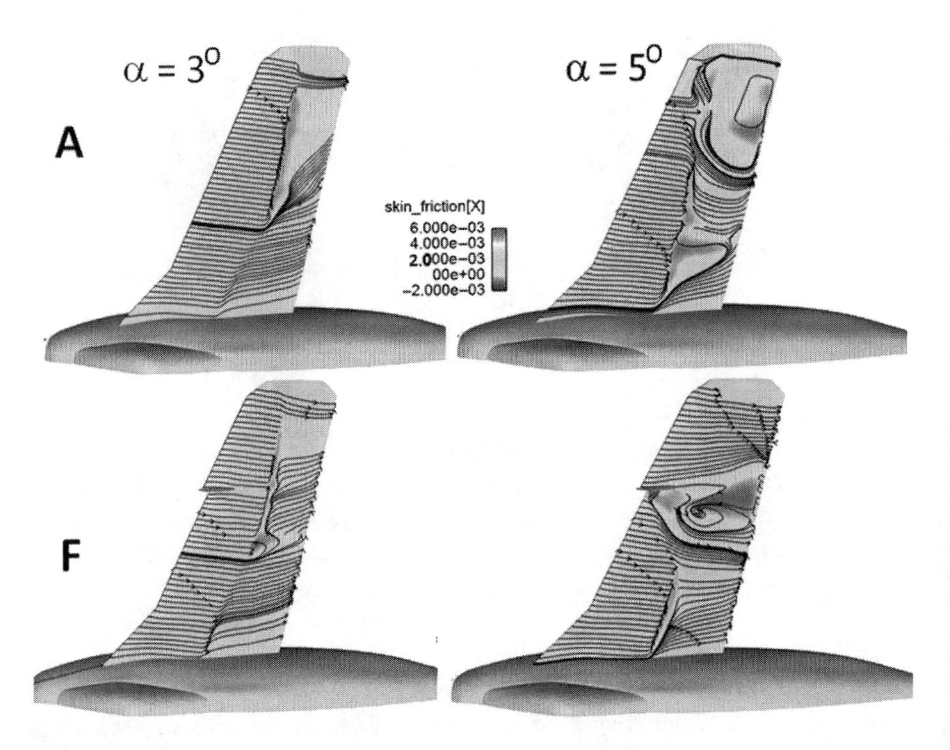

Figure 9.11 Component of skin-friction coefficient vector in x-direction $C_{f,x}$ and skin-friction lines on upper wing at 3° (left) and 5° (right).

Further illustration of the flow physics is obtained by considering the boundary layer and separation behavior (Figure 9.11). Recirculation is indicated by $C_{fx} < 0$. The shock–boundary layer interaction creates the separations at 3°. At 5°. The A-wing is stalled over most of the wing, but the F-wing separation is limited to the area just inboard of the fence.

Transonic Pitch-Up with Tip Stall: Sawtooth and Stall Fence Remedy

The usual rationale for boundary-layer fences is to block the decelerating spanwise boundary-layer flow; outboard of the fence, the boundary-layer flow will be streamwise again, and shielded from the decelerating layer from the wing root out to the fence. A vortex rolls over the top of the fence (Figure 9.6) and energizes the flow outboard. The Saab engineers call the devices "virveldelare" (vortex splitters) and attribute their function to stabilizing the vortical flow along the leading edge by splitting it. *Vortilons* are on the pressure side of the wing and act in a similar way but at higher angles of attack, where the leading-edge stagnation line has moved substantially below the leading-edge apex.

There is in this story a continuous change with leading-edge sweep. A straight high-*AR* wing behaves much like an airfoil; a small leading-edge radius will make a leading-edge separation bubble. A swept high-*AR* wing works pressure-wise like a cross section orthogonal to the leading edge, with a spanwise flow superposed. Hence, a leading-edge separation bubble becomes a leading-edge vortex. More sweep makes the vortex move downstream and become stronger. To do its work on the leading-edge flow, the vortex splitter need extend only through the suction peak and high-speed area close to the leading edge. This can be compared to the full chord fences seen, for example, on the MiG 15 and 17. The overall effect is to replace outboard separation with local separation, where the fence diverts the decelerating inboard leading-edge boundary layer chordwise to separate in the adverse pressure region.

9.3.3 Case Study III: The CRM Wing, a Minimum-Drag Design

The educational and research communities have been in need of a practical baseline swept-wing configuration to apply engineering-detailed wing-design principles to and then illustrate how these correspond to what mathematical optimization would determine. In other words, can mathematical techniques explore the design space and identify design directions that relate to what engineers discovered in the past 70 years of clinical experience since 1948, and perhaps go further and find aspects they were not aware of?

To see what is possible today for swept-wing design, we turn to the CRM wing.

CRM Swept-Wing Test Case

Responding to the need for aerodynamic design optimization benchmarks, the American Institute of Aeronautics and Astronautics (AIAA) Aerodynamic Design Optimization Discussion Group (ADODG) proposed six design optimization cases of various characters that could be used to compare methods. Case 4 addresses minimizing the drag of the CRM swept transport wing in transonic flight. The CRM configuration itself is known as the NASA CRM (https://commonresearchmodel.larc.nasa.gov/). It consists of wing, fuselage, and horizontal tail, and it is a neutral (and realistic) representative of a current transonic airliner (e.g. a Boeing or Airbus plane) that has become a broader benchmark for all types of CFD comparisons.

It is a new, publicly available and relevant baseline swept-wing geometry to which state-of-the-art CFD methods in general, and multidisciplinary optimization (MDO) design tools (cf. Chapter 1) in particular, can be applied.

NASA and Boeing designed the CRM model as a contemporary supercritical transonic wing suitable for a flight condition of $M_\infty = 0.85, C_L = 0.5$, and a chord Reynolds number of 40 million. It is defined as a stack of 21 airfoil sections given on the CRM website. This wing design exhibits fairly high performance over a reasonable neighborhood of the design point. The wing is not a single-operating point design. The result is a team effort described by J. Vassberg et al. [31], and it is a rather fully developed wing with trailing-edge break (so-called Yehudi), twist, thickness scheduling, etc., although few design details or explanations are described in the publication.

The initial geometry in the exercise is extracted from the CRM geometry. The geometry and specifications are given by the ADODG. The fuselage and tail are deleted from the original CRM, and the root of the remaining wing is moved to the symmetry plane.

The only requirement specified by the ADODG for the parameterization in Case 4 is that the planform must remain fixed and the changes in shape may only be made in the vertical (z) direction. See the Martins team website at http://mdolab.engin.umich.edu/content/aerodynamic-design-optimization-workshop for more details.

Problem Description

The ADODG currently defines the optimization problem as a 3D RANS-based drag minimization of the CRM wing in transonic flow and with single and multipoint conditions.

For the ADODG CRM wing geometry, Case 4.1 is the baseline single-point optimization. Results for this case have been reported by Martins and his team at the MDO Lab in the University of Michigan, and the resulting geometries and meshes are publicly available.

The shape design variables are the z-coordinate displacements of the free-form deformation (FFD) volume control points (i.e. allowed to move only in the vertical direction). In addition, each operating condition has an independent angle of attack variable. Equality constraints for the lift coefficient are enforced at each operating condition. A single pitching moment inequality constraint is enforced only at the nominal flight condition ($M = 0.85, C_L = 0.5$) for each optimization. Finally, two types of geometric constraints are imposed: the internal volume of the wing must be greater than or equal to the initial volume, and the thickness of the wing must exceed 25% of the initial thickness at any point.

CRM Wing Optimized for Minimum Drag

Since an in-depth description of the aerodynamic shape optimization software is beyond the scope of this book; we simply present some of the Martins team results for the CRM in Case 4 from Ref. [13]. Figure 9.12 presents results for the initial CRM

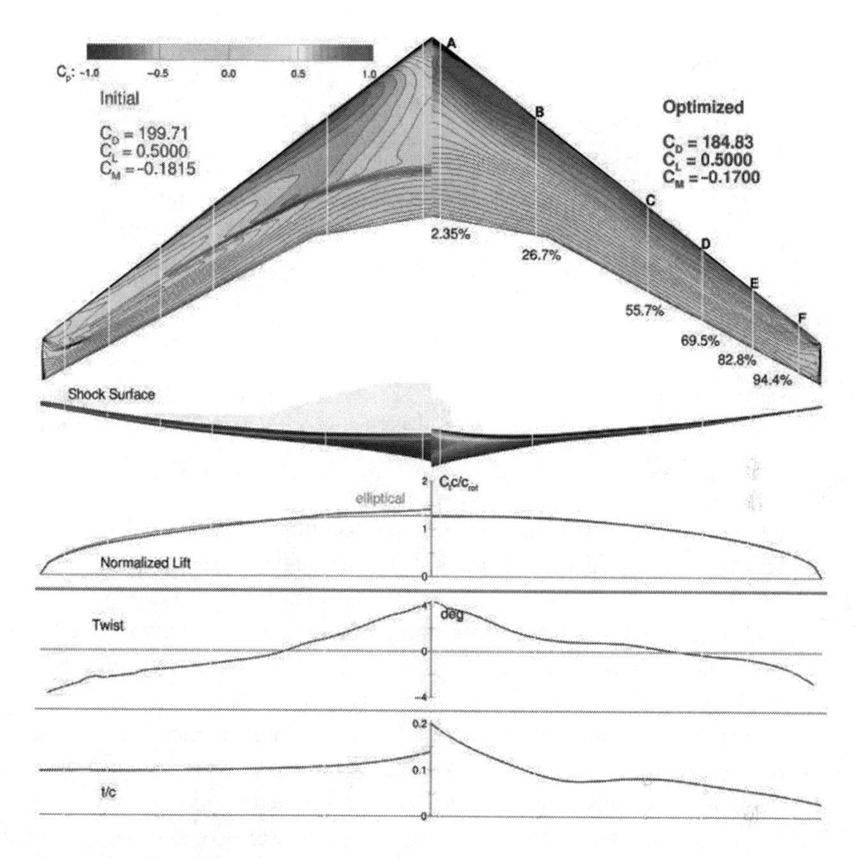

Figure 9.12 The optimized wing is shock-free and with 8.5% lower drag. (Courtesy of J. Martins [13], reprinted with permission)

wing shape on the left, while the optimized results are shown on the right. The planform view of the wing shows the C_p contours at the nominal operating condition, along with the corresponding drag coefficients. Just below the planform view, the front view also shows the C_p contours and a visualization of the shock surface, as well as the physical thickness variation in the wing. Below this, we plot the spanwise lift, twist, and thickness-to-chord ratio distributions. A reference elliptic lift distribution is also shown. The original wing (CRM) displays the classic isobar *unsweeping effect* of root and tip. Notice how the optimization procedure has modified the spanwise thickness and twist distributions to effectively straighten the isobars.

The optimization decreased the drag by 8.5% from 199.7 counts to 184.8 counts at the nominal flow condition. At the optimum, the lift coefficient target is met and the pitching moment is reduced to the lowest allowed value. The lift distribution of the optimized wing is much closer to the elliptical distribution, indicating an induced drag that is close to the theoretical minimum, achieved by an algorithm that fine tunes the twist distribution and airfoil shapes. The baseline wing has a near-linear

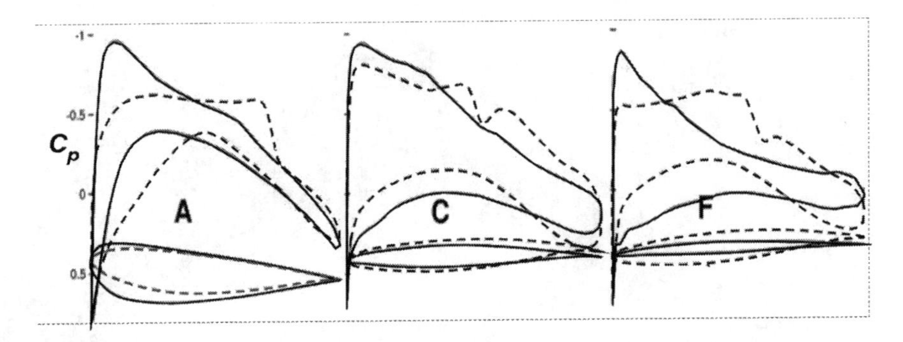

Figure 9.13 Cross-sectional shape and C_p distribution changes made at the root section A, mid-span section C, and tip section F. (Courtesy of J. Martins [13], reprinted with permission)

twist distribution. The optimized design has more twist at the root and at the tip and less twist near mid-wing. The overall twist angle only changed slightly from 8.06° to 7.43°.

The optimized thickness distribution is significantly different from that of the baseline. The optimizer chooses to increase the thickness at the root and decrease the thickness at the tip to maintain volume. The root t/c is over 20%.

The low thickness near the tip, however, would in practice incur a structural weight penalty. A stricter thickness constraint could lead to a more realistic design: an optimum design is only as good as the realism imposed by its constraints.

The condensation of very closely spaced pressure contour lines over a significant portion of the baseline wing indicates the shock wave formation. The optimized wing shows evenly spaced, parallel pressure contours, indicating a nearly shock-free solution at the nominal flight condition.

Chordwise Changes: Root, Mid-span and Tip Sections

Let us look in more detail at what the optimizer changed at the root (section A in Figure 9.13) by plotting the chordwise C_p distribution. A thickening of the nose and an increase in negative camber effectively removes the shock and produces a pressure distribution that is more similar to the midsection C.

The shock elimination can also be seen on the airfoil C_p distributions. The sharp increase in local pressure due to the shock becomes a gradual change from the leading edge to the trailing edge.

Tip Section

Another noticeable feature in the optimized wing is the sharp leading edge. The optimizer explores the weakness in the problem formulation. With a single-point optimization, there is no penalty for thinning out the leading edge. However, sharp leading-edge airfoils experience adverse performance at off-design conditions, since the flow is prone to separation at off-design angles of attack.

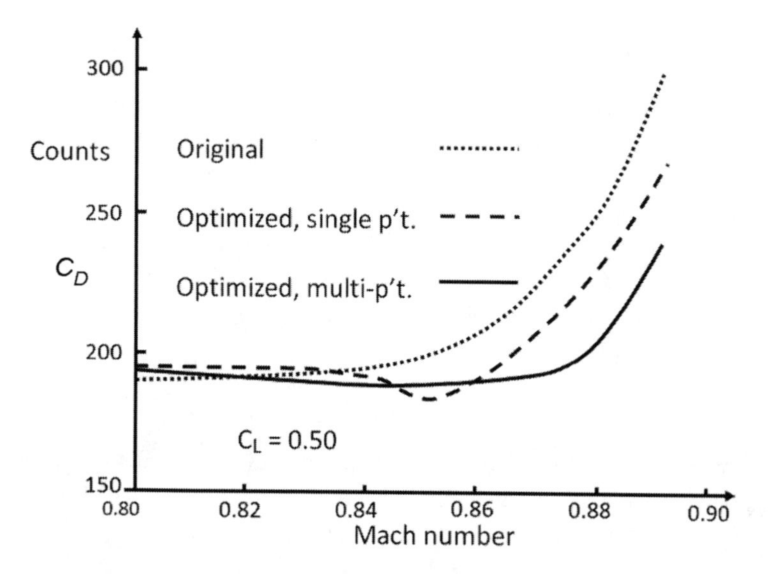

Figure 9.14 Mach sweeps of drag for the original and optimized CRM wings. (Courtesy of J. Martins [16], reprinted with permission)

The tip section F plot shows a severe slimming of the nose and a decrease in twist that effectively removes the shock and produces a C_p distribution that is more similar to the midsection C. The changes at both ends effectively mean that the pressure recovery slope from leading edge to trailing edge now is approximately constant along the span.

Multipoint Optimization of the CRM Wing

When a shape is optimized for a single operation point, its performance may degrade with even small flight-state deviations. This can be avoided by taking different flight states into the figure of merit (FoM), as is done in Martins et al. [16]. The results are summarized below.

Figure 9.14 shows how drag varies with Mach number for the original, the single-point optimized wing, and the multipoint optimized shape. The baseline has significant drag rise starting at $M0.85$ and steepening with increased Mach number, just like the airfoil drag rise curves we saw in Section 8.3. The minimization of drag at the nominal cruise condition only gives the lowest drag there, but even a small deviation worsens the drag substantially. That optimum is not robust with respect to the operating point, and this would be frowned upon by wing designers.

The multipoint optimization uses a weighted sum of drag at nine different M, C_L conditions. The weights were chosen to reflect the cost incurred at the particular condition as revealed by actual recorded Boeing 777 flights in service. The CRM wing is typical of similar aircraft. The resulting drag rise is gradual and monotonic and the price to pay in drag at nominal conditions is small: the Martins team multipoint optimized shape cost just 4 drag counts.

The following two lessons are to be learned.

- The parameterization of the shape must not allow very local shape changes leading to overall shapes with bumps. A single-operating-point wave drag can be mitigated by raising a bump at the shock position, but that is, of course, a bad design.
- The shape must be good for a range of operating conditions to avoid narrow drag buckets. Techniques for multi-operating-point optimization must be applied. Examples include forming a weighted average of results at different points or taking the worst case as FoM. It remains, of course, to investigate the robustness with respect to the weights.

Concluding Remark

Optimization cases such as these are of primarily academic interest and are not representative of industry aerodynamic shape optimization for an actual transonic aircraft configuration for several reasons.

1. The lack of the fuselage results in an incorrect lift distribution. In fact, the initial CRM when combined with the fuselage has a closer to elliptic lift distribution than the wing alone.
2. The analysis is performed at a wind tunnel Reynolds number of $5M$, as opposed to a flight Reynolds number of approximately $40M$. It is not known what effect this would have on the optimal design.
3. The lower bound on the thickness is too small to be practical. The optimized thickness distributions would result in considerable structural difficulties.
4. An aircraft wing is not rigid and deforms according to the actual flight loads. This bends and twists the wing, resulting in a different lift distribution, shock structure, and induced drag. Previous work by the Martins' team has investigated optimizations that address the shortcomings of these "aerodynamic-only" optimizations by performing aero-structural optimization, but this is beyond the scope of the ADODG benchmarks. Note that such aero-elastic deformation has been taken into account for the CRM case. The airfoil-stack definition published is in fact the $1g$ cruise flight shape for a reasonably sized structure in the baseline wing and not the jig shape.

9.4 Supersonic Slender-Wing Design

The configuration layout for supersonic flight of a military aircraft depends very much on the mission objectives. The basic design philosophy for a civil supersonic transport focuses, even more than for transonic flight, on minimizing shock waves that produce excessive wave drag, as we saw earlier in Figures 1.13, 2.21, and 3.1 and as demonstrated for airfoils in Figure 8.13. Consequently, the configuration tends to be slender with a long fuselage, slenderness ratio $s/\ell \lesssim 0.2$, and thin wings whose planform is hybrid and components carefully integrated. Although thin, the wings are highly constrained to provide the necessary depth and volume requirements with shaping for the deployment of high-lift devices.

9.4.1 Design Challenge: Overcome Drag

In order to design for supersonic flight with its large wave drag, we must first understand the origin of the drag, and from that understanding find ways to reduce it. As we have seen, both slenderness and thinness are key, hence thin wings of small *AR* with lift-enhancing coherent vortex separation represent the desired flow pattern. Remembering $M(L/D)$, the task then becomes to trade off
loss of cruise efficiency (i.e. larger D due to drag rise) with efficiency gained by supersonic cruise speed (i.e. larger M_{cruise}).
We shall study some of the fundamentals involved in such a trade-off in the next section. With these fundamentals in hand, two cases studies – the TCR and the Concorde – exemplify the design considerations.

9.4.2 Case Study IV: The TCR

Similarly to supersonic configurations, the near-sonic TCR is slender and its wings thin. This increases the critical Mach number above that for basic swept-wing concepts such as CRM, as Figure 1.9 suggestsed and the computations below substantiate.

At this point, however, we are getting ahead of our story. As explained in Section 1.1.5, the baseline TCR that Saab produced in its Cycle 1 design is a T-tail configuration. As we explain in Chapter 10, the T-tail TCR incurred excessive drag for trimming, so it was redesigned as a canard configuration, with the optimal version termed TCR-C15.

Here, we discuss some of the CFD investigations of TCR-C15 in trimmed flight. Since we are concerned about drag at high speed, it is appropriate to include all contributions to drag, including trim drag. The next chapter explains what is involved in the trim calculation; here, we focus on the surface pressure and shock wave development.

TCR-C15 Shock Patterns and Isobars: M*0.65–0.97*

The trim calculations will require equation solution using several canard deflection angle δ settings, but we begin first with $\delta = 0$.

Forces and Pitching Moment

Figure 9.15 plots lift C_L, drag as $C_D \times 10$, and pitching moment C_m to show the development with speed as the cruise speed is approached. We see that up to near $M_\infty = 0.97$, C_L increases more rapidly with M_∞ than C_D does. Close to $M1$, the lift-curve slope decreases, but shock stall has not yet set in.

Meanwhile, pitching moment $C_m < 0$ grows in magnitude, and increasingly so after $M_\infty = 0.85$, which of course must be trimmed by the canard, adding trim drag. The pitching moment change with Mach number is the well-known "tuck-under"

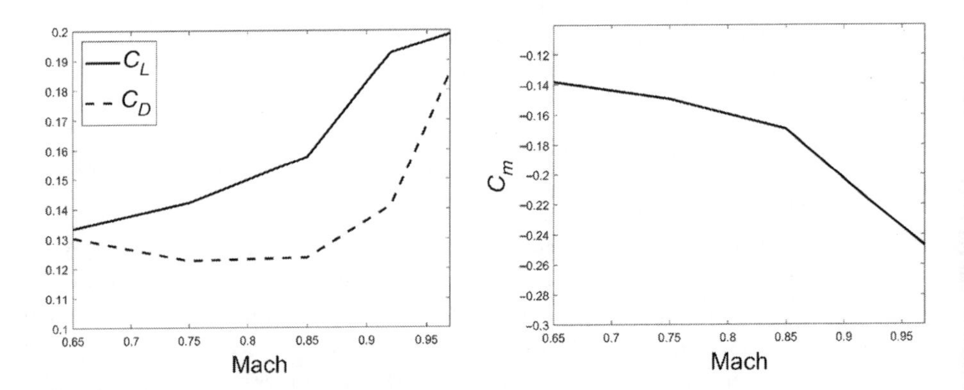

Figure 9.15 Mach-sweep computation, of C_L and $C_D \times 10$ (left); pitching moment C_m (right). EDGE RANS computation TCR-C15 configuration. (Courtesy of P. Eliasson [7], reprinted with permission)

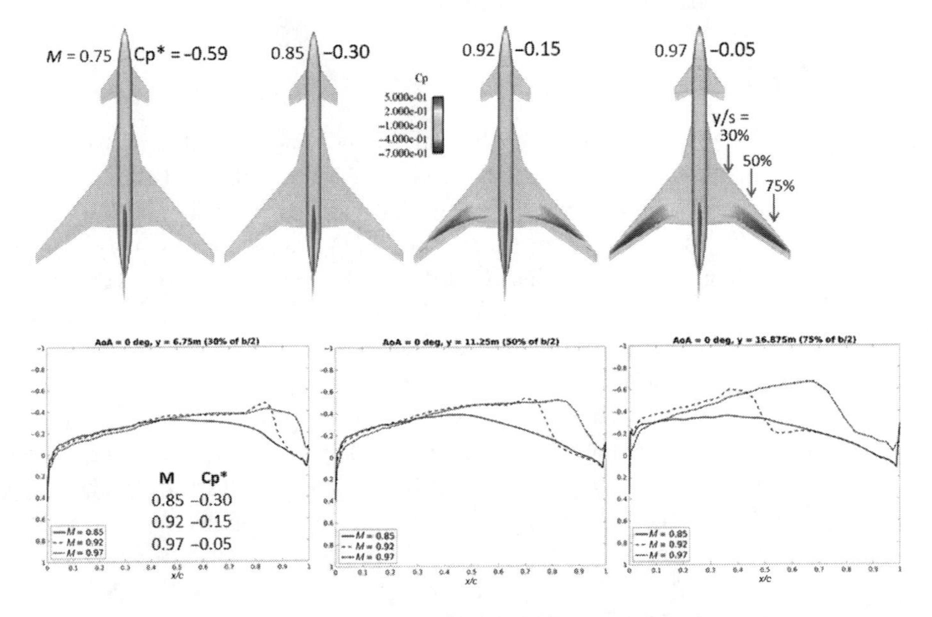

Figure 9.16 TCR-C15 transonic pressure distribution. (Top) Isobars on the upper surface; (bottom), C_p distributions plotted in three chordwise cuts; $\alpha = 0°$, $\delta = 0°$. AoA = angle of attack. (Courtesy of P. Eliasson [8], reprinted with permission)

phenomenon – the nose-down tendency caused by shock waves moving aft and shifting the overall pressure distribution. This trim problem saw mitigation in pitch-control systems, as we describe in the next chapter.

Surface Pressure

We must look at the upper surface pressure distributions and the shock patterns to understand better the increases in lift and drag. Figure 9.16 (top) also reveals that

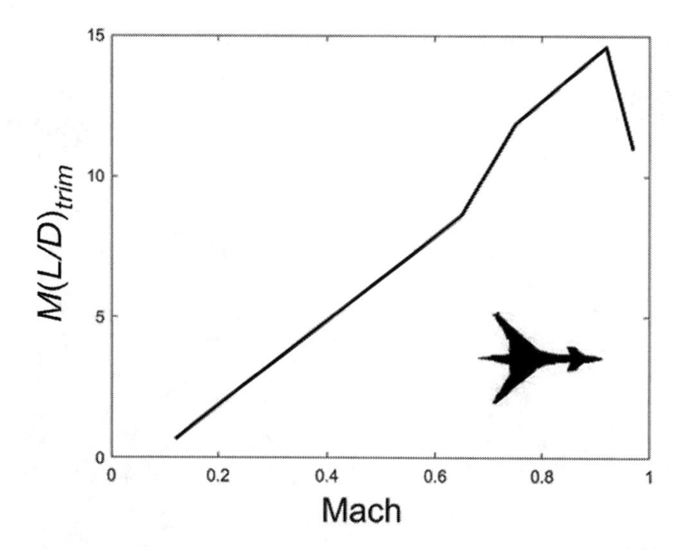

Figure 9.17 Cruise efficiency $M(L/D)$ in trimmed condition, aerodata from EDGE RANS computations.

the change in lift and pitching moment comes from the wing. The surface pressure distribution shows that the suction on the outer, hence more aft, part of the wing increases with increasing Mach number.

The chordwise upper surface pressure profiles show clearly how the shock and total lift move aft rapidly with M_∞ increasing from 0.85 to 0.97 – the tuck-under phenomenon.

Note that the shocks are not as swept as the leading edge, so there appears to be potential to reduce the wave-drag rise by further sweeping these shocks.

Cruise Efficiency

Figure 9.17 shows the TCR-C15 trimmed $M(L/D)$, which continues to rise after $M0.85$ and even $M0.92$, unlike what would be the case for the CRM wing. This comes about because even though the drag rises with increasing M_∞, the lift grows more rapidly.

We need to see what happens with drag if we increase the cruise speed beyond $M0.97$ to supersonic values. From the perspective of linear supersonic theory, Chapter 3 drew the conclusion that a subsonic leading edge results in lower wave drag. Below, we investigate the roles of sweep angle, slenderness, and thinness by comparing three classical planforms in supersonic flight. A detailed analysis should consider trimmed flight, including the trim drag, as with the TCR above. But the differences between the planforms are indicative even for a fixed angle of attack and no trim.

How Slenderness Affects Wave Drag

Chapter 1 discussed planforms very generally (e.g. Figure 1.11). Let us now study three of those planforms in more detail, all with $AR = 2$, the same wing area S,

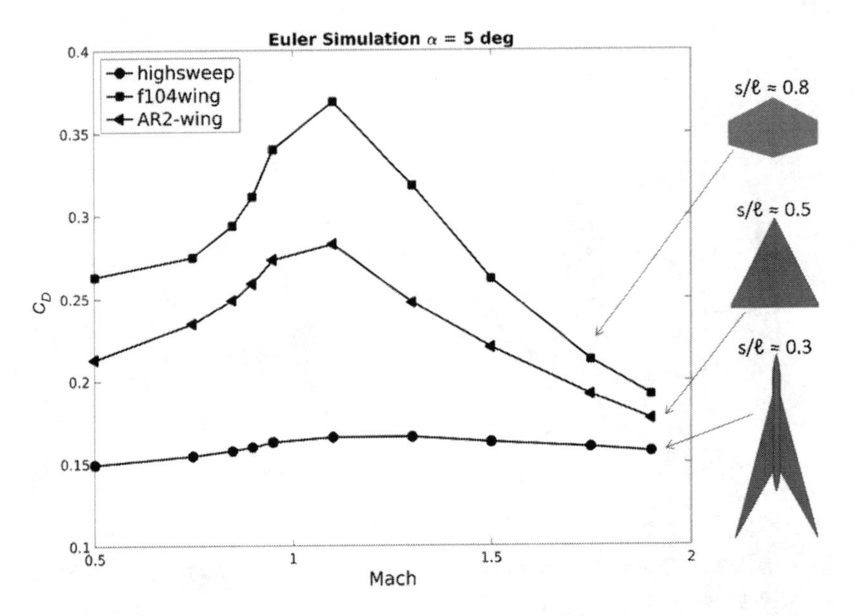

Figure 9.18 Inviscid drag for a Mach sweep at $\alpha = 5^\circ$ for the three supersonic wings. (Top) The Aerion F-104-type wing; (middle) a 63.4° sweep delta; (bottom) the "highsweep" body with wings swept significantly inside the Mach cone. EDGE Euler computations.

and thickness 4%. The first is a trapezoidal planform, similar to that of the F-104 Starfighter, with a circular arc airfoil; the second is a delta planform; and the third is a highly swept arrow wing. The last two have the symmetric 4% Royal Aircraft Establishment (RAE) airfoil. Further details of these three planforms are as follows.

1. An unswept wing of slenderness ratio $s/\ell \approx 0.8$ giving taper ratio 0.6 with supersonic leading and trailing edges. By intention, there is no fuselage; the F-104 fuselage gives subsonic leading edges at $M2$.
2. A delta wing of slenderness ratio $s/\ell \approx 0.5$ and also no fuselage, but because of the high sweep, it has a subsonic leading edge and a supersonic trailing edge.
3. A high-sweep arrow wing–body of slenderness ratio $s/\ell \approx 0.3$, with sweep to give subsonic leading and trailing edges.

In the following section, we draw some conclusions from Euler simulations. The Mach cone angle is 30° at $M2$. The delta planform has sweep angle $\Lambda = 63.4^\circ$ and is subsonic up to $M2.2$.

Figure 9.18 presents a Mach sweep from $M_\infty = 0.5$ to nearly 2 computed at $\alpha = 5^\circ$ for the three supersonic wings to see what each one can imply in terms of wave drag.

Sub- and Super sonic Leading/Trailing Edges

Chapter 3 described the effects of subsonic and supersonic leading edges for representative planforms shown in Figure 1.11. The computation confirms that for values of M_∞ in the sweep, the wave drag of a subsonic leading edge is lower than that of a supersonic edge; the higher the sweep, the lower the drag. The configuration should be slender enough to be inside the Mach cone rooted at the nose. Both the trapezoidal wing and the delta wing show dramatic maxima in C_D just past $M1$, while the high-sweep wing does not.

The Aerion Supersonic Business Jet

Aerion Corporation has been working on a small supersonic airplane design for a number of years. During that time, the design has evolved. It is expected to cruise at $M1.1$–1.2 without supersonic boom with a long-range cruise Mach number of 1.4. It has an essentially unswept wing, similar to the trapezoidal configuration in Figure 9.18.

The concept arose from the "natural laminar flow" idea, a fundamental property of compressible fluid mechanics. For an unswept wing at supersonic speed, the pressure on the wing falls continuously from the leading edge to the trailing edge. This means that the pressure gradient is favorable over the entire wing and, for a modest Reynolds number, the flow should stay laminar. The resulting low skin-friction drag could become an enabling technology. These considerations help to explain what was shown in Figure 3.23 and hold also for fighters such as the Saab Gripen (e.g. [22]).

9.4.3 Case Study V: A Concorde-like Configuration

On the slender delta-winged Concorde, the leading edge was made very sharp to provoke separation even at the low angles of attack required at cruising speed. It was also warped along its length in such a way as to ensure that the vortices grew evenly along the leading edge. Figure 9.19 gives a preliminary idea of the complexity of such a wing, although the details of camber, thickness, and drooped nose are difficult to capture in a photograph.

In addition to the leading-edge warp, the wing has spanwise variations in camber for reasons that will be explained later. The double-delta planform with high sweep inner section and lower sweep outer section was used in the fifties on the M 2 Saab J 35 and later on other high-speed transport design proposals.

Concorde was originally envisaged to fly with attached flow in cruise, but it was found in concept studies that the optimum cruise condition was obtained with a small amount of vortex flow that separates from the sharp leading edge, resulting in flow patterns as illustrated in Figures 2.18 and 2.19 and discussed in Section 1.2.2 and Figure 1.13, as well as in Figure 2.1. The high suction under the vortex gives an extra nonlinear lift contribution and thus reduces the incidence needed for a given lift (e.g. for takeoff and landing). It was further argued, and fully confirmed later, that the changes in flow with Mach number would be smooth and gradual. The leading edge would remain subsonic up to a high free stream Mach number. When supersonic, the shock waves would be highly swept and any shock-induced separation would still tend to create a free vortex layer.

Figure 9.19 Upper side of Concorde wing (left) (author's photo of Concorde at the Udvar–Hazy Center of Air and Space Museum); SUMO Concorde-like model (right).

The concept studies investigated many aerodynamic aspects, such as the following.

1. The effects of changes in leading-edge sweepback
2. The relative merits of planforms: delta, gothic, and ogee
3. The effects of cross-sectional area distribution and shape
4. The use of fore-and-aft camber to minimize the lift-dependent drag and trim changes
5. The low-speed characteristics and, in particular, the incidence/sideslip values beyond which the vortices would burst and a steady flow pattern would break down

Figure 9.20 defines the limitations within which one can/must design a slender wing for flight at about $M2$. It indicates the "region of no conflict" (i.e. where among all the trade-offs the requirements for the wing design can be satisfied) and thus the latitude allowed in determining the final wing form.

There are four low-speed limitations.

- The onset of Dutch-roll instability.
- The vortex breakdown at 5° sideslip
- The maximum cabin floor angle for comfort in takeoff and approach
- A required C_L for the approach

In addition, $\sqrt{M_\infty^2 - 1}\frac{s}{\ell} \leq 0.7,\ M_\infty \approx 2.2$ is needed for adequate cruise performance.

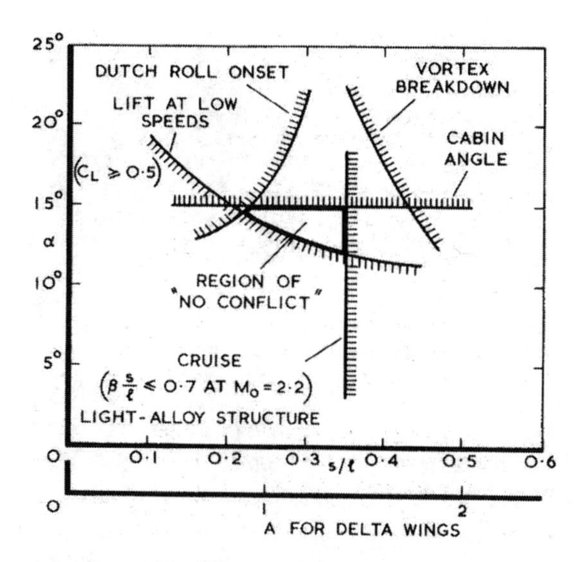

Figure 9.20 "Region of no conflict" indicates where the constraints are met and the requirements are satisfied by a slender delta wing. (From Küchemann [14], reprinted by permission from AIAA)

The wing of a supersonic aircraft has to satisfy the following two contradictory demands.

(1) Low wave drag
(2) Satisfactory low-speed performance

The first demand calls for a wing so slender that the leading edge is nominally subsonic. The second demand calls for a sufficiently high *AR* to achieve the needed low-speed performance. This, of course, goes together with a wave-drag increment in supersonic flight.

The double-delta wing in its inboard part is to meet the first demand, while the outboard part is to meet the second demand. The second delta, moreover, has the welcome effect of a small movement of the neutral point during the transition from sub- to super-sonic flight. This allows for a trim schedule with minimized trim drag. At $M2$, the leading edge of the Concorde wing can be considered as nominally subsonic about halfway to the tip and then, at the outboard delta, as nominally supersonic, and eventually, with increasing sweep at the wing tip, as subsonic again.

Computational Model of a Concorde-like Configuration

Our intent is to elucidate some of this parameter space with CFD computations on a Concorde-like configuration constructed from data published in the open literature and that comply with the region of no conflict.

The AIAA Concorde design book [24] and the excellent review by Wilde and Cormery [33] explain the geometry of the wing in enough detail to make an approximation to it, although the exact camber cannot be ascertained. Figure 9.21 shows the Concorde-like model we have constructed with SUMO.

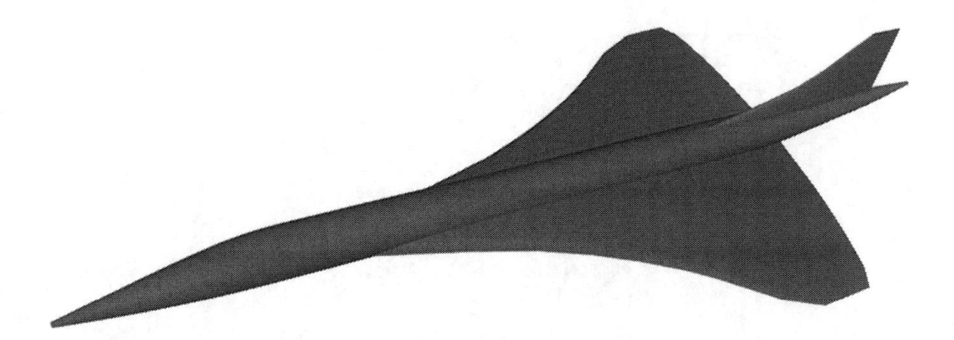

Figure 9.21 The Concorde-like model created and meshed with SUMO.

CFD Illustration of Some Concorde Aerodynamic Characteristics

The leading-edge vortices induce suction on the upper surface of the wing that remains constant along the chord, except for the region of the trailing edge. As the vortices grow with angle of attack, the proportion of the total lift contributed in this way increases, with the lift curve being nonlinear but monotone in the range appropriate to practical flight. This vortex flow is maintained at supersonic speed, although the effects of the vortices becomes less marked with increase in Mach number.

This section presents the results of Euler computations investigating some design questions for the wing. Among the design issues we investigate are the following.

- Cruise efficiency
- Controlled separation – nonlinear vortex lift
- Adequate lift for landing
- Drag penalty for drooped leading edge
- Upper surface isobars and shock waves

Cruise Efficiency

Concorde cruises at $M_\infty = 2$ and about $C_L = 0.1$, for which the angle of attack is roughly $\alpha = 5°$. Figure 9.22 presents a Mach sweep in inviscid flow at this incidence to indicate the overall cruise efficiency and confirms its rise toward $M2$. Steady straight and level flight needs moment balance provided by elevator deflection. Trim drag and viscous drag should be added to the wave drag computed, so this simple computation overestimates $M(L/D)$.

Controlled Separation: Nonlinear Vortex Lift

The flow pattern around a swept wing of very low *AR*, such as a delta of *AR* 2.0, is very different from that for a high-*AR*, moderately-swept wing. It is highly three-dimensional: "all tip effect" is a graphic way of describing it.

It presented an alternative to the traditional approach of treating flow separation as a phenomenon to be suppressed. Instead, it suggested that one could positively stimulate a separation pattern to generate a stable system of vortices that would exhibit regular growth with increasing angle of attack.

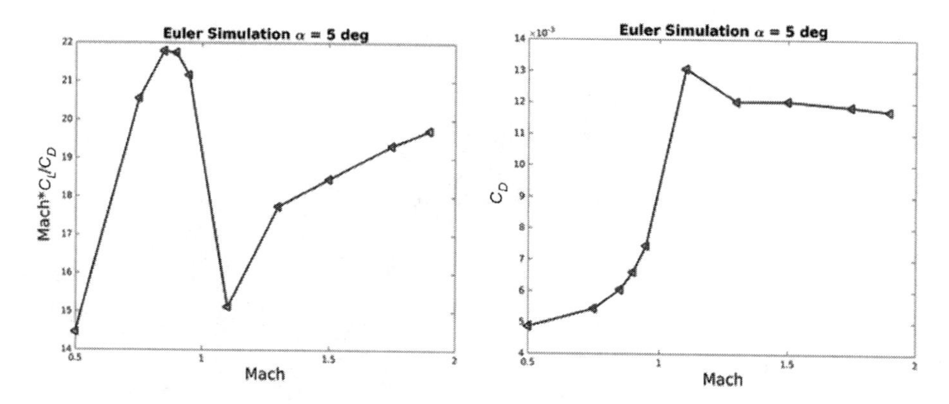

Figure 9.22 Cruise efficiency of the Concorde-like configuration. (Left) $M(L/D)$ and (right) C_D for Mach sweeps from 0.5 to 2. (Courtesy of M. Zhang, personal communication)

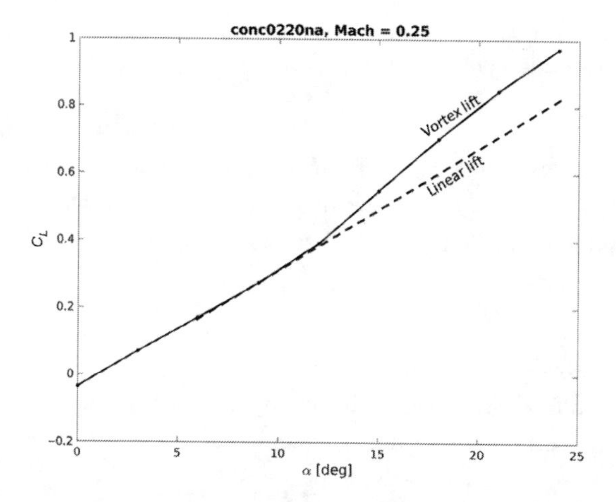

Figure 9.23 Nonlinear vortex lift computed in Euler simulation at low speed (e.g. landing).

Figure 9.23 shows the nonlinear vortex lift at low speed in inviscid flow. Estimating this vortex-lift enhancement was crucial, since it would determine whether the landing attitude would be acceptable.

Adequate Lift for Landing

As angle of attack increases, a slender delta produces lift only about half as vigorously as a modern swept-wing subsonic transport. The net result was that, just as cruise wave drag forced acceptance of much lower values of L/D than customary, in the absence of high-lift devices, takeoff and landing would have to be made with smaller lift coefficients. Roughly speaking, subsonic transports can count on a usable lift coefficient (without high-lift devices) of about 1.0, whereas slender deltas produce no more than 0.5. Even to generate such low lift, it looked as though a slender delta would have to be flown at angles of attack approaching 15°, which is very large

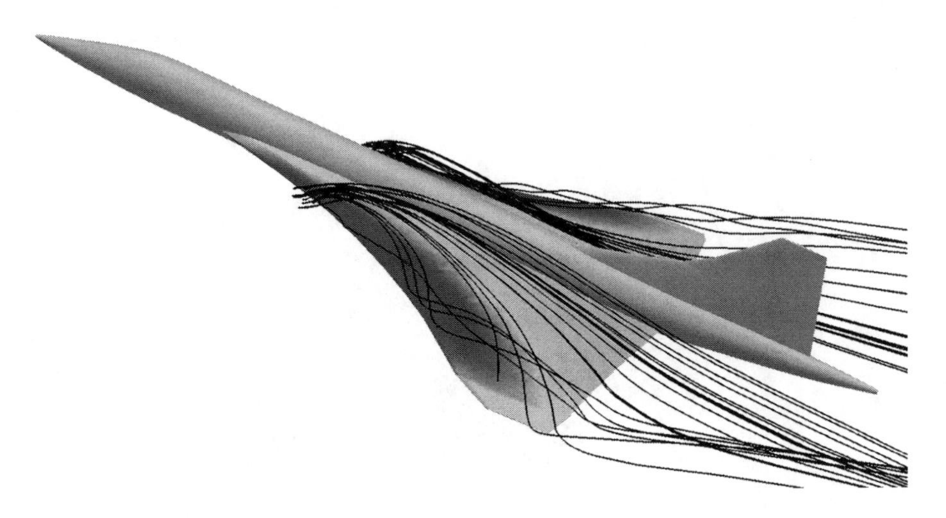

Figure 9.24 Vortex-flow features computed in Euler simulation, EDGE code. (Courtesy of M. Zhang, personal communication)

by subsonic standards. Nevertheless, early on in the Concorde deliberations, it was firmly established that, at angles of attack of this order, the lift was sufficient and could be trimmed longitudinally. Figure 9.24 shows the development of vortex flow over the Concorde.

Drag Penalty for a Drooped Leading Edge

Finally, considerable effort was expended reviewing the consequences of introducing changes to an existing design. For instance, reducing the leading-edge droop over the whole span would improve the supersonic drag.

The Concorde designers worked very hard to trade between sufficient droop at the leading edge for high lift during landing and the extra drag encountered at cruise. We have run two Euler computations for our Concorde-like configuration in cruise to assess the drag penalty for drooping the leading edge. The original wing was flattened to reduce its camber (i.e. the leading-edge droop). At cruise $C_L = 0.116$, the flattened wing C_D is 0.0106 and the original wing has $C_D = 0.0116$. The 10-drag-counts cruise-drag penalty for drooping the leading edge is only 10%.

Upper Surface Isobars and Shock Waves

Figure 9.25 presents iso-Mach curves for $M2$ flight over the upper surface to substantiate the absence of shock waves. In Figures 9.25 (right), the entire upper surface is supersonic, with Mach from 1.5 near-leading edge to 2.5 under the vortex and back to 2 at the trailing edge.

In Figure 9.25 (left), three chordwise plots of $C_p(x)$ at 24%, 40%, and 56% of half-spanwidth are shown.

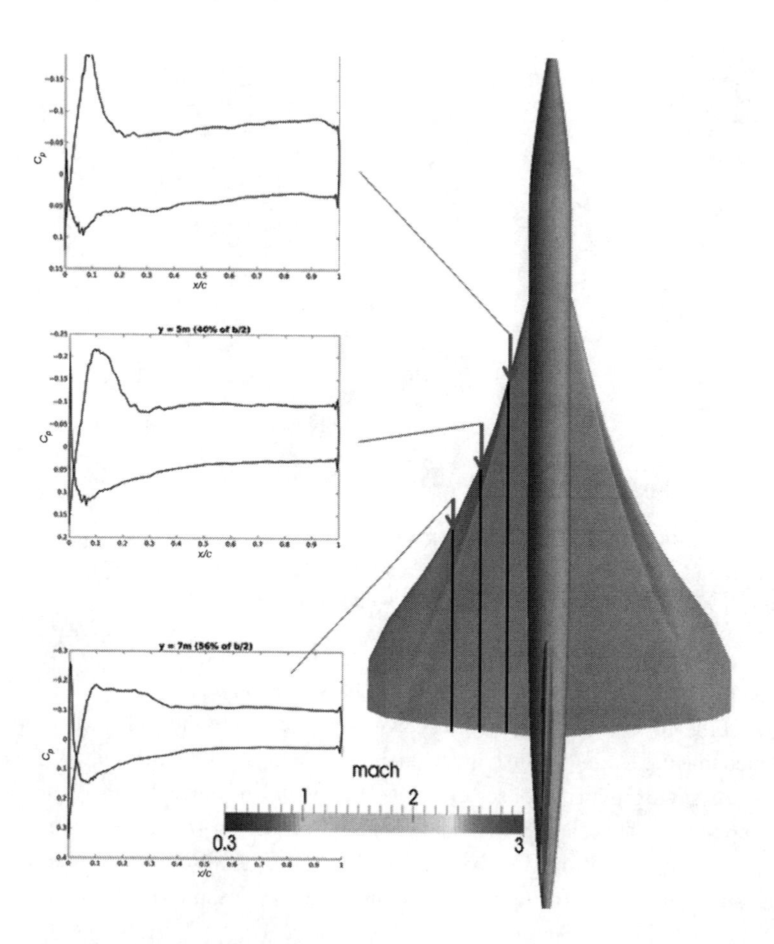

Figure 9.25 Mach distribution computed on wing upper side (right); chordwise C_p profiles (upper and lower side) at 24%, 40%, and 56% of half-span (left). (Courtesy of M. Zhang, personal communication)

Figure 9.25 shows how the suction peak widens outboard as the vortex expands and lifts off. This is actually a streamwise process along the swept leading edge. The suction peak is followed by an isentropic recompression (i.e. without shock) to the free stream velocity at about $M2$. The streamwise section of the wing has very slowly varying thickness. As expected for supersonic flow, pressure is almost constant on upper side and lower side, all the way to the trailing edge, which is supersonic, hence the pressure jump. The absence of shock waves over the wing proper gives low wave drag. Note the very sharp suction peak on the leading-edge pressure side, the result of the small leading-edge radius and the strong droop. It carries downforce, but this is only small, since it is so local.

As we have pointed out before, delta-wing flow fields are complicated and hard to estimate, so CFD is called to the rescue, and Euler computations with automatic mesh generation are quick. It remains, of course, for due diligence to assess the credibility of the inviscid flow model.

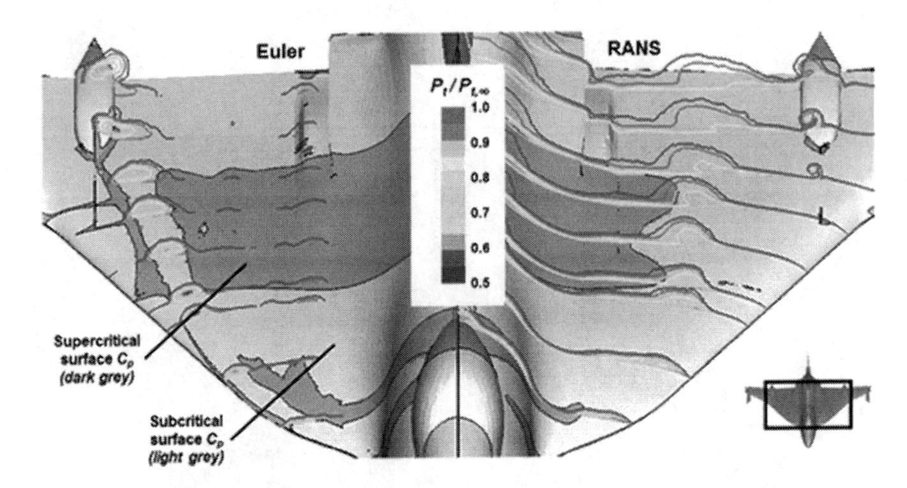

Figure 9.26 Comparison of Euler and RANS solutions on adapted grids for the F-16XL flight condition FC-70, $M_\infty = 0.97, Re_{cref} = 89M, \alpha = 4°$. EDGE code simulations. (Courtesy of S. Crippa, PhD dissertation [5], reprinted with permission)

Due Diligence: Aerodynamic Sharpness of the Leading Edge

All of the computations we have shown are solutions to the Euler equations. We have stated that the leading edge of the Concorde is "aerodynamically sharp." This is another way of saying that the onset of the vortex would be virtually the same in a RANS computation. This can only be confirmed by comparing solutions because there are no a priori criteria defining "aerodynamically sharp." A RANS computation confirmed the small difference from the Euler simulation on the same (RANS) grid.

Euler and RANS on F-16XL

A more thorough comparison of Euler versus RANS computation was carried out for the F-16XL to shed additional light on the questions of vortex location, sharpness of edges, and shock wave-boundary layer interaction.

Steady transonic vortex flow that does not undergo breakdown over the F-16XL at conditions $M_\infty = 0.97, Re_{cref} = 89 \times 10^6, \alpha = 4°$ is a challenging case of shock–vortex–boundary layer interaction. In his PhD thesis, S. Crippa [5] presents an interesting comparison between an inviscid solution and a RANS solution. This brings some insight into the role of the boundary layer in vortex and shock positions.

Figure 9.26 shows the substantial difference in vortex core strength between the Euler solution on the triply adapted "inviscid" grid and the RANS solution on a singly adapted "viscous" grid. Although the leading-edge section that joins the fuselage to the high-sweep section features a sharp leading edge, the high-sweep section features a decreasing radius toward the wing kink, which probably cannot be considered "aerodynamically sharp" at that point. If that is the case, the position of vortex onset in the Euler solution is grid dependent. For the Euler analysis, the discrete grid is responsible for the generation of vorticity and thus the refinement of the surface grid changes the predicted leading-edge vortex location, as Figure 9.26 suggests. By refining the blunt leading edge of the high-sweep section, the separation onset predicted by the inviscid method moves downstream.

Figure 9.26 also presents the topological difference in supercritical regions between the two computations. The supercritical region in the inviscid computation is clearly divided in the chordwise direction by the footprint of the leading-edge vortex. In the viscous result, the supercritical region extends from the symmetry plane to mid-span, whereby the weaker vortex has less impact on it. Most notably, although at $M_\infty = 0.97$ the shock is expected to be strong, the comparison indicates very weak shock–boundary layer interaction since there is little difference in the shock location terminating the supercritical regions in the Euler and RANS solutions. Even at its design cruise speed of $M_\infty = 0.85$, the shock location on the CRM wing computed in the Euler solution would be markedly different from that computed in the RANS solution. This leads one to believe that thin, slender configurations develop significantly different shock patterns from those for transonic swept wings such as the CRM.

9.5 Further Configuration Development

Wing design is never based on wing-alone benchmarks; it must also include wing–body calculations, because the flow in the body-junction region can strongly influence the aerodynamic behavior of a given wing. Furthermore, the configuration must be trimmed by balancing all moments by the tail plane or canard, a subject that is dealt with in the next chapter. The preceding sections have described Cycle 2 clean-wing design in some detail. The next step in the design procedure, Cycle 3, proceeds in many different directions to develop the configuration further. First among them is to join the fuselage to the clean wing. Adding the body brings interference effects such as interference drag, which occurs at both low and high speeds. Joining the wing root to the fuselage without smoothing the juncture creates acute angles between the surfaces. At low speed, the flow in the juncture region may separate before the trailing edge in a *juncture vortex*, which adds drag. Hirschel et al. [11] present details on the CRM wing–body juncture flow. High speed adds wave drag, and one must find the optimal wing–fuselage combination to minimize it. The area ruling discussed in Chapter 3 offers guidelines.

This design cycle typically results in a family of several feasible configurations meeting these requirements and competing with each other.

Modern Fighter Aircraft

The shapes of subsonic and transonic transport aircraft have converged, to a certain degree, to presumably optimum configurations (cf. Sections 9.2 and 9.3). The missions for airliners have been essentially the same since commercial air traffic began, which contributes to the convergence. Similarly, the design process has been streamlined and detailed along similar lines for all manufacturers. In fact, the aerospace historian John D. Anderson, Jr., in his book *The Grand Designers*. traces the systematic cycles in design back to the Wright brothers, later substantially refined by Frank Barnwell, master designer for the World War I UK aircraft company Bristol.

Such convergence is not at hand for modern fighter aircraft, particularly those that operate in the transonic and the (low) supersonic flight domain. The many different requirement profiles of such aircraft have led to quite different vehicle shapes, where signature aspects (radar, infrared) have also played a large role.

Current military aircraft seem to fall into the following two major configuration classes.

(1) The delta wing with a canard
(2) The hybrid wing (i.e. some combination of the trapezoidal wing with a highly swept strake)

European designs such as the Eurofighter Typhoon, the Dassault Rafale, and the Saab JAS 39 Gripen, which will be discussed below, favor the delta wing with a canard, while most of the current American and Russian fighter aircraft have a hybrid wing.

At high angle of attack, the highly swept strake develops a strong vortex system with a twofold effect. It produces a nonlinear lift increment, and also stabilizes the flow over the trapezoidal wing. Should the flow separate chaotically at high incidence on the outer part of the trapezoidal wing, the separated area will be much smaller compared to what it would be without the strake.

The delta wing with canard and the hybrid wing both have their merits. In some instances, one has advantages over the other. Which type is finally preferred depends on the mission requirements and on the assessment of the single mission elements. In addition, the experience of the design team, and even the product philosophy of the manufacturing company, may play a role in the decision, as illustrated in Hirschel et al. [11].

In an unusually candid exposé of the aerodynamic evolution of the Saab JAS 39 Gripen, Modin and Clareus and their colleagues [3, 12, 22] describe in detail the industrial design procedure leading to the final layout selection. We give here a synopsis of this multiyear industrial project to indicate the nature of the work involved and its magnitude.

9.5.1 Case Study VI: Cycle 3 Layout Selection of the JAS 39 Gripen

Layout Evaluation

As Figure 1.1 suggested, the initial layout studies at Saab included a variety of different layouts. Detailed design studies including structure and systems installation were carried out for a handful of the competing configurations, in particular the 2105 and 2102 in Figure 8.27, but also the 2107, among others. Their essential characteristics were analyzed, and the results formed the basis for the selection of a reference configuration for further refinements and the development of the final aircraft. Work soon concentrated on two alternatives: a close-coupled delta canard configuration 2105 and a more conventional aft-tail 2102. Although the choice of engine was not supposed to have any significant influence on the relative merits of the configurations, the studies adopted the F404 engine and set the wing loadings on these two layouts to give a common landing approach speed.

Modin and Clareus remark that

> It is a delicate task to make a fair comparative evaluation of the different merits of candidate airplane layouts.

This is because, in their selection, the technical results for the designs came out rather close. For example, the 2107 aircraft showed advantages in weight and some performance aspects. It had an unconventional air intake in the root of the dorsal fin. At high angles of attack the forebody blanketing might unacceptably distort the intake flow, and 2107 was dropped for technical risks.

To give some appreciation for this delicate selection task, we compare two major characteristics for the two remaining configurations, 2102 and 2105, as they were at the time when the layout selection was made in December 1980: high-speed and turn performance. For their further discussions of, among many other factors, instability level (e.g. static margin), weight, low-speed performance, etc., the reader is referred to Modin and Clareus [22].

High-Speed Performance: Area Ruling

As explained in Section 3.4.5, the wave drag at transonic speeds can be minimized by ensuring that the distribution of the aircraft total cross-sectional area along the length of the aircraft axis follows a smooth pattern such as that of a low-drag body of revolution. Let us now listen to Modin and Clareus describe their sophisticated contouring shaping of the Gripen by such area-ruling optimization.

Figures 9.27a and 9.27b show the plan views of Gripen prototypes 2102 without and 2105 with area ruling. Figure 9.27c plots the total cross-sectional areas of the two configurations in cuts perpendicular to the length axis (i.e. corresponding to $M1$). At $M1$, previous canard configurations produced an unfavorable saddle form. However, by careful local fuselage design, this tendency is not pronounced for the 2105, and has disappeared completely at $M1.1$. The maximum cross-sectional area of the 2105 is some 9% lower than that of the 2102.

The slope of the area distribution toward the aft end of the aircraft is of particular importance to supersonic wave drag. This favors the 2105 prototype because the absence of an aft tail and the forward position of the wing on the fuselage, necessary for the desired instability in pitch, makes it possible to obtain an aerodynamically clean aft end on the canard configuration with a favorable area distribution.

Figure 9.27d presents the zero-lift drag for the two configurations. The canard configuration is only slightly better at subsonic speeds, while at supersonic Mach numbers, the difference is quite significant.

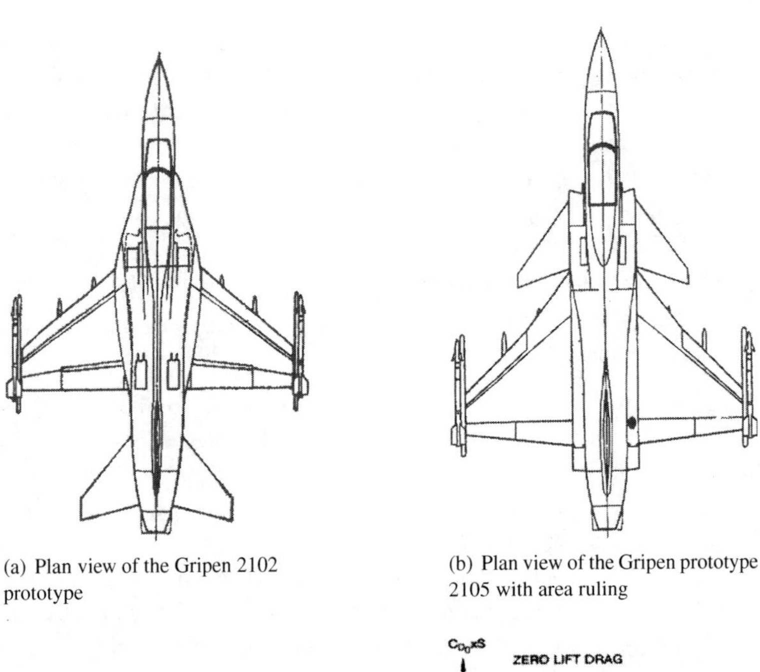

(a) Plan view of the Gripen 2102 prototype

(b) Plan view of the Gripen prototype 2105 with area ruling

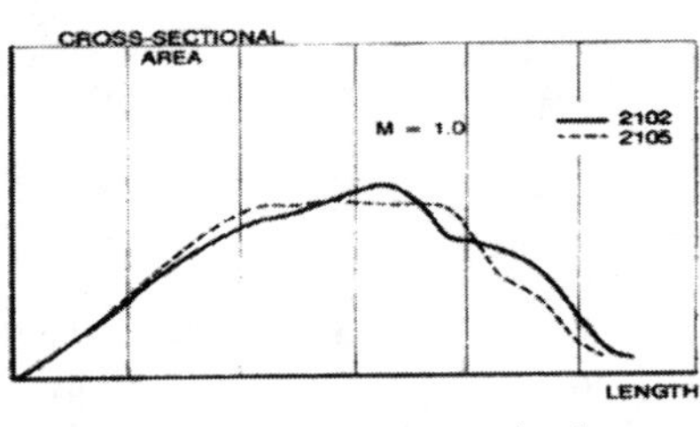

(c) Gripen prototype cross-sectional area

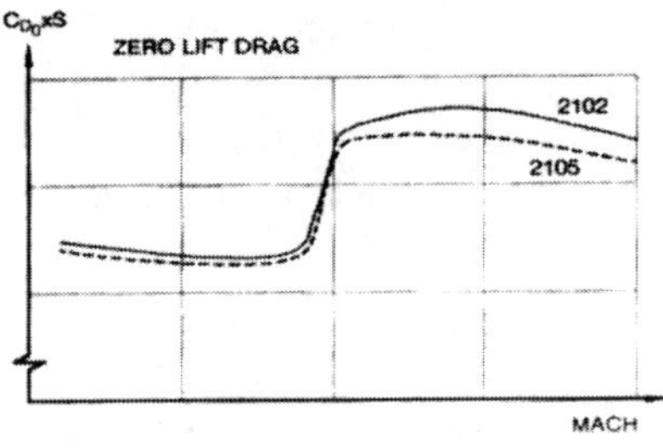

(d) Gripen prototype zero-lift drag

Figure 9.27 Variation of the cross-sectional areas of Gripen prototypes 2102 and 2105 and their effects, respectively, on zero-lift drag. (Courtesy of K. Nilsson [12], Saab, reprinted with permission)

Turn Performance

Important parameters for sustained turn performance are high Oswald efficiency factor *e* and, even more so, low spanloading. At low speed, both are more easily obtained on the (high-*AR*) wing-tail 2102 than on the 2105 delta canard configuration. There was a trade-off, however, made for high speed. Since the zero-lift drag is a dominant parameter at supersonic speed (see Figure 9.27d) to reduce drag, the 2105 canard camber was kept smaller than the optimum would be for subsonic maneuver. This trade-off in canard camber led to approximately 10% higher maneuver performance for the 2105 than for the 2102.

Final Selection

Many characteristics were taken into account, some of which are not directly related to aerodynamics, for selecting the final layout. In the technical evaluation, no single virtue alone was decisive in favor of the delta-canard configuration. However, when weighing all of the pros and cons when compared to the aft-tail contender, the canard airplane was found to be the best candidate to meet the stipulated requirements from both technical and economical points of view. After 2 years of intensive prototype development and evaluation, in December 1980, Saab selected the 2105 delta-canard concept. After another 8 years of further refinement and optimization, it took to the skies in 1988.

9.5.2 Topics in Configuration Development

The previous section indicated the breadth and depth of the industrial work required to develop the aircraft configuration (Cycle 3). The scope of this textbook allows us to just barely scratch the surface on this endeavor, and we do this on three specific topics concerning additions to the clean wing. The first is integration of the wing and fuselage and the second is wing tip design considerations. The third concerns adding high-lift devices to the clean wing, and we give one example of their analysis. The next chapter then looks at tail-plane development, matters of trim, etc.

Wing–Fuselage Integration

Junctions between surfaces are always sources of premature flow separation, often resulting in a *juncture vortex* (also termed a necklace vortex), leading to aerodynamic buffet and drag. The turbulent boundary layer developing on the forebody fuselage experiences a severe adverse pressure gradient where it encounters the wing surface. Consequently, the boundary layer separates from the fuselage just ahead of the wing leading edge. The resulting flow pattern is complex and fully 3D with a juncture vortex occurring in many cases, with a subsequent increase in drag.

In transonic cruise conditions, shock waves can have a strong impact on the flow structure, as we have seen in Section 9.3. What this means is that the fuselage shape must also be subject to drag minimization. The flow behavior at the wing–fuselage junction is also very sensitive to the geometry of the wing root, such as the sweep angle. Such a juncture vortex occurs on the CRM wing, and Hirschel et al. discuss this in depth in Ref. [11], to which the interested reader is referred.

Wing–Fuselage Fillet

Wing fillets (and tail fillets) are designed to fill the volume between surfaces where they meet at an acute angle to help keep the flow attached. The flow field around them can serve to induce extra lift on a wing in regions where the local velocities are otherwise low. Figure 9.28 illustrates the juncture flow and how a fillet helps to maintain attached flow.

Wing Tip Design

The shape of the wing tip influences the flow locally. The CRM planform is cut off streamwise at the tip and consequently the isobars are drawn forward, as shown on the left in Figure 9.12. The tip edge, however, also plays a role. A rounded tip edge

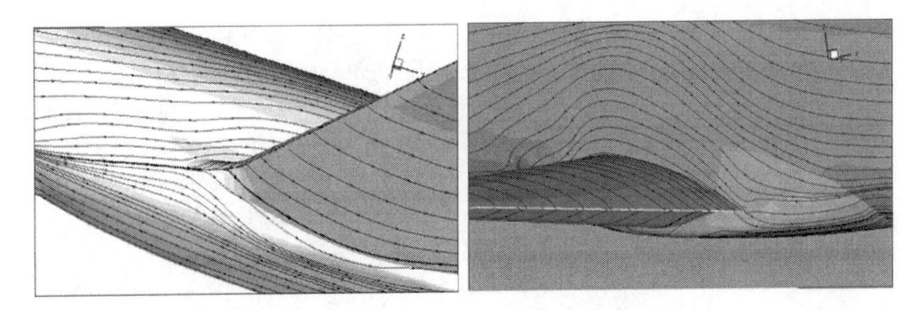

Figure 9.28 Fillets fill in and smooth over areas between surfaces that, intersecting at acute angles, trap and slow the air, causing unwanted separation and juncture vortices. (Courtesy of M. Zhang [34], EU project NOVEMOR, personnal communication)

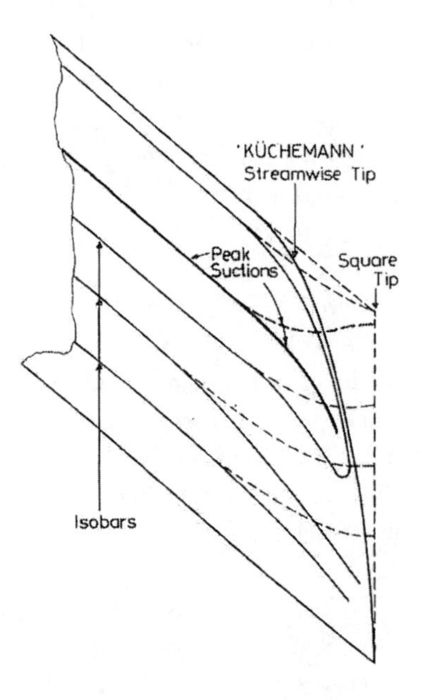

Figure 9.29 Küchemann wing tip – isobar patterns near the tips of swept-back wings. Dashed lines: constant sections and streamwise tip; solid lines: curved tip. (From [10], reprinted with permission)

facilitates the flow from the bottom of the wing to the top, while a sharp edge at the tip makes it more difficult and thus is equivalent to a larger span, with improved aerodynamic performance.

Figure 9.29 shows how giving the tip more sweep has the effect of sweeping the isobars downstream. Such curved tips were common on airliners in the 1990s (e.g. the Airbus A-321).

Winglets

Drag-reducing devices can also be integrated into the wing tip. Winglets are one or two cambered and twisted surfaces positioned in the rotating flow at the wing tip. They are

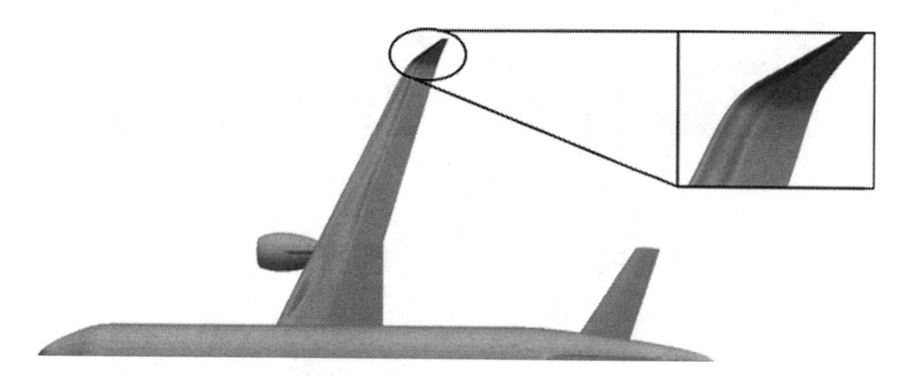

Figure 9.30 Example of the blended winglet design on the NOVEMOR reference wing; Mach = 0.78, $C_{L,wing} \approx 0.47$. (Courtesy of M. Zhang, PhD dissertation [34], reprinted with permission)

shaped to give a force with a positive forward component, which reduces the airplane drag. The shape is optimized for one flight condition, and in other conditions the drag reduction is less. The winglet is thus a useful device for airplanes spending most of their flying time cruising at a single defined flight state. Boeing, however, adopted a raked (highly swept) wing tip extension for the 767-400 in preference to the canted winglet that was proposed originally.

Blended Winglet Example

Current design philosophy seems to favor blending the winglet shaping into the overall wing design. This was the approach taken in the EU Project NOVEMOR for the design of a regional aircraft wing to cruise at $M_\infty = 0.78$ and $C_{L,design} = 0.47$. Figure 9.30 shows a blended winglet of about 10% of the main wing semi-span, designed to maintain swept parallel isobars at the tip and onto the winglet.

9.5.3 High-Lift Devices and Analysis

So far, we have been looking at configurations with clean wings, but high-lift devices are required for flight at low speed, such as at take-off and landing. The additional wing lift comes from the following.

- Increased airfoil camber
- Boundary-layer control due to re-energized or removed low-energy boundary layers
- Increments of the effective wing area

Trailing-edge flaps increase the camber and lift for a given incidence. Leading-edge flaps postpone leading-edge stall to a higher angle of attack, as we have seen in Chapter 8 for the 2D case.

Accurate computation of flows over high-lift devices is a challenging task. At the leading edge of the slat, there are effects of compressibility, unsteady cove flow, and the

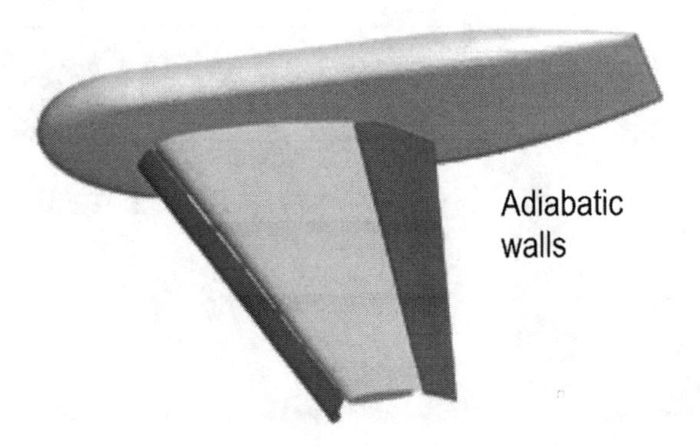

Figure 9.31 The NASA trapezoidal wing. Slat at 30° and flap at 25°. Brackets for slat and flap on underside not visible here. (Courtesy of M. Tomac, PhD thesis [30], reprinted with permission)

presence of a laminar separation bubble. On the main wing profile, confluent boundary-layer flow and adverse pressure gradient wakes are present. AIAA has recognized the importance of this aspect and has held three cycles of the High Lift Prediction Workshop, where the test case for the first cycle was the NASA trapezoidal wing.

Three-Element NASA Trapezoidal Wing

Rumsey et al. [25] and Slotnick [28] describe, and Figure 9.31 shows, the three-element NASA trapezoidal wing case for the high-lift prediction workshop (slat at 30°, flap at 25°) with brackets.

Compare Results with and without Transition Modeling

Transition modeling is known to be important in 2D multielement airfoil studies, so transition effects could be significant and therefore should be modeled. Figure 9.32 shows the forces and pitch moment computed with the $\gamma - \tilde{Re}_{\theta t}$ shear stress transport (SST) transition/turbulence model of Langtry and Menter [15] in the EDGE code, and it compares these with fully turbulent computations and with experiments. Transition modeling improves not only the absolute value of the forces, but also the angle of attack at which lift breaks down (Figure 9.32, left and center). The prediction of the pitching moment is also improved for 13–28° angle of attack.

Analyze Results

The $\gamma - \tilde{Re}_{\theta t}$ SST transition/turbulence model includes the solution of two additional transport equations for an intermittency parameter and a local transition-onset momentum-thickness Reynolds number. The effective intermittency regulates the production of turbulence kinetic energy (TKE) in the SST model.

Figure 9.33 indicates the laminar and turbulent areas by viewing the projected intermittency levels on the surface and iso-surface of elevated turbulent kinetic energy for an angle of attack of 28°.

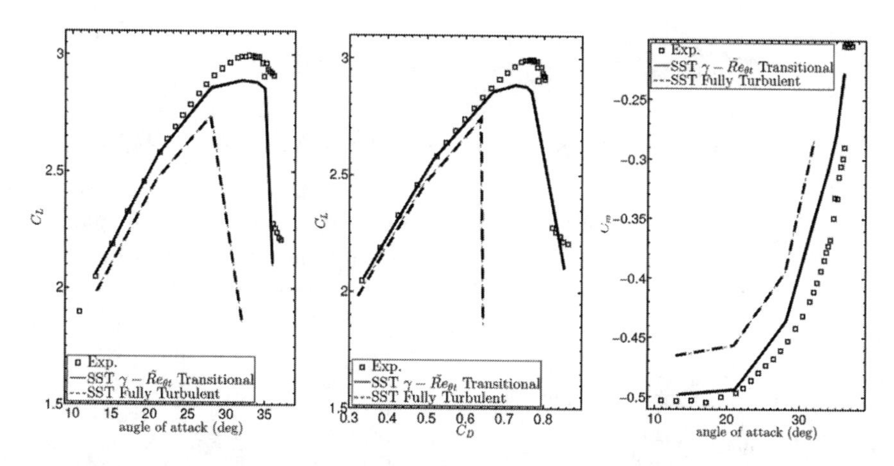

Figure 9.32 Integrated forces and moments computed with the EDGE code. NASA trap wing, $M = 0.2$, $Re_{mac} = 4.2M$, $\alpha = 28°$. (Left) $C_L(\alpha)$, (middle) drag polar, and (right) $C_m(\alpha)$. (Courtesy of M. Tomac, PhD thesis [30], reprinted with permission)

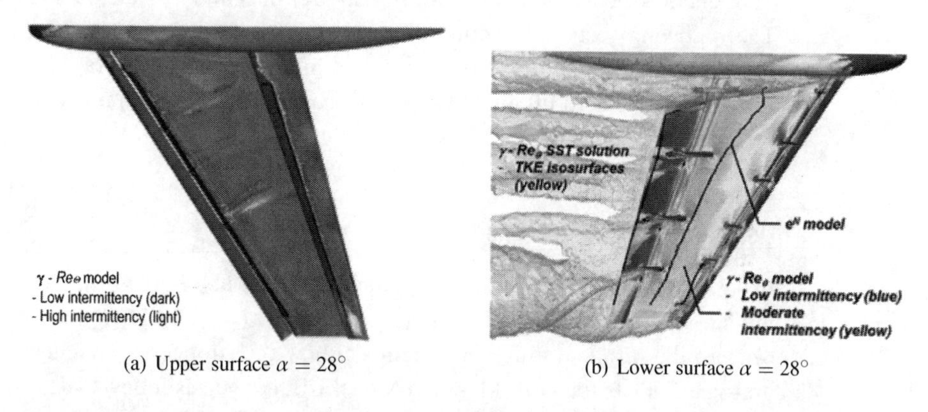

(a) Upper surface $\alpha = 28°$

(b) Lower surface $\alpha = 28°$

Figure 9.33 Comparison of approximate predictions of transition locations. Trap wing, $M = 0.2$, $Re_{mac} = 4.2M$, $\alpha = 28°$. (Courtesy of M. Tomac, PhD thesis [30], reprinted with permission)

Note that the levels have been adjusted to clearly indicate the laminar and transitional regions. The iso-surface of TKE has intentionally been removed from the upper surfaces on the left-hand side in Figure 9.33, since the intermittency works as a good indicator for the relatively clean flow on the upper side of the wing.

On the lower side, on the other hand, the flow is significantly more complex, and therefore, the transparent iso-surface is included as well. On the upper side, the results agree well with the computation by Eliasson et al. [6] that uses the database e^N method with the envelope approach in conjunction with the Spalart–Allmaras model to impose transition regions, indicated by a black line – this applies for all three elements.

On the lower side, the agreement is still good regarding the trailing-edge flap, except for the tripped boundary-layer flow due to the wakes from the brackets. On the main

Figure 9.34 Body of Model I-A, a Sears–Haack body with $r_{max}/l = 1/18$.

profile, on the other hand, the results disagree significantly. While the e^N database method has predicted a transition location approximately at mid-chord, the $\gamma - \tilde{Re}_{\theta t}$ SST results show mainly turbulent flow on the main profile. However, this demonstrates the state of the current transition model implementation in the EDGE code. Overall, this case indicates that reliable transition modeling is something that must become standard in the near future.

9.6 Wing–Body Mathematical Shape Optimization

The generic scheme for mathematical shape optimization was laid out in Section 1.3.3. The following sections describe such an exercise carried out as an MSc project at KTH by Dirk Franke [9]. The object is to minimize the wave drag of a wing–body configuration at zero lift and $M_\infty = 1.02$ using an optimization procedure to reshape the axisymmetric body.

9.6.1 Description of the Test Case

In 1957, McDevitt and Taylor [21] studied the effects of area ruling experimentally in wind tunnel tests. The wind tunnel conditions $M_\infty = 1.02$, $Re = 4M$ are chosen because the results at this Mach number show the strongest difference between the drag coefficients for Model I-A and Model I-B, defined as follows.

- Model I-A is a Sears–Haack body with a NACA64A008 wing. The wing is swept back 40° at the 40% chord line, with *AR* 6 and taper ratio 0.4. The body area distribution enters into the discussion of the results in Section 9.6.3.
- Model I-B is similar to Model I-A but with the body reshaped by area ruling to give the same area distribution, including the wing, as the Model I-A body alone.

Recall that the Sears-Haack body (Figure 9.34) is the axisymmetric shape with the lowest theoretical wave drag in supersonic flow for a given body length l and given volume $r/r_{max} = \left(1 - (1 - 2x/l)^2\right)^{3/4}$. "Theoretical" here means that wave drag is computed by the Prandtl–Glauert wave equation model in the far field (see Chapter 3). The nose is flat, yet the radius of curvature vanishes. McDevitt and Taylor found that Model I-B has much lower wave drag, but there is no reason to believe that it would be the optimum in a mathematical minimization procedure. This, of course, is precisely the point under investigation. The optimization modifies only the body, keeping it axisymmetric.

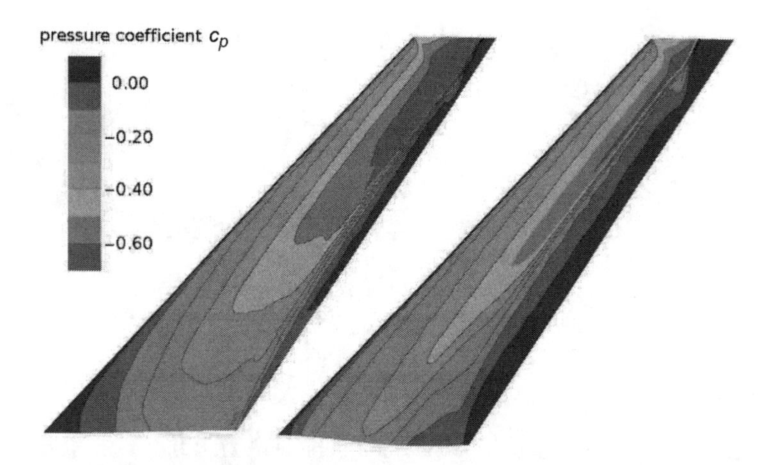

Figure 9.35 Pressure distributions C_p computed on the wing upper surface. (Left) Model I-A; (right) Model I-B. (Courtesy of D. Franke, MSc thesis [9], reprinted with permission)

The exercise is somewhat academic. Modern airliners do not have coke-bottle fuselages. Instead, area ruling has been applied to the whole configuration, including wing root, fairing, engine nacelles, and flap track fairings and is not visible to the naked eye.

Pressure Distribution on Models I-A and I-B

Figure 9.35 shows the computed pressure distributions on the wing upper surface. C_p^* is close to 0. The terminating shock is close to the trailing edge. For Model I-A (Figure 9.35, left), the isobars are strongly curved near the body for the whole chord length. Model I-B, with an area-ruled body, shows a significant improvement in isobar pattern with almost no curvature at the body. Outboard, the fuselage influence decreases, although the isobars on the wing are still not equal for both models. Close to the wing tip a shock cone appears, rooted at the wing tip leading edge. This establishes the two baselines for comparison with optimization results.

9.6.2 Drag Minimization by Optimization

The optimization problem is approached with Euler CFD at zero lift with Edge, so the only drag component left is wave drag. Referring to the definitions in Section 1.3.3, properties **P** are the coefficients C_L, C_D, and C_m. The figure of merit appearing as objective function FoM is thus C_D or, equivalently, total drag.

The constraints were set as inequality constraints on **P** to allow only minimal lift and moment coefficient excursions from those of the baseline.

In addition, the radial shape deviation from the baseline was limited by upper and lower bounds on the coefficients in the shape perturbation functions, which make up the parameter vector **X**. The shape functions were chosen as Hicks–Henne bumps (see Section 5.2.4).

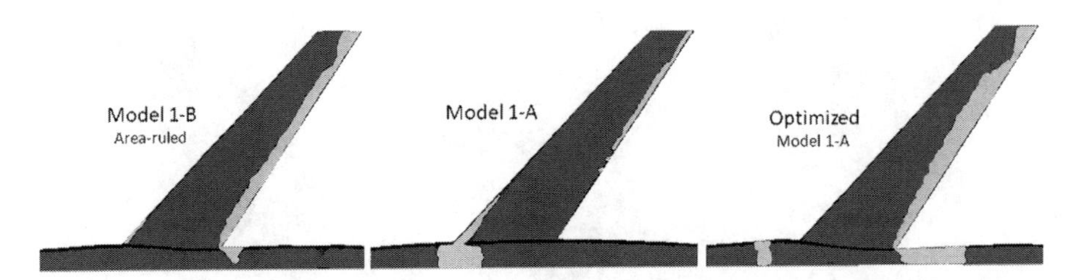

Figure 9.36 Mach-colored body and upper wing surfaces. (Left) Area-ruled Model I-B, (middle) original model I-A, (right) optimized from Model I-A. (Courtesy of D. Franke, MSc thesis [9], reprinted with permission)

The optimizer must leave the body untouched at $x/l \leq 0.4$, a surrogate for keeping the volume constant. This was implemented in the definition of the Hicks–Henne bumps.

The *shape* optimization problem now looks exactly like Eq. (8.1) in Section 8.6. It remains to detail how the shape modifications are transmitted to the computational mesh, which is all the CFD code sees.

The scheme is close to the one in Figure 1.17. A *surface* mesh on the baseline was deformed by a special-purpose "glue" code. The *volume* mesh was then adjusted by tools in the CFD code. Gridpoints on the axisymmetric body are allowed to move radially. The wing is a surface lofted linearly between the fixed root and tip, and gridpoints on it displaced by body deformation slide along the generators. These rules have to be amended at the junction to avoid large grid cell stretching ratios. Such details are a challenge to generalize. In this case, topology changes and infeasible shapes were ruled out by the bounds on the perturbations.

9.6.3 Results

The number of parameters was varied from four to seven and only the shape of the body was changed. The profile and dimensions of the wing are constant. Figure 9.36 shows initial and optimized Mach distributions, with Model I-A and I-B as baselines. The results with the two different baselines agree closely. The boundary between super- and sub-sonic domains is composed of shocks and Mach lines where the flow accelerates through $M1$.

Close to the leading edge, the supersonic flow accelerates. We recognize the shock wave on the wing, close to the trailing edge, and a further shock wave on the fuselage in front of the wing for both models. The shock wave on Model I-B compared to Model I-A is shifted aft.

Consider again Figure 9.36. Model I-B has a weaker shock on the fuselage. The high wave drag of Model I-A is also signaled by the unfavorable distortion of the isobars close to the fuselage, related to the shock wave in front of the wing. Further out, the shock wave on Model I-B is shifted toward the leading edge and weakened.

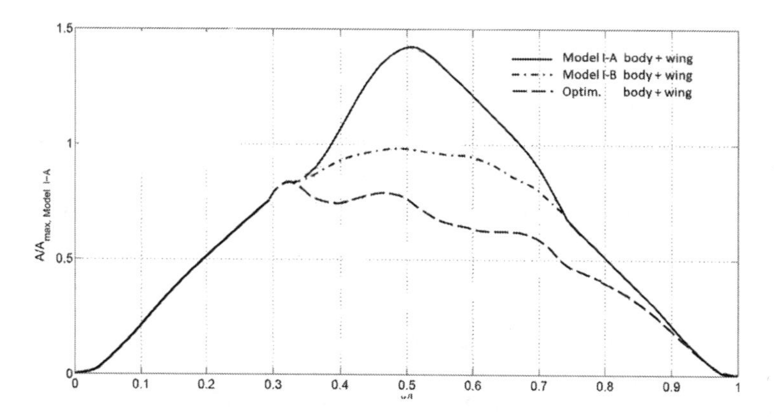

Figure 9.37 Cross-sectional areas of body and wing–body configurations, Models I-A and I-B and final optimized shape. (Courtesy of D. Franke, MSc thesis [9], reprinted with permission)

Cross-sectional Area Distribution

The final optimized shape with seven Hicks–Henne bumps has the smallest drag. Figure 9.37 shows the cross-sectional area distribution for Models I-A and I-B and the optimal shape.

Given either baseline, the optimization reduces the body cross section where it is allowed (i.e. aft of 40% of body length). This is in accordance with the area-rule concept. The contraction of the body is much stronger than on Model I-B. The volume is not constrained in the optimization, so these results might be expected.

Two spatial discretization schemes were tested with only small difference: 3% for the drag coefficient.

The optimization decreased the drag of Model I-A by about 60%. Model I-B is optimal by verbatim interpretation of the area rule. The optimization gave another shape, since the Sears–Haack area distribution is for axisymmetric bodies. It would remain to investigate the robustness of the optimal shape with respect to changes in constraints.

This concludes the study of isolated wing design. In the next chapter, we continue with the further development of the configuration, including trim balance and flying qualities.

9.7 Learn More by Computing

Gain hands-on experience of the computational tools for the topics in this chapter by working with the on-line resources. Exercises, tutorials, and project suggestions are found on the book website www.cambridge.org/rizzi. For example, unsweep the wing of Model I-A and compute the zero-lift drag at M 1.1 and compare to that of the original. Also compute Model I-A with a few small angles of attack to check the formula (Eq. 3.17). Software to compute many of the examples shown is available from http://airinnova.se/education/aero-dynamic-design-of-aircraft

References

[1] M. Blair. Evolution of the F-86. Technical report. AIAA-Paper-80-3039, 1980.

[2] R. Buschner. Pressure distribution measurements on a sweptback wing with jet engine nacelle. Technical report TM 1226. NACA, 1949.

[3] U. Clareus. Aerodynamiska för och sidoprojekt till jas 39 gripen. 2001. Flygteknik 2001 Congress (in Swedish).

[4] W. H. Cook. *The Road to the 707: The Inside Story of Designing the 707*. TYC Publishing Company, 1991.

[5] S. Crippa. *Advances in Vortical Flow Prediction Methods for Design of Delta-Winged Aircraft*. KTH School of Engineering Sciences, 2008.

[6] P. Eliasson, A. Hanifi, and S.-H. Peng. Influence of transition on high-lift prediction for the nasa trap wing model. Technical report. AIAA Paper 2011-3009, 2011.

[7] P. Eliasson, J. Vos, A. Da Ronch, M. Zhang, and A. Rizzi. Virtual aircraft design of transcruiser: Computing break points in pitch moment curve. Technical report. AIAA-2010-4366, 2010.

[8] P. Eliasson, J. Vos, A. DaRonch, M. Zhang, and A. Rizzi. Virtual aircraft design of transcruiser – computing break points in pitch moment curve. Presented at 28th AIAA Applied Aerodynamics Conference. American Institute of Aeronautics and Astronautics, June 2010.

[9] D. M. Franke. Interfacing and testing the coupling of the cfd solver edge with the optimization package `modeFRONTIER`. Diploma thesis, KTH School of Engineering Sciences, 2007.

[10] A. B. Haines. Computers and wind tunnels: complementary aids to aircraft design. *Aeronautical Journal*, 81(799): 306–321, 1977.

[11] E. H. Hirschel, A. Rizzi, and C. Breitsamter. *Separated and Vortical Flow in Aircraft Aerodynamics*. Springer, 2021.

[12] T. Ivarsson, G. Berseus, K. E. Modin, and U. Clareus. Definition av flygplan jas 39. In *Saab-minnen del 19*. Saab Veteran's Club, Link 2009, chapter 13 (in Swedish).

[13] G. K. W. Kenway and J. R. R. A Martins. Multipoint aerodynamic shape optimization investigations of the common research model wing. *AIAA Journal*, 54(1): 113–128, 2016.

[14] D. Küchemann. *Aerodynamic Design of Aircraft*. AIAA Education Series. AIAA, 2012.

[15] R. B Langtry and F. R. Menter. Correlation-based transition modeling for unstructured parallelized computational fluid dynamics codes. *AIAA Journal*, 47(12): 2894–2906, 2009.

[16] R. P. Liema, J. R. R. A. Martins, and Kenway G. K. W. Expected drag minimization for aerodynamic design optimization based on aircraft operational data. *Aerospace Science and Technology*, 63: 344–362, 2017.

[17] J. M. Luckring. Selected scientific and technical contributions of Edward C. Polhamus. Presented at 34th AIAA Aerodynamics Conference, AIAA Aviation Forum, Washington DC, June 2016.

[18] H. Ludwieg. Pfeilflügel bei hohen geschwindigkeiten. *Lilienthal-Ges. Ber.*, 127, 1940.

[19] F. Lynch. Commercial transports – aerodynamic design for cruise efficiency. In D. Nixon, editor, *Transonic Aerodynamics*, Volume 81 of *AIAA Progress in Aeronautics and Astronautics*. AIAA, 1982, pp. 81–144.

[20] W. H. Mason. AOE 4124 configuration aerodynamics, 2016. Available from www.dept.aoe.vt.edu/~mason/Mason_f/ConfigAero.html.

[21] J. B. McDevitt and R. A. Taylor. An investigation of wing-body interference effects at transonic speeds for several swept-wing and body combinations. Technical report. NACA RM A57A02. NACA, 1957.

[22] K. E. Modin and U. Clareus. Aerodynamic design evolution of the Saab JAS 39 Gripen aircraft. Presented at AIAA/AHS/ASEE Aircraft Design Systems and Operations Meeting, Maryland, MD, USA, September 1991.

[23] E. Obert. *Aerodynamic Design of Transport Aircraft*. IOS Press, 2009.

[24] J. Rech and C. Leyman. *A Case Study by Aerospatiale and British Aerospace on the Concorde*. AIAA Professional Study Series. AIAA, 2006.

[25] C. L. Rumsey, J. P. Slotnick, M. Long, R. A. Stuever, and T. R. Wayman. Summary of the first AIAA CFD high-lift prediction workshop. *Journal of Aircraft*, 48(6): 2068–2079, 2011.

[26] D. M. Ryle. *High Reynolds Number Subsonic Aerodynamics*. AGARD Lecture Series LS-37-70. AGARD, 1970.

[27] G. S. Schairer. Evolution of modern air transport wings. Technical report 80-3037. AIAA, 1980.

[28] J. P. Slotnick, J. A. Hannon, and M. Chaffin. Overview of the first AIAA CFD high lift prediction workshop. Technical report. AIAA Paper 2011-0862, 2011.

[29] A. M. O. Smith. Wing design and analysis – your job. In T. Cebeci, editor, *Numerical and Physical Aspects of Aerodynmic Flows*. Springer Science + Business Media, 1984, pp. 41–59.

[30] M. Tomac. *Towards Automated CFD for Enginerring Methods in Airfcraft Design*. KTH School of Engineering Sciences, 2014.

[31] J. Vassberg, M. Dehaan, M. Rivers, and R. Wahls. *Development of a Common Research Model for Applied CFD Validation Studies*. AIAA, 2008.

[32] F. Wänström. Rapport over studieresa till schweiz den 11–22 nov. 1945. Travel report, Saab Archive, November 1945 (in Swedish).

[33] M. G. Wilde and G. Cormery. The aerodynamic derivation of the Concorde wing. Presented at 11th Anglo-American Aeronautical Conference, London, England, September 1969.

[34] M. Zhang, R. Nangia, and A. Rizzi. Design and shape optimization of morphing winglet for regional jetliner. Technical report. AIAA 2013–4304, 2013.

10 Configuration Development and Flying Qualities

Stability and control comprised the single most critical requirement for flight.

The Wright brothers

Preamble

How the distribution of the aerodynamic forces and moments on the aircraft vary along the flight path determines its stability and the need for control with sustained authority. Previous chapters have dealt in some detail with aerodynamic performance (i.e. the mapping from shape to lift and drag for clean wings). It was assumed tacitly that the plane can be flown stably along the mission path, idealizing the plane as a mass point with lift and drag forces. Now it is time to address that issue more thoroughly. Flight simulation requires an airplane model responding to gravity, thrust, and realistic aerodynamic forces and moments: a six degree of freedom Newtonian *rigid* body model. The control system is responsible for transmitting pilot commands to control actuators, with authority ranging from autopilots to stability augmentation systems. The design is done with linearized models developed in control theory. Analysis and design require close interaction of the aerodynamicists with controls, structural mechanics, and flight mechanics experts.

The flight mechanical model is compiled from the aero-data and the mass and balance properties of the airframe. The aerodynamicist's task is to predict the aerodynamic forces and moments and express them in look-up tables of coefficients, and we explain how such tables can be populated.

Flight simulation models are used to assess flying qualities (i.e. how easy pilots will find it to fly the planned missions, quantified as pilot ratings). The ratings (or flying quality levels) can be predicted from the flight mechanics model without a pilot, but also flight simulation with a pilot in the loop is possible. Of particular interest are the stability properties that describe how well the aircraft recovers from external disturbances and how it reacts to commanded changes in flight attitude.

The response of the aircraft in steady flight to small disturbances can be represented as a superposition of a small number of flight natural modes, the quantitative properties of which provide the quantified flight-handling qualities.

The redesign of the baseline T-tail Transonic Cruiser to an all-moving canard configuration exemplifies these flight-dynamical concepts and brings them to life

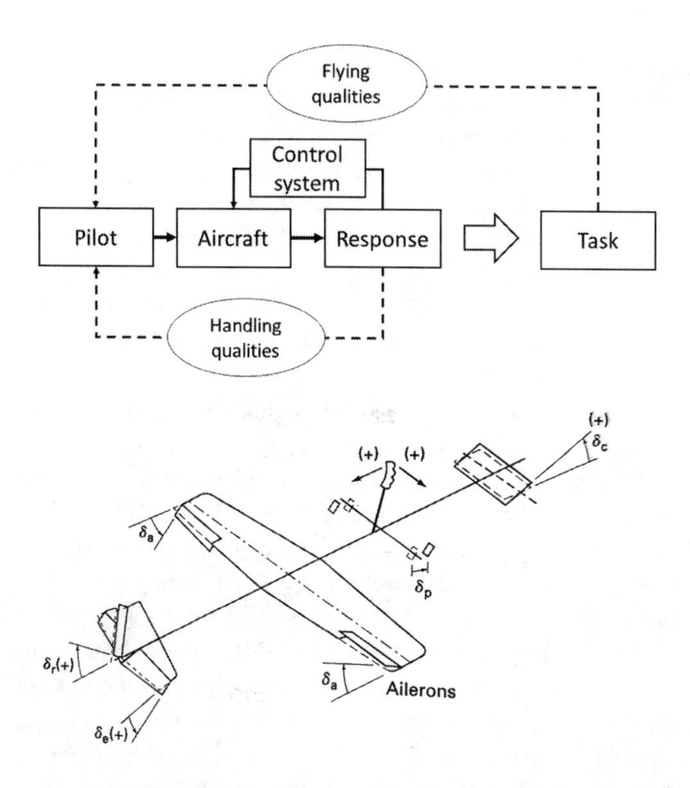

Figure 10.1 Flying and handling qualities of the airplane in response to pilot control actuation (top); control surfaces of conventional aircraft (bottom). (From R. Nelson [25], reprinted with permission)

in application. For example, with the aero-database computed, the transonic trim condition indicates that near Mach 1 the nose-down pitch moment increases strongly, requiring further canard deflection to trim. The canard has been correctly sized and has ample authority to trim at the highest flight speed.

10.1 Introduction

In his textbook [5], Cook describes the flying and handling qualities of an airplane as those properties that describe the ease and effectiveness with which it responds to pilot commands for executing a flight task. The signal-flow diagram in Figure 10.1 (top) shows the process involved in the pilot's perception of flying and handling qualities. The solid lines represent physical, mechanical, or electrical signal-flow paths, whereas the dashed lines represent sensory feedback information to the pilot. The pilot's perception of *flying qualities* comprises a qualitative description of how well the airplane carries out the commanded task. His or her perception of *handling qualities* is a qualitative description of the *adequacy* of the response to controls. For our purposes, these two are inseparable, and we call them flying qualities.

Assessment and Prediction of Flying Qualities

An aircraft with good flying qualities must obey the pilot's inputs precisely, rapidly, and predictably without unwanted excursions, uncontrollable behavior, or excessive physical pilot effort. Preferably, the aircraft should possess these desirable characteristics throughout its performance envelope. The National Advisory Committee on Aeronautics (NACA) World War II test pilot Melvin N. Gough put it in a slightly different form when he stated: "The flying qualities of an aircraft may be defined as the stability and control characteristics that have an important bearing on the safety of flight and on the pilot's impressions of the ease and precision with which the aircraft may be flown and maneuvered."

In the early days of aviation, evaluation of flying qualities was not possible until the flight tests of a prototype. That is quite late in the game to discover control problems. Major steps were taken between 1918 and 1938 toward the quantification of flying qualities, and a review is given in Ref. [36]. The Cooper–Harper ratings (Section 10.2.2) quantify pilot assessments. They enable such assessment by analysis of a linearized flight mechanical model and by flight simulation, so that potential problems can be addressed early in the design cycle. A few comments are in order.

Stick-fixed properties relate to the aircraft with fixed pilot-operated controls.

The stick-fixed dynamic properties give *predicted flying qualities*, exemplified by the characteristics for the "short-period" flight mode shown in Section 10.2.2.

With a pilot in the loop in a simulator or in flight tests, *assigned flying qualities* can be bestowed on the aircraft as information for future pilots.

George Hartley Bryan's classic book *Stability of Aviation* from 1911 [3] deals with flight stability by linearization of the equations of motion, and it pioneered the terminology of "derivatives," which since then has permeated the science of flight mechanics. The eigenmodes of the linearized dynamic model were analyzed very early on, particularly the *phugoid* motion by Lanchester in 1907 [19], as well as by Bryan. This is familiar to most of us from launching a paper dart too enthusiastically. It climbs, loses speed, drops the nose to dive and gather speed, climbs again, and the process repeats itself. The set of important modes is described in Section 10.2.1. The result of the mode analysis and the work leading up to the Cooper–Harper scales is that pilot ratings of flying qualities can be quantified by mathematical models of the aircraft. Problems can be identified and addressed in a systematic way, and properties of transient movement around straight and level flight can be quantified in charts such as those presented in Section 10.2.2.

When the configuration has flying-quality deficiencies, it becomes necessary to modify it. There are three ways to achieve this.

(1) Modify the aerodynamic design, the subject of this chapter.
(2) Modify the mass distribution.
(3) Introduce a flight control system to ameliorate the deficiency and optimize it for its purpose.

Control design is a discipline of its own, and it is beyond the scope of this book. However, it is essential to understand the basics of the relationship between the aerodynamics of the airframe and its stability and control characteristics. Here, we concentrate on the aerodynamic considerations given to configuring or redesigning the airframe layout.

10.1.1 Trim and Maneuver, Stability and Control

The configuration must be capable of straight and level unaccelerated flight in which the aircraft is said to be in *trim*. This usually requires a lift force to balance the moment in pitch with a *stabilizing surface* (e.g. the horizontal tail plane). Similarly, the vertical tail fin balances the yawing moment in flight with sideslip. A central question arises: Once in trim, how does the configuration respond to a wind gust that disturbs the equilibrium? Will the plane return to trim condition once the gust has passed? How long does it take?

This is the study of aircraft stability, a subject in the discipline of flight dynamics, for which the configuration developer must have some appreciation.

Similarly, the aircraft departs from trim conditions during a maneuver when the pilot deploys a control surface to initiate an acceleration. For example, actuating the elevator brings about a pitching moment, moving the rudder causes a yaw moment, and aileron deflection forces the aircraft to roll. All of these surfaces are part of the airframe and need to be properly sized. The characteristics of the motion produced by the deployed surface must be assessed, and the tools to do this are the subject of this chapter.

Flight Simulation to Identify Flying and Handling Qualities

The geometry of the lifting surfaces and their control surfaces determine the forces, and the inertial equations of aircraft motion provide the *link* between their actions on the aircraft and the aircraft's dynamic response. This system of nonlinear equations constitutes a mathematical model of that aircraft's behavior and can predict the aircraft's *stability and control characteristics* and hence can quantify its flying qualities so that the data can be used to further develop the configuration. Section 10.2.2 describes how desirable flying and handling qualities can be quantified. Section 10.1.2 details the flight-dynamic mathematical model, and Section 10.4 explains how to obtain the extensive information about the aircraft required by the model. Section 10.5.4 explores the sizing of control surfaces. In Section 10.5.1, the tools are brought to bear on the development of the transonic cruiser (TCR) concept for which the trim drag had to be reduced through a configuration change. Section 10.5.5 outlines cases with other applications of the tools, while Section 10.5.6 covers control issues found in flight test or wind tunnel campaigns.

10.1.2 Rigid Aircraft Flight Simulation Model

The aircraft is a body in motion under the influence of forces and moments, and its behavior can be expressed mathematically as a dynamic system. The classical textbooks by Etkin [9], Cook [5], and Nelson [25] show the development of the mathematical model from Newton's Second Law to the complete set of equations for 3D aircraft motion (Eq. (10.1)). They do not cover the production and compilation of force and moment data, but rather assume the data are available in the form of coefficients and their derivatives with respect to flight state. This text aims to help fill that gap.

The model is formulated in a body-fixed coordinate system with origin at the aircraft's center of gravity. This makes the inertia tensor constant; the downside is that coordinate rotations have to be performed to evaluate the right-hand sides, since forces are by convention recorded in a wind-fixed system. The inertial expressions for translation, rotation, and kinematical relationships are written in symbolic form as follows.

$$\begin{aligned} &Translation: && m\dot{\mathbf{V}} + \omega \times m\mathbf{V} = \mathbf{F}_{aero} + \mathbf{F}_{thrust} + \mathbf{F}_{gravity} \\ &Rotation: && \mathbf{I}\dot{\omega} + \omega \times \mathbf{I}\omega = \mathbf{M}_{aero} + \mathbf{M}_{thrust} \\ &Kinematics: && \dot{\mathbf{\Theta}} = \mathbf{L}(\mathbf{\Theta}, \omega) \end{aligned} \tag{10.1}$$

where $\mathbf{F}$ denotes aerodynamic (aero), propulsion (thrust), and gravity forces; $\mathbf{M}$ denotes aerodynamic and thrust moments around the center of gravity; $\mathbf{V}$ represents the velocity of the aircraft and ω its rotation rate vector $\omega = (p, q, r)$; m denotes its mass and $\mathbf{I}$ moments of inertia; $\mathbf{\Theta} = (\Psi, \theta, \Phi)$ is the set of Euler angles; and $\mathbf{L}$ is the Euler angle rates; refer to Figure 0.2 in the Nomenclature section.

Equations (10.1) constitute an evolution model with state vector $(\mathbf{V}, \omega, \mathbf{\Theta})$ that determines the instantaneous motion from forces and moments. There remains the issue of determining these from past and present flight states. The forces on the aircraft can be computed from the air stress tensor on the wetted surface. This requires knowledge of the relative air motion in detail. However, when we consider the aircraft flying into still air, the forces can only depend on the aircraft motion, both *instantaneous* as well as its *history*.

Force and Moment Data

The force and moment data can be determined by computational fluid dynamics (CFD), by wind tunnel experiment, and by flight tests. An example of the use of advanced CFD is given in Ref. [13]. In this text, we focus on CFD analysis that students can do on their own. The other two techniques are best done by organizations with trained staff and a large budgets. CFD can be relied on for attached flow cases, but significant separations with unsteady flow require very expensive time-resolved simulations with large-eddy simulation (LES) or similar techniques. Wind tunnel campaigns require a physical model that tolerates the forces in the tests, hence models for transonic speeds are expensive. Much attention to turbulence levels and the influence of the model mounting as well as shock reflections on wind tunnel walls is necessary. In addition, in most wind tunnels, flight Reynolds numbers cannot be matched. It follows that a fair amount of wind tunnel result correction is needed to extrapolate to free-flight conditions. Flight tests are, of course, the ultimate in fidelity, but significant manipulation of recorded accelerations and structural responses is

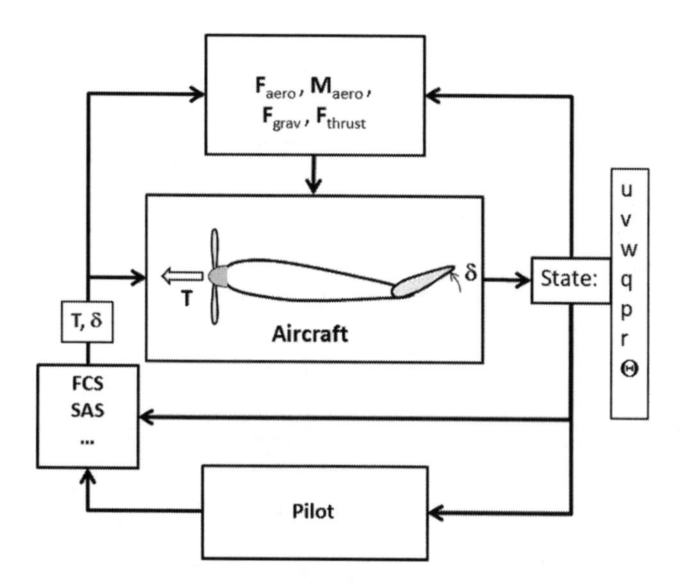

Figure 10.2 System diagram of aircraft, pilot, and control systems. SAS = stability augmentation system.

necessary to extract forces and moments. In addition, "pushing the envelope" – flying at extreme conditions to initiate stalls and spins, not to speak of dives at high speed – is dangerous.

A complete picture of the dynamic system includes the pilot and the flight control system (FCS). Mitigation of undesirable dynamics, such as overshoots and too slow damping of transients, is referred to as stability augmentation see (Figure 10.2). Hoeijmakers and Hulshoff [17], among others, have pointed out two ways to approach the estimation of the fluid–aircraft dynamic interaction using simulation.

Coupled Simulation

The *coupled-simulation method*, as shown for the Immelmann maneuver in Section 10.1.3, integrates the equations for the fluid flow simultaneously with the aircraft kinematic equations. This approach allows all aerodynamics to be included in the response of the aircraft, which can be augmented to include actuation of control surfaces and propulsion, as well as aero-elastic effects. In spite of its generality, however, at present the coupled-simulation method is too costly to be applied routinely, primarily because of the large number of simulations required to characterize the aircraft response over the full flight envelope (see Jou [18]).

Segregated Model

In the second approach, a range of aerodynamic behavior is encapsulated into pre-computed models (e.g. look-up tables) used in the simulation. The overall process to produce the tables by CFD (in machine learning terminology, the *training* of the model) and subsequent *use* by look-up in the flight simulator is sketched in Figure 10.3.

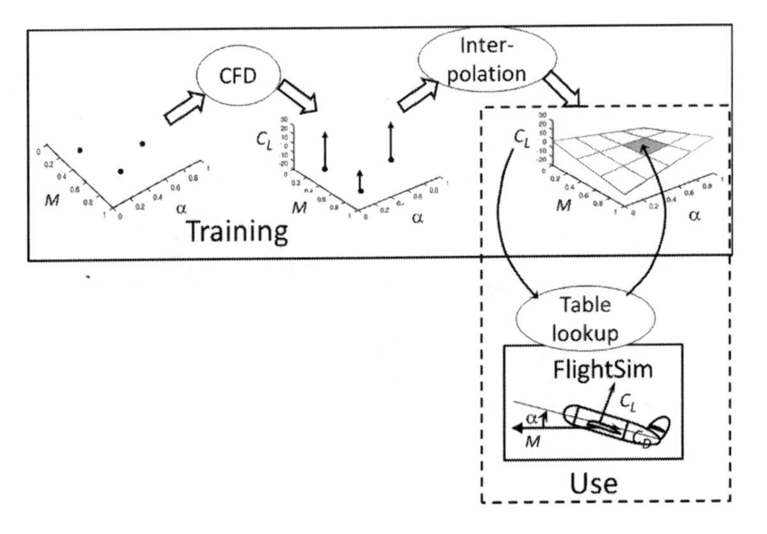

Figure 10.3 Production and use of look-up tables of force and moment coefficients.

Only $C_L(M, \alpha)$ is shown; the reader should imagine at least nine independent variables as input to the look-up and six force and moment results delivered.

Making such a reduced and more tractable model of the interaction of airflow and aircraft raises the fundamental issue of its mathematical structure.

Quasi-steady Model

A "quasi-steady" motion is characterized by immediate reaction of forces and moments to the aircraft state (i.e. airspeed, wind direction α, β, its rotation angular velocities, and throttle and control surface actuation). The air movement then adapts to the aircraft more quickly than the aircraft can react to forces, and the dependence on the history of the motion variables becomes small. Then the $\mathbf{F}, \mathbf{M}$ depend only on the current state, and the evolution equations in Eq. (10.1) become a set of ordinary differential equations.

Unsteady Motion

For quicker maneuvers, the state of the surrounding air must also come into play.

As an example, consider a single high-aspect ratio (*AR*) wing (see Chapter 3). The vortex sheet travels downstream and exerts forces on the wing, and the sheet "remembers" past states of the circulation around the wing. Another effect is the time delay of an air particle disturbed by the main wing before it influences the tail. A simple model is that the coefficients depend on a past value of, say, α, $\alpha(t - \tau)$, $\tau > 0$. For small delays τ we can approximate the following.

$$\alpha(t - \tau) \approx \alpha(t) - \tau\dot{\alpha}(t)$$

Thus, including a dependency on $\dot{\alpha}$ in the forces can account for some of the memory effect. There is a mathematical bend in the road to include such dependencies: $\dot{\alpha}$

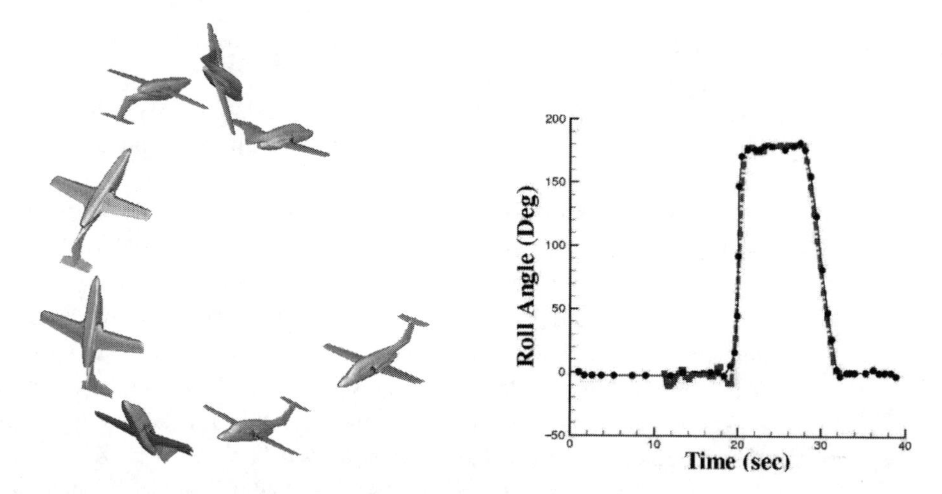

Figure 10.4 The simulated Ranger 2000 performs an Immelmann turn (left); roll angle simulated (squares) and in real flight (circles) (right). (Courtesy of M. Ghoreyshi [14], EU SimSAC project [29], reprinted with permission)

depends on accelerations $\dot{u}, \ldots$ (i.e. the very quantities to be computed from the forces). The system in Eq. (10.1) becomes a *differential-algebraic* system instead of a system of ordinary differential equations. However, we need not worry too much: a single differentiation of the algebraic relation between α and the velocities produces a nonsingular linear system to be solved for the accelerations. Trying to include higher derivatives in this way, however, will bring disaster.

The following two issues must be considered in computing the force tables by CFD.

(1) The computational cost is high since CFD runs must be performed for many parameter combinations. This will be discussed in Section 10.4.
(2) Varying the free stream direction, airspeed, and altitude is easy in a CFD analysis. How to manage the dependencies on p, q, r, and $\dot{\alpha}$, etc., is another matter, to be touched upon in Section 10.3.

10.1.3 Example Simulation: Immelmann Turn

Consider the sequence of still-pictures of a Ranger 2000 aircraft performing an *Immelmann turn* (Figure 10.4, left). An Immelmann turn consists of a half-loop followed by a half roll from inverted to normal flight attitude at the top of the loop. This maneuver was intended to provide the aircraft a rapid repeated run on a target, as well as a means to escape blast effects of its own weapon in low-altitude bombing. The maneuver is common in aerobatics. In the related "Herbst" maneuver, the climb should be so steep that the turn at the top is made almost in stall, which requires great pilot skills. The Immelmann turn is initiated with a pitch-up command at the entry, and then continues with the pilot pulling back on the stick such that the aircraft flies a loop, finishing with a 180° roll from inverted at the top of the loop. At the entry point, $V = 164$ m/second and the altitude is 2.98 km; at exit, at the top of the loop, the altitude is 3.71 km, with the heading being exactly the opposite.

The pictures were generated by a computer simulation of the maneuver in an actual flight test, solving Eq. (10.1) concurrently with computing the air movement by unsteady CFD.

The simulation considers the plane to be an essentially rigid body, with a few additional selected elastic degrees of freedom (DOFs), modeled by Newtonian mechanics. That accounts for $O(20)$ DOFs. The surrounding air was modeled by CFD for unsteady, compressible, inviscid flow, with 14 million grid cells. From this, we can infer that such simulations require significant computational resources; indeed, they are too expensive for the large number of design alternatives to be evaluated in the development of a new configuration. The agreement with the flight state is excellent; Figure 10.4 (right) shows the roll angle in simulation and in real life.

A comment may be suitable here: the plot shows roll angle suddenly increasing to half a turn at 20 seconds and then decreasing again more slowly at about 28 seconds. This is the commanded roll executed as part of the maneuver. The sudden jump is an artifact. In the vertical position, the craft executes a jump change in heading, say, from $-X$ to $+X$, which is a jump change in roll attitude.

But time histories of the control actuation (e.g. the rudder angles) disagree quite significantly. The explanation lies in how the "pilot actions" were selected in the simulation. The simulated pilot was tasked with using the controls to follow the real flight as closely as possible. Thus, the pilot was defined by the mathematical optimization problem to choose at each instant the control settings (with suitable restrictions) to minimize some measure of the overall agreement between real and simulated flight.

There are, of course, differences between the real physics and the mathematical model. The experiment shows that they are so small that the model can behave like the real thing when given reasonable, but not identical, control signals, that compensate for the differences. The lesson here is that *it is possible to predict the properties of the aircraft, such as "flying qualities," by computer simulation of a mathematical model.*

10.2 Stability of Aircraft Motion

In order to investigate aircraft motion, it is first necessary to establish that it can be brought into a condition of equilibrium (i.e. a condition of balancing forces and moments). The stability characteristics can be determined from equilibrium flight conditions. The aircraft is *statically stable* if, when the motion is disturbed, the forces and moments tend to restore equilibrium.

Dynamic stability describes the subsequent behavior after the initial response of the aircraft to the static restoring forces and moments.

The aircraft is piloted by human and/or computer, and one usually sees the piloting entity as a controller for the aircraft dynamical system, as in Figure 10.2.

We use a shorthand notation for the six DOF model Eq. (10.1).

$$\mathbf{A}\frac{d\mathbf{s}}{dt} = \mathbf{F}(\mathbf{s}, t, \boldsymbol{\delta}) \tag{10.2}$$

The explicit dependence on time represents externalities such as wind gusts. The flight-state vector $\mathbf{s}$ has $N = 9$ components as follows.

$$\mathbf{s} = (\boldsymbol{u}, \boldsymbol{\omega}, \boldsymbol{\Theta}) \tag{10.3}$$

The $\boldsymbol{\delta}$ is a set of controls: rudder, aileron, elevator, and thrust in a classical aircraft. The $\mathbf{A}$-matrix coefficient is a constant of the following form.

$$\begin{pmatrix} m\mathbf{E} & 0 & 0 \\ 0 & \mathbf{I} & 0 \\ 0 & 0 & \mathbf{E} \end{pmatrix}$$

$\mathbf{E}$ is a 3×3 unit matrix and $\mathbf{I}$ is the inertia matrix. Obviously, the model requires evaluation of the forces and moments on the aircraft from its flight state $\mathbf{s}$, present and past, including settings of control surfaces and throttles. Now, a rigid body has three translational and three rotational DOFs, giving 12 first-order ordinary differential equations. Equation (10.1) gives the translational speeds but not positions, so it needs three more kinematic relations such as Eq. (10.13) to describe the motion of the body in the world coordinate system. The forces that make up $\mathbf{F}$ are usually provided as nondimensional force coefficients. They are converted to forces and moments using reference quantities wing area S, wing span b, a wing chord c, often the mean aerodynamic chord MAC, and dynamic pressure q. In addition, the relative position of the center of gravity (CG) to the reference point for computed (or measured) moments must be supplied.

Reference Point for Moments and Static Margin

The *aerodynamic reference point* is a fixed point on the aircraft body. In principle, it can be defined anywhere as long as the moments are calculated with respect to that reference point. Issues arise when the model shall represent elastic deformations or movement of fuel, landing gear, deployment of flaps, etc., since such processes introduce changing inertias around the CG. In order to make the data recorded easy to interpret, in particular the C_m vs α-curve, it is useful to place the aerodynamic reference point where the influence of lift on moment around the CG is small.

As we shall see, a necessary condition for pitch static stability is that an increase in α that gives increased lift be accompanied by a change in moment tending to rotate the wing section nose down (i.e. $dC_{m,CG}/d\alpha < 0$). Thus we must consider the variation of $C_{m,CG}$ with CG position. We have for the moment coefficient at X, using the usual $xx_{,v}$-notation for differentiation of xx with respect to v,

$$\begin{aligned} C_m(X) &= C_{m,ref} - C_L \cdot \frac{X - x_{ref}}{c_{ref}} \\ C_m(X)_{,\alpha} &= C_{m,ref,\alpha} - C_{L,\alpha} \cdot \frac{X - x_{ref}}{c_{ref}} \end{aligned} \tag{10.4}$$

At the center of pressure (CP) $X = x_{cp}$ and at the aerodynamic center (AC) $X = x_{ac}$, defined in the Nomenclature, p. xxx, the left hand sides vanish, so

$$\text{Center of pressure:} \quad x_{cp} = x_{ref} + c_{ref} \cdot \frac{C_{m,ref}}{C_L} \tag{10.5}$$

$$\text{Aerodynamic center:} \quad x_{ac} = x_{ref} + c_{ref} \cdot \frac{C_{m,ref,\alpha}}{C_{L,\alpha}}$$

Thus,

$$C_{m,CG,\alpha} = -C_{L,\alpha} \cdot K_n, \; K_n = \frac{x_{ac} - x_{CG}}{c_{ref}}. \tag{10.6}$$

K_n is known as the *static margin* which helps quantify the degree of stability: $K_n > 0$, i.e. $x_{CG} < x_{ac}$ is necessary, but not sufficient for pitch stability. Note that x points aft, so this means CG *ahead* of AC.

The placement of the CG is a vital element of a safe-flying airplane. For static stability, the CG must always, under any condition of fuel loading, passenger loading, cargo loading, or flaps and landing gear retraction or extension be between a forward and an aft limit.

Trim Condition

The trim settings required for a steady desired speed, climb, and turn rate in general 3D motion can be obtained by solving the nonlinear algebraic system in Eq. (10.2) for the desired equilibrium state $\mathbf{s}^*$ as follows.

$$\mathbf{F}(s^*, \delta^*) = 0 \tag{10.7}$$

with the control and thrust settings $\boldsymbol{\delta}^*$ as unknowns. Of course, such a nonlinear system of equations may have no solutions; there might also be several control inputs to produce the same flight state.

10.2.1 Dynamic Stability and Eigenmodes of Motion

The following assumes some familiarity with mathematical analysis of nonlinear dynamical systems by approximation with linear, time-invariant systems of differential equations. This is treated in textbooks on differential equations and control theory. Examples are the classic text by Richard Bellman, *Stability Theory of Differential Equations* [1], the lighter *Introduction to Applied Mathematics* by Gilbert Strang, with a broader scope [33], and control theory books such as Ref. [15]. Older texts focus on single differential equations of high order and determinants, whereas modern treatment employs the matrix-vector approach used here and in control theory. The more mathematically oriented texts spend much effort on issues of "repeated roots" to the associated characteristic polynomials, but that is not necessary for the material here.

Small excursions from trimmed steady flight $\mathbf{s}^*$ are governed by the linearized set of differential equations for the perturbation $\delta s(t)$. The coefficients form the

Jacobian matrices **B**, **C** of partial derivatives of **F** with respect to the states **s** and $\boldsymbol{\delta}$: $B_{ij} = \frac{\partial F_i}{\partial s_j}, C_{ij} = \frac{\partial F_i}{\partial \delta_j}$ for the defined state of steady flight $s^*, \boldsymbol{\delta}^*$ when acted upon by disturbances to controls $\delta\boldsymbol{\delta}(t)$ and to the external influences, such as gusts, $\delta\mathbf{f}(t)$.

$$\mathbf{A}\frac{d\delta\mathbf{s}}{dt} = \mathbf{B}\delta\mathbf{s} + \mathbf{C}\delta\boldsymbol{\delta}(t) + \delta\mathbf{f}(t) \tag{10.8}$$

The matrix **A** is always nonsingular, so the existence and uniqueness of solutions are at hand. This is a linear set of ordinary differential equations with constant coefficients that for $\delta\mathbf{f}(t) = \delta\boldsymbol{\delta}(t) = 0$ admits exponentials $\delta\mathbf{s}(t) = e^{\lambda t}\mathbf{v}$ as solutions. They can be determined from the eigenvalue problem as follows.

$$[\mathbf{B} - \lambda\mathbf{A}]\mathbf{v} = 0 \tag{10.9}$$

which provides eigenvalues $\lambda_k = \eta_k + i\omega_k$ and eigenvectors $\mathbf{v}_k, k = 1, 2, \ldots$. In the generic case, there are N linearly independent eigenvectors and a general solution can be obtained by proper superposition of these, with coefficients $\{a_k\}$ chosen to satisfy the initial condition.

$$\delta\mathbf{s}(t) = \sum_{k=1}^{N} a_k e^{t(\eta_k + i\omega_k)}\mathbf{v}_k \tag{10.10}$$

ω_k is the angular frequency and η_k is a growth parameter for time evolution of the kth eigenmode $e^{t\lambda_k}\mathbf{v}_k$.

So, if $\eta < 0$, the mode is damped. If $\omega \neq 0$, the mode oscillates and there are two modes with the same angular frequency since complex eigenvalues must appear in complex-conjugated pairs. The corresponding vectors **v** are complex. The solution to the initial value problem in Eq. (10.8) is a superposition of all the modes and is of course still real.

Modes of Motion and Dynamic Stability

Consider a trajectory flown in a small part of a typical maneuver, such as a turn to a new heading or ascent from one flight level to another. The pilot makes the change from one equilibrium flight condition to another by exciting one or more of the modes and then terminates the maneuver by suppressing the modes. The modes of motion may also be excited by external perturbations. The properties of these modes of motion determine the *dynamic* stability, defined as the time history of the aircraft motion after being disturbed from trim. This means that aircraft flying and handling qualities can be described by a small number of parameters, the characteristics of the modes: frequency, damping, and time constant.

Frequency $f = \omega/(2\pi)$ is defined as the number of cycles per second.

Damping ratio ζ is a measure of the subsidence of the motion, defined for a real second-order system with two eigenvalues, λ_1, λ_2.

$$\zeta = -\frac{\lambda_1 + \lambda_2}{2\sqrt{\lambda_1\lambda_2}}$$

Here, there are more eigenvalues, and the use of ζ relies on connecting them in pairs. This is easy for oscillatory modes that have a pair of complex conjugate eigenvalues, in which case the following holds.

$$\zeta = \eta/\sqrt{\eta^2 + \omega^2}, |\zeta| \leq 1$$

When the eigenvalues are both real and nonpositive, the modes are strongly damped, $\zeta > 1$; if not, the system is *unstable*.

The *time constant* of the motion is the time it takes for the aircraft to return to equilibrium, say, with 1% of the initial perturbation amplitude, $T = \ln(0.01)/\eta$.

Damping of the aircraft modes of motion has a profound effect on flying qualities. With too little damping, the mode is too easily excited by inadvertent pilot inputs or by atmospheric turbulence. If there is too much damping, the aircraft motion following a control input is slow to develop, and the pilot may describe the aircraft as "sluggish." The aircraft mission again determines the optimum dynamic stability characteristics.

Important Modes of Motion

To exhaust the complete set of solutions for Eq. (10.10), the number of modes must equal the order of the system, but only the five modes described below are important. The remaining modes all decay very quickly. This is the standard set of modes commonly observed. Depending on the aircraft configuration, there may be other modes of significance, or combinations of these.

1. *Phugoid.* Analyzed already by Lanchester and Bryan, this mode is characterized by very slow oscillations in pitch angle and velocity (θ, V) and a nearly constant angle of attack, as explained in the discussion in the subsection on unsteady longitudinal motion. According to Ref. [9], the period and damping factor are roughly as follows.

$$T = \pi\sqrt{2}V/g \text{ and } \zeta = \frac{C_D}{\sqrt{2}C_L} \qquad (10.11)$$

   To excite this mode, begin with the airplane trimmed in level flight. Pull back slightly on the stick and maintain or increase elevator trim. This pitches the airplane up into a climb. As the airplane climbs, it loses speed and lift, causing it to gradually pitch downward and enter a dive. During the dive, the airplane gains speed and lift, bringing it back into a climb.

   The motion is usually of such a long period – about 93 seconds for a Boeing 747 – that it need not be highly damped for piloted aircraft. The period is independent of the airplane characteristics and altitude and depends almost only on the trimmed airspeed.

2. *Short period.* This mode is characterized by rapid oscillations in angle of attack about a nearly constant flight path with almost constant airspeed, as explained in the discussion in the subsection on unsteady longitudinal motion. The mode is probably best excited by rapidly deflecting the elevator. Its frequency and damping are very important in the assessment of aircraft handling. Usually it is

highly damped; for a Boeing 747, the period is about 7 seconds, while the time to halve the amplitude of a disturbance is only 1.86 seconds. The short-period frequency is strongly related to the airplane's static margin; in the simple case of straight line motion, the frequency is proportional to $\sqrt{-C_{m,CG,\alpha}/C_L}$.

Dynamic modes to describe the lateral motion of aircraft include the *roll subsistence mode* (often uninteresting), the *Dutch-roll mode*, and the *spiral mode*.

3. *Roll subsistence mode*. The roll mode consists of almost pure rolling motion and is generally a nonoscillatory motion showing how rolling motion is damped.
4. *Dutch-roll mode*. This mode is moderately fast side-to-side swaying of the aircraft. It involves oscillations in bank, yaw, and sideslip angles. In a flight-state display, a Dutch roll will be indicated by the velocity vector circle oscillating from side to side. The coupled roll and yaw motion is often not sufficiently damped for good handling. Transport aircraft often require active yaw dampers to suppress this motion. High directional stability (large C_n) tends to stabilize the Dutch-roll mode while reducing the stability of the spiral mode. Conversely, large rolling moment due to sideslip, $C_{l,\beta}$, stabilizes the spiral mode while destabilizing the Dutch-roll motion. Because sweep produces effective dihedral and because low-wing airplanes often have excessive dihedral to improve ground clearance, Dutch-roll motions are often poorly damped on low-swept-wing aircraft.
5. *Spiral mode*. This mode is very slow and may be stable or unstable. It can be observed by applying aileron or rudder to bank the airplane slightly in straight and level flight. With aileron and rudder reset to neutral, the aircraft will either return to level flight in a *stable spiral mode* or experience increasing bank angle accompanied by a dive: *spiral divergence*. The damping time to half-amplitude is usually given to assess the spiral. Similarly to the phugoid motion, the spiral mode is usually very slow and often not of critical importance for piloted aircraft. A Boeing 747 has a nonoscillatory spiral mode that damps to half-amplitude in 95 seconds under typical conditions, while many airplanes have unstable spiral modes that require pilot input from time to time to maintain heading.

The linearized model can also predict handling issues involving different DOFs, such as *adverse yaw*, a handling characteristic that may be experienced when rolling the airplane with ailerons. To execute a smoothly banked left turn, the pilot will drop the right and raise the left aileron to roll the airplane to the left. The resulting extra lift on the right is also likely to produce more drag, causing the airplane to yaw to the right: the *adverse yaw*. Thus, the pilot must counteract the adverse yaw to starboard by applying left rudder and possibly also elevator up to balance the reduction of lift acting vertically on the rolled wings.

10.2.2 Longitudinal Motion

Let us restrict the model to movement in a vertical plane. This will provide important information about the longitudinal behavior, the characteristics of pitching movements

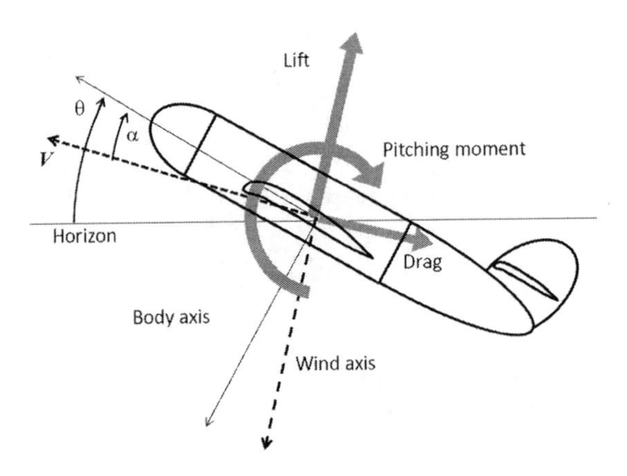

Figure 10.5 Velocity vector V, attitude angle θ, angle of attack α, and lift and drag forces.

(Figure 10.5). With the usual definitions of force and moment coefficients, Eq. (10.1) for the setup in Figure 10.5 collapses to the equations of 2D motion.

$$\begin{aligned} m\dot{u} + mqw &= QS\,C_x + T - mg\sin\theta \\ m\dot{w} - mqu &= QS\,C_z + mg\cos\theta \\ I_y\dot{q} &= QS\left(c_{ref}C_{m,ref} - (x_{ref} - x_{CG})C_z + (z_{ref} - z_{CG})C_x\right) \\ \dot{\theta} &= q \end{aligned} \tag{10.12}$$

where Q is dynamic pressure, m is the aircraft mass, x_{ref} is the *reference point* chosen for the pitching moment coefficient, x_{CG} is the x-station of the CG, and I_y is the moment of inertia around a y-parallel axis through the CG. C_x, C_z are $C_L(\alpha, \delta_e), C_D(\alpha, \delta_e)$ rotated through α.

Notes

We assume that the thrust vector passes through the CG and is aligned with the aircraft x-axis.

The flight mechanical body-fixed coordinate system is rotated 180° from the system we used for description of the aircraft geometry.

The moment that reacts to attitude rotation rate q is determined by the rotational increment $q(z - z_{cg}, -x + x_{cg})$ to velocity at (x, z). If forces and moments have been computed for rotation around another reference point, the effect is a contribution to translation velocity of the CG, $\Delta u = q(z_{ref} - z_{cg})$, $\Delta w = -q(x_{ref} - x_{cg})$, to be added to the CG velocity before evaluating forces and moments, but not in the kinematic relation to global position in Eq. (10.13).

The linearized system has only the phugoid and short-period modes.

To produce the trajectory in the earth-fixed (X, Z) system, Eq. (10.12) is augmented by the kinematic relations.

$$\begin{aligned} dX/dt &= U = -u\cos\theta - w\sin\theta \\ dZ/dt &= W = u\sin\theta - w\cos\theta \end{aligned} \tag{10.13}$$

Straight and Level Flight

Suppose the plane shall fly straight and level with airspeed $V = \sqrt{u^2 + w^2} = u$, which defines the dynamic pressure. Then $q = 0$, $\theta = \alpha$ and we must find the thrust T, angle of attack (and attitude) α, so

$$
\begin{aligned}
0 &= QSC_x + T - mg \sin\alpha \\
0 &= QSC_z + mg \cos\alpha \\
0 &= c_{ref} C_{m,ref} - (x_{ref} - x_{CG})C_z + (z_{ref} - z_{CG})C_x
\end{aligned}
\tag{10.14}
$$

With three parameters to choose, α, δ, and T, and three conditions, one for each of the three state variables u, w, and θ, this should work. It is possible to change the mass distribution (i.e. the location of the CG), by pumping fuel, but that is slow. The aerodynamics can be changed by deflecting the elevator. This primarily influences the moment and, to a lesser degree, overall lift and drag.

Roughly, then, for small α perform the following.

(1) Choose α so lift balances weight.
(2) Choose δ_e with this α so $C_{m,CG}$ vanishes.

Since δ_e also changes lift and drag, these steps must be iterated to solve the system of nonlinear equations. Finally, choose thrust T to balance the x-forces. The above indicates the weak influences that enable a solution by iteration such that there will be a unique solution. Of course, the Newton method will solve the equations in no time.

The glide slope with fixed elevator can be computed with $T = 0$.

Unsteady Longitudinal Motion: Pitch and Plunge

Now consider possible small excursions around straight and level flight at (almost) constant speed. Figure 10.6 illustrates two important motions, pitch and plunge.

(1) Pitch: the velocity vector direction (i.e. the aircraft attitude θ oscillates, while α is unchanged). This is similar to the phugoid motion.
(2) Plunge: the attitude θ remains fixed, but the plane climbs and dives slightly through changes in lift via varying α, similarly to the short-period mode of motion.

Textbooks on flight mechanics find the eigenvalues of the linearized system, derived analytically, by solving a quartic discriminant equation numerically. Experience is then brought in to show that the phugoid is primarily a pitching motion and the short period an oscillation in angle of attack. Now, the same information is encoded in the associated eigenvectors and shows this, as well as the associated small variations in velocities u, w, and hence altitude.

Stability Augmentation System

If the stick-fixed damping of a mode is very slow, the pilot must actively damp oscillations by control surface actuation, an undesirable contribution to pilot workload.

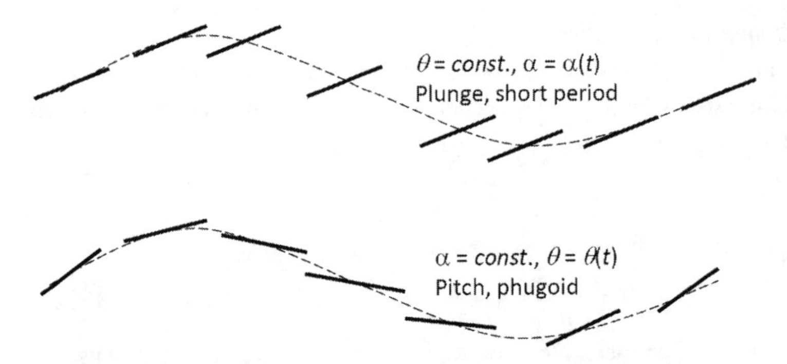

Figure 10.6 *Plunging motion* where α changes with time (top); *pitching* constant incidence motion where the attitude θ changes with time (bottom).

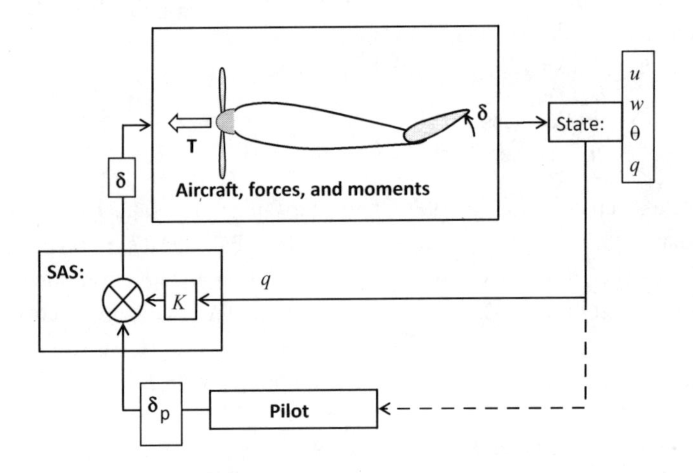

Figure 10.7 System with pitch rate feedback controller.

The problem can be mitigated by a feedback controller such as a *stability augmentation system* (SAS), exemplified by a "pitch damper" in Figure 10.7. It shall actuate the elevator to make the pitch follow the pilot's input; to this end, it senses the pitch rate by a gyro. The output from the pitch damper is added to the pilot's elevator command δ_p. The pitch rate feedback provides a contribution $kC_{m,e}\dot{\theta}$ in the moment equation. Judicious choice of the gain k is required to give a suitable damping rate of the short period; too much damping makes the aircraft unresponsive and sluggish, too little demands much pilot attention. Graphical characterizations such as those for the short-period mode relate properties of the stick-fixed flight modes to pilot evaluations.

Flight Simulation and Cooper–Harper Scale

Satisfactory flying qualities require the aircraft to be both stable and controllable. The optimum blend of stability and control for the defined missions should be

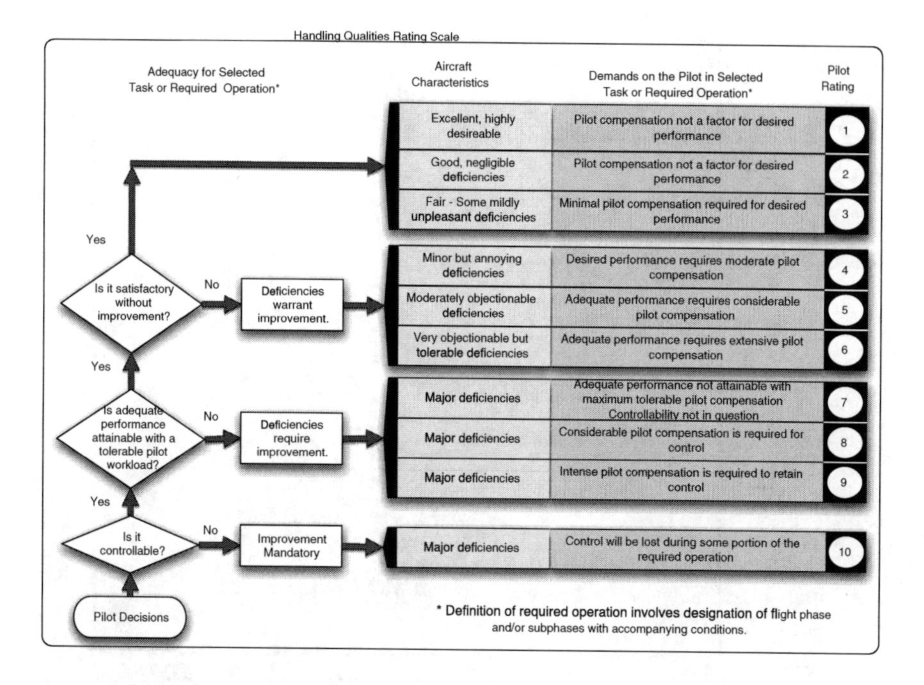

Figure 10.8 The Cooper–Harper scale. (From NASA report http://history.nasa.gov/SP-3300/fig66.htm, NASA public domain)

the aircraft designer's goal. What constitutes acceptable characteristics is often not obvious, and several attempts have been made to quantify pilot opinion on acceptable handling qualities. Subjective flying quality evaluations such as the Cooper–Harper ratings (Figure 10.8), are used to distinguish between "good-flying" and difficult-to-fly aircraft. Prediction of pilot ratings from the quantitative mathematical model is complex because human performance is also complex; subjective ratings must be turned into quantitative criteria. The link was established by characterization of airplane dynamics by numbers that describe a few modes of motion (Section 10.2.1); these numbers could be correlated with pilot ratings on the Cooper–Harper scale by interviews. The Cooper–Harper scale was developed by NACA and published in 1969 as Ref. [6] (Figure 10.8). Since its publication, it has become an accepted standard for the quantification of pilot evaluation of flying and handling qualities. The tests are performed by asking the pilot to fly a given "mission" such as a slow turn to a new heading, a dive and pull-up, acquiring a moving "target" by keeping the nose pointed at it, etc.

By correlating parameters of the trajectories flown with Cooper–Harper ratings, it was possible to produce iso-rating curves on charts of the motion parameters (see Section 10.2.1). The parameters relate to damping and frequency of aircraft oscillations, accelerations, etc. As an example, Figure 10.9 shows pilot ratings from "Satisfactory" to "Unacceptable" for the short period mode of motion. The mode

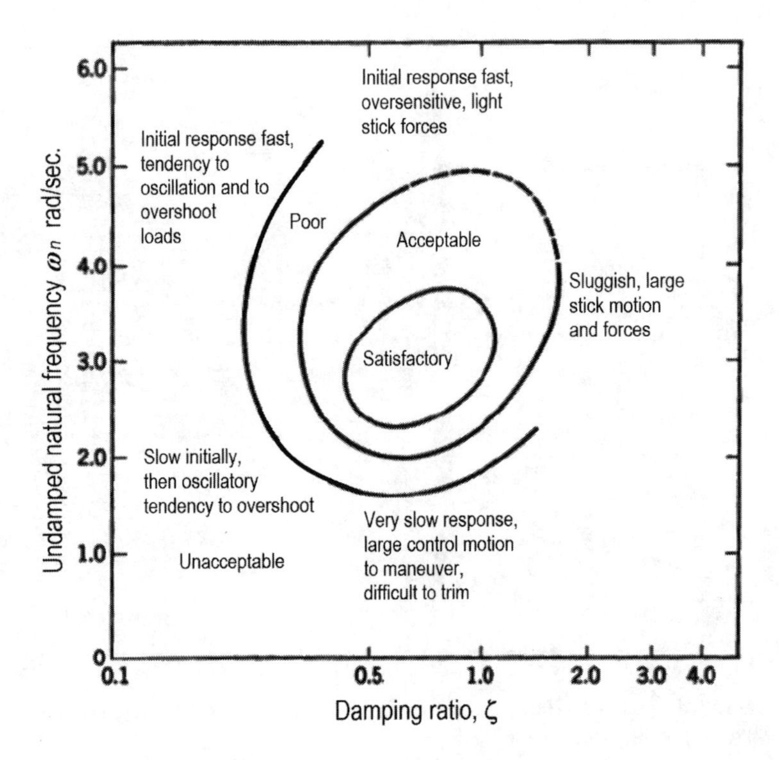

Figure 10.9 Short-period characteristics for different handling qualities. (From F. O'Hara [26], reproduced with permission)

is characterized by its "undamped" frequency $\omega_n = \sqrt{\omega^2 + \eta^2}$ (rad/sec) and damping ratio ζ (-). Note the log-scale for the damping ratio. Flight simulation also allows analysis of more violent maneuvers, for which the small-motion model is not applicable, and this can help further development of the flight control system to operate over the whole flight envelope.

10.3 Flight Simulation for Design Assessment

A generic flight simulator, let us call it FLIGHTSIM, integrates the system in Eq. (10.1) from the data supplied by the Aerodynamics (i.e. us), Weights and Balance, and Propulsion "offices." We shall discuss later how to produce the force and moment coefficient tables. Weights and balance means mass, inertia, and CG data, as well as the reference data used for defining the coefficients. A propulsion model is needed to provide realistic response (thrust) to throttle inputs and flight state; FLIGHTSIM would typically be equipped with an internal model with a few parameters, such as response time. For performance analysis, a more advanced engine model may be provided as an external software module. Similarly, the FLIGHTSIM package will include a

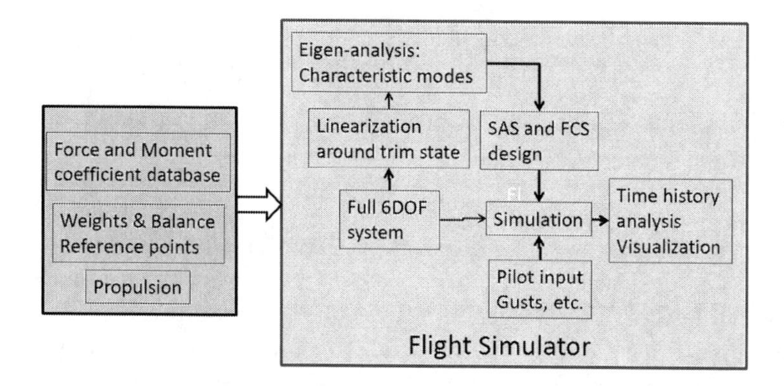

Figure 10.10 Generic flight simulator.

parametrized FCS for the user to supply the parameters, or even a design option for optimizing its parameters. The tasks for FLIGHTSIM are shown in Figure 10.10.

1. Stability analysis:
   a. Eigen-analysis of linearized model, stick-fixed, with and without FCS/SAS
2. Six DOF flight simulation:
   a. Test flights, including trim response
   b. Gusts
   c. Time history identification (nonlinear model)
3. FCS and SAS
4. Performance prediction
5. Miscellaneous: data review, results review, cross plots, etc.

Two typical examples from academia are: the static and dynamic stability analyzer (SDSA) by T. Grabowski from Warsaw University, a standalone C++ package [16], and the PHALANX code by M. Voskuijl from Netherlands Defence Academy (NLDA) built on MATLAB Multi-Body Dynamics and Aerospace Blockset Simulink packages [10].

Flight simulators focusing on realistic graphics are popular as, for example, dog-fighting games, and they enjoy commercial success. But the force and moment computations are often quite heuristic and it is hard to provide detailed aero-data of one's own, as would be required for use in design. FLIGHTGEAR (www.flightgear.org/) is a General Public License (GPL) open-source project that can use the United States Air Force (USAF) Digital DATCOM (see Section 1.1.4) to compute fidelity-level L0 aero-data for a limited range of aircraft geometries.

The aircraft manufacturers build very detailed flight mechanics models of each specific aircraft produced. The force and moment data are created by CFD, wind tunnel campaigns, and flight test. For each new feature on the aircraft, such as high-lift device settings, drop tank and armament attachment, amount of fuel, payload, etc., the aero-data must be updated.

10.3.1 Force and Moment Coefficients and Their Derivatives

The analysis of stability linearizes the dependence of forces and moments on the instantaneous state through a perturbation analysis. The derivatives of coefficients with respect to flight state (e.g. $C_{m,\alpha}$) provide much information to the flight dynamicist, since their values at trim enter directly into the linearized dynamics, as we have seen above. The values of the derivatives indicate how strongly a coefficient depends on the various independent variables. A large knowledge base has been built on this concept to help designers do their work. Books on flight mechanics describe forces and moments exclusively by such derivatives.

However, they are to be taken constant only for small perturbations to a given flight state. For flight simulation over a larger envelope, their values must be determined from the instantaneous flight state and change as the flight state changes.

There are, of course, many derivatives. The rigid aircraft has a state vector with three velocities, three angular velocities, and three attitude angles driven by three aerodynamic forces and three aerodynamic moments. It is not unreasonable to suppose that there are also purely aerodynamic forces due to accelerations associated with the flow disturbance as the airframe accelerates through it. Therefore, as a minimum in a mathematical model of our airplane, we might expect to need a derivative of each of the six forces and moments with respect to the states and to the controls, usually three surfaces and throttles. This is a total of $6 \times (9 + 4) = 78$ derivatives, many more than are used in most flying quality analyses (thankfully). The force and moment data sets needed for flight simulation can become very large, as discussed below.

There are three categories of derivatives as follows.

- *Static* derivatives: with respect to α and β, as well as the *quasi-steady* or *rotary* derivatives with respect to the rotation rates p, q, r.
- *Dynamic* or *unsteady* derivatives: could be numerous, but usually are limited to derivatives with respect to the time rate of change of α and β.
- *Control* derivatives: result from control actuation, usually limited to three control channels, and throttle.

Stability derivatives provide information about the *stiffness* and *damping* attributes of the dynamic system. For example, the *damping derivatives* characterize the variation of forces and moments with respect to angular rates.

We exemplify by the pitching moment coefficient C_m, with restriction to the 2D case for low-speed flight. Then, the coefficient depends only on angle of attack, pitching rate, angle of attack rate, and elevator deflection: $C_m = C_m(\alpha, q, \dot{\alpha}, \delta_e)$. The model is to be representative for small excursions around the trim state $\alpha = \alpha_0, q = \dot{\alpha} = 0$, $\delta_e = \delta_{e0}$ where the airplane flies straight and level.

$$\begin{aligned} C_m(t) = C_{m_0}(=0!) + C_{m,\alpha}(\alpha(t)-\alpha_0) + C_{m,q}q(t)\frac{c}{2V} &\quad \text{Static and quasi-steady} \\ +C_{m,\dot{\alpha}}\dot{\alpha}(t)\frac{c}{2V} &\quad \text{Dynamic} \\ +C_{m,\delta_e}(\delta_e(t)-\delta_{e0}) &\quad \text{Control} \end{aligned} \tag{10.15}$$

with δ_e the elevator deflection, c wing chord, $\dot{\alpha} = \frac{d\alpha}{dt}$, and V true airspeed.

Tabular Representation

The static and quasi-steady data can be represented as (a set of) multidimensional arrays of force and moment coefficients (or derivatives) versus six state parameters: angle of attack, Mach number, sideslip angle, and rotational velocity components.

A similar set of arrays is defined for three control channels: yaw, pitch, and roll. The aircraft may have more surfaces, but they will then have to be operated by the three control channels in suitable combinations. For instance, classical ailerons deploy antisymmetrically for roll control and the elevator is assigned to pitch. A tailless flying wing has elevons that deploy in coordination to provide both pitch and roll moments. There must be at least one other set of (e.g. spoilers), for yaw.

Finally, a table for unsteady derivatives, with dependence on i.e. $\dot{\alpha}, \dot{\beta}$, etc. can be supplied for different Mach and α. Such data are more involved to obtain, for slow maneuvers of less importance, and are often neglected.

Curse of Dimensionality for Tabular Models

In a representation of force and moment coefficients for a rigid aircraft there would be six tables of dimension nine or higher, depending on the number of control channels available, for example. With nine independent variables, and, say, each variable quantized with five steps, there are $5^9 \approx 2 \cdot 10^6$ different points in the discrete state space. As the JAS 39 Gripen example in Section 10.5.6 shows, five points may be far from the resolution necessary for a realistic model. The curse of high dimensionality makes a brute force approach for compiling the data infeasible. Even with exploitation of the tricks of the trade (see Sections 10.4.1–10.4.4), the population of the tables by CFD, wind tunnel campaigns, and flight tests is a serious undertaking and critical to the assessment of flying and handling qualities. The resulting aero-data set must be managed and updated over the lifetime of the configuration – usually decades – to accommodate additions and design modifications.

10.3.2 Decomposition into Lower-Dimensional Tables

Approximations and simplifications by exploitation of almost-linearity and weak interdependencies (see Eq. (10.16)), are used to lift the curse of dimensionality for the tabular representation. Thus, the data are usually approximated by sets of lower-dimensional tables. The SDSA model uses 3D tables (see Section 10.3.2). The tables can be used through linearization (see Section 10.3.1) or as they are by

interpolation and proper combination of results from the set of low-dimensional tables, noting that derivative × argument increment = function increment. As an example, consider now evaluation of the pitching moment coefficient, $C_m(t) = f(M(t), \alpha(t), \beta(t), q(t), p(t), r(t), \delta_e(t), \delta_a(t), \delta_r(t))$.
This function of nine variables is compiled from the 3D tables as follows.

$$\begin{aligned} C_m(t) = {} & f(M(t), \alpha(t), \beta(t), 0, 0, 0, 0, 0, 0) \\ & + f(M(t), \alpha(t), 0, q(t), 0, 0, 0, 0, 0) - f(M(t), \alpha(t), 0, 0, 0, 0, 0, 0, 0) \\ & + f(M(t), \alpha(t), 0, 0, p(t), 0, 0, 0, 0) - f(M(t), \alpha(t), 0, 0, 0, 0, 0, 0, 0,) \\ & + \dots \end{aligned} \tag{10.16}$$

which is exact if $f(.)$ is, actually, a sum of functions of three variables; in particular, if it is an affine function.
The evaluation in Eq. (10.15) for the 2D case can be carried out from the tabular representation as follows.

$$\begin{aligned} C_m(t) = {} & C_m(\alpha(t), q(t), 0, \delta_e) \text{ Static, quasi-steady, and control} \\ & + C_m(\alpha(t), q(t), \dot{\alpha}(t), 0) - C_m(\alpha(t), q(t), 0, 0) \text{ Dynamic} \end{aligned} \tag{10.17}$$

assuming a decomposition of the 4D table into two 3D ones, one vs $\alpha, q, \dot{\alpha}$ and one vs α, q, δ_e.

Caveat
The samples must be dense enough to catch the actual variations, even at the edges of the flight envelope. If the tables are too sparse, flight simulation will not match flight tests. There is always a risk of unpleasant surprises in flight tests probing the full envelope. Large-scale unsteady separated flow over a stalled configuration is hard to compute, and its effects cannot be captured accurately by a tabular representation limited to inclusion of the "unsteady" derivatives. Only when CFD can reliably predict separated flows will flight simulation be able to *predict* aircraft behavior in extreme situations. Nevertheless, it is very useful even without that capability to simulate trajectories with healthy flow that tend to the edges of the envelope. Control systems and pilot training can then be developed to deflect such trajectories from eventual loss of control. An example of the organization of the decomposed tables is given in Table 10.1.

The three control channels are the elevator (δ_e), the rudder (δ_r), and the aileron (δ_a). The tables are built up from 3D tables with α, M, and a third parameter: β, q, p and r, δ_e, δ_r, or δ_a (see Eq. (10.16)).

Look-Up by Interpolation
It remains to interpolate in a 3D table. This is fast if the independent variable set is a cartesian product of 1D grids and requires a search for the right interval, followed by interpolation to the eight corners of the 3D interval. The simplest scheme is a trilinear interpolant of the form $a_0 + a_1x_1 + a_2x_2 + a_3x_3 + a_{12}x_1x_2 + a_{23}x_2x_3 + a_{31}x_3x_1 + a_{123}x_1x_2x_3$ with exactly eight undetermined coefficients. The resulting function is

Table 10.1 A typical coefficient tabular format.

(a) Stability coefficients table

α	$Mach$	β	q	p	r	C_L	C_D	C_m	C_Y	C_ℓ	C_n
x	x	x	-	-	-	x	x	x	x	x	x
x	x	-	x	-	-	x	x	x	x	x	x
x	x	-	-	x	-	x	x	x	x	x	x
x	x	-	-	-	x	x	x	x	x	x	x

(b) Control coefficients table

α	$Mach$	δ_e	δ_r	δ_a	C_L	C_D	C_m	C_Y	C_ℓ	C_n
x	x	x	-	-	x	x	x	x	x	x
x	x	-	x	-	x	x	x	x	x	x
x	x	-	-	x	x	x	x	x	x	x

(c) Unsteady coefficients table

$Mach$	$C_{m,\dot{\alpha}}$	$C_{Z_{\dot{\alpha}}}$	$C_{X_{\dot{\alpha}}}$	$C_{Y_{\dot{\beta}}}$	$C_{\ell_{\dot{\beta}}}$	$C_{n_{\dot{\beta}}}$
x	x	x	x	x	x	x

continuous with discontinuities in derivatives. It is most suitably arranged as a sequence of seven linear interpolations. *Much* more work is needed to compute an interpolant with continuous derivatives, a piecewise tricubic function with 64 coefficients.

10.3.3 Weights and Balance, CG and Moments of Inertia

By suitable arrangement of the design layout and stabilizer (be it canard or tail), acceptable fore and aft limits of the center of gravity of the whole aircraft must be established. One of the tasks for the fuel system is to move the fuel to help keep the CG between the design limits. The distribution of the mass m is reflected in the position of the center of gravity (or centroid) $\mathbf{x}_{CG}$ and the moments of inertia matrix I around $\mathbf{x}_{CG}$,

$$\mathbf{x}_{CG} = \int \mathbf{x} dm/m, \ m = \int dm$$

and for $i, j = 1,2,3$

$$I_{ij} = (\sum_k a_{kk})\delta_{ij} - a_{ij}, \ a_{ij} = \int (x_i - x_{CG,i})(x_j - x_{CG,j})dm$$

To complete the model, m, $\mathbf{x}_{CG}$, and I must be predicted by the Weights and Balance Office. This is done by summing contributions from the components of the configuration, using formulas for simple shapes and applying data from existing aircraft. The book website has a tutorial on the basics applied to a toy balsa glider and a paper dart.

10.4 Building Aerodynamic Tables

Forces and moments are assumed to be functions of the instantaneous values of the velocities, control angles, and their rates. Some "dynamic" coefficients can be included, but significant dynamic effects of the inertia of the air for rapid maneuvers may be lost. For instance, a wind tunnel experiment with a wing section executing harmonic pitch will show different forces on the upstroke and downstroke at the same angle of attack – a hysteresis effect. This can be modeled for small-amplitude oscillation to some extent by the dynamic derivatives, but not for large-amplitude motion even without large-scale stall.

Estimation of Quasi-Steady and Unsteady Derivatives: Quasi-Steady Data

Quasi-steady effects are incorporated by taking into account local flow velocity increments from rotation. CFD codes use a method to solve the flow equations in a rotating coordinate system in which the aircraft is stationary. In a ground-fixed system, the aircraft CG flies a circle (yaw rate), a steady pull-up at the bottom of a looping (pitch rate; Figure 10.11), or a roll (roll rate). The centrifugal and Coriolis "forces" are absorbed by the computed pressure, so the summing up of forces on the aircraft can proceed as for straight and level flight. Note that the flight path is forced without concern for aircraft attitude or control surface deflections, as they might be necessary in a "real" flight. The rotation center is determined by the desired load factor n as follows.

$$r = \frac{V_\infty^2}{(n-1)g} \tag{10.18}$$

Recirculation of disturbances in the computational domain must be avoided by setting the center of rotation outside the computational domain.

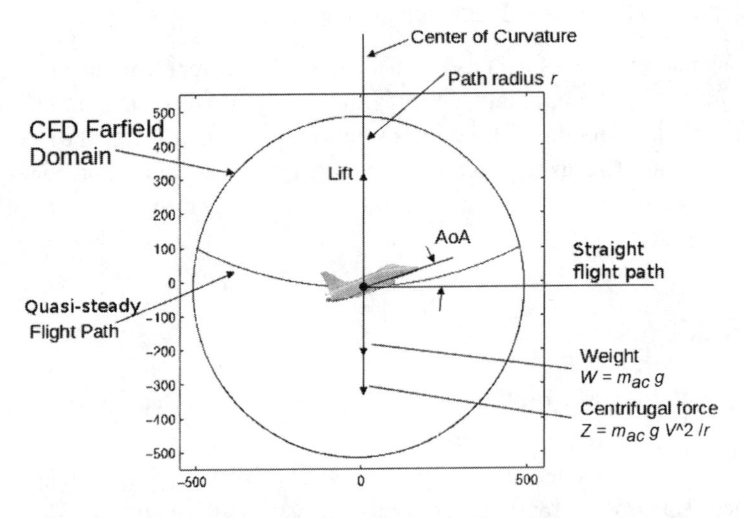

Figure 10.11 Simulation of quasi-steady pull-up. (Courtesy of M. Tomac, PhD dissertation [35], reprinted with permission)

Unsteady Derivatives
The wake trailing a lifting surface carries with it the circulation of all time up to "now," so an unsteady vortex lattice method (VLM) model can compute unsteady derivatives. An effect for aircraft with tails is the time delay it takes for a disturbance from the main wing to reach the tail. As such, these effects cannot be computed by steady CFD.

Control Derivatives
To accurately calculate the control surface forces, deflection of control surfaces must be incorporated into the CFD simulation. A number of techniques are in use.

Mesh Regeneration
A simple but expensive approach is to generate a new grid for each control deflection. The deformation/deflection can be applied to the surface model (the computed-aided design) by actual movement of the surfaces representing the control or by free-form deformation, as touched upon in Chapter 5. The former approach is necessary to model the opening and closing of gaps around the control device. The free-form deformation cannot make gaps appear, so the deflected surface will be connected to the stationary parts by a "web." Surface deformation is followed by surface and volume mesh regeneration. The free-form deformation can also be applied to the existing surface mesh, followed by volume mesh regeneration.

Gap Effects
As seen in Chapter 9 on high-lift devices, the gaps between the main element and the deployed surfaces are very important. One effect is that the flow through the leading-edge gap energizes the boundary layer on the suction side. Indeed, legendary aircraft designer Kelly Johnson put fixed *slots* just downstream of the leading edge on the Super Electra wing to allow higher angles of attack and higher lift without stall. The effects on control surfaces of airflow in gaps is illustrated in Section 10.5.5 on the Saab 340 elevator design.

Mesh Deformation
A deformation of the surface mesh can be propagated into the volume by a number of mechanisms ranging from movement along surface normals equilibrated by resistive spring forces, to solution of elliptic or parabolic partial differential equations for the coordinate perturbations. For resolution of boundary layers, the wall normal extent of the mesh cells is very small. It follows that even a very small deflection will move surface cells many cell thicknesses, invalidating ideas of small perturbations to existing cells.

Normal-Only Deflection: Transpiration Conditions
Inviscid flow models (i.e. VLM and Euler) implement the slip boundary condition by requiring the normal component of velocity to vanish. It is an old trick to apply this condition on moving surfaces by rotating the normal without changing the geometry.

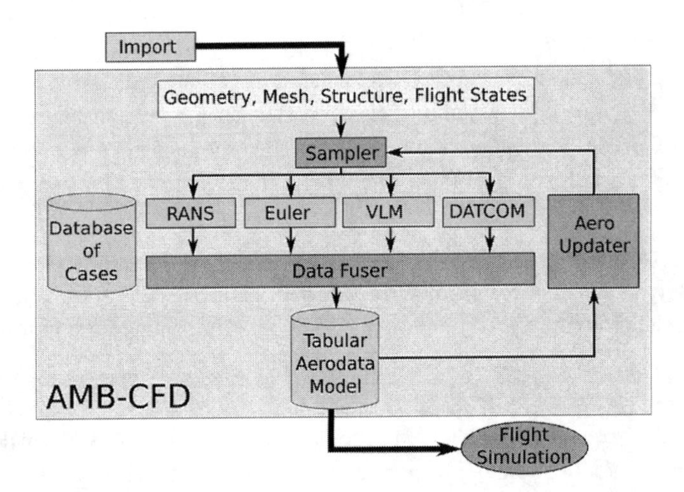

Figure 10.12 Architecture and functionality of a typical aerodynamic data set generator AMB-CFD.

The VLM code AVL [8] uses this, as does the EDGE solver. The limitation of the control deflection to small angles is the main disadvantage of this approach.

10.4.1 Efficient Population of the Tables

The cost of computing is problematic, and a brute force approach with a CFD calculation for every entry in the table extremely so. The entries can number in the thousands, or even more in late design stages. Fortunately, methods are available that can reduce the computational cost. An alternative to the brute force approach is proposed based on sampling, reconstruction, and data fusion of aerodynamic data. An introduction to such surrogate model techniques for different applications is given in Ref. [12]. The scheme here uses Kriging and Co-Kriging, named for South African mine engineer Danie Krige (1919–2013) and formalized in the 1960s by geostatistics pioneer Georges Matheron (1930–2000).

It is assumed that a low-fidelity model of the aircraft and suitable computationally cheap tools for computing its aero-data are available. A sketch of a particular table construction tool is shown in Figure 10.12.

The low-fidelity estimates are augmented with higher-fidelity data. Data fusion combines the two data sets into a single database. Once constructed, the look-up tables can be used even in real-time to fly the aircraft through the envelope covered in the tables.

10.4.2 Kriging and Co-Kriging

Aero-data F – now representing all coefficients used – at some state $\mathbf{s}$ must be approximated by, say, $\hat{F}(\mathbf{s})$ using a finite set $S = \{\mathbf{s}_i, f_i\}, i = 1, \ldots, N$. This is a "scattered data" interpolation problem, which will reappear in Chapter 11 and is a basic ingredient in big data management. First, the interpolator must be constructed in the *training* phase, and then used in the *use* or *evaluation* phase. The number of evaluations is so large that much effort can be spent in the training phase. It is natural to require the following of the scheme.

- It must work with "scattered data sets" S and not require sets to be, for example, tensor products as in the standard spline interpolation for shape representation.
- $\hat{F}$ is a smooth function.
- $\hat{F}$ returns (almost) exactly the observed values at the observation sites, $\hat{F}(\mathbf{s}_i) = f_i, i = 1, \ldots, N$.
- There is no exponential growth of the work in training and use with the dimensionality of $\mathbf{s}$.
- The scheme must evaluate $\hat{F}$ efficiently for large N.
- The set S can be extended in an incremental fashion.

The technique views F as a Gaussian stochastic field with stationary but unknown mean μ and unknown variance σ, and correlation between locations, $corr(F(\mathbf{s}_1), F(\mathbf{s}_2)) = \gamma(\mathbf{s}_1, \mathbf{s}_2)$ (the variogram), which is unity for coinciding points and decays with their inter-distance. Given these assumptions and parameter values to define $\gamma(.,.)$, $\hat{F}$ is chosen to maximize the likelihood of the actually observed data.

$$\hat{F}(\mathbf{s}) = \mu + \boldsymbol{\psi}^T \Psi^{-1}(\mathbf{f} - \mu \mathbf{1}) \tag{10.19}$$

where

$$\mathbf{1} = (1, 1, \ldots, 1)^T, \mathbf{f} = (f_1, f_2, \ldots, f_N)^T,$$

$$\psi_i = \gamma(\mathbf{s}, \mathbf{s}_i), \Psi_{ij} = \gamma(\mathbf{s}_i, \mathbf{s}_j), \; i, j = 1, \ldots, N$$

This guarantees the interpolation property and zero bias.

With a fixed variogram, this is an interpolation to the difference between the data and their mean. However, the variogram must be chosen to best represent the observed data, so the construction of the final Kriging estimator also requires optimization of likelihood over the variogram parameters (i.e. the Γ matrix). This is a nonlinear optimization task often solved by search algorithms that do not use gradients since differentiation is too complex.

Now, the force and moment data sets have "means" that vary very systematically, so Kriging is applied to the difference between the data and a standard regression model, such as a quadratic polynomial, which picks up the trends.

Training Phase

1. Construct a regression model R by, for example, least squares approximation, $R(\mathbf{s}_i) \approx f_i, \; i = 1, \ldots, N$, and record the values $r_i = R(\mathbf{s}_i)$.
2. Search the variogram parameter space.
For each candidate $\gamma(.,.)$
2.1. Construct the Ψ matrix.
2.2. Compute the maximum likelihood. This needs Ψ^{-1}.
3. Record the parameters and the Ψ matrix.

Evaluation/Use Phase

For each new site $\mathbf{s}$, compute the following.

$$\psi_i = \gamma(\mathbf{s}, \mathbf{s}_i), i = 1, \ldots, N$$
$$\hat{F}(\mathbf{s}) = R(\mathbf{s}) + \mu + \boldsymbol{\psi}^T \Psi^{-1}(\mathbf{f} - \mathbf{r} - \mu \mathbf{1}) \qquad (10.20)$$

The DACE MATLAB toolbox [22, 23] appeared in 2002 and won widespread acceptance for Kriging, as well as for engineering scattered data interpolation where there is no stochastic foundation for the model. DACE has been reimplemented in packages such as ooDACE [7], and similar tools are available also in Python [27]. Choice of the γ-function, particularly its decay rate with interpoint distance, can be critical. DACE and similar tools provide efficient algorithms for optimization of such method-internal parameters. It should be noted that Kriging works well when the distribution of sites $\mathbf{s}_i$ is uniform over the domain, but it has problems when the distribution density is anisotropic.

Co-Kriging

This technique uses f-values from a low-fidelity but cheap model f_c as a base, corrected by a few expensive high-fidelity values f_e. Co-Kriging exploits the correlation between fine and coarse model data to enhance the prediction accuracy. The ooDACE toolbox uses the autoregressive Co-Kriging model of Kennedy et al. [12]. Consider two sets of samples, $\{\mathbf{S}_c, \mathbf{f}_c\} = \{\mathbf{s}_c i, f_c(\mathbf{s}_c i), i = 1, \ldots, N_c\}$ and $\{\mathbf{S}_e, \mathbf{f}_e\} = \{\mathbf{s}_{ei}, f_e(\mathbf{s}_{ei}), i = 1, \ldots, N_e\}$.

First, a Kriging model $\hat{F}_c$ of the coarse data $\{\mathbf{S}_c, \mathbf{f}_c\}$ is constructed. Subsequently, the second Kriging model $\hat{F}_{co}$ is constructed on the residuals of the fine and coarse data $\mathbf{S}_e, \mathbf{f}_e - \rho \cdot \hat{F}_c(\mathbf{S}_e)$. The parameter ρ is estimated as part of the maximum likelihood estimate of the second Kriging model. Note that the configuration (the choice of the correlation function, regression function, etc.) of both Kriging models can be adjusted separately for the coarse data and the residuals, respectively. The resulting Co-Kriging interpolant is then defined similarly to Eq. (10.20).

Figure 10.13 shows the Co-Kriging approximation for a 1D example from Ref. [12] and the ooDACE MATLAB toolbox. The correlation function $\gamma(.,.)$ is Gaussian, the regression model $R(.)$ is linear, and $N_e = 4$ high-fidelity $\mathbf{s}_e, f_e$ samples are provided. Two of them are positioned at the borders of the parameter space to avoid extrapolation. Many low-fidelity predictions are generated by adding a smooth "error" to the target function. Co-Kriging produces an excellent approximation to the target using only four high-fidelity samples. The role of these samples is to correct the values of the low-fidelity samples, whose job is to provide information about the correlations. In contrast, Kriging applied to the four high-fidelity samples alone provides a poor representation.

10.4.3 A Kriging-based Framework for the Tabulation of Aero-Data

A framework such as that in Figure 10.12 for the generation of aerodynamic tables uses the following steps.

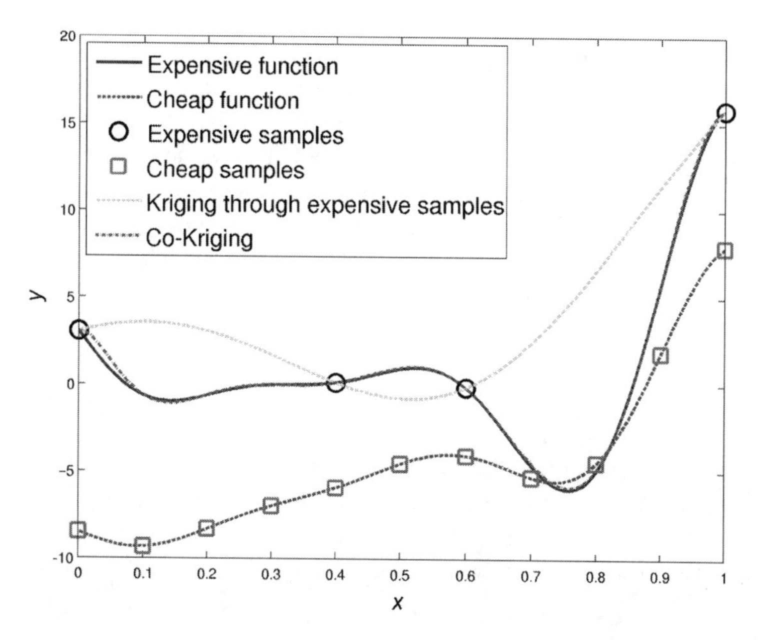

Figure 10.13 A one-variable Co-Kriging example from Ref. [12] and the DACE MATLAB toolbox.

1. The independent variables and their range are specified, and initial sampling is used to begin the procedure and to provide a quick overview of aerodynamic data throughout the parameter space by a low-fidelity method such as VLM.
2. A surrogate model $\hat{F}_c$ based on Kriging is generated.
3. The parameter space is iteratively refined by *adaptive sampling*: sample points are added at untried locations to improve the accuracy and verify the robustness of the surrogate model.
4. When the low-fidelity surrogate model is complete (i.e passes an error estimate test), high-fidelity data are generated and Co-Kriged. Feedback on expected variance of the prediction, etc., is used to select new high-fidelity samples. This is an "engineer-in-the-loop" process (see Section 10.4.2).

VLM is accurate only for low-speed attached flow. For higher speeds, a Level 2 (Euler CFD) prediction that models compressible flow is needed; for cases with boundary-layer interaction or significant flow separation, a Level 3 prediction by Reynolds-averaged Navier–Stokes (RANS) or detached-eddy simulation (DES) is called for. The engineers know that only sufficiently high M_∞ needs Euler CFD and only α approaching and exceeding stall needs RANS (or DES). The number of expensive hi(gher)-fidelity samples N_e can thus be limited by restricting them to the subdomains where the predictor $\hat{F}_c$ is known to be inaccurate.

The Co-Kriging predictor $\hat{F}_{co}$ correction will automatically decay with distance from the high-fidelity set. Of course, the engineer should check that low-fidelity predictions match the high-fidelity ones in the domain in which he or she believes low-fidelity to be accurate.

Adaptive Sampling

Using the stochastic field model, Kriging can estimate the variance of its predictor. It therefore makes sense even for the low-fidelity predictor to start with a coarse set **S**, find domains where the variance estimate is large, and augment **S** by new samples there. Such design of experiment (DOE) rules using statistical methods have been instrumental in the design and interpretation of experimental campaigns. Classical texts in this area are, for example, Box et al. [2] and the groundbreaking book by Fisher, in its ninth (!) edition [11]. Kriging can provide feedback to the sampling scheme with the aim of producing the cheapest predictor with prediction errors satisfying required confidence levels.

Data Fusion

Data fusion is the combination of aero-data produced by different sources into a single surrogate model. Kriging can use data sets **S** fused by simple amalgamation, or results with higher fidelity can be Co-Kriged into a single Kriging predictor. The tools must allow backtracking so that the pedigree of a model is certain at all times.

10.4.4 Co-Kriging Use Case

This section shows the fused aerodynamic coefficients from analysis by low-fidelity VLM (L1) and high-fidelity Euler CFD (L2) on a reference configuration. The results come from the AGILE project [4].

Figure 10.14 (left) shows the fused $C_L(M, \alpha, \delta)$ results as surface over the M–α plane, with elevator deflection $\delta = 0^\circ$ over the flight envelope, dots for low-fidelity samples and crosses for Euler samples. The Co-Kriging predicts the nonlinear behavior at higher angles of attack, as the high-fidelity samples indicate.

The surrogate model is reliable with max(RMSE) $= 0.048\%$. The final M, α sample sets for building the surrogate models are shown in Figure 10.14 (right) with 22×3

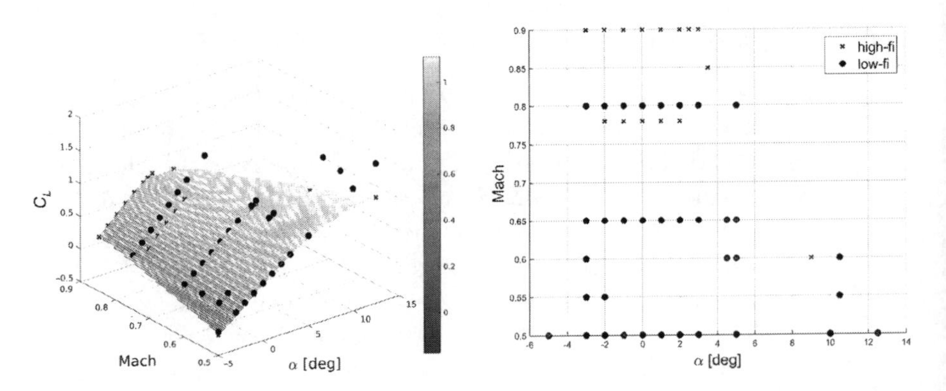

Figure 10.14 Coefficient surface with low-fidelity (dots) and high-fidelity (crosses) data points (left); final set of low-fidelity and high-fidelity samples in angle of attack $\times$ Mach plane (right). (Courtesy of M. Zhang, PhD dissertation [24], reprinted with permission)

high-fidelity samples and 35 × 3 low-fidelity samples for elevator deflections −3, 0, and +3 degrees. For each elevator deflection, the samples are at the same angle of attack and Mach locations.

10.5 Applications: Configuration Design and Flying Qualities

10.5.1 Wing Positioning

The position of the CG relative to the aircraft aerodynamic center is crucial, and proper axial positioning of the lifting surfaces is necessary for good longitudinal stability. Wing dihedral influences the lateral and roll dynamics. Two studies drawn from the SimSAC Project exemplify the trade-offs involved in laying out the lifting surfaces. We discuss first wing location on the Transonic Cruiser (TCR) and then optimization of wing and tail locations for the Alenia SMJ jet airliner. Subsonic tail-last (not canard) airplanes are generally balanced to bring their CGs near the wing-alone aerodynamic center, which is near 25% of the MAC at the wing center. The tail contribution shifts the aircraft aerodynamic center further aft and gives a suitable static margin.

Tailless airplanes, on the other hand, must have the CG ahead of the wing-alone aerodynamic center to be inherently statically stable. If the wing is swept back, it can be trimmed at a reasonably high lift coefficient with trailing-edge up elevon deflections. The degree of static stability desired and the maximum lift coefficient obtained are interrelated.

The aerodynamic center of canard airplanes is considerably ahead of the 25% point of the wing MAC. The canard configuration offers trimming with an upload, which means that the main wing needs to produce less lift, reducing the total induced drag. It must also be set back enough to provide suitable static margin.

10.5.2 Improving TCR Trim Characteristics

L0 design tools (e.g. the USAF DATCOM; see Chapter 1), predicted that the baseline TCR in Figure 10.15a would require excessive elevator deflections to trim, and this section describes the low-speed rigid-aircraft trim analysis that was carried out to confirm this hypothesis and what reconfigurations were considered. With the baseline T-tail geometry, aerodynamic data were computed using the Tornado L1 VLM code. The flight simulator SDSA produced trim and dynamic stability analysis that confirmed the handbook prediction of 20° tailplane deflection at sea-level speed of 160 m/s ($M \approx 0.47$) and angle of attack of 3.5°. A tailplane deflection of 18° was found for trim.

The configuration should be redesigned to reduce control surface deflection and its associated trim drag. Moreover, it was found from L1 aero-elastic modeling (see Chapter 11) that the T-tail was prone to flutter, thus providing further reason to reconfigure the baseline. Figure 10.15 summarizes the evolution of the design for pitch control from baseline T-tail through baseline with moved main wing, three-surface configuration, to the final all-moving canard.

(a) The Saab baseline T-tail configuration

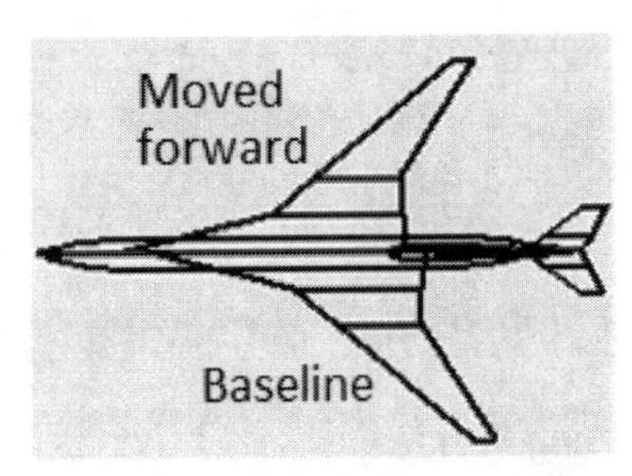

(b) Moving baseline wing forward 5% fuselage length

(c) Reconfigure T-tail baseline to three-surface layout

(d) Reconfigure T-tail baseline to a basic canard layout TCR-C

Figure 10.15 TCR design evolution: (a) baseline T-tail configuration from Saab; (b) wing moved forward to reduce trim drag; (c) first reconfiguration to a three-surface layout; (d) second reconfiguration to a basic Canard layout, TCR-C-basic displayed, further developed to final TCR-C15. (From EU project SimSAC [29], reprinted with permission)

Wing Repositioning

Moving the main wing further forward 5% of the fuselage length (Figure 10.15b) reduced the trim requirement to approximately 13°, which is still excessive, so just moving the wing was not a viable solution to the trim-drag problem, nor to the flutter problem.

Three-Surface Configuration

A three-lifting surface configuration with canard, main wing, and horizontal tail was considered next (see Figure 10.15c), but was rejected due to its complexity and the lack of available design guidelines. It should be noted that three lifting surfaces provide an extra DOF so the aircraft can be trimmed while optimizing for another variable, such as drag.

All-Moving Canard

The final configuration considered was equipped with an *all-moving canard* (Figure 10.15d).

A canard configuration has several advantages. It can trim the configuration by lift, reducing the induced drag compared to a tail surface. It also offers improved stall response. If designed correctly, the canard stalls before the main wing and induces a nose-down pitching moment that reduces the angle of attack and mitigates main-wing stall.

Table 10.2 TCR-C: selected configurations trimmed at 160 m/s at sea level. AoA = angle of attack.

	$x_W(-)$	$x_C(-)$	S_C (m^2)	AoA α (○)	Canard δ (○)	Static margin (%)
TCR-C2	0.26	0.13	65	2.4	6.6	4.42
TCR-C8	0.26	0.017	47	2.2	6.6	4.64
TCR-C15	0.26	0.12	72	3.2	5.9	3.12
TCR-C17	0.26	0.017	65	2.2	4.2	−2.65

Longitudinal stability is a key driver for sizing the canard, and the critical flight state is in low speed, since in transonic flight the aerodynamic center shifts aft, increasing the aircraft stability margin.

Sizing the TCR Canard Layout

Trade studies were carried out with VLM, keeping the planform and area of the main wing fixed while varying the following three design parameters: the apex position of the main wing x_W/ℓ, the apex position of the canard wing x_C/ℓ, and the canard area S_C, where ℓ is the fuselage length. The static margin K_n is given as a fraction of $MAC = 11.88m$.

Four of the most promising designs are given in Table 10.2, which shows the angle of attack, canard deflection δ required to trim, and static margin K_n at 160 m/s at sea level.

TCR-C2 and TCR-C15 have the canard wing located foremost in the fuselage to maximize its moment arm to the CG, and TCR-C8 and TCR-C17 have the canard in the nose cone. The latter option interferes with systems housed in the nose cone, such as the weather radar. In the former option, the doors cannot be located in the fore fuselage.

Some general conclusions can be drawn from Table 10.2. When the canard is moved forward, its moment arm increases. The force it needs for a given moment around the CG is smaller, so a smaller deflection is required. However, the aircraft aerodynamic center also moves forward, impairing longitudinal stability. Featuring a larger distance between canard and main wing, the TCR-C17 requires a smaller canard deflection for trim but is statically unstable in pitch.

Since the product of arm and area appears in the moment, increasing the canard area gives a similar effect, as can be seen when comparing the TCR-C8 and the TCR-C17 with its larger canard area. The TCR-C15 was chosen as the most appropriate configuration: the reduction in control surface deflection at trim is balanced by a tolerable reduction in stability.

Traditional Sizing by Tail Volume

Raymer [28] defines the canard volume coefficient as follows.

$$c_c = \frac{\ell_c S_c}{MAC \cdot S}$$

where ℓ_c is the canard moment arm. He suggests a volume coefficient of approximately 0.1, taking into account only the exposed canard area.

The TCR-C15 has a canard moment arm of 28 m, an exposed canard area of 60 m^2, a mean aerodynamic chord of 11.77 m, and an extended wing area of 489 m^2. The resulting canard volume coefficient is 0.29. With the main wing geometry being frozen, the volume coefficient can be reduced by decreasing the distance between the canard and the main wing or by reducing the canard area.

Both alternatives would result in poor trim ability, the opposite to the goal of the exercise. One concludes that the control surface volume rule of thumb is inadequate for such an unconventional design as the canard TCR. VLM tools proved useful for approaching unconventional configurations that are scarcely mentioned in aircraft design literature.

Transonic Trim Changes Due to Compressibility

Background: Tuck-under

Longitudinal trim changes when accelerating or decelerating near and even through Mach 1 cause a nose-down tendency called *tuck-under*. A severe problem for the Lockheed P-38 and its contemporaries, it was investigated at length in the UK in the early 1940s through high-speed dives. The tuck-under made pull-up from a steep dive very slow, so the dives must allow substantial altitude reserves. Several nonlinear factors come into play in the transonic flight regime. One is the change in $C_{L,\alpha}$ that results from shock aft movement, as demonstrated for an airfoil in Chapter 6. The effect on $C_{m,\alpha}$ is also related to the rearward movement of the aerodynamic center.

Such nose-down longitudinal trim change near Mach 1 was an issue for the wide-delta-winged Douglas F4D-l Skyray, later called the F-6, used by the US Navy in the 1950s. It had fully powered elevon controls but no Mach trim change compensation. Mach trim could be more easily installed on horizontal tails that, like those on contemporary jetliners, can be made all-moving in addition to carrying the elevator. The McDonnell Douglas F4 Phantom from the late 1950s had a sophisticated powered Mach trim system with force feedback to the pilot.

The TCR-C15 Transonic Shift of Aerodynamic Center

Often given in percentage of the mean aerodynamic chord MAC, the static margin K_n was defined in Section 10.2 by Eq. (10.6). With the CG x_{CG} at 38.33 m from the nose, the values for K_n are given in Table 10.3, which shows the movement of the aerodynamic center and changes to static margin with increasing M_∞.

A static margin over 32% indicates that at high speed the TCR-C15 is very stable with slow response to canard deflection. This illustrates one of the design problems associated with transonic flight, namely how to accommodate the large shift in neutral point from low to high speed. One possible answer to this is to relax the longitudinal stability requirement at low speed. This would result in the need for stability augmentation and would also present certification challenges.

Table 10.3 TCR-C15: aerodynamic center (AC) and static margin (K_n) for increasing M_∞, where $x_{CG} = 38.33$ m and $MAC = 11.77$ m.

M_∞	AC (m)	$K_n(\%)$
0.12	38.90	4.70
0.65	39.93	13.63
0.85	40.62	19.46
0.97	42.12	32.20

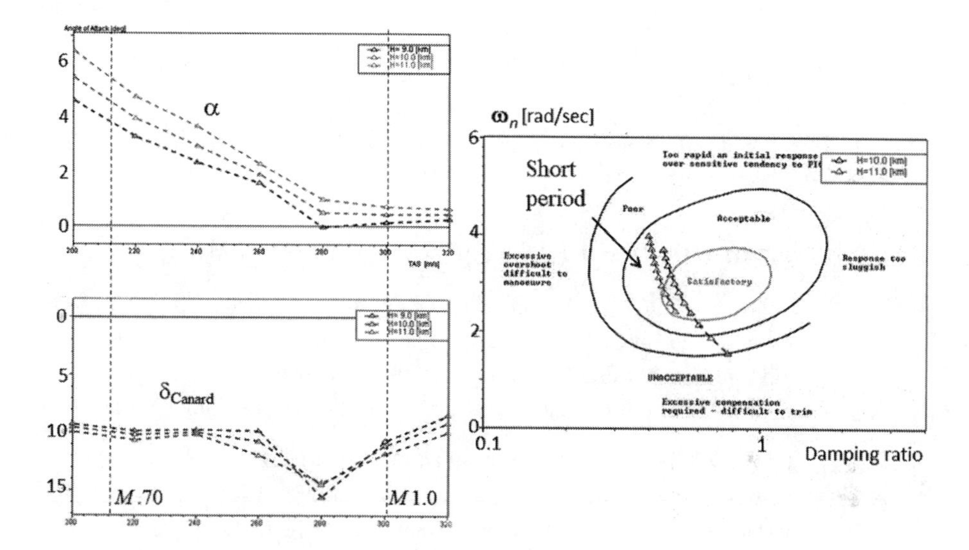

Figure 10.16 TCR-C15 trim conditions and short-period mode properties at transonic speed. (From EU project SimSAC [30], reprinted with permission)

Transonic Trim and Stability

Figure 10.16 presents the trim conditions and associated SDSA-predicted short period mode properties at transonic speed and cruise altitudes. The deflection angles remain fairly constant until the speed of sound is approached, where there is a significant increase in canard deflection angle δ from 10° to 15° before it returns back to approximately the same deflection at higher speeds. This manifestation of tuck-under is controllable. Cruise at $M_c = 0.97$ at 11 km altitude requires a deflection of approximately 12°. The short period mode characteristics are good. The phugoid (not shown) is strongly damped, as can be expected from the rough approximation in Eq. (10.11), showing a damping ratio proportional to D/L. The TCR cruise D/L is large compared to current airliners traveling at slightly lower Mach numbers.

With increasing Mach number, the lift and pitching moments of the wing change dramatically. In particular, a large jump in the upper surface suction from $M0.85$ to $M0.92$ occurs on the outer wing panel, behind the CG, thus contributing a pitch-down moment. This is the source of the tuck-under seen in the "dip" in the canard deflection in Figure 10.16.

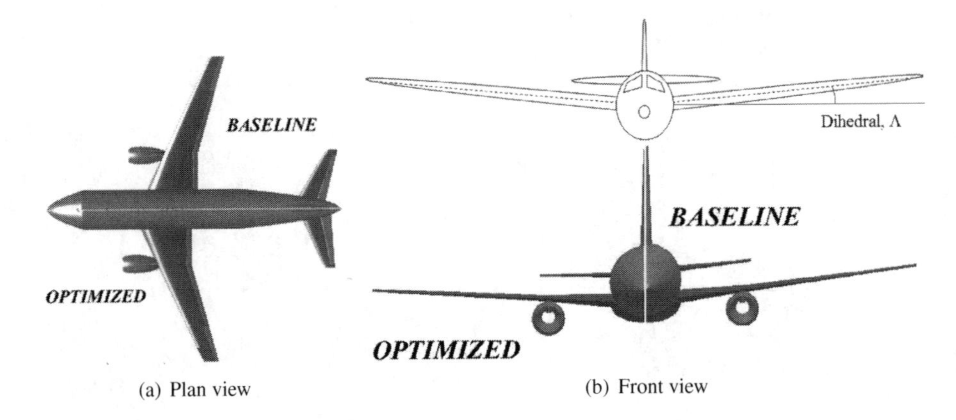

Figure 10.17 Comparison of optimized and baseline configurations for SMJ. (From EU project SimSAC [29], reprinted with permission)

10.5.3 Improved Dutch-Roll Characteristics for the Alenia SMJ

SimSAC studied the Alenia 70-seat Regional Commuter configuration concept SMJ. At high speed and altitude, the baseline was not compliant with JAR 23 rules for Dutch roll and had too high elevator deflections for trim. The exercise was to improve its stability and control characteristics by variations of positional parameters as follows.

1. Wing dihedral angle – reduced from 7.25° to 3.0°
2. Wing longitudinal position – moved ahead 2% of fuselage length
3. Horizontal tail dihedral angle – reduced from 6.0° to 0°
4. Incidence of the horizontal tail – changed from 0° to −3°

The configurational variations concluded in the new configuration shown in Figure 10.17.

10.5.4 Tail and Canard Layout and Assessment

Tail and canard surfaces stabilize the aircraft and provide moments needed for maneuver and trim. They add wetted area and structural weight and should be as small as possible. Straight-tapered tail surfaces with hinge lines all at constant percentage chord have become the norm for most aircraft. In general, the control surfaces are sized based on the required control power (i.e. their contributions to pitch moment $\Delta C_m/\Delta\delta_e$ and yaw moment $\Delta C_n/\Delta\delta_r$).

Horizontal Tail

Horizontal tails are generally used to provide trim and control over a range of conditions such as takeoff rotation, approach trim, and nose-down acceleration near stall. These conditions must be considered for all allowed CG positions, such as takeoff rotation with the CG in its most forward position for aircraft with nose gear and most aft position for craft with tail gear.

Tail surfaces are normally loaded downward in cruise, for some planes by as much as 5% of the aircraft weight. If stability requirements are relaxed with the application of active controls, the size of the tail surface and/or the magnitude of tail download can be reduced. The weight as well as wetted area are decreased, giving improved performance. Control power usually determines the sizing of the tail.

Vertical Tail

For older and current aircraft, stability requirements for Dutch-roll modes were an issue in sizing the vertical tail. For aircraft with fly-by-wire flight controls, vertical tails are not sized for Dutch roll, provided that the tail is sized to give the control system sufficient authority to stabilize the airplane. For this purpose, the rudder should be designed to return the aircraft from a 10° sideslip disturbance at any altitude. The critical control sizing constraint in straight and unaccelerated flight for twin-engine planes is that aerodynamic moments must balance engine thrust with one engine inoperative, creating windmilling drag, and the other engine giving maximal thrust.

10.5.5 Control Surfaces

Elevator Sizing for the AGILE Design Campaign 1 Configuration

Some very large airplanes are cruise trim critical. The tail is sized to be buffet free or below drag divergence at dive Mach number. Drag divergence is used as an indication of the likelihood of elevator control reversal because of high airloads and possible aero-elastic effects at transonic speed when strong shock appears on the suction side of the stabilizer. Further elevator deflection increases the loading and may cause structural deformation of the elevator and eventually control response reversal.

A specific example is the Design Campaign 1 (DC1) exercise, a A320-like configuration from the AGILE project. The participants in the project computed the aerodynamic coefficients and derivatives with L1 TORNADO and L2 SU2 tools, including elevator deflections, and fused the results in an aero-database for longitudinal flight simulation. Elevator deflections are implemented by the SU2 free-form deformation (FFD) box to deflect the mesh by rotation around the elevator hinge line. Angles up to about 5° are tolerated without unacceptable mesh cell deformation.

TORNADO usually took less than a minute per VLM analysis on a modern laptop, and the SU2 Euler solver took around 4 minutes on a 32-core workstation for the reference aircraft. The computational cost ratio is at least 128 (provided that the SU2 parallel computing speed-up is linear). Thus, even very dense L1 samples are affordable.

Prediction of Flying Qualities from Fused Data

The participants analyzed the flight performance and flight dynamics with PHALANX and compared results obtained with L1 + L2 fused data by Co-Kriging to that obtained with entirely L1 data. The PHALANX MATLAB tool makes extensive use of the Simulink © platform, its toolboxes and the Simscape © environment for simulating physical systems.

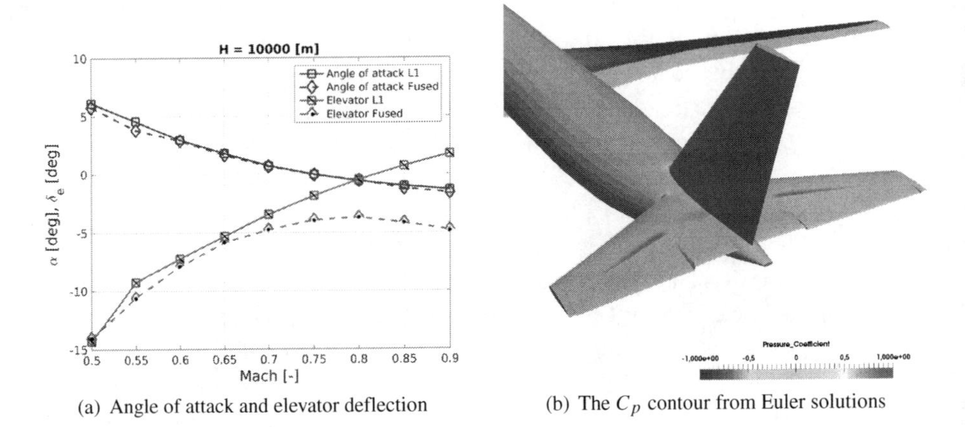

(a) Angle of attack and elevator deflection (b) The C_p contour from Euler solutions

Figure 10.18 (a) Angle of attack and elevator deflection for L1 and L1 + L2 fused data for trimmed flight at 10 km altitude as a function of Mach number; (b) C_p contours on the horizontal tail from SU2 Euler solutions, $M = 0.78$, $\alpha = 0°$ with elevator deflection $\delta = 4°$. The elevator deflection is modeled by FFD deformation of the mesh. (From AGILE EU project, M. Zhang [4], private communication)

With the necessary input data for weight, inertia, and engine thrust, the simulation model assessed the trim condition (e.g. the prediction of flight envelope limits and power required as a function of Mach number) to determine the flying qualities for various flight conditions.

Figure 10.18a shows the angle of attack and elevator deflection in trimmed flight for the whole range of Mach numbers at 10 km altitude. The L1 + L2 and L1 databases give similar predictions for the angle of attack. For subcritical speeds, the L1 and L1 + L2 elevator deflections agree well, as expected, but deviate at higher Mach. In the transonic speed regime, the aerodynamic center on the swept wing moves aft, an effect not modeled by VLM. The moment around the CG decreases, needing less elevator lift for trim. Notice the shock wave in the C_p-contours on the upper side of the elevator in Figure 10.18b, and the "web" between the elevator and tailplane.

Rudder Sizing for Asymmetric Twin-Engine Configurations

For twin-engine aircraft, controlled flight with one engine inoperative is a requirement for rudder authority. To reduce the rudder deflection needed to counteract the axisymmetric thrust, Raymer [28] proposes an asymmetric concept shown in Figure 10.19, inspired, Raymer says, by Burt Rutan's asymmetric Boomerang. The issue in question is: Does the rudder size of the concept have sufficient authority in one-engine-out operation? In L2 mode, EDGE can compute the aerodynamics of control surface deflection through the transpiration boundary condition (see Section 10.4), with flow tangency to the deflected surface prescribed on the nondeflected surface. This method avoids mesh deformation, so one mesh can be used for all of the deflection cases.

The propeller is simulated as a "very short" nacelle in SUMO to make the grid resolve the propeller disk. A specified mass flow through the disk adds momentum, which

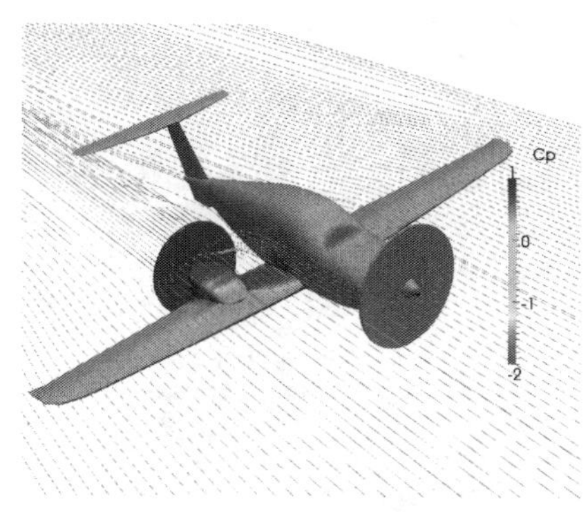

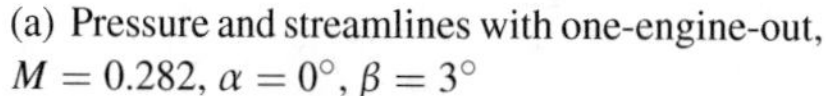
(a) Pressure and streamlines with one-engine-out, $M = 0.282$, $\alpha = 0^\circ$, $\beta = 3^\circ$

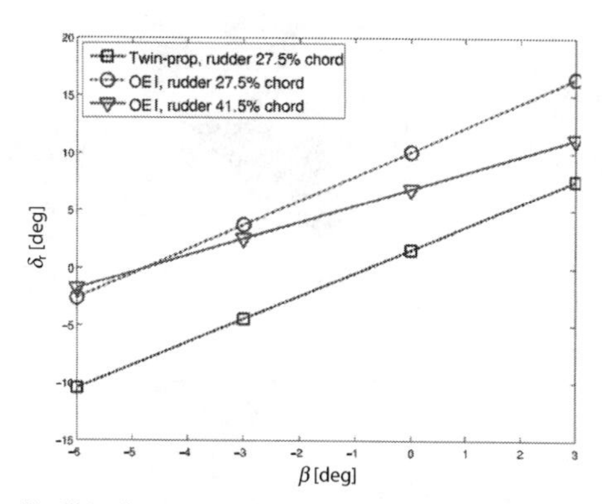

(b) Sideslip angle β versus rudder deflection δ_r in yaw trim

Figure 10.19 Aerodynamic moments and forces computed by L2 Euler EDGE. OEI = one engine inoperative. (Courtesy of M. Zhang, PhD dissertation [24], reprinted with permission)

provides the pressure jump across it. The propeller thrust was chosen to balance the cruise drag, and Figure 10.19a visualizes the surface pressure and streamlines with nose propeller inoperative.

The analysis predicts that, with both engines operating, the asymmetric aircraft trims without sideslip with rudder deflection about 1.6° (square-symbol line in Figure 10.19b) to counteract the inherent nose-left yaw. Should the front engine then fail, the off-center thrust yaws the aircraft further to the left. The trim condition then is the round-symbol line in Figure 10.19b. Zero-yaw flight then requires a rudder deflection around 10°.

Such deflection for trim is problematic because compensation for a strong lateral gust would require additional yaw-moment authority, and rudder stall then becomes an uncomfortable possibility. There are two ways to enhance rudder authority: one is to enlarge the whole tail fin, thus reducing the magnitude of C_n at $\beta = 0$; another is to increase the rudder fraction of the tail to increase the magnitude of control derivative $C_{n,\delta}$. Resizing the rudder to 50% larger than the baseline (27.5% chord to 41.5% chord) improves the trim authority. Zero sideslip now requires 7° deflection instead of 10° (triangle-symbol line in Figure 10.19b), which is consistent with linear aerodynamics.

Aileron Time-to-Roll Simulation for the Saab 2000

Figure 10.20 (top) defines the hinge moment coefficient C_h and the roll moment coefficient C_ℓ resulting from aileron deflection δ_a.

The ailerons should have the following.

(1) Sufficiently large influence $C_{\ell,\delta}$ on the roll moment C_ℓ with a nearly linear relationship between control force and aircraft response throughout its operating range.
(2) Acceptably small hinge moment C_h so that the wheel force is not too large.

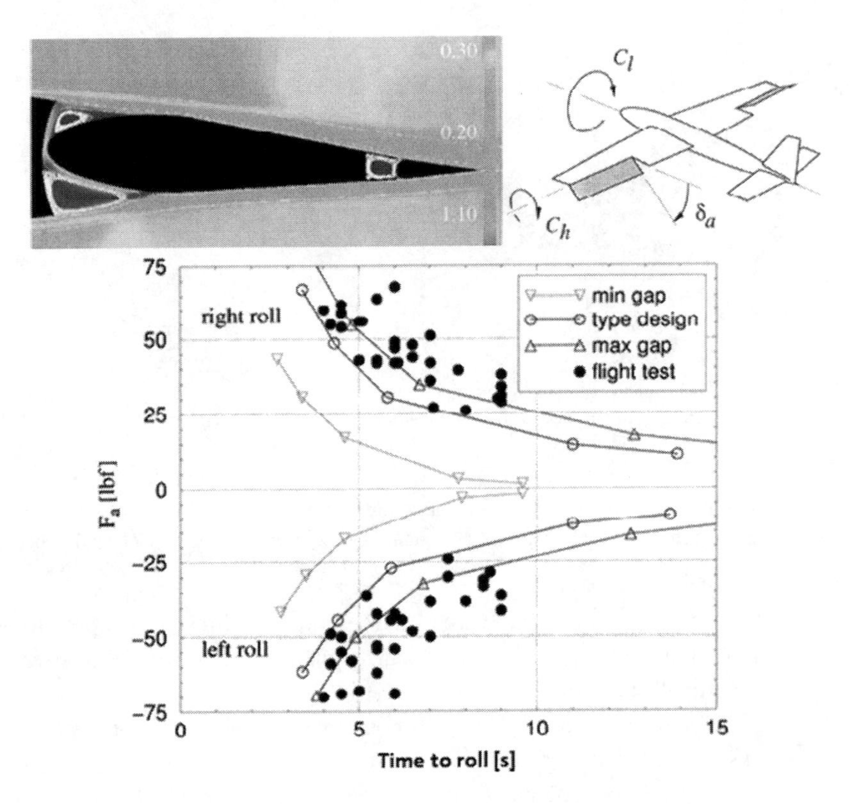

Figure 10.20 Hinge moment coefficient C_h and roll moment coefficient C_ℓ at aileron deflection δ_a (top right). Computed Mach number distribution and streamlines around the Saab 2000 aileron (top left) $\delta_a = 0°, M_\infty = 0.298, Re = 10.7 \times 10^6$. Control wheel force F_a vs time-to-roll-60° at 139 m/s, (bottom). (Courtesy of Erkki Soinne, PhD thesis [31], reprinted with permission)

The flow-field plot shows a 2D idealization of flow around the aileron. The study focuses on the gaps between aileron and main wing element and between aileron and its trailing-edge tab. The wing of a production aircraft is much more complex, involving many details, including actuation and linkages that must pass requirements for flight certification. Prediction of pilot wheel forces is valuable in aircraft aerodynamic design. Prediction of the aero-forces is a demanding task because the sensitivity of the flow to the size of the gap between control surface and main wing element and the Reynolds and Mach numbers makes it difficult to obtain an accurate CFD solution in the vicinity of the slots and gaps.

E. Soinne in his dissertation [32] examined the aileron deflection wheel force of the Saab 2000 commuter aircraft in a detailed analysis combining CFD and flight-mechanical simulations. The regulations require that a roll of 60° from a steady 30° banked turn can be executed in not more than 7 seconds for airspeeds from minimal climb speed to maximum operating speed.

The effects of aileron gap size on the wheel forces were computed with 2D RANS computations on two aileron sections. The inboard section was taken through the aileron tab (see Figure 10.20) and the outboard section was taken through the aileron exterior hinge. The values at the two sections were weighed together to make a 3D aileron effectiveness and a hinge moment coefficient. Computations were made with the minimum, nominal, and maximum gap sizes resulting from normal manufacturing tolerances. With the computed aerodynamic data, which included dynamic derivatives, Soinne carried out simulations of the airplane's mechanical control system. The Saab in-house six DOF system FORMIC takes the aerodynamic hinge moment to the control wheel, accounting for nonlinearities due to friction, backlash, centering cams, disconnect and spring units, and aileron differential. Figure 10.20 (bottom right) shows the wheel force as a function of the time needed to roll the required 60° at the maximum operating airspeed of 139 m/s. The effects of the tolerances are most prominent at high dynamic pressure (i.e. high speed). The wheel force F_a required of the pilot increases in faster rolls to around 50 lbf for a 7-second time to roll. Wheel force values due to all allowed production tolerances, measured in flight testing of a production aircraft, are shown to verify the modeling.

10.5.6 Aerodynamic Problems Discovered in Flight Testing

It happens (all too frequently) that aerodynamic problems surface during or just before flight testing a prototype vehicle. Problems arise for all sorts of reasons, and remedial actions take either of two directions: to adapt the control laws in the FCS to counteract the undesirable effects or to change the aerodynamic shape to avoid the problem. A slope of a moment curve with respect to either α or β that changes abruptly or even changes sign is very difficult for the control laws to treat. It is better to modify the flight shape in such cases, and we present two such examples.

UCAV in Sideslip Suffers Break in the $C_{\ell,\alpha}$ Curve

Saab has flight tested a subscale unmanned combat aerial vehicle (UCAV) demonstrator called SHARC (Swedish Highly Advanced Research Configuration). The main objectives of the SHARC program were to demonstrate strike missions, low-signature features, high survivability and autonomy, and to test its airworthiness. To this end, SHARC has a flat body with chined sides and engine air intake between twin fins to reduce the radar cross section. Wind tunnel tests on the candidate configuration revealed undesirable roll and yaw moment dependence on sideslip. The results here appear in Ref. [21]. Figure 10.21a shows rolling moment C_ℓ following a V-curve versus angle of attack α at sideslip angle $\beta = 15°$ and with the slope break at $\alpha = 15°$.

Flow visualizations and pressure distributions are presented at an angle of sideslip of 15°. Figures 10.21c and 10.21d compare the solutions for angles of attack of 12° and 15°. The side wind coming from the left of the UCAV (right sides of Figures 10.21c and 10.21d) highly affects the formation of the body vortices. This is explained by a modified effective sweep angle of the chines of the UCAV: higher sweep on the leeward side and lower sweep on the windward side. The two body vortices thus have

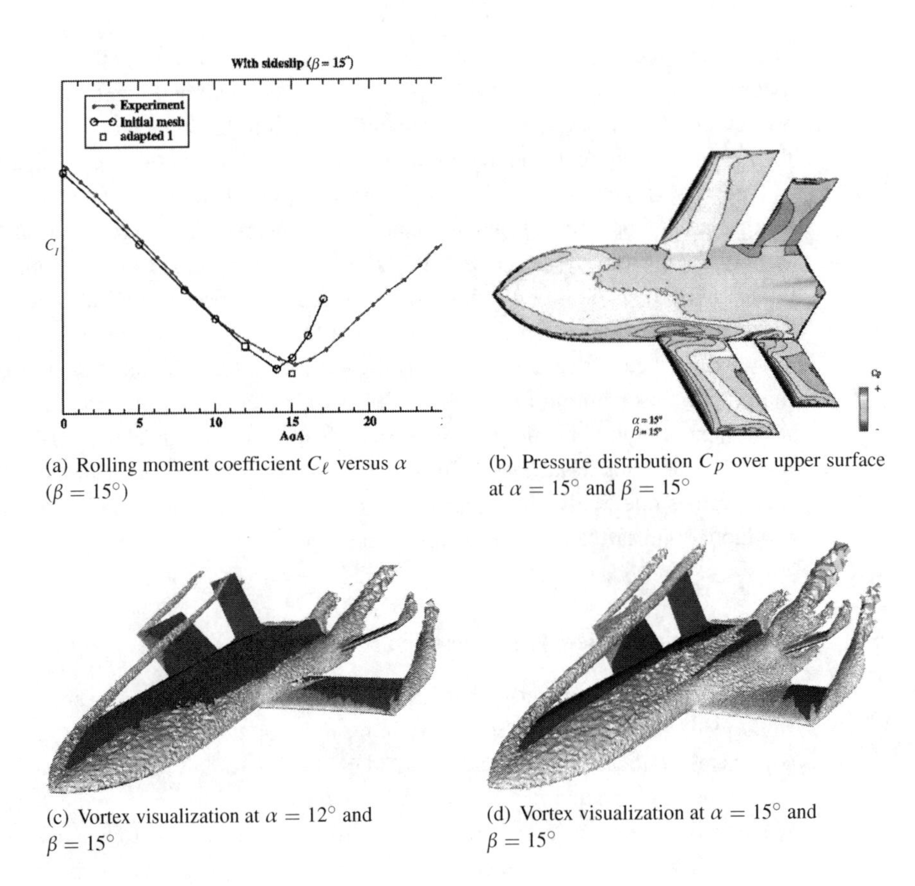

(a) Rolling moment coefficient C_ℓ versus α ($\beta = 15^\circ$)

(b) Pressure distribution C_p over upper surface at $\alpha = 15^\circ$ and $\beta = 15^\circ$

(c) Vortex visualization at $\alpha = 12^\circ$ and $\beta = 15^\circ$

(d) Vortex visualization at $\alpha = 15^\circ$ and $\beta = 15^\circ$

Figure 10.21 UCAV flow fields at $\alpha = 12^\circ$ and 15° at sideslip $\beta = 12^\circ$. (Courtesy of Y. LeMoigne, PhD thesis [20], reprinted with permission)

completely different positions. On the windward side, the vortex remains close to the wing and forms more downstream than in straight flight. On the leeward side, the vortex forms closer to the nose and lifts off the body.

The asymmetry makes the leeward vortex hit the tail. At 12° in Figure 10.21c, the vortex is mainly under the tail but hits the wing leading edge. As the angle of attack increases the vortex moves up, hitting the tail at 15° (Figure 10.21d) and passing over it for higher angles. The windward vortex starts to lift off too, but at a location near the wing, which creates a very complex separated flow on the rear part of the UCAV at $\alpha > 15^\circ$. With 15° of sideslip, convergence of the steady-state iterations was indecisive, indicating flow unsteadiness.

The asymmetry is also seen in the pressure distribution. At $\alpha = 12^\circ$, the leeward vortex is rather weak and there is no trace of it on the body nor on the wing (see Figure 10.21b). As the angle of attack increases, the footprint on the fuselage becomes more pronounced. On the windward side, the low-pressure region below the vortex is present even at 12°, but the lowest pressures are found close to the leading edge of the wing where the vortex lifts off. The tail on the windward side also shows the low-pressure mark of the vortex, already seen in the results without sideslip. The interaction of the leeward vortex with the tail has a large influence on the stability of the vehicle, as is shown in the next paragraph.

Prediction of the Aerodynamic Characteristics

The V-shape rolling moment curve with sideslip is predicted rather well by computations (see Figure 10.21a). The decreasing rolling moment until 15° is certainly due to the windward vortex getting stronger and generating suction on the port side, while above that angle less suction is present, particularly in the rear part of the vehicle. The effect of the leeward vortex on the tail above 15° can also be a contributing factor to the sudden change in slope at that angle. In this case too, the numerical results follow the experimental curve until 14°. A finer mesh improves it until at least 15°, where the minimum C_L is reached.

The strong nonlinearity of the moment coefficient certainly complicates the design of the control laws. It is difficult to explain in detail, but the vortex hitting the tail plays an important role. Analysis of the results of the wind tunnel campaign led to a change in tail layout before the configuration could proceed to flight testing.

Saab JAS 39 Gripen Cock's Comb Mitigates Break in $C_{n,\delta_{canard}}$

Control systems can be designed to mitigate undesirable variations of forces and moments with flight state and control surface deflections. The Gripen fighter is, like other fighters of its generation, unstable in pitch; this improves turn performance and can be controlled by the fly-by-wire system. There was some indication early on about nonsmooth yaw behavior for certain canard deflections at high angles of attack, and it was decided that the problem, if any, would be solved by adjustment of the control laws after enough aerodynamics data had been obtained. Initial flight tests were uneventful, but in a following test the prototype crashed on approach, and on national television, following serious pitch oscillations. Thus, the control laws had to be revisited. The revision following the crash included a detailed wind tunnel investigation, which revealed a very large, sudden dip in a moment over a few degrees of deflection. This was too much for the control system to handle, and the aerodynamicists were tasked with smoothing the behavior by shape modification. A very small shoulder strake, dubbed the Cock's Comb, was installed above and behind the canard to stabilize the vortex flow. The plot in Figure 10.22 was obtained from the published memoirs of Saab engineer Ulla Teige [34]. It shows the yaw moment curve at angle of attack $\alpha = 28°$ and sideslip $\beta = 10°$ before (solid line with the serious dip) and after (dashed line), the modification. The caption in Swedish claims that "The Cock's Comb makes a difference – also for aircraft!" Indeed, shoulder strakes also appear on Gripen's generation colleagues, the Eurofighter and Panavia Tornado.

In an elementary way, we have completed the study of the flying qualities of the rigid aircraft. What remains is to determine the shape of the wing under load during flight, the topic of the next chapter.

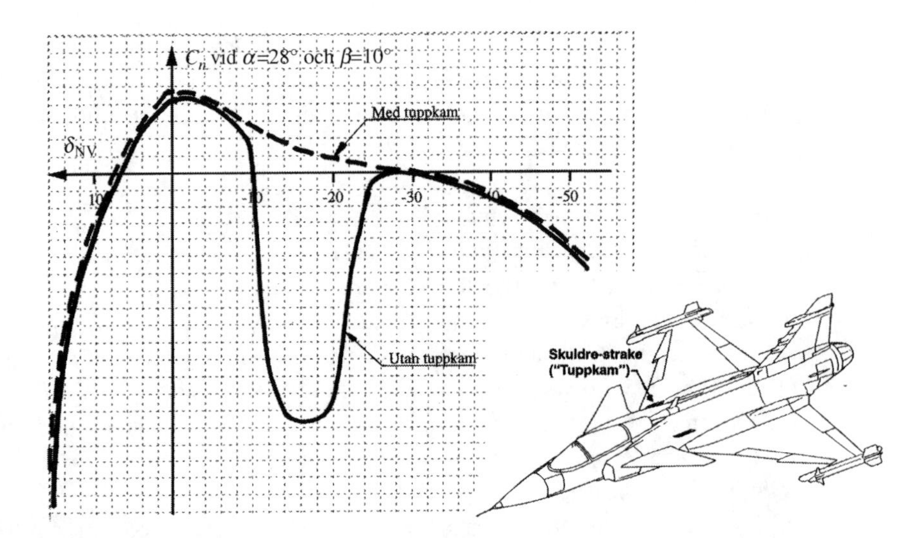

Figure 10.22 Yaw moment vs canard deflection, without and with Cock's Comb on the Saab JAS 39 Gripen. The Cock's Comb is just aft of the canard trailing edge. (Courtesy of K. Nilsson [34], Saab, reprinted with permission)

10.6 Learn More by Computing

Gain hands-on experience of the computational tools for the topics in this chapter by working with the on-line resources. Exercises, tutorials, and project suggestions are found on the book website www.cambridge.org/rizzi. For example, use the VLM tools to analyze the TCR low speed stability as measured by the static margin. Software to compute many of the examples shown is available from http://airinnova.se/education/aero-dynamic-design-of-aircraft

References

[1] R. Bellman. *Stability Theory of Differential Equations*. McGraw-Hill, 1953. Reprinted Dover Publications, Inc., 1969.

[2] G. E. P. Box, W. G. Hunter, and J. S. Hunter. *Statistics for Experimenters: Design, Innovation, and Discovery*, 2nd edition. Wiley, 2005.

[3] G. H. Bryan. *Stability in Aviation. An Introduction to Dynamical Stability as Applied to the Motions of Aeroplanes*. Macmillan and Company, Ltd, 1911.

[4] P. D. Ciampa and B. Nagel. AGILE the next generation of collaborative MDO: Achievements and open challenges. Presented at 2018 Multidisciplinary Analysis and Optimization Conference, June 2018. AIAA-Paper-2018-3249.

[5] M. V. Cook. *Flight Dynamics Principles: A Linear Systems Approach to Aircraft Stability and Control*. Elsevier Aerospace Engineering Series, 3rd edition. Elsevier, 1997.

[6] G. E. Cooper and R. J. Harper. The use of pilot rating in the evaluation of aircraft handling qualities. Technical report TN D-5153. NASA, 1969.

[7] I. Couckuyt, T. Dhaene, and P. Demeester. ooDACE toolbox: A flexible object-oriented kriging implementation. *Journal of Machine Learning Research*, 15: 3183–3186, 2014.

[8] M. Drela. *Flight Vehicle Aerodynamics*. MIT Press, 2014.

[9] B. Etkin and L. D. Reid. *Dynamics of Flight – Stability and Control*. John Wiley and Sons, 1995.

[10] T. Fengnian and M. Voskuijl. Automated generation of multiphysics simulation models to support multidisciplinary design optimization. *Advanced Engineering Informatics*, 29: 1110–1125, 2015.

[11] R. A. Fisher. *The Design of Experiments*, 9th edition. Macmillan, 1971.

[12] A. I. J. Forrester, A. Sobester, and A. J. Keane. *Engineering Design via Surrogate Modelling: a Practical Guide*, Volume 226 of *Progress in Astronautics and Aeronautics*. John Wiley & Sons, Ltd, 2008.

[13] M. Ghoresyshi, K. J. Badcock, A. DaRonch, D. Vallespin, and A. Rizzi. Automated CFD analysis for the investigation of flight handling qualities. *Mathematical Modelling of Natural Phenomena*, 6(3): 166–188, 2011.

[14] M. Ghoreyshi, D. Vallespin, A. DaRonch, K. J. Badcock, J. Vos, and S. Hitzel. Simulation of aircraft manoeuvres based on computational fluid dynamics. Presented at AIAA Guidance, Navigation and Control Conference, Toronto, Ontario, 2010. AIAA-2010-8239.

[15] T. Glad and L. Ljung. *Control Theory*. Taylor & Francis, 2000. Reprinted by CRC Press, 2010.

[16] T. Goetzendorf-Grabowski, D. Mieszalski, and E. Marcinkiewicz. Stability analysis in conceptual design using SDSA tool. Presented at Special Session of the AIAA AFM Conference, Toronto, 2010. AIAA.

[17] H. W. M. Hoeijmakers and S. J. Hulshoff. Numerical simulation of the unsteady aerodynamic response of a complete aircraft. Presented at ICAS Congress, Melbourne, 1998.

[18] W.-H. Jou. A systems approach to CFD code development. Presented at ICAS Congress, Melbourne, 1998.

[19] F. W. Lanchester. *Aerodynamics, Constituting the First Volume of a Complete Work on Aerial Flight*. A. Constable and Co., Ltd, 1907.

[20] Y. LeMoigne. *Adaptive Mesh Refinement and Simulations of Unsteady Delta-Wing Aerodynamics*. PhD thesis, Aeronautical and Vehicle Engineering, KTH Royal Institute of Technology, 2004.

[21] Y. LeMoigne. Adaptive mesh refinement sensors for vortex flow simulations. In P. Neittaanmäki, T. Rossi, K. Majera, and O. Pironneau, editors, *Proceedings of the ECCOMAS 2004 Congress*. University of Jyväskylä Press, 2004.

[22] S. N. Lophaven, H. B. Nielsen, and J. Søndergaard. DACE, a Matlab Kriging toolbox. version 2.0. Technical report. Technical University of Denmark, 2002.

[23] S. N. Lophaven, H. B. Nielsen, and J. Søndergaard. Aspects of the Matlab toolbox DACE. Technical report. Technical University of Denmark, 2002.

[24] Z. Mengmeng. *Contributions to Variable Fidelity MDO Framework for Collaborative and Integrated Aircraft Design*. Doctoral thesis, Engineering Sciences, KTH Royal Institute of Technology, 2015.

[25] R. C. Nelson. *Flight Stability and Automatic Control*, 2nd edition. WCB/McGraw-Hill, 1998.

[26] F. O'Hara. Handling criteria. *Journal of the Royal Aeronautical Society*, 71(676): 271–291, 1967.

[27] F. Pedregosa et al. Scikit-learn: Machine learning in Python. *Journal of Machine Learning Research*, 12: 2825–2830, 2011.

[28] D. P. Raymer. *Aircraft Design: A Conceptual Approach*, 6th edition. American Institute if Aeronautics and Astronautics, 2018.

[29] A. Rizzi. Modeling and simulating aircraft stability and control – the SimSAC project. *Progress in Aerospace Sciences*, 47(8): 573–588, 2011.

[30] A. Rizzi, P. Eliasson, T. Goetzendorf-Grabowski, J. B. Vos, M. Zhang, and T. S. Richardson. Design of a canard configured transcruiser using ceasiom. *Progress in Aerospace Sciences*, 47(8): 695–705, 2011.

[31] E. Soinne. *Aerodynamic and Flight Dynamic Simulations of Aileron Characteristics*. PhD thesis, Dept. Aeronautics, KTH, 2000.

[32] E. Soinne. Aerodynamically balanced ailerons for a commuter aircraft. *Progress in Aerospace Sciences*, 37: 497–550, 2001.

[33] G. Strang. *Introduction to Applied Mathematics*. Cambridge-Wellesley Press, 1986.

[34] U. Teige. Aerodata - vad är det? In *Saabminnen Del 22* SAAB Veteran's Club, 2012, pp. 92–97 (in Swedish).

[35] M. Tomac. *Towards Automated CFD for Enginerring Methods in Airfcraft Design*. Doctoral thesis, KTH School of Engineering Sciences, 2014.

[36] W. G. Vincenti. *What Enginees Know and How They Know It: Analytical Studies from Aeronautical History*. Johns Hopkins University Press, 1990.

11 Airload–Structure Interactions and Aero-Elastic Effects

> The traditional approach ... for aircraft with fairly slender high aspect ratio wings was to recognize that the structure is beam-like and then to represent the major aircraft components ...by beams lying along reference axes positioned at, for example, the locus of shear centres [i.e flexural axis].
>
> Jonathan Cooper[1]

Preamble

Up to now, we have assumed that the aircraft is rigid, making all of our aerodynamic computations on its flight shape. This chapter addresses how the manufactured (jig) shape deforms to the flight shape.

All wings are elastic and deform under load into the flight shape. The overall objective of this chapter is to show how the wing shape can be determined for a given flight condition. The aerodynamicist participates with computation fluid dynamics (CFD) and delivers the aero-loads in a multidisciplinary team analyzing this interaction problem. With the aero-loads in hand from CFD computations, the structural engineer can provide a *sized* structural model of the particular wing in the iterative process of the *aero-elastic loop* to find the deformation to the flight shape. In keeping with the scope of this book, we consider the aero-structural teamwork in the early stages of aerodynamic design. The loop is exemplified by static aero-elastic analyses in low-subsonic airflows, such as static wing deformation, divergence, and control reversal.

In Section 11.4, this chapter's overview of the fluid–structure interaction is shown at the software level in a *partitioned* approach of a *loosely coupled, modular framework*, which we call AeroFrame. It illustrates the interactions of the two disciplines CFD and computational structural mechanics, the work done by the team members, and the required data exchange.

[1] *Introduction to Aircraft Aeroelasticity and Loads* [19], with permission of John Wiley & Sons, Ltd.

In two case studies, Section 11.5 shows AEROFRAME in action on the following.

(1) A simple model in a wind tunnel undergoing wing divergence and control reversal
(2) Determination of the flight shape of an unmanned aerial vehicle in pull-up and push-down maneuvers

In both cases, the structures expert in our team has been Aaron Dettmann, who was studying for his MSc thesis [6].

11.1 Introduction

Aero-elasticity is concerned with the interaction of airflow with flexible structures. Load-carrying structures for aerospace applications are light weight, flexible designs. Although designed to sustain the loads encountered during operation, they may undergo quite considerable elastic deformations.

The wings will bend and twist when subjected to aerodynamic loads. Depending on structural and flow characteristics, this change in shape and loading may lead to *stability* problems, which are categorized as static or dynamic (i.e. involving oscillations).

Dynamic phenomena, such as flutter and buffeting, involve unsteady aerodynamic loads and structural dynamics. The unsteady loads continuously feed energy into the structure, causing it to be accelerated in an oscillating motion. Hence, not only elastic properties but also the mass distribution influence the response.

11.1.1 Aeroloads Deform Flight Shape and Affect Aerodynamics

The deformations on an airliner with high-aspect ratio (*AR*) wings can be large; in cruise, the wing tips of a Boeing 787 Dreamliner bend up around 10 ft. under aerodynamic and gravity forces, and up to 26 ft. (!) in a 2.5*g* pull-up at maximum takeoff weight (see Figure 11.1, top).

As introduced briefly in Section 1.4.2, the *aero-load data* are used to assess the aircraft structure. The analysis requires a mechanical model (e.g. a finite-element (FE) description of the structure) to translate the forces and moments into deformations and stresses.

Stresses are useful for estimating service life, fatigue, failure modes, etc. The deformations also must be taken into account for predicting actual cruise and maneuver performance, at least to the point of deciding *if* the aero-elastic shape change is significant. Next-generation airliners may well have higher-*AR* wings, since these reduce the drag, so wing deformation in flight becomes a more important issue, as it always has been for high-performance gliders and high-altitude aircraft such as the Lockheed-Martin U-2 Spyplane.

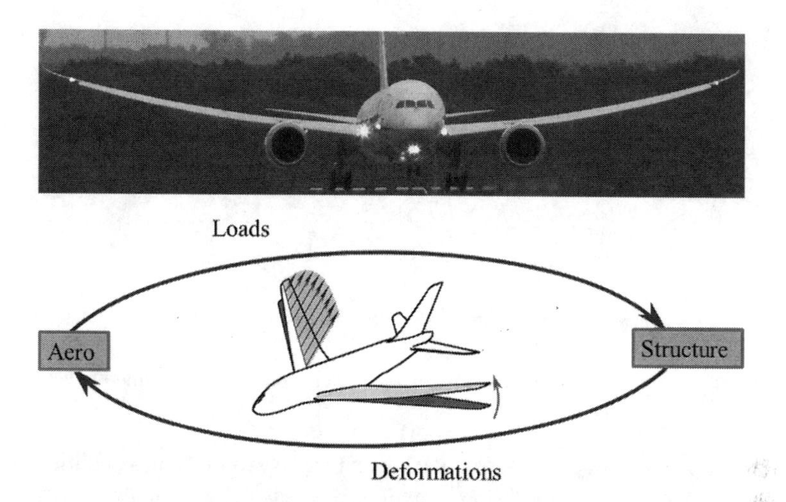

Figure 11.1 The B-787 Dreamliner wing deforms visibly in flight (top). (Courtesy of M. Bogdan, Cargospotter YouTube channel, reprinted with permission.) An elastic wing deforms under load (bottom). (Courtesy of A. Dettmann, MSc thesis [6], reproduced with permission)

The trend to larger-span wings is countermanded by the increased structural weight needed to support the increased wing bending moment, and the design of such airframes requires close integration of structural and aerodynamic design.

Both static and dynamic aero-elastic effects are usually deleterious. "Wing divergence," "buffeting," "flutter," and "aileron reversal" are effects that bound the flight envelope and must be guarded against.

For example, the World War I Fokker D-8 monoplane suffered a number of wing failures thought to be caused by weakness of the wing structure. The story is related in Ref. [14]. The military required the wing structure to be stiffened, yet the first modified airplane crashed on its first sortie with wing failure. Loading the wing with sandbags, Fokker discovered that the wing tip incidence increased significantly as the wing bent under load. In flight, the wing tip twist increased its airload, leading to more twist, etc., and structural failure occurred at sufficiently high speed.

How could this happen? Figure 11.2 helps to answer this question. *Wing loading* has been computed in Chapter 3 to assess wing bending moments, but the wing also twists under load. The lift on a wing section can be represented by a force acting approximately at the 1/4-chord point. For most wing section structures, the cross-section flexural axis is behind the 1/4-chord, so the section twists nose up under the lifting force (Figure 11.2, left.) This, in turn, increases the lift force, with more twist, etc. The structure must be stiff enough to limit the aero-elastic bending and twist for all flight states in the envelope.

The D-8 modification had strengthened the *rear* wing beam spar. This also moved the flexural axis aft, giving more arm to the lift force, and the net effect was that torsional, not bending, deformation grew to structural failure.

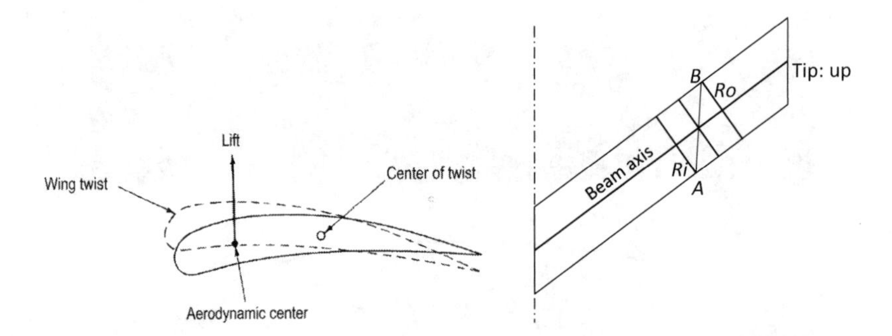

Figure 11.2 A wing section twists nose up under aero-load (left); bending of swept wing (right).

Most deflections under load have a negative impact on flying qualities, but by judicious analysis and clever design, it may be possible to alleviate the negative effects by "aero-elastic tailoring," adapting the *stiffness distribution* of the load-carrying structures and the jig shape of the wing.

Bending and Torsion of Swept Wings

For purely geometric reasons, a forward-swept wing has a stronger, and a backward-swept wing a weaker divergence tendency than a straight wing. This can be seen by imagining the beam representing the wing and a stream-parallel strip on the wing surface (Figure 11.2, right). When the wing bends tip up (toward the reader), the outboard rib **Ro** comes up higher than the inboard one **Ri**, so the point **B** is higher than **A**. If the wing is swept forward (i.e. flies toward the top of the page), **B** is on the leading edge and the effect is a nose-up twist of a stream-parallel strip with increased lift. The opposite is the case for a backward-swept wing.

11.1.2 The Aero-elastic Loop: Fluid–Structure Interaction

Computational models provide tools to analyze aero-elastic effects and thereby help identify and improve problematic designs. If aero-elastic analysis tools are built from *separate* aerodynamic (i.e. computational fluid dynamics (CFD) and computational structural mechanics (CSM) models, which is a fairly common approach, the two disciplines need to exchange deformation and load data.

Figure 11.3 shows how the structures and aerodynamics interact via the surface shape of the aircraft. It also outlines a notional concept with the following steps for separate CFD and CSM solvers to interact to solve the fluid–structure interaction problem.

Start from an undeformed state (e.g. jig shape of wing), compute aerodynamic loads, and subsequently map them onto a structure model to obtain the deformations. Then update the shape of the aerodynamic model to reflect these deformations, which in turn changes the airloads, etc.

The exchange of loads and deformations is implemented in a computational *aero-elastic loop*.

Running it can produce one of two outcomes: convergence to a structural *equilibrium* with small deformations, showing that the structure is sufficiently stiff to balance the aerodynamic forces, or continued growth of deformations to become uncomfortably large.

The latter outcome shows only that this particular iterative way of solving the coupled problem fails. Other algorithms may do better. They usually start from small dynamic pressure when the iteration is guaranteed to converge and increase it in steps, just as the sweeps for angle of attack and Mach number are set up in MSES (Chapter 7). In this way, the growth of deformations with dynamic pressure is obtained.

In steady, level flight as well as quasi-steady flight maneuvers, such as steady climb and descent, pull-up, or turns, the aerodynamic loads can be considered steady. But in other scenarios, the load may change quickly (e.g. in a strong gust). Then the deformation occurs rapidly, so that inertial forces come into play, requiring dynamic analysis of structures, and ultimately also of the flow.

With aerodynamic and gravity loads known, the aircraft designer can decide on the size (i.e. strength, stiffness) of the major structural components. This defines the overall stiffness and a part of the mass distribution. But loads depend on mass distribution, so again a coupled problem arises, calling for some solution of equations. The critical loads are found on the edges of the flight envelope. Indeed, the intended envelope may have to be restricted as the structural design evolves together with the mass distribution.

11.1.3 Aero-elastic Scenarios

Keeping the aero-elastic loop in mind (Figure 11.3), the following discussion looks at a number of aero-elastic scenarios. They are either *static*, involving only steady aerodynamic and gravitational forces, or *dynamic*, with time-varying forces and flow motions along with airframe inertial effects.

Static Deformation

Aerodynamic loads and structural response may converge to a static aero-elastic equilibrium where airloads on the deformed shape balance internal elastic forces. It is

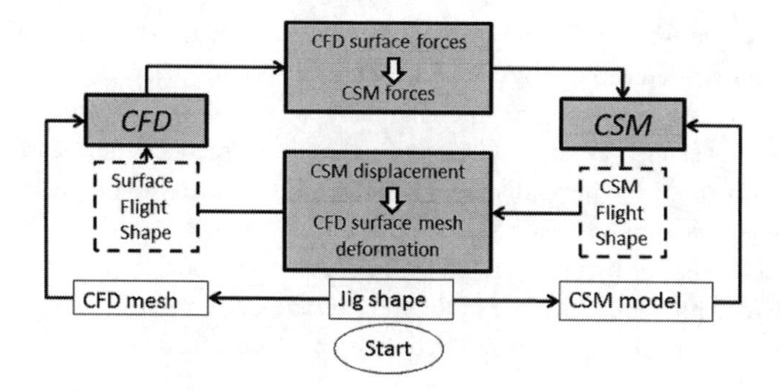

Figure 11.3 Notional diagram of the fluid–structure interaction indicating the flow of information in an *aero-elastic loop*.

quite possible – indeed, it is common – that the forces on the loaded and deformed structure are larger than on the undeformed structure. When the aerodynamic forces overcome the maximal structural-restoring force, runaway wing deformation in twist and bending called *aero-elastic divergence* occurs, and the structure fails, as on the Fokker D-8 monoplane. The main factors influencing this phenomenon are the dynamic pressure and wing torsional stiffness. Forward-swept wings are particularly sensitive to divergence, as we saw above. The conventional practice is to *design the flight shape* of the aircraft (i.e. the aero-elastically deformed configuration). The manufactured shape (jig shape) is determined later on by calculating backward from the desired deformed configuration using the corresponding known loading condition. Obviously, this process requires that there is a single well-defined design load case corresponding to a certain flight condition (e.g. cruise) for which this jig-shape transformation can be performed. In fact, wings are designed, optimized, and built such that their deformed flight shape is a minimum drag configuration. This is precisely what Martins et al. [10] have done for the Common Research Model (CRM) wing.

Control Efficiency and Reversal

Since wings are flexible, applications of control surfaces have different effects, depending upon the flight speed. Thus, one must estimate how wing twist and control effectiveness comply with the intended flight envelope.

When a trailing edge control surface down is applied, the resultant extra lift occurs close to the trailing edge, aft of the flexural axis, causing a moment to deflect the section nose down. This reduces the lift obtained through deflection of the flap and reduces the *control efficiency*. The actual net lift increment drops with increasing speed – the efficiency declines – until it vanishes at the *reversal speed*. Beyond this speed, application of the control will have the *opposite* effect to that intended, called *control reversal*. Although not necessarily disastrous, this phenomenon is undesirable, as control of the aircraft is poor around the reversal speed when the control surface is ineffective.

11.1.4 Objectives and Scope of the Chapter

The overall objective of this chapter is to show how the wing shape at a given flight condition can be determined by a multidisciplinary team analyzing the interaction problem in the aero-elastic loop. Applications to static aero-elastic analyses in low-subsonic airflows, such as static wing deformation, divergence, and control reversal, show the loop at work. Unsteady aerodynamic loads, effects of compressibility, and structural dynamics are not treated, although the framework is general and such effects could be accounted for by higher-fidelity CFD and CSM models and suitable solution algorithms. For broader and more in-depth coverage of aero-elastic phenomena and computational tools, the reader is referred to textbooks devoted to aero-elasticity (e.g. [7, 14, 19]).

The aircraft design aero-structural teamwork is carried out in a software framework requiring significant computational resources.

In contrast to this, and in keeping with standard aero-elastic curricula, we first consider a model that is simple enough that an analytic aero-elastic model can be established. The static solution can be computed in an iterative fashion, and it turns out that the wing diverges exactly when the iteration does.

The generalization to the static aero-elasticity of a configuration with linear flow and structural models is also given, and the dynamic problems are touched upon.

This chapter's overview of the fluid–structure interaction is shown, at the software level, in a *partitioned* approach because this illustrates the interactions of the two disciplines of CFD and CSM, the work done by the team members, and the required data exchange. A *loosely coupled, modular framework* is proposed with clear lines for disciplines to act on data describing the shared state of the forces and deformations.

In the same way as a sized baseline configuration must be produced (Chapter 1) for clean-wing aerodynamic design to proceed, *structural sizing* needs to be carried out by the structural engineer. He or she determines the specific parameters in the CSM model to adequately bear the worst-case aero-loads and gravity loads, as well as other loads created by gusts or bumps on the runway.

For a deeper study of the structural design of aircraft, the reader is referred to specialist texts (e.g. [5, 9, 13]).

Finally, two case studies apply the AEROFRAME framework to the aero-elastic loop. The two examples treated are a wing model in a wind tunnel undergoing divergence and control reversal (Section 11.5.1) and determination of the flight shape of a unmanned aerial vehicle (UAV) in pull-up and push-down maneuvers (Section 11.5.2). In both cases, the structures expert in our team has been Aaron Dettmann, who was studying for his MSc thesis [6]. This thesis should be consulted for further CSM details.

11.2 Model of Wing Section in Torsion

To progress further, we need to develop the aero-elastic loop (Figure 11.3), into a specific computational or analytical model. The example here is the classic textbook example (e.g. in Cooper [19]): a twisting wing section in a wind tunnel (Figure 11.4). The section is free to rotate around a twist axis constrained by a torsion spring. As is the case for conventional wings, the twist axis is behind the aerodynamic center. When you consider the implications for wing design, bear in mind the following limitations of this model.

- Divergence of a real wing involves both twist and bending.
- The aerodynamics is modeled by the "coefficients" usually employed for stability analysis and flight simulation. It models high angle of attack and unsteady flows quite crudely and ignores stall and hysteresis effects.
- The spring-and-damper model is a rough approximation to the damping provided by a real structure.

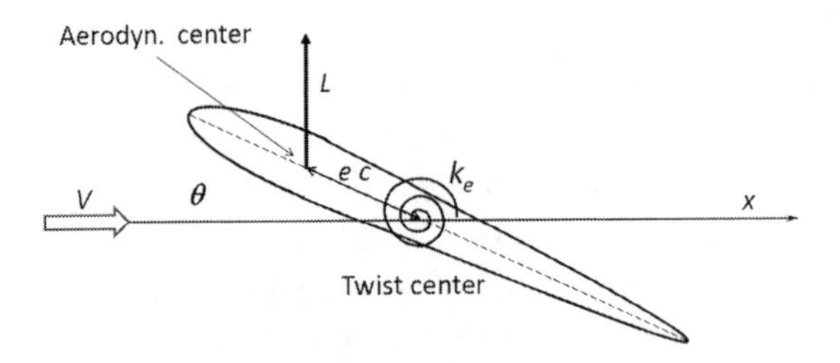

Figure 11.4 The single degree of freedom twisting airfoil.

Although much practical aero-elastic analysis is carried out with linear models such as this one, the student should not read too much into its quantitative predictions. Its purpose is just to illustrate the concepts. A more realistic development, albeit one that is more abstract, comes in the next section.

Consider the symmetric, rigid chord c airfoil in Figure 11.4 with incidence θ, *eccentricity* ec between the lift acting on the aerodynamic center and *flexural* (twist) axis, and *torsional stiffness* k_e (Nm/radian) per unit span. The oncoming stream is parallel to the x-axis. The moment due to the aerodynamic lift, proportional to the dynamic pressure $q = 1/2\rho V^2[N/m^2]$, is as follows.

$$M = qec^2(c_L(\theta, \dot{\theta})\cos\theta + c_D(\theta, \dot{\theta})\sin\theta) \tag{11.1}$$

Now M less the spring-restoring moment accelerates the twist. Include a structural damping term, $D > 0$, to complete the model as a nonlinear harmonic oscillator. The lift and drag coefficients include dependence on the pitching rate (i.e. an aerodynamic damping term; see Section 10.1.2 for a discussion). Set the spring for zero moment at $\theta = \theta_0$ to give a nonvanishing lift force at small dynamic pressures, and the structural model becomes as follows.

$$J\ddot{\theta} + D\dot{\theta} + k_e(\theta - \theta_0) = M \tag{11.2}$$

with J being the moment of inertia per unit span around the twist center.

Equations (11.1) and (11.2) describe the fluid–structure interaction problem as an oscillator driven by a force depending nonlinearly on the position and velocity, so bifurcations and nonuniqueness of solutions are to be expected. An analysis limited to small angles of attack θ is still very informative. Neglect drag and approximate $c_L(\theta, \dot{\theta}) \approx a\theta + b\dot{\theta}$,

$$a = c_{L,\alpha}(\alpha = 0, \dot{\alpha} = 0), \quad b = c_{L,\dot{\alpha}}(\alpha = 0, \dot{\alpha} = 0)$$

in the formulas below. The validity of this approximation is discussed in Etkin [8], for example and is found to be dubious; however, one result of this is that $b > 0$ for small Mach numbers and slow section movements. For convenience, define the parameter $p = ec^2$, and the dynamic system becomes, to first order in θ, the following linear oscillator.

$$J\ddot{\theta} + (D - pqb)\dot{\theta} + (k_e - pqa)\theta - k_e\theta_0 = 0. \tag{11.3}$$

11.2.1 Statics

Equilibrium of Eq. (11.3) (i.e. $\dot{\theta}, \ddot{\theta} \equiv 0$) gives the following linear equation.

$$(k_e - pqa)\theta = k_e\theta_0 \tag{11.4}$$

Consider its solution for increasing q, $\theta^* = \frac{\theta_0}{1-q/q^*}$, and $q^* = \frac{k_e}{pa}$ only when $q < q^*$, predicting divergence at $q = q^*$, a well-known result of *wing section torsional divergence*. Note that θ is defined differently in the Cooper reference above, hence the difference in the formula.

The *staggered iteration* of the aero-elastic loop, suggested in Figure 11.3, computes equilibria by alternating aerodynamic ("CFD") and structural ("CSM") computations. It can be applied to solve Eq. (11.4) as follows.
For $n = 0, 1, \ldots$

- "CFD" computes the moment by Eq. (11.1), given the deflection θ_n:
$M_n = pqa\theta_n$.
- "CSM," (Eq. 11.4) computes the deflection produced by M_n: $\theta_{n+1} = \theta_0 + \frac{M_n}{k_e}$.

This process converges only for $q < q*$ and diverges for $q > q*$ and predicts the divergence dynamic pressure correctly. In general, the process may diverge before the wing. But *if* the iteration converges, the dynamic pressure is below divergence.

11.2.2 Dynamics

Equation (11.3) is the standard constant coefficient oscillator equation, with stability properties independent of the equilibrium. It is *stable* for
$D - pqb > 0$ and $k_e - pqa > 0 \Leftrightarrow q < q*$.
With increasing dynamic pressure q, the instability occurs when
$q > D/pb$ or $q > q*$, whichever comes first.
The latter is the *static divergence limit*: if $q > q*$, the divergence is a nonoscillatory blow-up, no matter the damping.

The former is purely dynamic and oscillatory with exponentially growing amplitude, and one is tempted to call it *flutter*. The analog is dubious, since real lifting surface flutter usually is a persistent oscillation with amplitude growing with dynamic pressure. Such a process needs the proper nonlinearities, such as hardening springs, which are included in the model, and the above is too crude a description for that.

11.3 Aero-Elastic Configuration Model

The overview here gives the mathematical formulation of the problem to be solved but does not go into numerical details. The formulation is valid for any CFD and CSM model, but it assumes familiarity with elements of structural mechanics such as the direct stiffness formulation. Specifics of the beam model used in AEROFRAME are given in Section 11.3.4, and the CFD vortex lattice method (VLM) model was treated in Chapter 3.

11.3.1 Statics

We wish to determine the deformation for a given dynamic pressure, as well as the critical dynamic pressure for static divergence. The latter is an eigenvalue problem related to the former, and we proceed to show how it appears for linear flow and structure models with a finite number of degrees of freedom. The matrices appearing in the description below all depend on the deformations, and neglecting them makes the relations correct only to almost the first order. In particular, the standard procedure is to compute the stiffness matrix and the downwash matrix for the undeformed structure and to take into account the effect of displacements on the flow tangency only. In VLM, this is the free stream scalar product with change in surface unit normal, $\boldsymbol{V}_\infty \cdot \delta\hat{\boldsymbol{n}}$. But several variations on the theme of including the effects of deformation in the matrices are possible.

Let the displacements of the structure be $\boldsymbol{\delta}^S$. The resulting displacement vector of the aerodynamic shape is $\boldsymbol{\delta}^A = \boldsymbol{L}_D^{S\to A}\boldsymbol{\delta}^S$. The matrix $\boldsymbol{L}_D^{S\to A}$ transfers *displacements* (translations and rotations) from the structure to the aerodynamic model. It is constructed by a specific scheme adapted to the structural and aerodynamic shape degrees of freedoms, to be discussed in detail in Section 11.3.3. A similar matrix $\boldsymbol{L}_F^{A\to S}$ transfers *forces* from aerodynamic degrees of freedom to structures. The aerodynamic forces on the deformed surface, computed by a linear flow model for dynamic pressure q, become $\boldsymbol{F}^A = \boldsymbol{F}_0 + q\ \boldsymbol{AIC}\ \boldsymbol{\delta}^A$, where $\boldsymbol{F}_0 = q\boldsymbol{h}_0$ are the forces on the undeformed shape, also proportional to q, and $\boldsymbol{AIC}$ is the matrix of aerodynamic influence coefficients from displacement to force. This can be computed for the VLM model, for example, with the matrices already in use there (see Chapter 3). The structural response to forces is also assumed to be linear, characterized by the stiffness matrix $\boldsymbol{K}$: $\boldsymbol{K}\boldsymbol{\delta}^S = \boldsymbol{L}_F^{A\to S}\boldsymbol{F}^A$, all rearranged into the following.

$$(\boldsymbol{K} - q\boldsymbol{B})\,\boldsymbol{\delta}^S = \boldsymbol{L}_F^{A\to S}\boldsymbol{F}_0 \tag{11.5}$$
$$\boldsymbol{B} = \boldsymbol{L}_F^{A\to S}\ \boldsymbol{AIC}\ \boldsymbol{L}_D^{S\to A}$$

If the structural model is suitably restrained, as in the wing in wind tunnel example in Section 11.5.1, $\boldsymbol{K}$ becomes nonsingular. For free flight, the rigid body modes form the null space of $\boldsymbol{K}$ and must be projected out, as is done in the application to the UAV OptiMale, for example (Section 11.5.2). It follows that for small dynamic pressures there is a unique solution. When q approaches the smallest real positive eigenvalue q^* of $\boldsymbol{K} - q\boldsymbol{B}$, the solution blows up.

The iteration $\boldsymbol{K}\boldsymbol{u}_{n+1} = q\boldsymbol{B}\boldsymbol{u}_n + \boldsymbol{L}_F^{A\to S}\boldsymbol{F}_0$ will converge for small enough q. Whether it will for *all* $0 < q < q^*$ is dubious.

Iterative algorithms for finding $\boldsymbol{\delta}^S$ and q^* that use only matrix-vector products ($\boldsymbol{K u}$, $\boldsymbol{K}^{-1}\boldsymbol{u}$ and $\boldsymbol{B u}$) can be implemented in the aero-elastic loop without forming the matrix $\boldsymbol{K} - q\boldsymbol{B}$; an example is given in Section 11.4.1. Depending on the algorithm, it is possible in such scenarios to include the higher-order effects of matrix dependence on deformations.

11.3.2 Aero-elasto-dynamics by the Doublet Lattice Model

It would be straightforward to extend a restrained structural model with inertia and damping terms in an attempt to simulate slow aero-elastic dynamics. But unless dynamics of the air are also modeled (in the wing section example by the appearance of $b = c_{L,\dot{\alpha}}$), the coupled model cannot flutter and is of little use.

The doublet lattice method (DLM), described in Rodden's textbook [14], has become the standard approach to low-speed modeling of dynamic aero-elasticity. It is formulated in the frequency domain using downwash functions for unsteady potential flow. The $\boldsymbol{AIC}$ matrix becomes dependent on the frequency ω of the harmonic motion. There are two parameters to vary – the nondimensional *reduced frequency* $k = b\omega/V$ and dynamic pressure q – in the search for combinations that give unbounded solutions.

11.3.3 Coupling Aero and Structural Models via Mesh

The aero-model computes surface forces on a surface deformed by displacements to the structure, and the structural model computes deflections from loads on the structure. A detailed analysis of actual stresses is possible only when all pieces have been designed late in the design cycle. However, approximate deflections can be computed with much simpler models. We quote from Rodden's book:

> The idealization of the structure of a lifting surface as an elastic axis to simplify aero-elastic analyses is a traditional approach that is still useful in preliminary design.

Figure 11.5 shows the arrangement of the elastic axis and transfer of aerodynamic forces and deformations by rigid connectors, as described by Rodden,

> ...a swept-back elastic axis connected by rigid rods to the aerodynamic control points in the aero-elastic analysis. The rigid rods are connected to the elastic axis in directions appropriate to the internal chordwise structure: streamwise, if the ribs are streamwise, or perpendicular to the elastic axis if the ribs are swept.

The iterations in the aero-elastic loop are terminated by a tolerance test on the difference between successive approximations to the structure deformation $\boldsymbol{\delta}^S$.

Aero-elastic Coupling: Transfer of Displacements and Forces

The coupling of the fluid and structure solvers constitutes the central aspect of the partitioned solution approach. A meaningful physical representation of the whole fluid–structure system requires *coupling boundary conditions* expressing that deformations and loads must be equal at the fluid–structure interface in accordance with Newton's

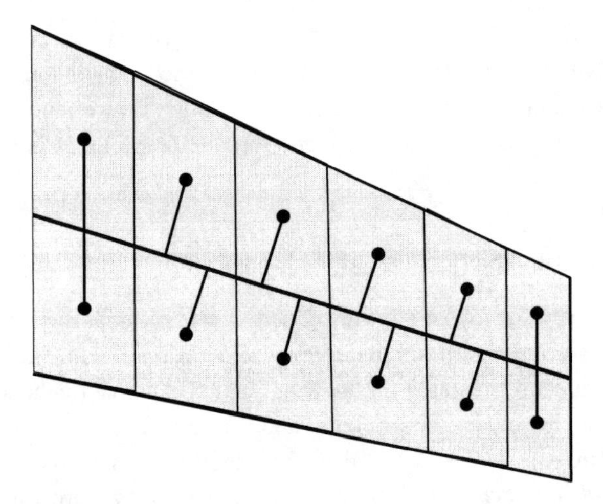

Figure 11.5 Swept wing planform, beam elastic axis, and rigid connectors to aero-points. (After Rodden [14])

third law. In addition, *all* aerodynamic loads must be transferred from the fluid to the structure domain, and the work done by the fluid on the surface must equal the work done by the transferred forces on the structure model.

When both CFD and CSM models possess a common outer surface (the "skin"), the mappings are 3D scattered data interpolations from one set of sites on the surface to another set.

A set of beams modeling the load-bearing structure is very useful and special. The AeroFrame CFD and CSM model is illustrated in Figure 11.6. While the VLM uses a mesh of lamina, the beam model uses a mesh of line segments. It is obvious that there is a mismatch between the models. The beam model is not even topologically close to the skin, nor to the VLM lamina. The aero-forces on the skin must be mapped to forces and moments acting on the beams. Similarly, the deformations of the beams – translations as well as rotations – must be transferred to surface deformation.

Work Conservation in Force Transfer

Consider a set of points $\boldsymbol{x}_i^A$ on the surface S, with aero-forces $\boldsymbol{F}_i^A$ on the surface S, and a set of points $\boldsymbol{x}_i^S$ on the structure with transferred forces $\boldsymbol{F}_i^S$. The surface points are displaced under load by $\boldsymbol{\delta}_i^A$. The work done on the surface and structure is as follows.

$$W_A = \sum_i \boldsymbol{F}_i^{A*} \boldsymbol{\delta}_i^A = W_S = \sum_i \boldsymbol{F}_i^{S*} \boldsymbol{\delta}_i^S$$

The transfer is (usually) a linear map, say:

$$\boldsymbol{F}_i^S = \sum_j \boldsymbol{T}_{ij}^{A \to S} \boldsymbol{F}_j^A$$

The matrix of all the $\boldsymbol{T}_{ij}$ defines the map $\boldsymbol{L}_F^{A \to S}$ (Section 11.3).

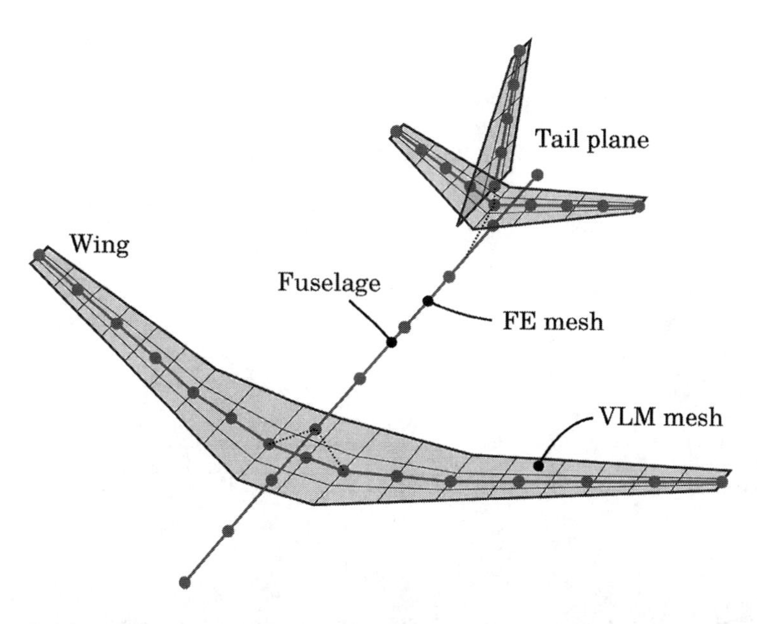

Figure 11.6 Concept of joining the VLM with the FE beam model. Aircraft components use separate models and discretizations for the aerodynamics and the structure analyses. For some components (e.g. the fuselage), there may not be a corresponding CFD mesh. (Courtesy of A. Dettmann, MSc thesis [6], reprinted with permission)

When S is close to the surface of the structural model, which must then contain shell or plate elements, the map is essentially a scattered-data interpolation. The equality of work requires simply that

$$\delta_j^A = \sum_i \boldsymbol{T}_{ij}^{A \to S*} \delta_i^S$$

Once the force transfer map is defined, the displacement transfer follows

$$\boldsymbol{T}_{ij}^{S \to A} = \boldsymbol{T}_{ji}^{A \to S*}$$

as the transpose of the force transfer matrix.

Forces and Displacements for Beam Models

Beams are 1D objects, and transfer of surface pressures to forces and moments acting "on" the beams is nontrivial. Similarly, transfer of the beam deformations to surface normal displacements must be defined. Neither process can be accomplished by some general scheme for scattered data interpolation. The development above must be adapted to this situation.

A point on a beam element must be assigned to a surface "element" in such a way that the whole surface is covered when all points on the beams are traversed. Usually, the beams are loaded by discrete point forces and moments applied to the nodes. Figure 11.7 shows a beam node P and its assigned surface element ΔS to which it is connected by the vector $\mathbf{l}$. The load on node P is then $\boldsymbol{F}^S = \boldsymbol{F}^A$ and the moment

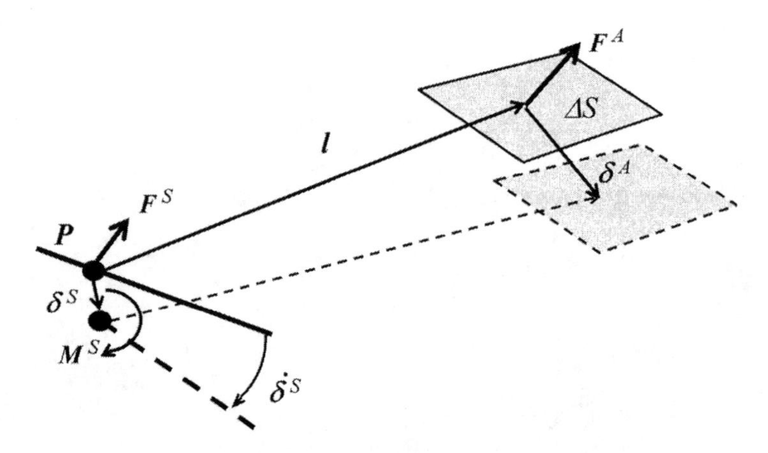

Figure 11.7 Forces and displacements for a beam node and its assigned surface element.

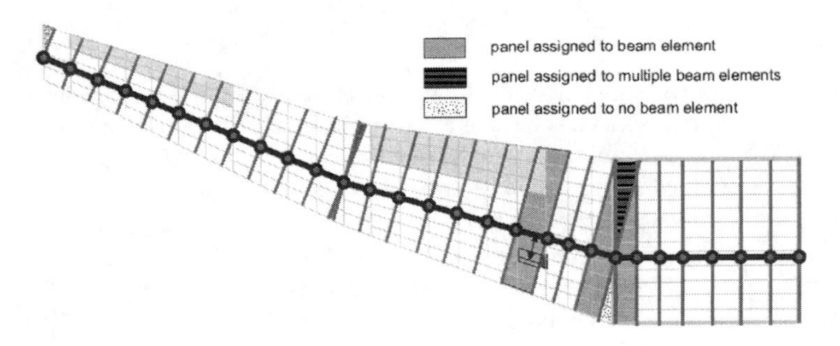

Figure 11.8 Assignment of VLM panels to a wing beam. (Courtesy of K. Seywald, MSc thesis [16], reprinted with permission)

$\boldsymbol{M} = \boldsymbol{l} \times \boldsymbol{F}^A$. The work of the aero-forces is $dW_A = \mathbf{F}^A \cdot \boldsymbol{\delta}^A$, and the work on the structure is $dW_S = \boldsymbol{F}^S \cdot \boldsymbol{\delta}^S + \boldsymbol{M} \cdot \dot{\boldsymbol{\delta}}^S$. We see then that work conservation requires $\boldsymbol{\delta}^A = \boldsymbol{\delta}^S + \boldsymbol{l} \times \dot{\boldsymbol{\delta}}^S$. The displacement of the assigned surface element is that of a body rigidly connected to P as it translates and rotates.

Assignment of a surface to the beam is easy when it can be swept out by curve segments, each assigned to a point on the beam. Usually, to be consistent with the development of the beam model, segments should be orthogonal to the beam. This is how wing structures with spanwise spars and ribs are commonly built. However, this does not work when the beam has kinks (see Figure 11.8). The shaded areas are either not covered by the sweep or are multiply covered, so ad hoc rules are applied to make the transfer one-to-one. A common alternative is to choose the segments chordwise. Then only the wing tip may need special attention. Another technique is to assign beam load sites as the closest point on the beam to the aero-force points.

11.3.4 Beam CSM Model

Aircraft are often made up of long and slender structures such as the wings and the fuselage. For low-fidelity analyses, such structures are reasonably modeled by 3D beam theory, which is the approach taken here with a FE formulation. Beam models are computationally inexpensive and are widely used for aero-elastic analyses, often in combination with the VLM and DLM as in Ref. [14].

Euler–Bernoulli beam theory, also known as classical beam theory, is used. The Euler–Bernoulli beam model is well covered in the literature, thus only a brief overview will be given here.

The partial differential equation for bending of the beam's (straight) elastic (x-)axis in the xz-plane is as follows.

$$\frac{\mathrm{d}^2}{\mathrm{d}x^2}\left(EI_y\frac{\mathrm{d}^2u_z}{\mathrm{d}x^2}\right)+\frac{\mathrm{d}m_y}{\mathrm{d}x}+q_z=0 \tag{11.6}$$

where u_z is the z-displacement, E is Young's modulus of the material, I_y is the surface moment of inertia around y of a cross section orthogonal to the elastic axis, $q_z(x)$ is transversal loading, and m_y is the applied moment per unit length. Examples are given in Section 11.5.1. The standard discretization of a 3D beam is a two-node element with 12 degrees of freedom, the three displacements u_x, u_y, and u_z and rotations Θ_x, Θ_y, and Θ_z of each node. The rotations are assumed small and define the end tangents, and the third angle is the torsion of the beam (Figure 11.9).

$$\Theta_y=-\frac{\mathrm{d}u_z}{\mathrm{d}x}\text{ and }\Theta_z=\frac{\mathrm{d}u_y}{\mathrm{d}x} \tag{11.7}$$

The element combines the bending stiffness in the xz-plane (Eq. (11.6)) with bending in the xy-plane, dilation along x, and torsion as follows.

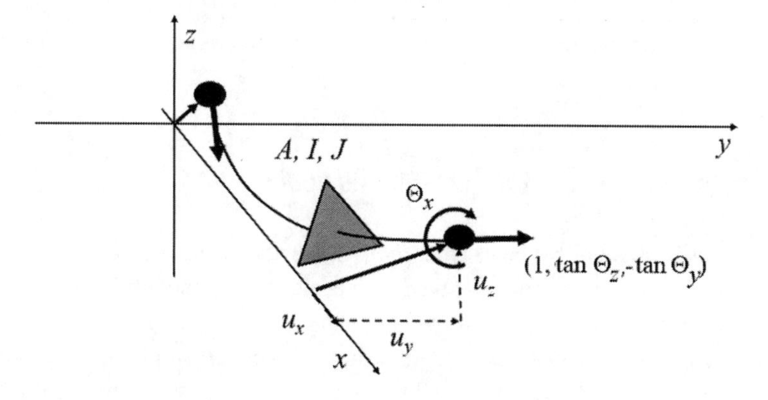

Figure 11.9 Two-node, 12 degrees of freedom 3D beam element.

$$\frac{d^2}{dx^2}\left(EI_z\frac{d^2u_y}{dx^2}\right)+\frac{dm_z}{dx}-q_y=0$$
$$\frac{d}{dx}\left(AE\frac{du_x}{dx}\right)+q_x=0 \tag{11.8}$$
$$\frac{d}{dx}\left(GJ\frac{d\Theta_x}{dx}\right)+m_x=0$$

where A is the cross-section area, $G = 1/2\ E/(1+\nu)$ the shear modulus, I_y and I_z are the area moments of inertia about the local y- and z-axes, and J is the torsional stiffness parameter.

The theory makes use of significant simplifications and assumptions. It is assumed that a normal plane section *remains plane* and *normal* to the deformed beam axis. The material is assumed isotropic with a Hookean linearly elastic relation between stresses and strains. For materials such as fiber composites or aluminum, this assumption implies a restriction to small strains. For small displacements, the model becomes linear, but it can be extended to cover geometric nonlinearity (i.e. large displacements), still with small strains. Since axial forces do not contribute to any bending loads, global buckling phenomena cannot be represented.

Deformations of wings with realistic geometries have been found to differ no more than $\pm 5\%$ between high-fidelity FE shell models and simple beam models. The differences were mainly attributed to shear deformations and the effects of restrained warping. Both effects are reflected by the FE shell model but not by the beam model.

Equations (11.6) and (11.8) are completely uncoupled as a result of the simplifying assumptions, so bendings about the y- and z-axes are uncoupled (no skew bending). Assembly of elements, forces, and constraints in the direct stiffness formulation produces the following linear system.

$$\boldsymbol{K}\delta^S = \boldsymbol{F}^S \tag{11.9}$$

where $\boldsymbol{K}$ is the stiffness matrix and $\boldsymbol{F}^S$ is the load vector, computed from the applied forces and moments.

11.4 Modular Framework for Aero-Elastic Loop

One can distinguish between *monolithic*, as in the wing section model above, and *partitioned* methods to compute the aero-elastic loop. A monolithic method compiles and solves the governing equations including both fluid and structural dynamics in a single system of equations, while partitioned methods are based on separate software modules for the CSM and CFD domains that interact in the aero-elastic loop environment.

A major advantage of the loosely coupled partitioned approach is that it maintains clear lines that separate the quite different models, solution methods, etc., from each

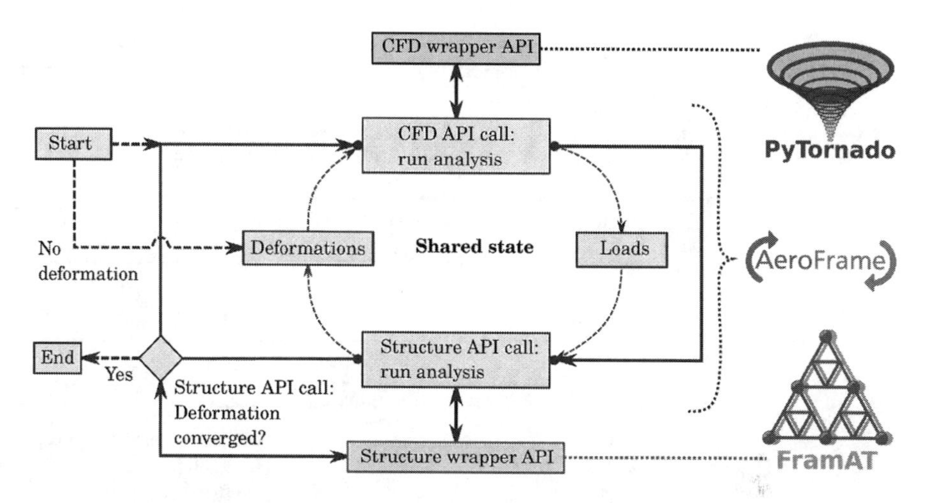

Figure 11.10 Conceptual implementation of the aero-elastic loop to find static equilibria. (Courtesy of A. Dettmann, MSc thesis [6], reprinted with permission)

other (fluid and structure) so that well-established modeling, discretization, and solution methods can be applied interchangeably. The approach enables "plug-and-play" with CFD and CSM modules.

CFD software was presented in earlier chapters. The FE technology for computational structural mechanics is well covered in many textbooks (e.g. [2, 4, 17]). The mapping of forces and deformations depends on the specifics of the CFD and CSM/FE models and requires some data processing by wrappers and application program interfaces (APIs).

11.4.1 The AeroFrame Aero-Elastic Modular Framework

The computational framework AeroFrame is illustrated in Figure 11.10. It provides the VLM code PyTornado for the CFD domain and the Euler–Bernoulli beam FE code FramAT for the CSM domain. The solvers can access the shared loads and deformation data to operate on a common aircraft model in a solution procedure.

Starting from an undeformed state, a first CFD analysis is performed. The computed loads are shared within the framework. Subsequently, a structure analysis proceeds by reading the shared loads and applying them to the structure model. The computed deformations are again shared within the framework and are accessible to the CFD analysis; thus, the flow field for a deformed aircraft can be computed, etc.

The loop of alternating aerodynamic and structure analyses including the sharing of loads and deformations is continued until the structural deformation has converged.

Let us show how the GMRES algorithm of Saad and Schultz [15] can be run in the aero-elastic loop to solve $(\boldsymbol{K} - q\boldsymbol{B})\boldsymbol{u} = \boldsymbol{b}$ or $\boldsymbol{u} - \boldsymbol{K}^{-1}\boldsymbol{B}\boldsymbol{u} = \boldsymbol{K}^{-1}\boldsymbol{b}$ (see Section 11.3). The matrix inverse is just a notational shorthand for the solution of a linear system with $\boldsymbol{K}$ as coefficients. $\boldsymbol{K}$ is sparse and its inverse is not to be computed. The algorithm builds and uses the Krylov subspace (i.e. $\boldsymbol{U} = (\boldsymbol{u}^n), n = 0, 1, \ldots N$ where

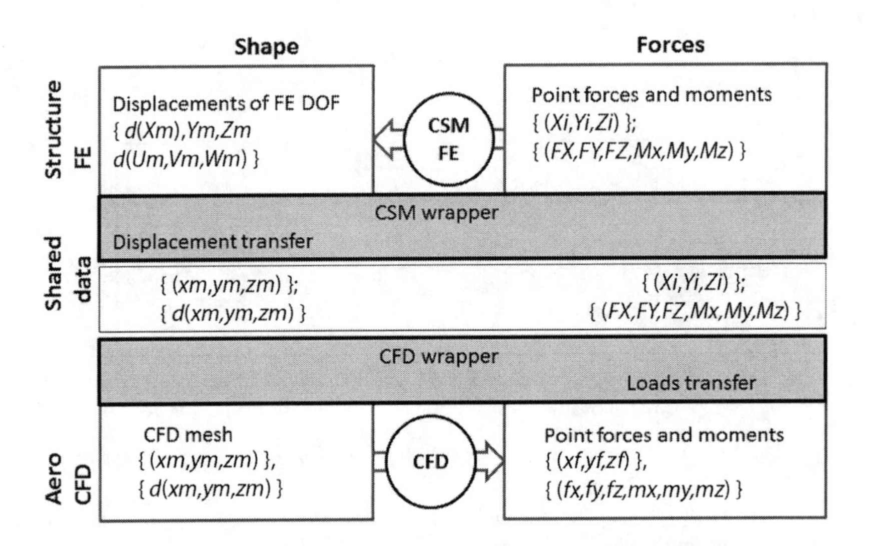

Figure 11.11 An example of data shared and wrapper tasks. DOF = degree of freedom.

$\boldsymbol{u}^{n+1} = \boldsymbol{u}^n - \boldsymbol{K}^{-1}\boldsymbol{B}\boldsymbol{u}^n$ and $\boldsymbol{u}^0$ is an initial guess, say $\boldsymbol{K}^{-1}\boldsymbol{b}$). However, $\boldsymbol{K}^{-1}\boldsymbol{B}\boldsymbol{u}^n$ is the result of a CFD computation with displacements $\boldsymbol{u}^n$ followed by a CSM analysis (i.e. one iteration of the basic cycle). The $\boldsymbol{u}^n$ are small displacements and can be scaled to be small without changing the results of the algorithm. In addition, eigenvalue problems (e.g. to find q^*) are solved efficiently by the Arnoldi algorithm using the Krylov subspace using openly available software (see [11]).

An example of the data shared and the tasks of the wrappers is shown in Figure 11.11.

The data computed by a module are shared so other modules can access them without knowing how they were computed. The CFD wrapper in this example has been made responsible for transferring the surface forces in the CFD mesh to the structural model point forces and moments. It must know the location of the FE points, and if follower loads are used, the FE points will change in every iteration, so they must be reexported and reread. Similarly, the FE wrapper is responsible for transferring the FE degrees of freedom displacements to the CFD mesh. It needs to know only the target, the CFD mesh points, which will change, and it exports the increments as responses to their coordinates.

Each CSM and CFD code needs its own wrapper, written to access and export the data in the defined format. The wrapper also provides scripting for running the analyses, managing input and output files, control variables for execution, etc.

There remains the issue of the format of the data. AeroFrame can share data in the CPACS format (Section 5.1.1), which is capable of storing all data for an aircraft. It also has a specific "bare bones" data format for the VLM and beam CSM solvers currently implemented.

To enable plug-and-play with CFD and CSM codes, the shared data format should follow a widely accepted standard, such as the CFD General Notation System (CGNS) [1], for exchanging mesh-based data. CGNS provides both the data format definition and libraries for manipulating such data with bindings for C, FORTRAN 90, FORTRAN 77, PYTHON, and MATLAB. Most CFD codes have CGNS interfaces, but CGNS has not become popular in the FE community to the same extent.

Implementation

The CFD and CSM tools are developed as stand-alone tools that know nothing about the other tools. The actual core of AEROFRAME is a framework that houses the shared data and executes scripts to run the modules through their defined APIs. It needs to know no implementation details about the modules. The modular design of the framework, closely related to the idea of loose coupling, enables the CFD and structure tools to be interchangeable. In other words, any solver module exposing the same API can be plugged in.

11.5 Case Studies: Elasto-static Wing Effects

Two case studies verify how well the aero-structural AEROFRAME framework with its low-fidelity CFD TORNADO and CSM FRAMAT modules carries out the aero-elastic loop to study static wing deformation and predict the critical airspeeds for divergence and control reversal in the first case of an academic example of a wing clamped to a wind tunnel wall. The second case is a more realistic aircraft situation – a UAV in high-altitude flight – to which the software is applied to predict the flight shape in two flight conditions. The two cases require aero-structural teamwork. The discussion, however, goes into details of just the aerodynamic aspects of the work – the CSM work is sketched only briefly. Full details of the CSM tasks, including structural sizing of the wing in both case studies, can be found in Aaron Dettmann's MSc thesis [6].

11.5.1 Case Study I: Divergence and Control Reversal for a Clamped Wing with Ailerons

A KTH course on aero-elasticity includes wind tunnel experiments for a simple flat-plate wing model and collection of experimental data for flutter, control reversal, and divergence. The experimental data are compared with AEROFRAME computations to demonstrate that a static aero-elastic wind tunnel experiment can be simulated. An experienced user can set up the simulation with little effort.

Figure 11.12 shows the rectangular planform of semi-span 1600 mm and chord 350 mm with ailerons of chord 60 mm all along the trailing edge. Hung vertically from the roof of the wind tunnel test section, the wing is clamped at the root and is subjected to a low-speed airflow with free stream density ρ and airspeed V_∞. Effects of gravity are negligible due to the vertical hanging.

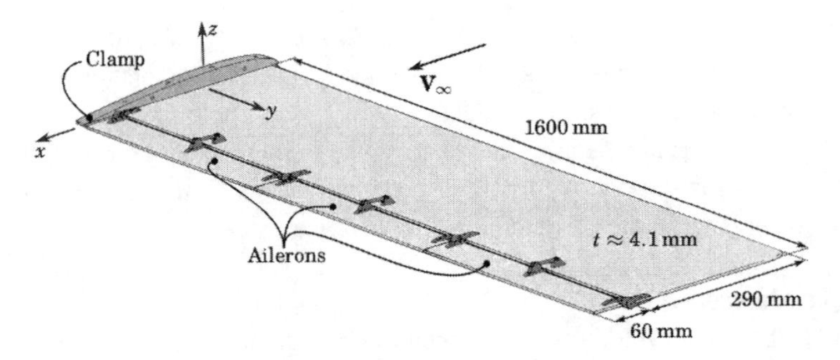

Figure 11.12 The wind tunnel model. (Courtesy of A. Dettmann, MSc thesis [6], reprinted with permission)

The wing structure, including the aileron, is made of a glass fiber–epoxy composite plate with constant thickness $t \approx 4.1$ mm and material density ϱ. The fibers are oriented in the spanwise and chordwise directions, such that no bending–torsion coupling results due to the composite material. It is further assumed that the material is linearly elastic with Young's modulus E and shear modulus G in the spanwise direction, and that the influence from material damping can be neglected.

The aileron is split into three identical sections along the span, attached to the wing with an (almost) moment-free metal hinge axis. The aileron sections will not deform elastically for small deflections of the main wing plate. They contribute to inertial and aerodynamic forces but not to the wing stiffness. Further assume that the deformation of the plate cross section can be neglected.

The elastic axis of the beam with bending stiffness EI and torsional stiffness GK is placed along the shear center of the wing. I is the area moment of inertia and K is the torsion constant of the wing section.

Dettmann determined the basic material properties of the glass fiber-reinforced polymeric (GFRP) composite plate in ground vibration tests. Initial guesses for the Young's modulus E, the shear modulus G, and the density ϱ were adjusted so that a FE beam model with only wing normal bending and torsion degrees of freedom would match measured eigenfrequencies for a clean GFRP reference plate and for the wing model (Table 11.1).

Comparison of Experiment and Simulation Results

For the divergence simulation, an angle of attack of 0.1° was set, simulating a small initial perturbation, and all aileron deflections δ_a were held to 0°. For the reversal analysis, all aileron deflections δ_a were set to 1°, 5°, and 10° with zero angle of attack on the main element. The simulations were run for successively increasing airspeeds V_∞ up to the point of divergence and reversal. Figure 11.13a plots the deflection u_z and the twist Θ_y at the wing tip versus airspeed V_∞. The divergence simulation shows that the wing deformation rate du_z/dV_∞ increases with increasing airspeed. Between 24 and 25 m/s, the wing tip deflection exceeds 10% of the wingspan.

Table 11.1 Estimated material parameters of the wind tunnel model. The total mass m_{total} includes the plate material as well as all hinges and joints, but not the clamp shown [6].

Young's modulus **E**	Shear modulus **G**	Density ϱ	m_{total}
32.5 GPa	5.1 GPa	1960 kg/m^3	4.856 kg

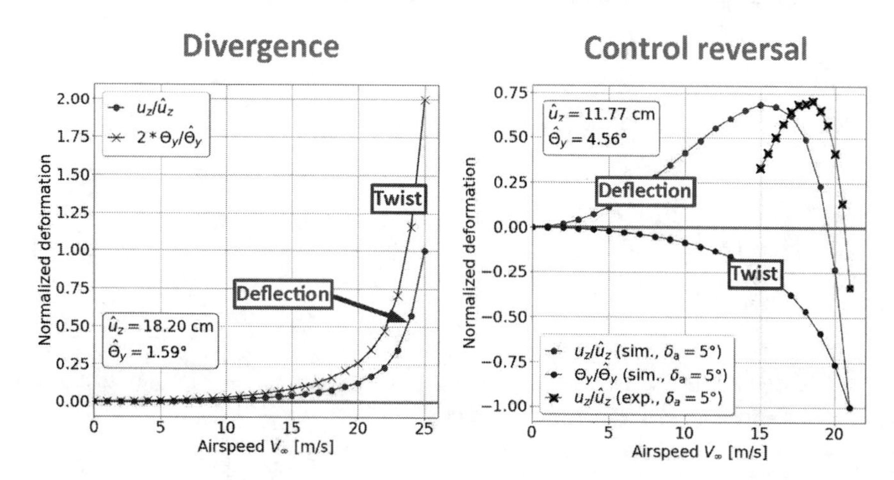

Figure 11.13 Divergence and control reversal of the wind tunnel model. Normalized wing tip deflection and twist computed for divergence and control reversal of the wind tunnel model (sim. = simulation, exp. = experiment). (Courtesy of A. Dettmann, MSc thesis [6], reprinted with permission)

In a similar way, Figure 11.13b plots the tip deformation in the control reversal simulation. For all three aileron settings δ_a, the tip deflection u_z and its twist Θ_y around the spanwise axis change in a similar manner. With increasing flow velocity, the wing initially bends up, $u_z > 0$, and twists leading-edge up. The deflection reaches a maximum, and with further increased airspeed decreases to almost no deflection. This defines the control reversal speed around 19.5 m/s. Higher airspeed bends the wing down. The divergence speed was close to 23.5 m/s, so control reversal happens before divergence.

Figure 11.14 indicates how the wing deformation is accompanied by a loss of control efficiency. An aileron deflection increment $\Delta\delta_{\text{a}}$ produces a roll moment coefficient increment ΔC_ℓ, so the relevant control derivative is $C_{\ell,\delta_{\text{a}}}(q) \approx \frac{\Delta C_\ell}{\Delta\delta_{\text{a}}}$ (see Chapter 10). The *control efficiency* is defined as $\frac{C_{\ell,\delta_{\text{a}}}(q)}{C_{\ell,\delta_{\text{a}}}(0)}$. Here, $C_\ell \equiv M_x/(q_\infty S_{ref} b_{ref})$, where M_x is the roll moment at the wing root, q_∞ is the freestream dynamic pressure, $S_{ref} = 0.56$ m^2, and $b_{ref} = 1.6$ m. Close to the airspeed associated with zero tip deflection, aileron inputs will have no effect, and at higher speed, the ailerons even produce reversed moments (Figure 11.14). The curve was produced from AeroFrame analyses for $\delta_{\text{a}} = 1°$ and $\delta_{\text{a}} = 5°$. Figure 11.13b also shows the tip deflection measured in the experiment for the reversal experiment with an aileron

Table 11.2 Simulated and measured critical speeds (strip: Theodorsen strip theory).

Critical speed	Strip	VLM	Experiment
Divergence speed (m/s)	20.6	25.0	23.5
Reversal speed (m/s)	18.9	20.5	≥21

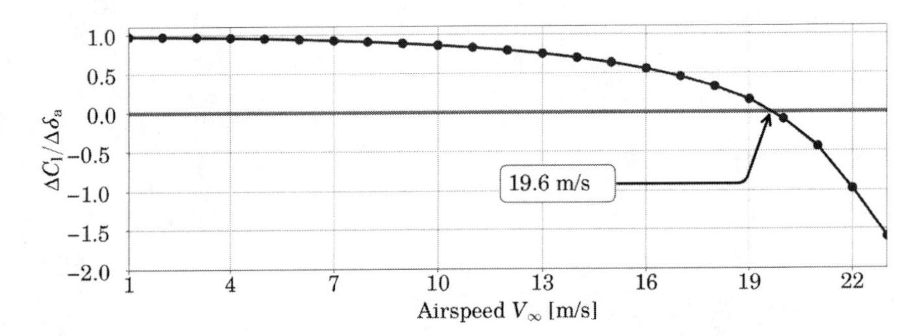

Figure 11.14 Loss of control efficiency with increased airspeed. (Courtesy of A. Dettmann, MSc thesis [6], reprinted with permission)

setting of 5° (squares). The deflection data come from optical observation of three markers with some uncertainty in the calibration, leading to an unknown offset in the ordinate. Nevertheless, the data clearly show a deformation behavior similar to the simulation, and at similar airspeeds.

Due to the fairly complex load situation in control reversal, finding the equilibria requires multiple iterations. All loops are initialized based on the undeformed state. In the case of the reversal simulation, the first guess is an upward deflection and a negative twist independent of the airspeed. Only thereafter is the solution iteratively improved until the final, possibly negative deflection is found. The critical reversal airspeed was concluded to be greater than or equal to 21 m/s.

In the course of the experiment, divergence and reversal speeds were estimated using the Theodorsen strip theory and a three degrees of freedom FE beam model. Theodorsen strip theory is a linear model in which local aerodynamic loads are computed based on the local angle of attack of a chordwise wing strip. The structure model used the same material parameters and the same bending and torsional stiffness data as AeroFrame. Table 11.2 sums up the critical speeds predicted by the Theodorsen model and AeroFrame and observed in the experiment.

In contrast to the VLM, strip theory does not consider any effects due to the finite span, and it no doubt overestimates the load distribution at the wing tip. Therefore, it seems reasonable that the results based on the VLM model are closer to the experimental results.

The AeroFrame results agree at least qualitatively with experimental evidence. The predictions of the critical speeds are well within the accuracy limits that could be expected with the experimental setup and the simplicity of the computational model.

Figure 11.15 The OptiMale UAV. (Courtesy of D. Charbonnier, AGILE EU project [3], reprinted with permission)

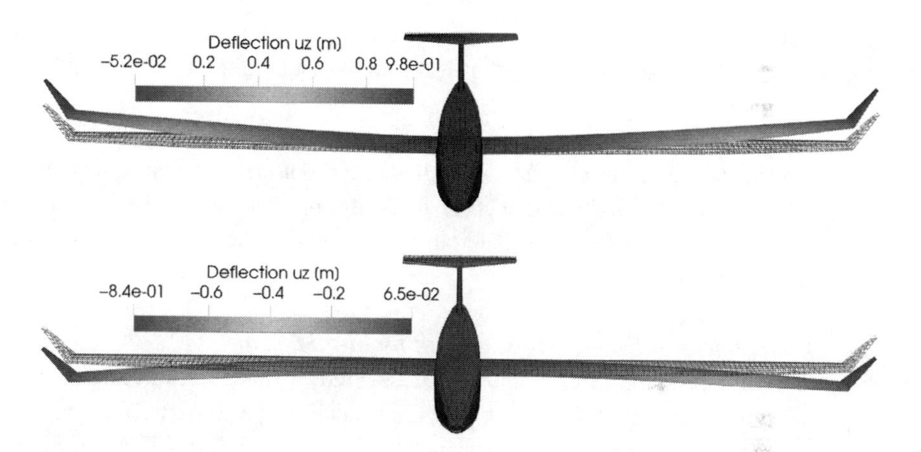

Figure 11.16 Deformation (u_z) in $3g$ pull-up and $-1.5g$ push-down as computed using high-fidelity models in AGILE. (Courtesy of A. Dettmann, MSc thesis [6], reprinted with permission)

11.5.2 Case Study II: OptiMale Long-Endurance UAV Flight Shape

The OptiMale aircraft, initially developed during the German AeroStruct research project, is a medium-altitude long-endurance (MALE) UAV in a conventional low-wing configuration, with T-tail, two rear-mounted turbofan engines, and two external wing-mounted fuel tanks (Figure 11.15). Cruising at 15 km or higher altitude, the reference model has a span of about 33 m and a high *AR* of approximately 20. The maximum takeoff mass is 10 t.

This sections summarizes our aero-structural analysis of the OptiMale UAV, demonstrating the capabilities of AeroFrame for a more complete, fully elastic aircraft model. The AGILE project performed a partitioned aero-elastic analysis with higher-fidelity Euler equation CFD and detailed FE CSM to obtain a minimum-weight design for OptiMale [12]. Figure 11.16 displays the deformations in the $3g$ pull-up and the $-1.5g$ push-down produced by the AGILE software framework, similar to AeroFrame. We shall compare predictions from AeroFrame with these.

Table 11.3 Estimated material parameters of the OptiMale structural model. The total mass m_{total} includes the entire beam model.

Young's modulus **E**	Shear modulus **G**	Density ϱ	m_{total}
68.9 GPa	26.0 GPa	2800 kg/m^3	10,000 kg

CSM Model Generation

The AGILE study defined two flight maneuvers to be most critical for the structure, namely a quasi-steady pull-up at a load factor of $n_z = 3$ and a push-down at $n_z = -1.5$. Both maneuvers were performed with a Mach number $M = 0.367$ at sea-level conditions. Based on the force equilibrium for a rigid aircraft, one can estimate the required lift coefficients C_L for these quasi-static flight maneuvers (at lowest/highest points).

Dettmann used the aero-structural sizing tool dAEDalus by Seywald [16] to estimate the required beam properties. This tool sizes the main wing of an aircraft by the aerodynamic and inertial loads during critical quasi-steady flight maneuvers. Table 11.3 lists the material parameters for the beam properties used in the simulations.

Treatment of Free-Flying Aircraft by Inertia Relief

Unlike the wind tunnel model in Case Study I, the forces accelerate a free-flying aircraft as well as deform it. In the FE formulation, the global stiffness matrix $\boldsymbol{K}$ would be singular with a null-space spanned by rigid body motions. Such combined "free-to-fly" and "free-to-flex" dynamic systems can be treated by the "inertia relief" method described in more detail in Ref. [18]. The general idea is to keep the acceleration terms

$$\boldsymbol{M}\ddot{\boldsymbol{\delta}}^S \tag{11.10}$$

in the force balance equation with $\boldsymbol{\delta}^S(t)$ taken as a rigid-body motion of all of the degrees of freedom, and $\boldsymbol{M}$ as the mass matrix of the FE model. This gives six new unknowns – the rigid body accelerations – that can be computed from zeroth and first moments of the force balance equation, remembering that all elastic forces vanish for rigid-body motion. The static equilibrium equation between the aerodynamic and gravity loads, the inertia loads (which are the acceleration terms in Eq. (11.10)), and the internal elastic loads can then be solved for the deformations. The FE model is constrained as usual to eliminate rigid-body motions; the reaction forces at the constraints should vanish.

The inertia relief method was applied in the analyses of the pull-up/push-down flight where the aircraft's acceleration, and therefore inertial loads, are known from the nature of the maneuvers. The global angle of attack and the elevator deflection were calibrated to achieve the required trim state. Pitch moment contributions from the lifting surfaces were considered, but not from other components such as the engines.

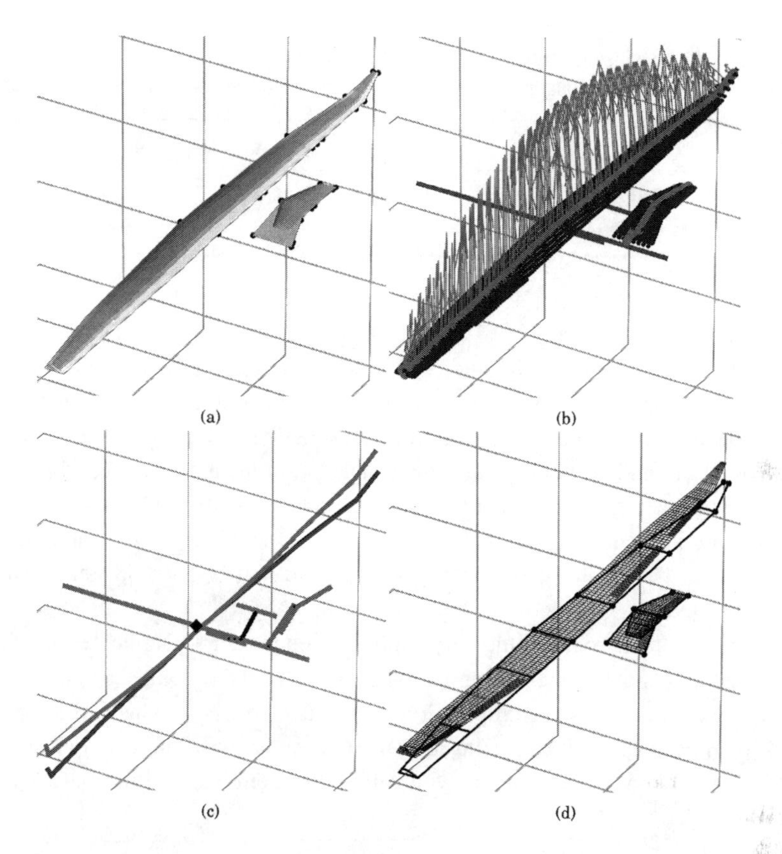

Figure 11.17 The OptiMale UAV: (a) pressure distribution in the undeformed state, (b) aero-loads transferred to beam nodes (nodal moments and inertia loads omitted), (c) deflection of the FE beam model, and (d) deformed VLM mesh. (Courtesy of A. Dettmann, MSc thesis [6], reprinted with permission)

Comparison with High-Fidelity Models

With the fully assembled beam model for the aircraft structure in hand, AEROFRAME simulated the pull-up and push-down maneuvers at trimmed flight conditions. Figure 11.17 shows data for the $3g$ pull-up. The pressure distribution computed with TORNADO is shown in Figure 11.17a. The discretized load distribution was mapped onto the beam model (Figure 11.17b). The structure deformation was computed with FRAMAT (Figure 11.17c) and subsequently translated into a VLM mesh deformation (Figure 11.17d). Table 11.4 presents a comparison of the wing tip deformations in the pull-up and the push-down maneuvers.

AEROFRAME predicts deformations of a similar magnitude to those from the AGILE higher-fidelity model (cf. Figure 11.16). The deflection estimate for the pull-up differs by 20 cm. The estimate for the push-down differs by only 2 cm. There are multiple reasons that may account for these differences, the most obvious being the different models to represent the aerodynamics and the structure. The most difficult part of this case study was the generation of a somewhat realistic beam and mass distribution model.

Table 11.4 Computed deformations in low- and high-fidelity analyses. In the AeroFrame model, the deformations were computed with and without consideration of gravity and inertial loads.

Model	u_z ($n_z = 3$)	u_z ($n_z = -1.5$)
High-fidelity (AGILE)	98 cm	−84 cm
Low-fidelity (with mass)	118 cm	−82 cm

The OptiMale study is rather unusual in the sense that a detailed geometric and structural description for the studied aircraft was already available, and the low-fidelity analysis using VLM and beam theory was a verification and validation exercise.

The aircraft conceptual design process normally begins with a coarse definition of the aircraft and low-fidelity analysis to size and refine a baseline configuration. Once initial size parameters have been estimated, higher-fidelity models would be developed for subsequent design cycles. In this scenario, it is of course desirable to have low-fidelity models that are reasonably accurate so that changes in later design cycles remain small.

This completes our rudimentary exposition on the aerodynamic design of aircraft in the last four chapters, beginning with the airfoil, then the isolated clean wing, elementary development of the configuration, its flying qualities, and lastly the determination of its flight shape. What remains is trying to rerun some of our examples yourself. Go to the book's website, immerse yourself in the exercises and tutorials, and try your hand at it.

11.6 Learn More by Computing

Gain hands-on experience of the computational tools for the topics in this chapter by working with the on-line resources. Exercises, tutorials, and project suggestions are found on the book website www.cambridge.org/rizzi. For example, study how the aero-elastic loop iterations gradually change the shape of the rectangular wing (Figure 11.12) measured in the windtunnel. Software used to compute many of the examples shown is available from http://airinnova.se/education/aero-dynamic-design-of-aircraft

References

[1] CGNS home page. Available from https://cgns.github.io/WhatIsCGNS.html

[2] K. J. Bathe. *Finite Element Procedures*, 2nd edition. Prentice Hall, 2014.

[3] P. D. Ciampa and B. Nagel. AGILE the next generation of collaborative MDO: Achievements and open challenges. Presented at 2018 Multidisciplinary Analysis and Optimization Conference, June 2018. AIAA-Paper-2018-3249.

[4] R. D. Cook, D. S. Malkus, M. E. Plesha, and R. J. Witt. *Concepts and Applications of Finite Element Analysis*, 4th edition. Wiley, 2002.

[5] J. Cutler. *Understanding Aircraft Structures*, 3rd edition. Blackwell Science, Ltd, 1999.

[6] A. Dettmann. *Loosely Coupled, Modular Framework for Linear Static Aeroelastic Analyses*. Diploma thesis, KTH School of Engineering Sciences, 2019.

[7] E. H. Dowell. *A Modern Course in Aeroelatsicity*. Solid Mechanics and Its Applications, 3rd edition. Kluwer Academic Publishers, 1995.

[8] B. Etkin and L. D. Reid. *Dynamics of Flight – Stability and Control*. John Wiley and Sons, 1995.

[9] D. Howe. *Aircraft Loading and Structural Layout*. Aerospace Series. Professional Engineering Publishing, Ltd, 2004.

[10] G. K. W. Kenway and J. R. R. A. Martins. Multipoint aerodynamic shape optimization investigations of the Common Research Model wing. *AIAA Journal*, 54(1): 113–128, 2016.

[11] R. B. Lehoucq, D. C. Sorensen, and C. Yang. *ARPACK Users' Guide: Solution of Large-Scale Eigenvalue Problems with Implicitly Restarted Arnoldi Methods*. SIAM, 1998.

[12] R. Maierl, A. Gastaldi, J.-N. Walther, and A. Jungo. Aero-structural optimization of a MALE configuration in the AGILE MDO framework. In *ICAS 2018*, 2018. Available from www.agile-project.eu/cloud/index.php/s/p55SINI9eJBJYoP.

[13] T. H. G. Megson. *Aircraft Structures for Engineering Students*. Elsevier Aerospace Engineering Series, 4th edition. Elsevier, Ltd. 2007.

[14] W. P. Rodden. *Theoretical and Computational Aeroelasticity*, 1st edition. Crest Publishing, 2012.

[15] Y. Saad and M. B. Schultz. GMRES: A generalized minimal residual algorithm for solving nonsymmetric linear systems. *Siam Journal on Scientific and Statistical Computing*, 7(3): 856–869, 1986.

[16] Klaus Seywald. *Impact of Aeroelasticity on Flight Dynamics and Handling Qualities of Novel Aircraft Configurations*. Dissertation, Technische Universität München, München, 2016.

[17] I. M. Smith, D. V. Griffiths, and L. Margetts. *Programming Finite Element Computations*. 5 edition, 2014.

[18] J. Wijker. *Mechanical Vibrations in Spacecraft Design*. Springer Berlin Heidelberg, Berlin, Heidelberg, 2004.

[19] J.R Wright and J.E Cooper. *Introduction to Aircraft Aeroelasticity*. John Wiley & Sons, Ltd., 2008.

Index

Printed in the United States
by Baker & Taylor Publisher Services